# Landsenses in Green Spaces

# Landsenses in Green Spaces

Editors

**Jiang Liu**
**Xinhao Wang**
**Xin-Chen Hong**

Basel • Beijing • Wuhan • Barcelona • Belgrade • Novi Sad • Cluj • Manchester

*Editors*

Jiang Liu
School of Architecture and
Urban-Rural Planning
Fuzhou University
Fuzhou
China

Xinhao Wang
School of Planning, DAAP
University of Cincinnati
Cincinnati
United States

Xin-Chen Hong
School of Architecture and
Urban-Rural Planning
Fuzhou University
Fuzhou
China

*Editorial Office*
MDPI
St. Alban-Anlage 66
4052 Basel, Switzerland

This is a reprint of articles from the Special Issue published online in the open access journal *Forests* (ISSN 1999-4907) (available at: www.mdpi.com/journal/forests/special_issues/YMM8OIPKUD).

For citation purposes, cite each article independently as indicated on the article page online and as indicated below:

Lastname, A.A.; Lastname, B.B. Article Title. *Journal Name* **Year**, *Volume Number*, Page Range.

**ISBN 978-3-7258-0602-7 (Hbk)**
**ISBN 978-3-7258-0601-0 (PDF)**
**doi.org/10.3390/books978-3-7258-0601-0**

Cover image courtesy of Jiang Liu

# Contents


**About the Editors** . . . . . . . . . . . . . . . . . . . . . . . . . . . . . . . . . . . . . . . . **vii**

**Preface** . . . . . . . . . . . . . . . . . . . . . . . . . . . . . . . . . . . . . . . . . . . . . . **ix**

**Jiang Liu, Xinhao Wang and Xinchen Hong**
Landsenses in Green Spaces
Reprinted from: *Forests* **2024**, *15*, 333, doi:10.3390/f15020333 . . . . . . . . . . . . . . . . . **1**

**Jiang Liu, Yi-Jun Huang, Zhu Chen and Xin-Chen Hong**
Multi-Scale Effects of Landscape Pattern on Soundscape Perception in Residential Green Spaces
Reprinted from: *Forests* **2023**, *14*, 2323, doi:10.3390/f14122323 . . . . . . . . . . . . . . . . . **9**

**Xuan Guo, Jiang Liu, Zhu Chen and Xin-Chen Hong**
Harmonious Degree of Sound Sources Influencing Visiting Experience in Kulangsu Scenic
Area, China
Reprinted from: *Forests* **2023**, *14*, 138, doi:10.3390/f14010138 . . . . . . . . . . . . . . . . . **28**

**Lei Luo, Qi Zhang, Yingming Mao, Yanyan Peng, Tao Wang and Jian Xu**
A Study on the Soundscape Preferences of the Elderly in the Urban Forest Parks of
Underdeveloped Cities in China
Reprinted from: *Forests* **2023**, *14*, 1266, doi:10.3390/f14061266 . . . . . . . . . . . . . . . . . **49**

**Dongxu Zhang, Xueliu Liu and Wei Mo**
Comparison of Soundscape Evaluation in Forest-Type and Urban-Type Han Chinese
Buddhist Temples
Reprinted from: *Forests* **2023**, *14*, 79, doi:10.3390/f14010079 . . . . . . . . . . . . . . . . . **75**

**Yuting Yin, Yuhan Shao, Huilin Lu, Yiying Hao and Like Jiang**
Predicting and Visualizing Human Soundscape Perception in Large-Scale Urban Green Spaces:
A Case Study of the Chengdu Outer Ring Ecological Zone
Reprinted from: *Forests* **2023**, *14*, 1946, doi:10.3390/f14101946 . . . . . . . . . . . . . . . . . **103**

**Yu Wei and Yueyuan Hou**
Forest Visitors' Multisensory Perception and Restoration Effects: A Study of China's National
Forest Parks by Introducing Generative Large Language Model
Reprinted from: *Forests* **2023**, *14*, 2412, doi:10.3390/f14122412 . . . . . . . . . . . . . . . . . **125**

**Bingzhi Zhong, Hui Xie, Tian Gao, Ling Qiu, Heng Li and Zhengkai Zhang**
The Effects of Spatial Characteristics and Visual and Smell Environments on the Soundscape of
Waterfront Space in Mountainous Cities
Reprinted from: *Forests* **2023**, *14*, 10, doi:10.3390/f14010010 . . . . . . . . . . . . . . . . . **151**

**Yanyan Cheng, Ying Bao, Shengshuai Liu, Xiao Liu, Bin Li, Yuqing Zhang, et al.**
Thermal Comfort Analysis and Optimization Strategies of Green Spaces in Chinese
Traditional Settlements
Reprinted from: *Forests* **2023**, *14*, 1501, doi:10.3390/f14071501 . . . . . . . . . . . . . . . . . **169**

**Junyi Li, Ziluo Huang, Dulai Zheng, Yujie Zhao, Peilin Huang, Shanjun Huang, et al.**
Effect of Landscape Elements on Public Psychology in Urban Park Waterfront Green Space: A
Quantitative Study by Semantic Segmentation
Reprinted from: *Forests* **2023**, *14*, 244, doi:10.3390/f14020244 . . . . . . . . . . . . . . . . . **185**

**Shan Shu, Lingkang Meng, Xun Piao, Xuechuan Geng and Jiaxin Tang**
Effects of Audio–Visual Interaction on Physio-Psychological Recovery of Older Adults in
Residential Public Space
Reprinted from: *Forests* **2024**, *15*, 266, doi:10.3390/f15020266 . . . . . . . . . . . . . . . . . . . . . **206**

**Guodong Chen, Jiayu Yan, Chongxiao Wang and Shuolei Chen**
Expanding the Associations between Landscape Characteristics and Aesthetic Sensory
Perception for Traditional Village Public Space
Reprinted from: *Forests* **2024**, *15*, 97, doi:10.3390/f15010097 . . . . . . . . . . . . . . . . . . . . . **223**

**Yuanping Shen, Qin Wang, Hongli Liu, Jianye Luo, Qunyue Liu and Yuxiang Lan**
Landscape Design Intensity and Its Associated Complexity of Forest Landscapes in Relation to
Preference and Eye Movements
Reprinted from: *Forests* **2023**, *14*, 761, doi:10.3390/f14040761 . . . . . . . . . . . . . . . . . . . . . **246**

**Tian Liu, Bingyi Mi, Hai Yan, Zhiyi Bao, Renwu Wu and Shuhan Wang**
Spatiotemporal Distribution Analysis of Spatial Vitality of Specialized Garden Plant Landscapes
during Spring: A Case Study of Hangzhou Botanical Garden in China
Reprinted from: *Forests* **2024**, *15*, 208, doi:10.3390/f15010208 . . . . . . . . . . . . . . . . . . . . . **262**

# About the Editors

**Jiang Liu**

Dr. Jiang Liu is a Professor at the Department of Landscape Architecture at the School of Architecture and Urban-Rural Planning at Fuzhou University, China. His research focuses on multi-sensory landscape experience and mechanism, soundscape planning and design, and landsenses ecology. He has published over 50 journal papers in peer-reviewed journals. He is the PI of three national funded research projects, the director of Environmental Psychology and Health Innovation Research Team, and vice head & co-founder of the "Key Lab of Digital Technology in Simulation and Prediction of Territorial Space" at Fuzhou University.

**Xinhao Wang**

Dr. Wang, AICP, is a professor of Planning at the School of Planning, College of DAAP, University of Cincinnati. (UC) He holds a Ph.D. in City and Regional Planning from the University of Pennsylvania, a Master of Community Planning from the University of Rhode Island, a Master of Science in Geo-Sciences and a Bachelor of Science in Geography from Peking University, Beijing, China. He teaches courses in geographic information systems (GIS), environmental planning, statistics and research methods. His research interests are in the areas of environmental planning and GIS application in planning (e.g., analysis of the spatial pattern of water quality indicators; the relationship between land use and water quality; GIS-based water quality modeling and flooding analysis; spatial distribution of human health and environmental indicators; and resilience). Dr. Wang's publications and presentations can be found in various planning and environmental journals, proceedings and conferences. Prior to UC, he worked as a research associate, planning consultant and senior GIS analyst. Dr. Wang is the Co-Director of the Joint Center of Geographic Information Systems and Spatial Analysis at UC.

**Xin-Chen Hong**

Dr. Xin-Chen Hong received his Ph.D. in Landscape Architecture from Fujian Agriculture and Forestry University (including a visiting process at University of British Columbia). He is now an associate professor at the School of Architecture and Urban-Rural Planning at Fuzhou University, China. Dr. Hong's research interests are social sensing and computational social science. His recent projects include the National Natural Science Foundation of China, and the Program of Humanities and Social Science Research Program of Ministry of Education of China. Dr. Hong has published more than 60 articles in peer-reviewed scientific journals, including two ESI High Cited Papers. He serves as reviewers of Nature Communications, Ecological Indicators, Urban Ecosystems, Frontiers in Public Health, etc.

# Preface

In recent years, there has been a growing emphasis on the importance of multi-sensory perception in the planning, design and management of green spaces. Recognized as essential ecological infrastructure, this approach aims to shape landscape experiences effectively, fostering high-quality environments and ultimately enhancing human well-being. Grounded in the landsense ecology theory, landsenses underscore the integration of human perception from sensory and psychological dimensions into ecological environmental research. This approach stands as a pivotal methodological and technical strategy for advancing the development of green spaces within the context of building smart and resilient cities.

We are privileged to have served as Guest Editors for this Special Issue, and we extend our gratitude to all authors and reviewers for their outstanding contributions and unwavering support. Additionally, we would like to express our appreciation to the *Forests* editorial team for their enthusiastic dedication and expert editing. We are sincerely thankful for the contributions made by our colleagues and scholars from various institutions. Furthermore, we acknowledge the generous funding provided by the National Natural Science Foundation of China (no. 52378049, no. 52208052), and the Humanities and Social Science Research Program of Ministry of Education of China (no. 21YJCZH038), which supported many of the research endeavors included in this issue.

By prioritizing multi-sensory perception in the planning, design and management of green spaces, landscape professionals aim to create environments that not only appeal to our visual senses, but also engage us on a deeper, more holistic level. These efforts not only contribute to the creation of high-quality landscapes, but also promote human well-being by fostering connections with nature and providing opportunities for relaxation, rejuvenation and sensory exploration.

**Jiang Liu, Xinhao Wang, and Xin-Chen Hong**
*Editors*

*Editorial*

# Landsenses in Green Spaces

**Jiang Liu [1], Xinhao Wang [2] and Xinchen Hong [1,*]**

[1]  School of Architecture and Urban-Rural Planning, Fuzhou University, Fuzhou 350108, China; jiang.liu@fzu.edu.cn
[2]  School of Planning, University of Cincinnati, Cincinnati, OH 45221, USA; wangxo@ucmail.uc.edu
*  Correspondence: xch.hung@outlook.com

## 1. Introduction

Green spaces, serving as crucial ecological infrastructure, offer numerous ecological system services and enhance human well-being, particularly in densely built environments. Theories and technical approaches for greening high-density agglomerations are progressively adopting a multidisciplinary approach [1]. The impact of landscape experience on human mental and/or physical well-being has garnered growing attention in the fields of green space and public health [2,3]. The domains of research dedicated to the planning, design, and management of green spaces emphasize the significance of multi-sensory perception, guided by traditional visual perception, in shaping landscape experiences to create high-quality landscapes [4–6]. The term "landsenses" is derived from "landsenses ecology". As a recently emerging scientific discipline grounded in ecological principles and an analytical framework encompassing natural elements, physical senses, psychological perceptions, socio-economic perspectives, process risk, and related aspects, landsenses ecology integrates landscape ecology with people's vision and social needs. It concentrates on land-use planning, construction, and management aimed at sustainable development [7]. Landsenses emphasize the incorporation of human perception from sensory and psychological dimensions into ecological environmental research. Within this framework, we posit that the theory advanced by landsenses ecology not only offers an effective avenue for investigating the relationship between humans and the environment, but also serves as a crucial methodological and technical approach for the development of green spaces in the context of constructing smart and resilient cities.

This Special Issue in *Forests* explores the role of landsenses in green spaces. It is comprised of 13 papers involving multi-sensory studies conducted in green spaces. This collection contains works in seven research fields:

(1)  Mechanisms of multi-sensory interaction and their effects;
(2)  Indicators for landsenses characteristics of green spaces;
(3)  Landsenses with cultural and regional significance;
(4)  Theoretical and technical approaches for landsenses creation;
(5)  Innovative application of the Internet of Things and multi-source data in green space studies;
(6)  Social perception, machinery perception and virtual reality;
(7)  Planning, design and management of green space based on landsenses ecology.

## 2. Summary of Articles Included in the Special Issue

Green spaces play a crucial role in promoting sustainable urban environmental management and enhancing social well-being. These spaces not only deliver numerous ecosystem services, but also positively influence the mental and physical health of urban residents, as well as encourage social interaction. Our research collections concentrate on diverse urban green spaces, encompassing urban forests, residential green areas, scenic zones, urban waterfront green spaces, botanical gardens, traditional villages, and more.

**Citation:** Liu, J.; Wang, X.; Hong, X. Landsenses in Green Spaces. *Forests* **2024**, *15*, 333. https://doi.org/10.3390/f15020333

Received: 20 January 2024
Revised: 30 January 2024
Accepted: 4 February 2024
Published: 8 February 2024

The quality of soundscapes significantly influences life quality and visitor experiences. Our research collections delve into the intricacies of soundscape quality, concentrating on the perception of soundscapes and examining influential factors across dimensions such as landscape patterns, environmental functional characteristics, visual perception, and visitation experiences. Moreover, our research collections extend to the development of predictive models for soundscape perception, which are founded upon the acquisition of sufficient data. Liu et al. explored a new perspective on the interrelationships between soundscape perception and landscape pattern on a multi-scale [8]. The authors confirmed that the scale effect of landscape patterns can affect soundscape perception based on 30 residential green spaces. They found that the multi-scale patterns of vegetation and buildings play more critical roles in forming soundscapes in residential areas. Guo et al. explored the relationships between the harmonious degree of sound sources (SHD) and visiting experience indicators [9]. They suggested that natural sounds were the most influential sound source and visual landscape perception, while human sounds and mechanical sounds both had significant positive effects on soundscape perception. There is an indirect relationship between the SHD of sound sources and the evaluation of comprehensive impression. Luo et al. analyzed the soundscape preferences of elderly residents in underdeveloped cities in China for urban forest parks and the relationships between the soundscape preferences and landscape features [10]. They found that the most influential factors affecting the soundscape preferences of the elderly include the length of time spent in the waterfront environment, the time spent in the forest park, and the importance of road signs. Zhang et al. conducted comparative research to investigate subjective soundscape evaluations between typical forest-type and urban-type Han Chinese Buddhist temples [11]. They found that respondents in forest-type temples preferred natural sounds, while respondents in urban-type temples preferred Buddhism-related man-made sounds. Yin et al. predicted individual-scale soundscape perception in large-scale urban green spaces (UGSs) based on environmental visual, aural, and functional characteristics [12]. Prediction results suggested that people's perceived soundscape satisfaction increased as the distance from the ring road increased, and it gradually reached its highest level in the green spaces stretched outside the ring road.

Multisensory integration can convey comprehensive information, thereby enhancing the stereoscopic and richness experience of environmental quality. Our research collections focus on the mechanisms of multisensory interactions, including visual, auditory, tactile, olfactory, gustatory, and thermal perceptions. Furthermore, the research explores the impact of the objective landscape environment on the psychological dimensions of the public. Wei et al. investigated the influence of sensory perception of forests on visitors' restoration effects from a multidimensional and multisensory perspective [13]. They utilized a generative large language model to address the dilemma posed by traditional self-report scale measures and revealed that different sensory quantities (sight, hearing, touch, and taste) have varying effects on visitor restoration. Zhong et al. explored the influence of spatial characteristics and visual and smell environments on the soundscape of waterfront space in mountainous cities (WSMCs) [14]. They found that $L_{Aeq}$ and the normalized soundscape difference index (NDSI) are more affected by spatial characteristics, and the soundscape comfort degree (SCD) is more affected by visual and smell environments in WSMCs. Meanwhile, they summarized the recommended values of spatial characteristics and visual and smell environment indicators. Cheng et al. analyzed the factors affecting the thermal comfort of green spaces [15]. They found that water and greening coverage are the primary factors affecting the thermal comfort of spaces. Increasing water area and creating multi-level greening spaces are effective measures to improve the thermal comfort of green spaces in the settlement. Li et al. revealed the influence of different landscape elements in urban park waterfront green spaces on public psychology and behavior [16]. Landscape elements have significant different contributions to the four experience dimensions, i.e., emotional, cognitive, psychological, and behavioral. The spatial element contributes most significantly to public's psychological response. Focusing on a special group, Shu et al.

investigated how audio–visual interactions in common public spaces within Chinese urban residential areas might have restorative effects on older adults [17]. Their results indicate the importance of establishing residential areas that incorporate both natural elements and diverse activities, encompassing both auditory and visual stimuli, to support the well-being and healthy aging of older adults in Chinese residential settings.

Moreover, a thorough comprehension of public aesthetic perception and preferences is crucial for crafting top-notch landscape planning and design. This understanding helps meet the expectations of residents and visitors while accomplishing diverse objectives, including environmental optimization, community interaction, and the preservation of regional characteristics. Our research collections explore the complexity of the relationship between landscape design, landscape features, and perceived preferences. Chen et al. expanded the landscape characterization system for the public space of the traditional village by integrating multiple dimensions [18], including landscape spatial form, visually attractive elements of the landscape, and landscape color. Results indicated that the public preferred a scenario with a high proportion of trees, relatively open space, mild and uniform color tones, suitability for movement, and the ability to produce a restorative and peaceful atmosphere. Shen et al. combined objective and subjective landscape complexity to investigate the effects of landscape design intensity on preference and eye movement [19]. They suggested that the significant relationship between objective or subjective landscape complexity, or preference and eye movement metrics, was dependent on landscape types. Liu et al. examined the spatiotemporal distribution patterns of spatial vitality and explored the correlation between plant landscape characteristics and spatial vitality [20]. They suggested that to establish vibrant specialized plant landscapes, managers and planners involved in the planning and design process should prioritize a comprehensive consideration of and respect for the visual aesthetics and functional needs of visitors.

## 3. The Researchers' Perspectives on Landsenses in Green Spaces

To delve deeper into the contributors' perspectives on landsenses in green spaces, an open-ended question was introduced: "What should be the primary focuses and challenges for landsenses research in public spaces?" Selected contributors were invited to share brief comments, and their comprehensive responses are presented below.

### 3.1. Personal Perspective 1

The primary research focus of landsenses lies in the mechanisms of multi-sensory experiences, emphasizing the need to address the significance of these issues in the research agenda. Identifying the sensory dimension(s) that contribute most to the visiting experience and understanding their interconnections in specific contexts pose complex challenges, with variations in different scenarios. A key challenge in landsenses research for green spaces is the integration of qualitative and quantitative approaches to offer more applicable and instructive theoretical outcomes for planning and design practices. Additionally, the evolving field of landsenses ecology holds significant potential in human settlement science, requiring further clarification and enrichment of its theoretical framework and methodology through additional research.

(Prof. Dr. Hui Xie, Chongqing University)

### 3.2. Personal Perspective 2

In landsenses research, it is best to use the power of contemporary science and technology, but also take into account the historical sense of the place and other special requirements and achieve a certain flexible variability. At the same time, at the macro and micro levels, these studies should pay attention to people's feelings and the healthy ecological development of the whole region. Traditional research methods should also be combined with artificial intelligence or big data, so that the true reliability of research results can be guaranteed. At present, the biggest challenge lies in the lack of due attention to the academic community and the insufficient participation of personnel.

(Prof. Dr. Dongxu Zhang, Guangzhou University)

### 3.3. Personal Perspective 3

As a senior landscape and tourist cross-discipline, I think the landscape perception of public space should be more focused on interaction with space and people, such as the material, color, light, sound and other sensory experience. It should more attention should be paid to deeper psychological responses and spiritual needs, such as belonging, happiness, security, authenticity, and other sociological properties. I think the challenge of public space landscape perception research should be the measurement of human objective physical indicators, such as ergonom measuring, corticol measurement, etc. Today, in the face of the rapid development of artificial intelligence, more technological intelligent detection methods and evaluation methods should be advocated.

(Assoc. Prof. Dr. Jian Xu, South China University of Technology)

### 3.4. Personal Perspective 4

Landsenses research in public spaces, rooted in the principles of landscape ecology, aims to achieve sustainable land use planning, construction, and governance. The focus of this research includes a comprehensive examination of natural elements, physical and psychological perceptions, socio-economic factors, processes, and risks. Investigating elements such as light, heat, water, and societal dynamics is crucial for creating public spaces that cater to diverse needs. However, landsenses research faces challenges, including the necessity for interdisciplinary collaboration across ecology, psychology, sociology, and urban planning. The integration of diverse data sources, standardization, privacy concerns, resource constraints, and community engagement are significant hurdles. Overcoming these challenges is essential to ensure that landsenses research contributes to the development of public spaces that are sustainable, inclusive, and attuned to the varied experiences and preferences of the community.

(Associ. Prof. Qunyue Liu, Fujian University of Technology)

### 3.5. Personal Perspective 5

The focuses of landsenses research could be put on the correlation and interaction between human perception, psychology, emotion, behavior and the urban and built environment. Due to the development of digital technologies such as the Internet of Things (IoT) and Artificial Intelligence (AI), the extensive and in-depth application of these technologies has not only provided landsenses research with the use of diversified sensing sensors, but also offered technical possibilities for the processing, analysis and application of the massive amounts of information.

The future challenges lie in the need to evolve quantitative research and to build more advanced perceptual systems, so as to analyze, compare and optimize different types of urban public spaces, among which the public space of healthy communities should be closely monitored because of its fundamental role in the process of healthy city construction and the realization of the goal of healthy and sustainable development. Exploring the construction of healthy communities, from the perspective of landsenses ecology, is really conducive to the further application of landsenses ecology to the practice of sustainable development and residential environment construction, offering support for the successful realization of health and sustainable development goals.

Researchers can explore ways to create a healthy community public space system integrated with multiple landscape perception elements including natural elements, physical sensory elements and psychological sensory elements. In a full use of the ecosystem services with all types of natural elements, health service facilities and ecological infrastructure are combined to form composite health and welfare service facilities, so that natural factors detrimental to the population health are eliminated and avoided, the natural elements beneficial to the community health are introduced and optimized, and the accessibility and sense of belonging influencing psychological health are provided. Simultaneously, the comfort of the physical perception of the crowd could be improved to avoid the discomfort or even impact on human senses and mental health caused by excessive stimulation of the human senses. Both normal performance of human senses and comfort are promoted through the creation of good physical perception.

(Associ. Prof. Xiao Liu, South China University of Technology)

### 3.6. Personal Perspective 6

Landscape research in public spaces is dedicated to creating urban environment that are both aesthetically pleasing and functional, focusing on meeting diverse user needs, promoting environmental sustainability and social–cultural activities, ensuring safety and convenient access, and striving to enhance the health of residents. Challenges faced include limited budgets, the need to adapt to environmental changes, and the efficient use of emerging technologies, all of which require interdisciplinary collaboration and innovative strategies to ensure the long-term development and maximization of social value in public spaces.

(Assoc. Prof. Dr. Yuhan Shao, Tongji University)

### 3.7. Personal Perspective 7

Landsenses research in public spaces should not only emphasize individual users' multi-sensory perceptions and diverse cultural backgrounds, but also understand the social interaction pattern importance of creating inclusive and diverse public spaces. Additionally, attention should be given to constructing and maintaining green infrastructure in public spaces for sustainable ecosystem services. However, landsenses research in the context of rapid urbanization faces challenges. How to improve limited urban spaces? How to ensure public spaces serve equally diverse societal groups? How to balance cultural differences in expectations and needs from the human collective? Landsenses ecology covers a broad range of topics, and therefore, in landsenses research, it is essential to define its conceptual framework and specific research fields clearly. This will showcase the unique contributions of landsenses ecology. I am looking forward to witnessing its distinctive role in the future.

(Ph.D. Candidate Zhu Chen, Leibniz University Hannover)

### 3.8. Personal Perspective 8

As we all know, the information conveyed by an objective environment provides multisensory stimulation and subsequently has an impact on human physical and mental health. As the main place for urban residents to relax, entertain, and unwind, the landsenses of public spaces also have a significant impact on the urban residents' physical and mental health, including the special landscape structure, landscape composition, and landscape pattern characteristics. The changes in landscape characteristics of public spaces, such as sky view factors, building height, vegetation area, water ratio, etc., may also change the public's perception attitude to some extent. Therefore, using semantic segmentation and virtual reality as technical support, the landscape features of urban public spaces have been analyzed and quantified, exploring the impact of landscape elements, facility elements, natural elements, and construction elements on the psychological, emotional, cognitive, and behavioral dimensions of urban residents, which provides data support for exploring the impact of landscape features on public psychology.

However, the landscape features of public spaces are diverse and complex, and it is particularly important to select appropriate indicators to reflect the landscape features of public spaces. Similarly, the landsenses of urban residents in public spaces are also influenced by many factors, making it equally challenging to select appropriate landsense indicators. Additionally, how to more accurately quantify the changes in landscape features of public spaces and the changes in physical and mental indicators of urban residents remains a huge challenge in landsenses research in public spaces.

Therefore, in future research, the development and innovation of technical means for quantifying landscape features of public spaces will need to be continuously carried out. In addition, obtaining objective and accurate landsenses of urban residents in public spaces is also crucial for future research, as the data obtained from the questionnaire are subjective and cannot objectively reflect the landsenses of urban residents. Moreover, we will continue to explore a scientifically reasonable landscape feature evaluation system for landsenses research in public spaces.

(Ph.D. Candidate Junyi Li, Fujian Agriculture and Forestry University)

### 3.9. Personal Perspective 9

Landsenses research in public spaces focuses on understanding the sensory experiences of individuals and the interaction of these experiences with different environmental elements. This understanding is crucial for creating or modifying spaces to enhance human experience. A major challenge is the inherent complexity and subjectivity of human sensory perception and psychological cognition. The task is therefore to develop methods to accurately quantify and measure these perceptions and experiences. This is essential if they are to be effectively linked to various environmental factors, thereby facilitating informed and practical design decisions in the development of public spaces.

(Ph.D. Candidate Xuan Guo, Leibniz University Hannover)

In summarizing the contributors' perspectives, the primary focuses of landsenses research in public spaces can be generalized as follows:

- **Application of Technologies:**

Utilizing technologies such as AI, IoT, and big data to enhance the understanding of public spaces.

- **Interdisciplinary Collaboration:**

Encouraging collaboration across different disciplines to uncover the multi-sensory mechanisms influencing the visitor experience.

- **Creating Inclusive and Diverse Public Spaces:**

Designing public spaces that cater to the needs of a diverse range of users.
As for the challenges identified:

- **Integration of Diverse Data Sources:**

Addressing the complexity of integrating data from various sources.

- **Combining Theoretical Results into Practice:**

Bridging the gap between theoretical research outcomes and practical implementation.

- **Efficient Use of Emerging Technologies:**

Ensuring the effective and ethical utilization of emerging technologies in landsenses research.

- **Lack of Appropriate Indicators of Perceptual Process:**

Developing suitable indicators to measure the perceptual processes involved in public space experiences.

- **Insufficient Attention from Academic Community and Public Awareness:**

Addressing the need for increased attention and awareness from both the academic community and the public.

- **Adaptation to Environmental Changes:**

Responding to the challenges posed by environmental changes to ensure the relevance and effectiveness of landsenses research in evolving contexts.

## 4. Conclusions

Since the emergence of the landsenses ecology theory, it has not only been serving as an effective avenue to delve into the relationship between humans and the environment, but also representing a crucial methodological and technical approach for the development of green spaces within the framework of constructing smart and resilient cities. While initially advocated by researchers in the field of ecology, it has shown significant potential in driving the systematic reorganization of traditional knowledge and accomplishing structural upgrades.

There is a comprehensive interaction between ecological processes and human perception in landsenses ecology. Grounded in the fundamental principles of ecology, landsenses ecology delves into the connection between natural elements and human physical perception and psychological cognition. This approach provides a rational means to integrate ecological processes with human perception. By focusing on changes in human perception at various scales during landsenses creation, it facilitates achieving a balance between the supply and demand of the natural ecosystem and the human socio-economic system. As evident in our research compilation, the studies span various scales, ranging from macro levels like urban ecological zones to micro levels like residential green spaces.

Landsenses ecology promotes transitioning from scattered to integrated multi-source data optimization. The data underpinning landsenses ecology consist of extensive information on ecology, its associated dynamic processes, and human psychological and physical perception. The implementation strategy involves multidisciplinary system reorganization and multi-scale spatial optimization management to achieve the processing, analysis, and application of landsenses data through the melioration model. It advocates employment of diverse technical methods to acquire and integrate scattered data information from various sources. As highlighted by many contributors to this compilation, the incorporation of quantitative approaches is crucial in advancing landsenses research and developing more sophisticated perceptual systems. A promising avenue for achieving this lies in the integration of diverse data sources, facilitated by digital technologies like the Internet of Things (IoT) and Artificial Intelligence (AI).

In the era of sophisticated spatial practice and rapid advancements in science and technology, landsenses ecology advocates a shift in spatial creation thinking from static to dynamic. Landsenses creation serves as the primary method in landsenses ecology practice, seeking to imbue one or more ecological visions into a medium through suitable forms of expression. The objective is to make these visions accessible to individuals and other entities. At its core, landsenses creation aims to facilitate resonance among people and foster shared behaviors. It conceptualizes all environments as integral components of a holistic landsenses system. The systematic methodology applied to physical perception and the comprehensive approach to psychological cognition in landsenses ecology research impart ecological characteristics to human perception information, forming the foundational data for landsenses creation. Nevertheless, uncovering the complete panorama of the human physical perception and psychological cognition process remains a significant journey. Our collection highlights a predominant research focus on soundscape within the auditory perception, indicating the need for increased attention to other sensory and cognitive dimensions.

Despite the global acceptance of the concept of "landsenses" still being in its early stages, its theoretical framework is evolving through the continuous efforts of researchers. This Special Issue aims to garner attention from scholars worldwide, shedding light on the ongoing developments and significance of landsenses in the realm of research.

**Author Contributions:** Conceptualization, J.L., X.W. and X.H.; writing—original draft preparation, J.L.; writing—review and editing, X.W., X.H. and J.L. All authors have read and agreed to the published version of the manuscript.

**Funding:** This project was supported by the National Natural Science Foundation of China (52378049, 52208052), the Humanities and Social Science Research Program of Ministry of Education of China (Grant No. 21YJCZH038), and Fujian Natural Science Foundation, China (2023J05108).

**Conflicts of Interest:** The authors declare no conflict of interest.

## References


1. Du, P.; Al-Kodmany, K.; Ali, M.M. *The Routledge Handbook on Greening High-Density Cities: Climate, Society and Health*; Routledge: Abingdon, UK, 2024.
2. Guo, X.; Liu, J.; Albert, C.; Hong, X.-C. Audio-visual interaction and visitor characteristics affect perceived soundscape restorativeness: Case study in five parks in China. *Urban For. Urban Green* **2022**, *77*, 127738. [CrossRef]
3. Hong, X.-C.; Cheng, S.; Liu, J.; Dang, E.; Wang, J.-B.; Cheng, Y. The Physiological Restorative Role of Soundscape in Different Forest Structures. *Forests* **2022**, *13*, 1920. [CrossRef]
4. Hong, X.-C.; Liu, J.; Wang, G.-Y. Soundscape in Urban Forests. *Forests* **2022**, *13*, 2056. [CrossRef]
5. Liu, J.; Yang, L.; Xiong, Y.-C.; Yang, Y.-Q. Effects of soundscape perception on visiting experience in a renovated historical block. *Build. Environ.* **2019**, *165*, 106375. [CrossRef]
6. Chen, Z.; Zhu, T.-Y.; Liu, J.; Hong, X.-C. Before Becoming a World Heritage: Spatiotemporal Dynamics and Spatial Dependency of the Soundscapes in Kulangsu Scenic Area, China. *Forests* **2022**, *13*, 1526. [CrossRef]
7. Zhao, J.-Z.; Liu, X.; Dong, R.-C.; Shao, G.-F. Landsenses Ecology and Ecological Planning Toward Sustainable Development. *Int. J. Sustain. Dev. World Ecol.* **2016**, *23*, 293–297. [CrossRef]
8. Liu, J.; Huang, Y.-J.; Chen, Z.; Hong, X.-C. Multi-Scale Effects of Landscape Pattern on Soundscape Perception in Residential Green Spaces. *Forests* **2023**, *14*, 2323. [CrossRef]
9. Guo, X.; Liu, J.; Chen, Z.; Hong, X.-C. Harmonious Degree of Sound Sources Influencing Visiting Experience in Kulangsu Scenic Area, China. *Forests* **2023**, *14*, 138. [CrossRef]
10. Luo, L.; Zhang, Q.; Mao, Y.; Peng, Y.; Wang, T.; Xu, J. A Study on the Soundscape Preferences of the Elderly in the Urban Forest Parks of Underdeveloped Cities in China. *Forests* **2023**, *14*, 1266. [CrossRef]
11. Zhang, D.; Liu, X.; Mo, W. Comparison of Soundscape Evaluation in Forest-Type and Urban-Type Han Chinese Buddhist Temples. *Forests* **2023**, *14*, 79. [CrossRef]
12. Yin, Y.; Shao, Y.; Lu, H.; Hao, Y.; Jiang, L. Predicting and Visualizing Human Soundscape Perception in Large-Scale Urban Green Spaces: A Case Study of the Chengdu Outer Ring Ecological Zone. *Forests* **2023**, *14*, 1946. [CrossRef]
13. Wei, Y.; Hou, Y.-Y. Forest Visitors' Multisensory Perception and Restoration Effects: A Study of China's National Forest Parks by Introducing Generative Large Language Model. *Forests* **2023**, *14*, 2412. [CrossRef]
14. Zhong, B.; Xie, H.; Gao, T.; Qiu, L.; Li, H.; Zhang, Z. The Effects of Spatial Characteristics and Visual and Smell Environments on the Soundscape of Waterfront Space in Mountainous Cities. *Forests* **2023**, *14*, 10. [CrossRef]
15. Cheng, Y.; Bao, Y.; Liu, S.; Liu, X.; Li, B.; Zhang, Y.; Pei, Y.; Zeng, Z.; Wang, Z. Thermal Comfort Analysis and Optimization Strategies of Green Spaces in Chinese Traditional Settlements. *Forests* **2023**, *14*, 1501. [CrossRef]
16. Li, J.; Huang, Z.; Zheng, D.; Zhao, Y.; Huang, P.; Huang, S.; Fang, W.; Fu, W.; Zhu, Z. Effect of Landscape Elements on Public Psychology in Urban Park Waterfront Green Space: A Quantitative Study by Semantic Segmentation. *Forests* **2023**, *14*, 244. [CrossRef]
17. Shu, S.; Meng, L.; Piao, X.; Geng, X.; Tang, J. Effects of Audio–Visual Interaction on Physio-Psychological Recovery of Older Adults in Residential Public Space. *Forests* **2024**, *15*, 266. [CrossRef]
18. Chen, G.; Yan, J.; Wang, C.; Chen, S. Expanding the Associations between Landscape Characteristics and Aesthetic Sensory Perception for Traditional Village Public Space. *Forests* **2024**, *15*, 97. [CrossRef]
19. Shen, Y.; Wang, Q.; Liu, H.; Luo, J.; Liu, Q.; Lan, Y. Landscape Design Intensity and Its Associated Complexity of Forest Landscapes in Relation to Preference and Eye Movements. *Forests* **2023**, *14*, 761. [CrossRef]
20. Liu, T.; Mi, B.; Yan, H.; Bao, Z.; Wu, R.; Wang, S. Spatiotemporal Distribution Analysis of Spatial Vitality of Specialized Garden Plant Landscapes during Spring: A Case Study of Hangzhou Botanical Garden in China. *Forests* **2024**, *15*, 208. [CrossRef]

**forests**

**MDPI**

*Article*

# Multi-Scale Effects of Landscape Pattern on Soundscape Perception in Residential Green Spaces

Jiang Liu [1,2], Yi-Jun Huang [1,2], Zhu Chen [1,3,*] and Xin-Chen Hong [1,2]

[1] School of Architecture and Urban-Rural Planning, Fuzhou University, Fuzhou 350108, China; jiang.liu@fzu.edu.cn (J.L.); yijunhuang_fzu@126.com (Y.-J.H.); xch.hung@outlook.com (X.-C.H.)
[2] Fujian Key Laboratory of Digital Technology for Territorial Space Analysis and Simulation, Fuzhou University, Fuzhou 350108, China
[3] Institute of Environmental Planning, Leibniz University Hannover, Herrenhäuser Str. 2, 30419 Hanover, Germany
* Correspondence: chen@umwelt.uni-hannover.de; Tel.: +86-18659133048

**Abstract:** Soundscape quality in green spaces of residential areas directly contributes to residents' quality of life. It has close relationships with landscape characteristics, which should be considered in landscape planning and design processes in residential areas. Accordingly, this study proposed a new perspective on the interrelationships between soundscape perception and landscape pattern on multi-scale, based on a case study of 30 residential green spaces in Fuzhou, China. Percentage of Landscape (PLAND), Patch Density (PD), Landscape Shape Index (LSI), and Patch cohesion index (COHESION) were utilized to represent the landscape pattern of vegetation, buildings, and roads in the residential areas. Soundscape perception was interpreted using the sound dominant degree (SDD) of sound sources and overall soundscape quality. The examined spatial scales range from 20 m to 180 m, with concentric circles spaced 20 m apart for each sampling point. Correlation analyses indicated that most landscape indices of vegetation and buildings were correlated with these soundscape perception indicators, while limited landscape indices of roads were associated with them. Based on the multi-scale landscape indices, multiple linear regression models for the SDD of sound sources and overall soundscape quality were established, confirming that the scale effect of landscape patterns can affect soundscape perception. Expressly, results indicated that these models were chiefly influenced by the landscape indices at a scale less than 120 m, but the scale effect of landscape pattern on the SDD of birdsong, pleasantness, and quietness was not so evident. Furthermore, we found that the number of explanatory variables may somewhat affect the model performance. The overall interpretability of these landscape indices for the SDD of sound sources was better than that of overall soundscape quality, implying the complexity of the latter. This study offers a fresh insight into the relationship between landscapes and soundscapes at varying scales. The findings can provide useful information for the promotion strategies of landscapes and soundscapes, especially in residential green spaces.

**Keywords:** landscape pattern; sound dominant degree; soundscape quality; residential area; green space; scale effect

**Citation:** Liu, J.; Huang, Y.-J.; Chen, Z.; Hong, X.-C. Multi-Scale Effects of Landscape Pattern on Soundscape Perception in Residential Green Spaces. *Forests* **2023**, *14*, 2323. https://doi.org/10.3390/f14122323

Academic Editor: Panteleimon Xofis

Received: 24 October 2023
Revised: 15 November 2023
Accepted: 21 November 2023
Published: 27 November 2023



## 1. Introduction

With rapid urbanization and population growth, noise pollution has become a pertinent issue worldwide. The World Health Organization (WHO) has stated that noise pollution is one of the primary threats to human health and well-being, and a primary cause of deteriorating urban environmental quality [1]. Decision-makers have considered the quality of the acoustic environment as a crucial component of environmental impact assessment and policies [2]. The Environmental Noise Directive (END) [3] demands member countries assess noise and manage key urban functional areas. Many European countries like Ireland have completed the noise mapping [2,4]. In China, the evaluation of the impact of the acoustic environment started relatively late. However, the established assessment approaches only consider controlling the noise levels, which is not equal to improving the quality of the acoustic environment. This is because human perception of the acoustic environment is also affected by the perceived sound sources, the psychological perception of the user, and other non-acoustic factors [5,6]. The soundscape is "an acoustic environment as perceived or experienced and/or understood by a person or people, in context" as defined in ISO 12913-1 [7]. This concept emphasizes the relationship between individuals, environment, and sound, and takes a more comprehensive approach to improve the quality of the acoustic environment [8–11]. Soundscape research aims to shift the focus from equating the acoustic environment with noise to sound as a resource. Currently, multiple approaches are available for exploring soundscape perception, typically through collecting human perception data along with physical and psychoacoustic information regarding the acoustic environment and context [12]. The definition and utilization of various data collection methods have been outlined in ISO 12913-2 [13], including questionnaires, soundwalk, and interview guidelines. Among these methods, the questionnaire has been commonly used for gathering subjective responses from a large sample across various scales [12,14].

The soundscape in the green spaces of residential areas is especially essential for human benefits, because these green spaces are the core areas for the daily life of urban dwellers. The residential green spaces provide residents with places for leisure, recreation, and socialization [15], and offer opportunities for inhabitants to access the natural environment and experience nature-based pleasure conveniently. Furthermore, studies have found the interrelationships between the quality of residential green spaces with children's health [16], residents' body mass index [17], and psychological health [18,19]. The soundscape serves as a key that unlocks the world, offering both normal-sighted and visually impaired inhabitants meaningful experiences and memories [20]. The poor acoustic environment in these areas can negatively affect residents' physiological and psychological states [21]. Moreover, excessive exposure to noise pollution can lead to health risks such as sleep disorders, cardiovascular diseases, increased stress and anxiety, and cognitive impairment in children [22,23]. The soundscapes in residential green spaces are indispensable for providing good environmental quality and reducing the annoyance caused by air pollution and noise [24]. However, they have not received attention as much as the soundscapes in other green spaces like urban parks or forests [25–28].

Many studies have proven that soundscape perception has interrelationships with landscape features [29,30], such as landscape aesthetics, spatiotemporal dynamics, and biodiversity [11]. Notably, landscape patterns representing landscape structures and ecological processes have been recognized as one of the most essential characteristics affecting soundscape perception [31]. Existing soundscape studies generally explored the relationship between landscape patterns and soundscape perception at only one scale. However, the landscape spatial patterns are differentiated across various scales [32]. This difference caused by scale effects may also affect soundscape perception directly or indirectly. The research of acoustic ecology has started to pay attention to this aspect, and the explored scales ranged from local to regional. At the local scale, it has been shown that the strength of the relationship between the acoustic entropy, the centroid, and skewness with vegetation and topographic features is influenced by the scale effect of landscape, especially at 25 m and 50 m [33]. Acoustic metrics also respond to the percentage of natural vegetation cover at different scales, and scale effects can eliminate redundancy in acoustic metrics. Especially at the 100 m scale, the surrounding landscapes are more likely to influence acoustic features [34]. At the regional scale, one study found a strong association between acoustic indices and habitat structure and quality between 1.5 km and 3 km [35]. In addition, the scale effect of landscape patterns may also indirectly affect soundscape composition. For example, tree cover at 20 m and 500 m scales has a direct and significant impact on bird species richness [36], which may indirectly affect the perceived intensity and diversity of birdsongs as well as soundscape restoration in the environment [37,38].

Nevertheless, the scale effects of landscape spatial features on soundscape perception are still under-explored, and relevant research was rarely set in residential green spaces. Accordingly, this study aims to explore the relationships between soundscape perception and landscape pattern at different scales, and then examine the scale effect of landscape pattern on soundscape perception. To this end, this study took 30 green spaces within 20 residential areas in Fuzhou, China, as a case study. These green spaces served as the center to create nine equally spaced concentric circles ranging from 20 to 180 m, representing different spatial scales. This study can provide essential data references and empirical evidence for soundscape planning and management in residential areas.

## 2. Research Method

### 2.1. Study Area

This study selected the residential areas on the south side of Jinshan Avenue in Fuzhou, China, as the case study sites, trying to minimize the influence of site location factors and the biases of research results (Figure 1a). The selection of residential areas was based on the following principles: (1) the building time of the residential area should be between ten and fifteen years and have relatively complete facilities; (2) the internal green spaces of the residential area should be well planned with diverse and representative characteristics; (3) the green spaces in the residential area should be away from city roads to avoid abundant external noise interference. Ultimately, 20 residential communities were selected and 30 representative green space samples were chosen for the acoustic data collection on site, labeled from SP1 to SP30 (Figure 1b).

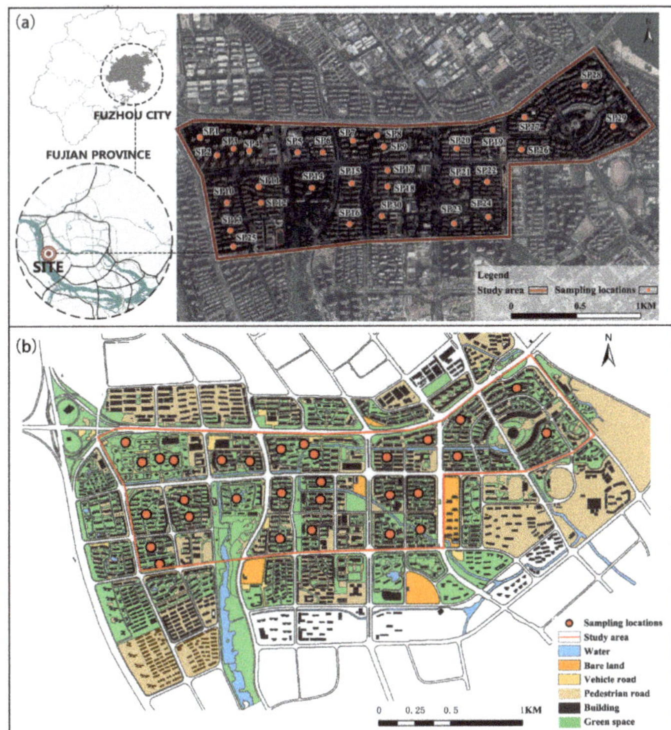


**Figure 1.** (**a**) Location and (**b**) land use types of the study area.

This study analyzed the landscape pattern based on three land use types: buildings (BD), vegetation (VT), and roads, including vehicular roads (VR) and pedestrian roads

(PR). The previous study indicated that the landscape pattern of these three land use types significantly determines the landscape spatial characteristics within the region [39]. The category of typical sound sources in the study area is outlined in Table 1. The classification and identification are based on the pilot study in the study area and adapted from relevant literature [40,41].

**Table 1.** Category of typical sound sources.

| Primary Sound Category | Sound Source | Abbreviation |
| --- | --- | --- |
| Natural Sound | Bird song | BS |
| | Leaves rustling | LR |
| Human Sound | Surrounding speech | SS |
| | Children playing | CP |
| Traffic Sound | External traffic | ET |
| | Internal traffic | IT |

A nine-scale circular buffer zone centered with each sampling site was created to reflect different spatial scales. The scale radius was set from 20 m to 180 m. The scales cover 20 m, 40 m, 60 m, 80 m, 100 m, 120 m, 140 m, 160 m, and 180 m. The set size refers to previous studies [30,32,34] and China's Standard for urban residential area planning and design [42]. We further assigned these scales into three relative ranks for better analysis and understanding (Table 2).

**Table 2.** Relative scale ranks for the study area.

| Relative Rank | Scale |
| --- | --- |
| Small scale | 20 m, 40 m, and 60 m |
| Medium scale | 80 m, 100 m, and 120 m |
| Large scale | 140 m, 160 m, and 180 m |

*2.2. Data Collection and Measurement*

2.2.1. Landscape Spatial Data

The status quo of the three land use types within the study area was vectored in ArcGIS10.2, including buildings, vegetation, and vehicular and pedestrian roads, based on Google satellite imagery of the study area, topographic maps of residential areas provided by the Natural Resources Bureau of Cangshan District, as well as street view maps, and on-site investigations. Subsequently, the nine-scale concentric circular zone of each sampling site was used to extract the corresponding landscape types.

This study selected four landscape metrics from two aspects, landscape composition and spatial pattern, deduced from the literature exploring the relationship between landscape features and soundscape perception [26,30,31]. The indices included Percentage of Landscape (PLAND), Patch Density (PD), Landscape Shape Index (LSI), and Patch cohesion index (COHESION), as shown in Table 3. Fragstats4.2 was employed to calculate such landscape indices (see Supplementary Materials).

2.2.2. Soundscape Perception Data

The soundscape perception data was obtained through a questionnaire survey conducted in November 2021 for 15 non-rainy days. Data were collected between 8:00–11:00 am and 3:00–6:00 pm. These two periods were the active times for most residents, which was practical for engaging participants. In addition, these timeframes can relatively ensure that we capture diverse and dynamic natural, human, and traffic sounds in the morning and afternoon. The questionnaire consisted of two parts: (1) demographic, social, and behavioral characteristics of the participants, including gender, age, education level, activity frequency, and purpose of visit; and (2) participants' evaluation of sound sources and overall soundscape quality of the green spaces. This study was approved by the Institutional Review Board (IRB) of Fuzhou University. All subjects were informed of the purpose and content of the questionnaire before the survey. They were able to quit this survey at any time without any reason if they wanted to

**Table 3.** Land use types and selected landscape indices.

| Land Use Type | Landscape Index | Abbreviation | Explanation |
|---|---|---|---|
| Vegetation | Percentage of landscape index of vegetation. | PLAND_VT | The relative proportion of vegetation patches in the entire landscape. |
| | Patch density of vegetation. | PD_VT | The number of vegetation patches per unit area reflects the intensity of patch density. |
| | Landscape shape index of vegetation | LSI_VT | The degree of regularity of the shape of vegetation patches. |
| | Patch cohesion index of vegetation. | COHESION_VT | The connectivity of vegetation patches. |
| Building | Percentage of landscape index of building. | PLAND_BD | The relative proportion of building patches in the entire landscape. |
| | Patch density of building. | PD_BD | The number of building patches per unit area. |
| | Landscape shape index of building. | LSI_BD | The degree of regularity of the shape of building patches. |
| | Patch cohesion index of building. | COHESION_BD | The connectivity of building patches. |
| Road | Percentage of landscape index of vehicular road. | PLAND_VR | The relative proportion of vehicular road patches in the entire landscape. |
| | Patch density of vehicular road. | PD_VR | The number of vehicular road patches per unit area. |
| | Landscape shape index of vehicular road. | LSI_VR | The degree of regularity of the shape of vehicular road patches. |
| | Patch cohesion index of vehicular road. | COHESION_VR | The connectivity of vehicular road patches. |
| | Percentage of landscape index of pedestrian road. | PLAND_PR | The relative proportion of pedestrian road patches in the entire landscape. |
| | Patch density of pedestrian road. | PD_PR | The number of pedestrian road patches per unit area. |
| | Landscape shape index of pedestrian road. | LSI_PR | The degree of regularity of the shape of pedestrian road patches. |
| | Patch cohesion index of pedestrian road. | COHESION_PR | The connectivity of pedestrian road patches. |

The sound source perception was evaluated by perceived occurrences of sounds (POS) with a 5-point scale (1—never, 2—occasionally, 3—normal, 4—often, 5—frequently), and perceived loudness of sounds (PLS) also scored by a 5-point scale (1—very weak, 2—weak, 3—normal, 4—strong, 5—very strong). The overall soundscape quality assessment consisted of the following 6 adjectives: pleasant, harmonious, vibrant, comfortable, eventful, and quiet [25,43], which were also rated on a 5-point scale (from 1-strongly disagree to 5-strongly agree). Furthermore, we also calculated the sound dominant degree (SDD) based on POS and PLS, as shown in Equation (1), which refers to the dominance of a specific sound source [44].

$$SDD_{Ji} = POS_{Ji} \times PLS_{Ji} \tag{1}$$

in this equation, POS denotes perceived occurrences of individual sounds, and PLS denotes the perceived loudness of individual sounds. Similarly, j represents the jth sample, and i represents the ith sound source.

### 2.3. Statistical Analysis

On the 30 survey sites, 350 questionnaires were distributed, with 338 returned. After eliminating questionnaires with incomplete or false information, there were 308 valid questionnaires, resulting in an effective rate of 91.1%. The reliability and validity of the collected questionnaire were examined, with the results indicating a Cronbach's alpha coefficient of 0.751, indicating acceptable reliability. The Kaiser-Meyer-Olkin (KMO) value is 0.796, more significant than the recommended 0.7 threshold. In addition, the significance value of the Bartlett test was 0.000, which is less than the standard 0.05 level, suggesting that the data had good validity. The demographic characteristics of the respondents are presented in Figure 2.

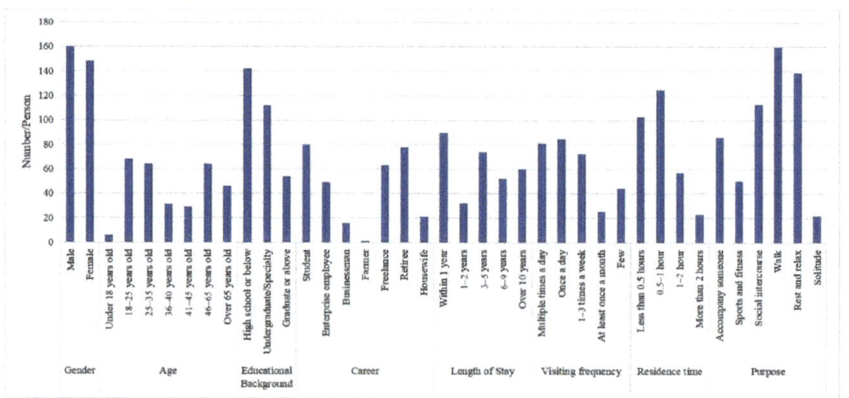


**Figure 2.** Statistical sample information ($n$ = 308).

The processed landscape index and soundscape perception data were statistically analyzed in SPSS 26. Spearman's rho correlation analysis was applied to explore the significant relationship between landscape index and soundscape perception of different landscape types at multiple scales. Stepwise multiple linear regression analysis was employed to identify further the key landscape indices affecting soundscape perception. The SDD was set as the dependent variable, and landscape indices that show significant correlations with SDD were used as independent variables. The collinearity diagnostic rule was applied to ensure no collinearity issue among the independent variables, as indicated by the variance inflation factor (VIF) being less than 10.

## 3. Results

### 3.1. Basic Analysis

#### 3.1.1. Sound Source Perception

The POS and PLS in each sampling point were calculated through statistical analysis of the questionnaire data, as shown in Figure 3. Regarding POS, birdsongs generally had a higher value in natural sounds, while the value of leaves rustling was lower. To some extent, the sounds of surrounding speech and children playing are highly similar. The differences in the PLS values for natural and artificial sounds were identical to their POS. Specifically, the PLS of bird song and surrounding speech were relatively high. Additionally, internal traffic sounds were generally perceived as higher in PLS than external ones.

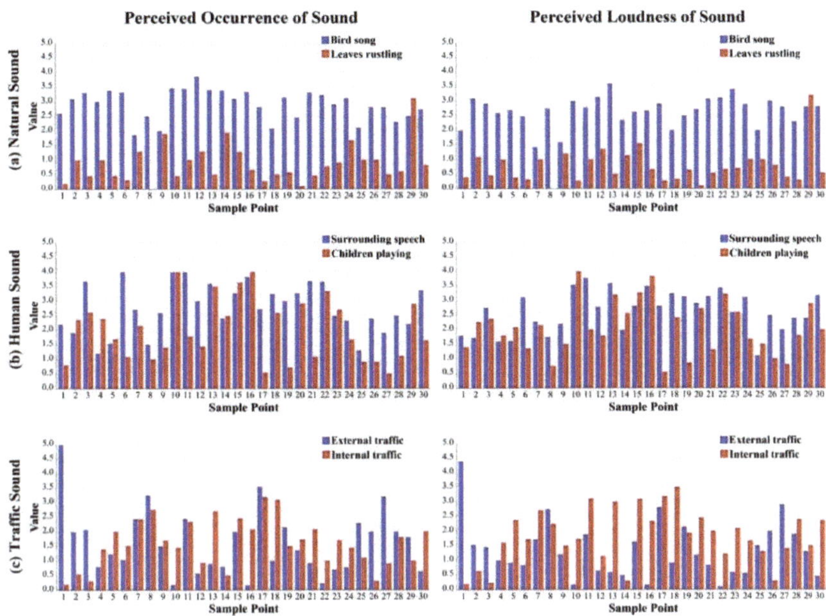

**Figure 3.** Mean Frequency and Intensity Map of Sound Sources.

### 3.1.2. Principal Component Analysis of Overall Soundscape Quality

Principal Component Analysis (PCA) was performed on the six soundscape perceptual attributes to synthesize the main character of overall soundscape quality in the study area. Referring to ISO standards and the characteristics of residential areas [7,45], the following three common factors were extracted: soundscape pleasantness, soundscape eventfulness, and soundscape quietness (Table 4). This outcome is similar to the results of previous studies [45]. The cumulative variance contribution rate of the three factors was 73.873%, indicating that they can capture relatively comprehensive information on the overall soundscape quality.

**Table 4.** PCA results of soundscape perception factors.

| Common Factor | Factor | Component | | |
|---|---|---|---|---|
| | | 1 | 2 | 3 |
| F1 (soundscape pleasantness) | pleasant | 0.879 | | |
| | harmonious | 0.729 | | |
| F2 (soundscape eventfulness) | eventful | | 0.840 | |
| | vibrant | | 0.747 | |
| F3 (soundscape quietness) | quiet | | | 0.934 |
| | comfortable | | | 0.632 |

Note: Extraction method: principal component analysis. Rotation method: Kaiser normalized varimax method. Rotation converged in 4 iterations.

### 3.2. Correlations between Landscape Indices of Different Land Use Types and Soundscape Perception Indicators

#### 3.2.1. Vegetation

Figure 4 shows many correlations between landscape indices of vegetation and sound source SDD and soundscape quality. The SDD of natural sounds is only related to LSI_VT and PD_VT. Specifically, the SDD of both bird song and leaves rustling positively correlates with LSI_VT at all examined scales except for 20 m. Furthermore, the SDD of bird song also has a positive relation to PD_VT at the scale from 40 m to 180 m. The SDD of human sounds has correlations with all landscape indices of vegetation. The SDD of both human sounds only correlates with PD_VT at 40 m–100 m, and the SDD of surrounding speech has more correlations at 120 m–180 m. The SDD of surrounding speech presents negative correlations with PLAND_VT and COHESION_VT ranging from 40 m to 180 m; however, apart from the scale of 100 m with the former. The SDD of children playing showcases two correlations with LSI_VT at 80 m and 100 m. Similarly, the SDD of external and internal traffic also exhibits significant relationships with the four indices, all of which are negative. Notably, only LSI_VT correlates with the SDD of these two traffic sounds. The SDD of external traffic correlates with PD_VT at 40 m–160 m. Both PLAND_VT and COHESION_VT are related to the SDD of internal traffic at 40 m–120 m and 40 m–100 m, separately. Regarding the correlations between the soundscape quality indicators and the landscape indices, soundscape pleasantness and eventfulness are positively related to LSI_VT at 80 m, 100 m and 140 m, and 80 m–160 m, respectively. PLAND_VT and COHESION_VT only have relationships with soundscape eventfulness at 60 m to 180 m and 80 m to 180 m, respectively. Nevertheless, soundscape quietness only correlates with COHESION at the 20 m scale among all indices.

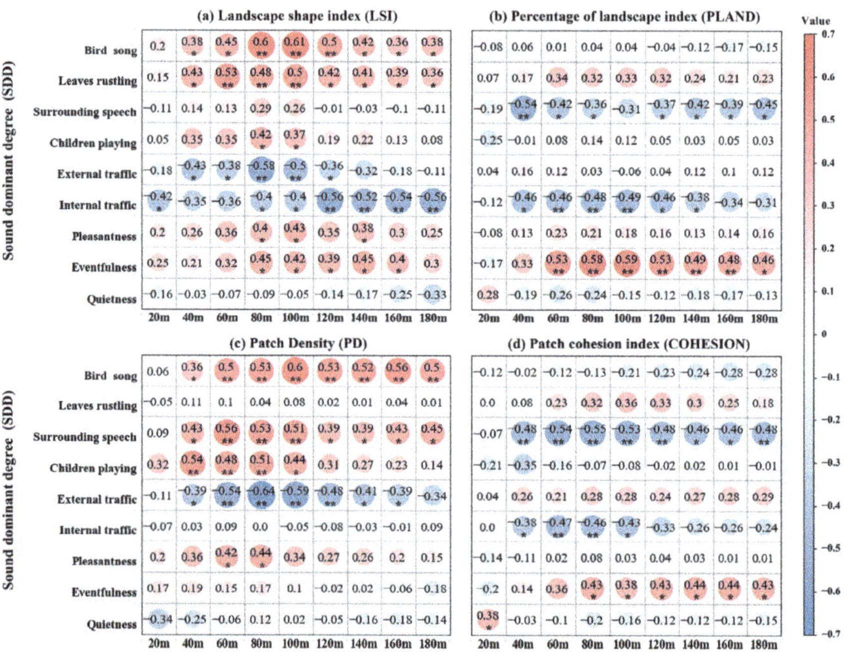

**Figure 4.** Correlations between landscape indices of the vegetation and the sound dominant degree (SDD) of sound sources and overall soundscape quality (* indicates $p < 0.05$, ** indicates $p < 0.01$).

### 3.2.2. Buildings

Figure 5 shows the correlations between the landscape indices of buildings and the SDD of sound sources and overall soundscape quality. LSI_BD has a relatively weaker correlation with the SDD of sound sources than other landscape indices, especially for the relations with the SDD of natural sounds. Apart from LSI_BD, all the building landscape indices exhibit significant positive correlations with the SDD of bird song at the scale of 140 m–180 m. However, the SDD of leaves rustling is unrelated to the four landscape indices at all tested scales. The relationships between the SDD of surrounding speech and the four building landscape indices are significant, mainly at medium and large scales. The SDD of children playing is found to have very weak correlations with the building indices, which appear only on one scale of each index. In addition, negative correlations are located between the SDD of external traffic and PLAND_BD and PD_BD at 120 m–140 m and COHESION_BD at 140 m–180 m. The correlations between the SDD of internal traffic and the four indices are found generally at small and large scales. The soundscape quality components are found to have rare correlations with the building landscape indices. Soundscape pleasantness and soundscape quietness are not associated with any of these indices. Correlations between soundscape eventfulness and the indices are chiefly below 100 m but not too much.

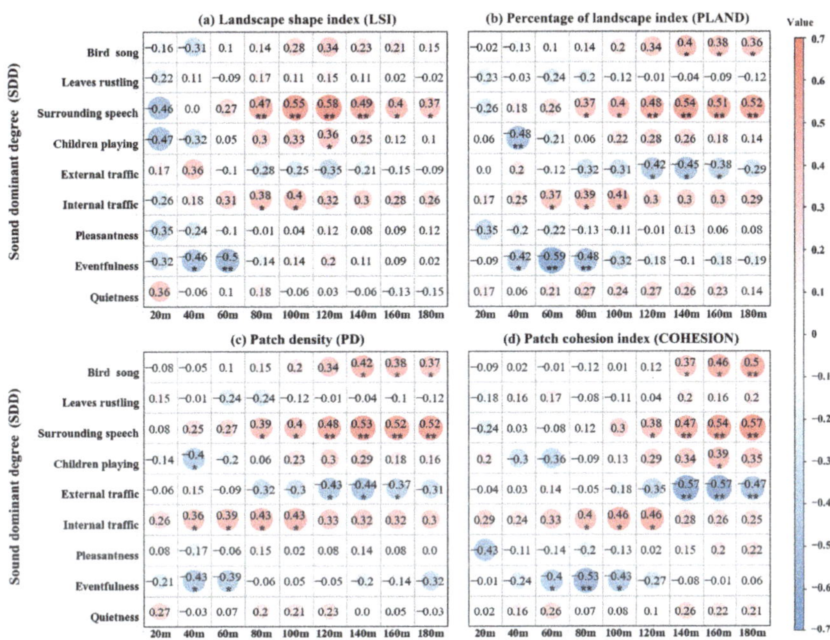

**Figure 5.** Correlations between building landscape indices and the sound dominant degree (SDD) of sound sources and overall soundscape quality (* indicates $p < 0.05$, ** indicates $p < 0.01$).

3.2.3. Roads

Figures 6 and 7 indicate that only a few landscape indices representing vehicular and pedestrian roads at specific scales significantly correlate with the SDD of sound sources. Compared to vehicular landscape indices, pedestrian landscape indices showcase more correlations with the SDD of sound sources and soundscape quality. Notably, PLAND_PR exhibits the most correlations with the SDD of surrounding speech at each scale ranging from 40 m to 180 m and with the SDD of internal traffic from 80 m to 180 m except for 160 m. Likewise, COHESION_PR has continuous correlations with the SDD of leaves rustling at the scale ranging from 20 m to 80 m. Soundscape pleasantness presents only two correlations with vehicular road landscape indices, including PLAND_VR and COHESION_VR; both are found at 80 m. Soundscape eventfulness is related to PLAND_PR at most scales among the three components, from 80 m to 140 m. Interestingly, the four landscape indices of pedestrian roads are all positively correlated with soundscape quietness at the 20 m scale. However, no significant correlation was found between the landscape indices of vehicular roads and soundscape quietness.

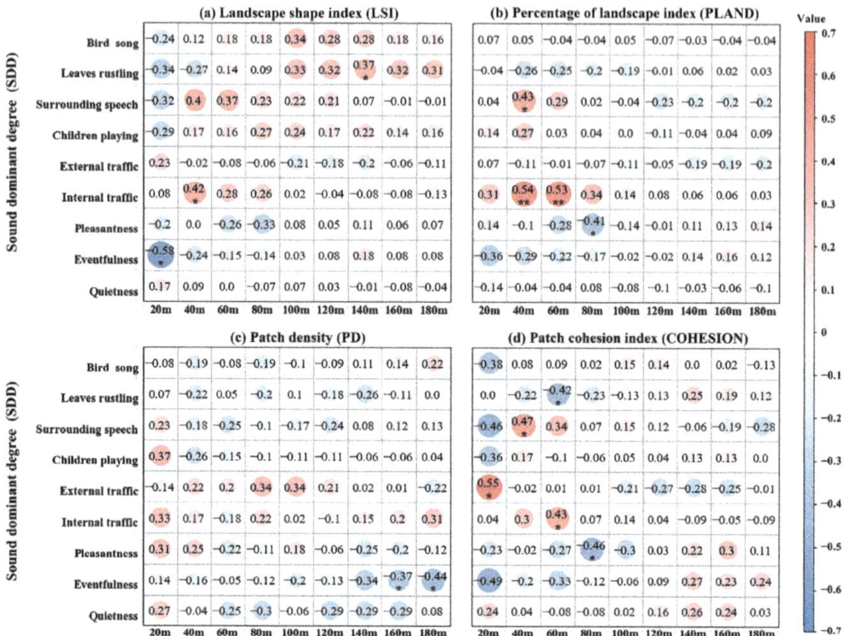

**Figure 6.** Correlations between vehicular road landscape indices and the sound dominant degree (SDD) of sound sources and overall soundscape quality (* indicates $p < 0.05$, ** indicates $p < 0.01$).

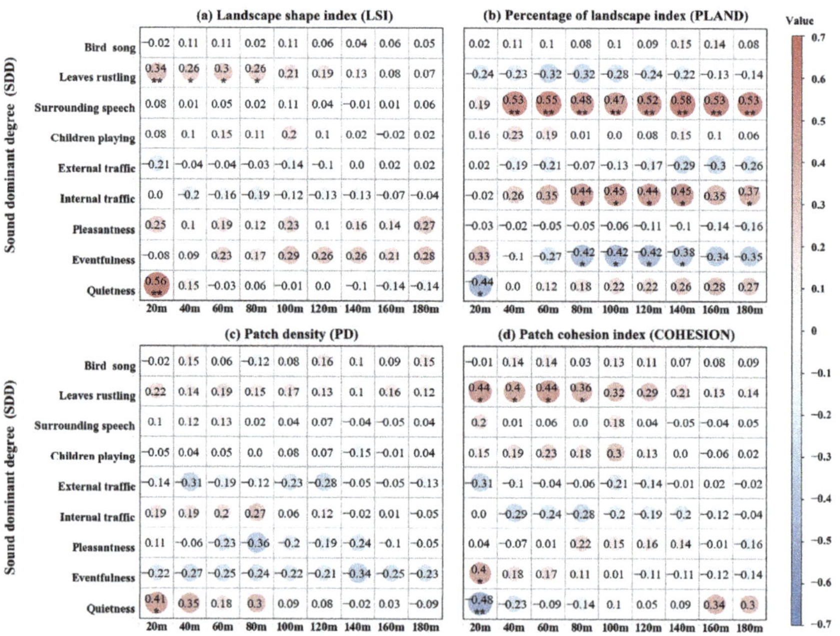

**Figure 7.** Correlations between pedestrian road landscape indices and the sound dominant degree (SDD) of sound sources and overall soundscape quality (* indicates $p < 0.05$, ** indicates $p < 0.01$).

### 3.3. Regression Models for Soundscape Perception Utilizing Multi-Scale Landscape Indices

#### 3.3.1. Models for the Sound Dominant Degree of Sound Sources

The results of stepwise multiple linear regression analysis are shown in Table 5. Most independent variables are at medium scales (80 m, 100 m, and 120 m), followed by the indices at small scales (20 m, 40 m, and 60 m). The number of significant landscape indices at large scales (140 m, 160 m, and 180 m) is the least. Only 100 m-PD_VT has a considerable impact on the SDD of bird songs. The SDD of surrounding speech is affected by landscape indices of vegetation and buildings at medium and large scales. PD_VT and PD_BD at 80 m, and LSI_BD at 160 m have positive effects on it, while PD_VT negatively affects it at 120 m. The 40 m-COHESION_BD has a negative impact on the SDD of children playing, but its intensity (Beta) is higher than the other two significant variables, 40 m-PD_VT and 120 m-LSI_BD. The SDD of external and internal traffic is influenced by building landscape indices that are 140 m-COHESION_BD and 100 m-LSI_BD, respectively. There are no landscape indices found to have a significant influence on the SDD of leaves rustling. However, according to the results of the correlation analysis below, there were correlations between the SDD of leaves rustling and landscape indices of vegetation and roads.

**Table 5.** Multivariate stepwise regression models for the sound dominant degree (SDD) of sound sources using multi-scale landscape indices.

| Dependent Variable | Independent Variable | Beta | VIF | t | $R^2$ | F |
|---|---|---|---|---|---|---|
| Bird song SDD | 100 m-PD_VT | 0.53 | 1 | 3.311 ** | 0.256 | 10.963 ** |
| Surrounding speech SDD | 80 m-PD_VT | 1.082 | 2.973 | 7.736 ** | 0.849 | 23.013 ** |
|  | 80 m-PD_BD | 0.672 | 1.571 | 6.607 ** |  |  |
|  | 160 m-LSI_BD | 0.418 | 1.156 | 4.797 ** |  |  |
|  | 120 m-PD_VT | −1.046 | 3.862 | −6.564 ** |  |  |
| Children playing SDD | 40 m-PD_VT | 0.396 | 1.207 | 2.86 ** | 0.540 | 12.367 *** |
|  | 40 m-PLAND_BD | −0.535 | 1.291 | −3.739 ** |  |  |
|  | 120 m-LSI_BD | 0.478 | 1.496 | 3.104 ** |  |  |
| External traffic SDD | 20 m-COHESION_VR | 0.460 | 1.012 | 2.703 ** | 0.598 | 11.429 ** |
|  | 140 m-COHESION_BD | −0.619 | 1.012 | −3.632 ** |  |  |
| Internal traffic SDD | 100 m-LSI_BD | 0.363 | 1.003 | 2.242 ** | 0.399 | 8.646 ** |
|  | 120 m-LSI_VT | −0.586 | 1.003 | −3.620 * |  |  |

Note: * indicates $p < 0.05$, ** indicates $p < 0.01$, *** indicates $p < 0.001$.

#### 3.3.2. Models for the Overall Soundscape Quality

The regression models for the overall soundscape quality indicators show that not many landscape indices have a significant relationship with them (Table 6). The scale effect of the landscape pattern on overall soundscape quality does not seem to be obvious, especially for pleasantness and quietness. Both pleasantness and quietness are only affected by one metric, 80 m-PLAND_VR and 20 m-PD_PR, separately. Eventfulness is negatively affected by 60 m-PLAND_BD and 100 m-COHESION_VT, while positively affected by 140 m-PLAND_VT.

**Table 6.** Multivariate stepwise regression models for overall soundscape quality using multi-scale landscape indices.

| Dependent Variable | Independent Variable | Beta | VIF | t | $R^2$ | F |
|---|---|---|---|---|---|---|
| Pleasantness | 80 m-PLAND_VR | −0.506 | 1 | −2.988 ** | 0.227 | 8.927 ** |
| Eventfulness | 60 m-PLAND_BD | −1.023 | 1.828 | −5.431 ** | 0.728 | 13.492 ** |
|  | 100 m-COHESION_VT | −0.948 | 2.647 | −4.182 * |  |  |
|  | 140 m-PLAND_VT | 0.467 | 2.242 | −2.238 * |  |  |
| Quietness | 20 m-PD_PR | 0.406 | 1 | 2.354 * | 0.135 | 5.543 * |

Note: * indicates $p < 0.05$, ** indicates $p < 0.01$.

## 4. Discussion

### 4.1. Typical Landscape Characteristics Associated with Soundscape Perception

The correlation results in vegetation indicated that LSI_VT and PD_VT significantly impacted soundscape perception at 80 m and 100 m scales. In contrast, other landscape indices had a relatively weaker effect on sound source perception. A possible reason is that landscape types and spatial configurations at scales lower than 80 m are relatively simple. People's perception of landscape elements weakens after over the scale of 100 m. On the impact of the SDD of different sound sources, LSI_VT, PLAND_VT, and PD_VT, they strongly promoted the SDD of bird song. This result indicates that the common birds in residential areas prefer vegetation with relatively complex and irregular shapes, similar to previous research [46,47]. However, some scholars have found that birds survive in regularly shaped urban green spaces [48], highlighting the uniqueness and value of studying the soundscapes of residential green spaces. In contrast, the above three landscape indices negatively impacted the SDD of surrounding speech and internal transportation noise, which may be attributed to the relatively complex vegetation structures that help to reduce anthropogenic noise [49]. Regarding overall soundscape quality, both LSI_VT and PD_VT had substantial impacts on soundscape pleasantness at the 80 m scale, while LSI_VT, PLAND_VT, and COHESION_VT had significant implications on soundscape eventfulness at the 100 m scale. This result may be due to the promoting effect of bird songs on soundscape pleasantness [50,51], and people were more likely to perceive various sound sources in vegetation at this scale. Based on such, it is suggested that the scale from 80 m to 100 m is more suitable for landscape creation by establishing a diverse and well-structured vegetation, which can not only enhance the perception of positive sound sources and weaken the impact of negative sound sources, but also improve the overall soundscape quality.

For the relationship between the landscape indices and the SDD in building land use, the SDD of surrounding speech and internal traffic had the strongest positive correlation with PD_BD and PLAND_BD. This finding indicates that in densely built areas, sound propagates perhaps through multiple reflections from walls [52,53], because the buildings in high-density urban areas can obstruct the free propagation of road traffic noise, leading to lower noise levels within residential areas and the formation of sound shadow zones [54,55]. This suggests that the layout of buildings can be utilized in residential planning to create sound shadow zones for improving acoustic environment quality. For example, high-density or high-rise buildings can be arranged on the side of external roads or residential areas combined with commercial building layouts to form strong sound shadow zones. In comparison, PD_BD and PLAND_BD had positive effects on the SDD of internal traffic sound at the scale of 100 m–120 m, which may be due to the increase in building area and density leading to the proximity of roadways to vegetation, thereby enhancing the perception of traffic sound. For overall soundscape quality, there was no significant correlation between the various landscape indices of buildings and soundscape pleasantness, while LSI_VT, PLAND_VT, and COHESION_VT were all positively correlated with soundscape eventfulness. However, we found that at the scale of 60 m–80 m, the size of building patches increases, and patch clusters become more aggregated, leading to soundscape eventfulness decreases. This result may be due to the relatively low area percentage of vegetation at that scale, thereby reducing the attractiveness of the place to birds and negatively impacting bird abundance [56], further affecting soundscape eventfulness. Additionally, LSI_BD negatively affected soundscape eventfulness at 40–60 m, possibly due to the complex architectural forms that affect sound propagation, decreasing soundscape eventfulness. These findings indicate that the pattern of buildings in residential areas significantly impacts sound perception. Therefore, increasing the height and density of buildings near urban roads is advisable while reducing the distance between buildings along the streets. These strategies are beneficial in creating good sound shadow areas, thereby reducing the permeability of noise. Moreover, designing relatively diverse forms of individual buildings can also effectively minimize external traffic sound, and adopting a simple arrangement method in the overall layout can promote the perception of positive soundscapes inside the residential area.

Regarding the analysis of road land use, results showed that the PLAND was more closely related to the SDD of surrounding speech and internal traffic compared to other landscape indices. The relationship between the SDD of bird song and landscape indices of roads was not as significant. This finding differs from other studies that indicated a negative correlation between bird abundance and road exposure [57,58]. Regarding overall soundscape quality, PLAND_VR and PLAND_PR had significant negative effects on soundscape pleasantness and eventfulness, respectively. These findings indicated that visual exposure to residential roads, including vehicle and pedestrian roads, may decrease the perception of overall soundscapes. From the perspective of acoustics, the sound reflection effect of road surfaces could also be a reason that increases the difficulty of perceiving natural sounds, thus reducing the overall soundscape quality. This study found a significant correlation between the quietness of soundscapes and the landscape indices of pedestrian roads, particularly at a scale of 20 m. The PD_PR and LSI_PR were found to positively impact the quietness, consistent with previous research [59]. These findings suggest that the landscape pattern of pedestrian roads is a crucial factor affecting the quietness of residential areas at the 20 m scale. Therefore, increasing the complexity of road forms, such as meandering and scattered footpaths, can be a helpful design strategy to help create a quiet environment in residential areas. In addition, previous studies have shown that paving materials can also impact the quality of soundscapes, with grass having more silent proposed background noise, making it more popular than gravel [60]. Therefore, in addition to adjusting the layout of pedestrian roads, sound-absorbing materials such as grass and wooden boards can also be considered paving materials to optimize the sound environment further [61].

*4.2. Scale Effects of Landscape Pattern on Soundscape Perception*

The SDD models reveal that most landscape indices with significant impacts are at small to medium scales, namely less than or equal to 120 m. However, the SDD of bird song appears unaffected by scale differences in landscape pattern, as it has only one explanatory variable at 100 m. The SDD is primarily influenced by landscape indices related to vegetation and buildings at various scales, with the only significant road landscape index being PD_VT. Vegetation landscape indices considerably impact the SDD of bird song, surrounding speech, and internal traffic at medium scales. This suggests that structurally rich vegetation is more effective at attracting birds and people and reducing internal traffic noise at medium scales [51]. However, the 120 m-PD_VT has a negative impact on surrounding speech, which we speculate might be due to residents within the surveyed neighborhoods preferring to visit nearby vegetation compared to those further away. The building landscape indices significantly affect all the SDD of sound sources at various scales except for birdsongs. Building landscape indices positively affect surrounding speech (primarily from adults) at medium and large scales, indicating that human activities are influenced by building density and complexity. This phenomenon was also observed in previous research [62]. Interestingly, the SDD of children playing is negatively impacted by PLAND_BD at a small scale (20 m), in contrast to the surrounding speech. We speculate that children's activities within the surveyed residential areas may often occur without adult supervision, as guardian behavior is associated with child behavior [63]. Furthermore, the explanatory variables are at medium and small scales, indicating that the activity range of children may be smaller than that of adults. The SDD of traffic noise model results show that building landscape indices at medium and large scales also influence the propagation of internal and external traffic noise to some extent. Specifically, buildings that are both distant from vegetation (greater than or equal to 100 m) and densely built help mitigate the transmission of external traffic noise inward. Conversely, internally uniform building forms within the residential areas assist in controlling internal traffic noise [64]. The road landscape index COHESION_VR only significantly impacts the SDD of external traffic at a small scale (20 m), indicating that adjacent urban roads outside the residential areas contribute significantly to noise pollution in internal green spaces. Therefore, in residential planning, it is crucial to address areas near urban roads on the periphery of residential areas to mitigate the negative impact of external roads on the residential soundscapes. Additionally, at a medium scale (120 m), LSI_VT has a negative impact on internal traffic noise. This suggests that designing structurally diverse vegetation within residential areas at a 120 m distance can effectively reduce internal traffic noise intensity. Such design can be implemented through plant arrangements within green spaces by enriching plant species or structural combinations such as tree-shrub-grass plant configuration [65].

Similarly, soundscape quality is primarily influenced by landscape indices at small to medium scales. This suggests that landscape design within a radius of 100 m or more minor can more effectively enhance the overall soundscape quality of residential green spaces. Among the three components, only eventfulness is influenced by landscape indices at different scales (Table 6), including 60 m-PLAND_BD, 100 m-COHESION_VT, and 140 m-PLAND_VT. This indicates that, at small scale, soundscape eventfulness is mainly affected by building density, while at medium and large scales, it tends to be influenced by vegetation structures and areas. This is because, on the one hand, vegetation can create favorable habitat conditions, attracting various sound-producing organisms, such as birds and insects, to inhabit them [66]. On the other hand, low-density building layouts provide more space for sound propagation, reducing the frequency of sound reflections on building surfaces [67]. Only 80 m-PLAND_VR has a significant negative impact on pleasantness (Beta = −0.506). Combining this result with the findings of the external traffic SDD model, we found that adjacent external roads to the surveyed residential areas may generate significant traffic noise, thereby diminishing the inside soundscape quality. This is consistent with previous research findings [64,68]. However, for urban residential area planning, the proportion of external roads is often challenging to alter. Therefore, to ensure or enhance the pleasantness of the residential internal soundscape, special attention should be paid to controlling the areas near the edges of residential neighborhoods adjacent to urban roads. For example, this can be achieved by increasing vegetation density in these areas or constructing water features to create masking effects through water sounds [69,70], reducing the perceived intensity of external noise. Quietness is only positively influenced by 20 m-PD_PR, indicating that moderately increasing the density of pedestrian roads at a small scale can promote soundscape quietness. This finding suggests that, for instance, a modest increase in the density of pedestrian roads or paths with varied forms within the interior or adjacent residential green spaces can allow residents to experience a quieter sound environment.

In addition, we found a correlation between the number of explanatory variables and the $R^2$ of the regression models for the SDD and overall soundscape quality indicators. Regression models with multiple significant independent variables exhibited better fit than models with only one independent variable. Furthermore, landscape pattern indices at different scales demonstrated more substantial explanatory power for the SDD than for overall soundscape quality. The analysis results confirm our research hypothesis that landscape spatial characteristics at different scales significantly influence soundscape perception. However, our analysis results also indicate room for improvement in the performance of regression models. Among the SDD models, except for the SDD model of bird song (which had only 25.6% of explained variance), the other models exhibited relatively good fits. Based on previous research, we speculate that this could be due to additional factors affecting the SDD, such as landscape diversity [51]. High landscape diversity in green spaces can provide better habitat quality, attracting more birds and enhancing the richness and perceptual intensity of birdsongs. However, landscape diversity, such as Shannon's Diversity Index (SHDI) or Simpson's Diversity Index (SIDI), is typically measured at the landscape scale [71]. Thus, this landscape feature may influence the SDD of birdsong at larger scales. The SDD model of surrounding speech achieved the best fit with an $R^2$ of 84.9%. This suggests that combining landscape structural characteristics of medium- or large-scale vegetation and buildings can effectively predict the SDD values of surrounding speech. Among the three regression models for overall soundscape quality indicators, only the eventfulness model exhibits good explanatory power at 72.8%. In contrast, the models for pleasantness and quietness have less satisfactory fits at 22.7% and 13.5%, respectively. This indicates that influencing factors and processes may be more complex than individual sound perception for the overall sound quality. Moreover, other factors besides landscape spatial characteristics may play a more decisive role. Previous research found that sound characteristics and cultural background influence soundscape quality. The factors, such as natural sound occurrence, sound preferences, noise intensity, and people's cultural and cognitive backgrounds, can significantly impact people's perception of soundscape pleasantness and quietness [11]. Therefore, the results of this study confirm that only incorporating landscape spatial characteristics at different scales may be insufficient to fully explain the variability in soundscape quality [30].

*4.3. Limitations and Future Study*

The present study has several limitations. Firstly, it only focused on the residential areas of a typical southern coastal city, Fuzhou. However, the soundscape perception and its effects on human well-being may vary across locations [72,73]. Thus, the results may not fully reflect the acoustic characteristics of residential areas in other regions. In addition, this study only considered the landscape elements within the residential areas. Still, it did not include the impact of the urban land cover types outside residential areas, such as urban roads. The perception of sound sources near the boundaries of residential areas can be affected by external noise sources outside the residential area. To overcome such shortcomings, follow-up studies could supplement typical case studies in different regions. In addition, researchers could consider the impact of urban land cover types on the soundscape of residential areas, select sampling points as far away as possible from external noise sources, or conduct time-segmented studies. Additionally, this study only reflected the soundscape characteristics in one season, and future research could also investigate and compare the acoustic environments of residential areas during different seasons. Furthermore, the participants in this study were not trained before the survey. They were randomly selected and asked in the field. This might affect the integrity of their responses to some extent because some of them may not recognize the sound types and features in their daily life. This offers the opportunity for further improvement in future research. Also, we encourage a combination of acoustic measurement with the questionnaire method in the subsequent investigation. Some interesting results could be found if the acoustic features (e.g., spectral contents [74]) and human responses are explored and compared simultaneously. Finally, the dimensions of indicators in this study were considered only in terms of spatial scale, shape, and aggregation degree. In the future, other indicators, such as the proximity index, should be included to enrich the dimensions of the indicators.

## 5. Conclusions

This study analyzed the relationship between soundscape perception and landscape indices at multiple scales within 30 residential green spaces in Fuzhou, China, which offers an innovative insight into soundscape planning and management in urban green spaces. Soundscape perception was captured from the SDD of the sound source and overall soundscape quality indicators. Landscape indices relating to vegetation, buildings, and roads ranged from 20 m to 180 m. Results showed that most landscape indices at different scales were correlated with the SDD and overall soundscape quality in vegetation and building land use. In contrast, fewer landscape indices in road land use were related to the soundscape perception. This indicates that the multi-scale patterns of vegetation and buildings play more critical roles in forming soundscapes in residential areas. Regression models illustrated that the SDD and overall soundscape quality were affected primarily below the 120 m scale. This suggests that landscape planning and design strategies for promoting soundscape quality can be more useful within this scale. However, the scale effects of landscape patterns on the SDD of birdsong, pleasantness, and quietness seemed insignificant. Furthermore, we found that the multi-scale landscape indices can better explain SDD variance than overall soundscape quality. This means that the components determining soundscape quality are more complex. They can hardly be explained only by landscape spatial features, and therefore variables such as sound features and context factors should also be included for interpreting the soundscape quality. We further argue that such variables may account for more important positions than landscape spatial patterns according to the model's explanatory ability. We are confident that our findings can help planners better understand the useful scales for landscape planning and management to improve soundscape perception in residential areas. Moreover, such results can also advance the state of knowledge regarding the relationships between landscapes and soundscapes. This study serves as helpful data support and empirical evidence for urban soundscape planning and design and related studies in the future.

**Supplementary Materials:** The following supporting information can be downloaded at: https://www.mdpi.com/article/10.3390/f14122323/s1.

**Author Contributions:** Conceptualization and methodology, J.L. and Z.C.; formal analysis, Y.-J.H. and Z.C.; investigation, J.L., Y.-J.H. and X.-C.H.; writing—original draft preparation, J.L., Y.-J.H. and Z.C.; writing—review and editing, J.L., Z.C. and X.-C.H.; supervision, J.L. All authors have read and agreed to the published version of the manuscript.

**Funding:** This research was funded by the National Natural Science Foundation of China (52378049, 52208052), and the Program of Humanities and Social Science Research Program of Ministry of Education of China (Grant No. 21YJCZH038).

**Data Availability Statement:** Data available on request due to restrictions, e.g., privacy or ethical.

**Acknowledgments:** The authors appreciate the help from Zhen Huang in field investigation.

**Conflicts of Interest:** The authors declare no conflict of interest.

# References

1. World Health Organization; Regional Office for Europe. *Environmental Noise Guidelines for the European Region*; World Health Organization: Geneva, Switzerland; Regional Office for Europe: Copenhagen, Denmark, 2018.
2. Murphy, E.; King, E.A. Strategic environmental noise mapping: Methodological issues concerning the implementation of the EU Environmental Noise Directive and their policy implications. *Environ. Int.* **2010**, *36*, 290–298. [CrossRef] [PubMed]
3. European Union. Directive 2002/49/EC of the European Parliament and the Council of 25 June 2002 relating to the assessment and management of environmental noise. *Off. J. Eur. Communities* **2002**, *2002*, L189.
4. King, E.A.; Murphy, E.; Rice, H.J. Implementation of the EU environmental noise directive: Lessons from the first phase of strategic noise mapping and action planning in Ireland. *J. Environ. Manag.* **2011**, *92*, 756–764. [CrossRef] [PubMed]
5. Raimbault, M.; Dubois, D. Urban soundscapes: Experiences and knowledge. *Cities* **2005**, *22*, 339–350. [CrossRef]
6. Zhang, M.; Kang, J. Towards the Evaluation, Description, and Creation of Soundscapes in Urban Open Spaces. *Environ. Plan. B Plan. Des.* **2007**, *34*, 68–86. [CrossRef]
7. *ISO 12913-1*; 2014 Acoustics—Soundscape—Part 1: Definition and Conceptual Framework. ISO: Geneva, Switzerland, 2014.
8. Kang, J. From dBA to soundscape indices: Managing our sound environment. *Front. Eng. Manag.* **2017**, *4*, 184–192. [CrossRef]
9. Abrantes, E. Perceiving, Navigating and Inhabiting: Performance design through sonic strategies. *Perform. Res.* **2021**, *26*, 31–38. [CrossRef]
10. Aletta, F.; Kang, J. Towards an urban vibrancy model: A soundscape approach. *Int. J. Environ. Res. Public Health* **2018**, *15*, 1712. [CrossRef]
11. Chen, Z.; Hermes, J.; Liu, J.; von Haaren, C. How to integrate the soundscape resource into landscape planning? A perspective from ecosystem services. *Ecol. Indic.* **2022**, *141*, 109156. [CrossRef]
12. Mancini, S.; Mascolo, A.; Graziuso, G.; Guarnaccia, C. Soundwalk, Questionnaires and Noise Measurements in a University Campus: A Soundscape Study. *Sustainability* **2021**, *13*, 841. [CrossRef]
13. *ISO/TS 12913-2*; 2018 Acoustics—Soundscape—Part 2: Data Collection and Reporting Requirements. ISO: Geneva, Switzerland, 2018.
14. Tarlao, C.; Steffens, J.; Guastavino, C. Investigating contextual influences on urban soundscape evaluations with structural equation modeling. *Build. Environ.* **2021**, *188*, 107490. [CrossRef]

15. Mygind, L.; Clark, G.M.; Bigelow, F.J.; Fuller-Tyszkiewicz, M.; Knibbs, L.D.; Mavoa, S.; Flensborg-Madsen, T.; Bentsen, P.; Lum, J.; Enticott, P.G. Green enrichment for better mind readers? Residential nature and social brain function in childhood. *J. Environ. Psychol.* **2023**, *88*, 102029. [CrossRef]

16. Ahmed, S.M.; Mishra, G.D.; Moss, K.M.; Mouly, T.A.; Yang, I.A.; Knibbs, L.D. Association between residential greenspace and health-related quality of life in children aged 0–12 years. *Environ. Res.* **2022**, *214*, 113759. [CrossRef] [PubMed]

17. Cummins, S.; Fagg, J. Does greener mean thinner? Associations between neighbourhood greenspace and weight status among adults in England. *Int. J. Obes.* **2012**, *36*, 1108–1113. [CrossRef] [PubMed]

18. Wang, Z.P.; Wang, W. An empirical study on the impact of green spaces in residential areas on the mental health of residents under COVID-19. *Landsc. Archit. Front.* **2020**, *8*, 46–59. [CrossRef]

19. Xiao, Y.; Miao, S.; Zhang, Y.; Chen, H.; Wu, W. Exploring the health effects of neighborhood greenness on Lilong residents in Shanghai. *Urban For. Urban Green.* **2021**, *66*, 127383. [CrossRef]

20. Mediastika, C.E.; Sudarsono, A.S.; Kristanto, L.; Tanuwidjaja, G.; Sunaryo, R.G.; Damayanti, R. Appraising the sonic environment of urban parks using the soundscape dimension of visually impaired people. *Int. J. Urban Sci.* **2020**, *24*, 216–241. [CrossRef]

21. Guo, L.H.; Cheng, S.; Liu, J.; Wang, Y.; Cai, Y.; Hong, X.C. Does social perception data express the spatio-temporal pattern of perceived urban noise? A case study based on 3,137 noise complaints in Fuzhou, China. *Appl. Acoust.* **2022**, *201*, 109129. [CrossRef]

22. Singh, D.; Kumari, N.; Sharma, P. A review of adverse effects of road traffic noise on human health. *Fluct. Noise Lett.* **2018**, *17*, 1830001. [CrossRef]

23. Tobías, A.; Recio, A.; Díaz, J.; Linares, C. Health impact assessment of traffic noise in Madrid (Spain). *Environ. Res.* **2015**, *137*, 136–140. [CrossRef]

24. Dzhambov, A.M.; Markevych, I.; Hartig, T.; Tilov, B.; Arabadzhiev, Z.; Stoyanov, D.; Gatseva, P.; Dimitrova, D.D. Multiple pathways link urban green- and bluespace to mental health in young adults. *Environ. Res.* **2018**, *166*, 223–233. [CrossRef] [PubMed]

25. Guo, X.; Liu, J.; Chen, Z.; Hong, X.-C. Harmonious Degree of Sound Sources Influencing Visiting Experience in Kulangsu Scenic Area, China. *Forests* **2023**, *14*, 138. [CrossRef]

26. Liu, J.; Kang, J.; Behm, H.; Luo, T. Effects of landscape on soundscape perception: Soundwalks in city parks. *Landsc. Urban Plan.* **2014**, *123*, 30–40. [CrossRef]

27. Brambilla, G.; Gallo, V.; Zambon, G. The Soundscape Quality in Some Urban Parks in Milan, Italy. *Int. J. Environ. Res. Public Health* **2013**, *10*, 2348–2369. [CrossRef] [PubMed]

28. Hong, X.C.; Cheng, S.; Liu, J.; Dang, E.; Wang, J.B.; Cheng, Y. The Physiological Restorative Role of Soundscape in Different Forest Structures. *Forests* **2022**, *13*, 1920. [CrossRef]

29. Matsinos, Y.G.; Mazaris, A.D.; Papadimitriou, K.D.; Mniestris, A.; Hatzigiannidis, G.; Maioglou, D.; Pantis, J.D. Spatio-temporal variability in human and natural sounds in a rural landscape. *Landsc. Ecol.* **2008**, *23*, 945–959. [CrossRef]

30. Liu, J.; Kang, J.; Luo, T.; Behm, H.; Coppack, T. Spatiotemporal variability of soundscapes in a multiple functional urban area. *Landsc. Urban Plan.* **2013**, *115*, 1–9. [CrossRef]

31. Liu, J.; Kang, J.; Behm, H.; Luo, T. Landscape spatial pattern indices and soundscape perception in a multi-functional urban area, Germany. *J. Environ. Eng. Landsc. Manag.* **2014**, *22*, 208–218. [CrossRef]

32. Wu, J. Effects of changing scale on landscape pattern analysis: Scaling relations. *Landsc. Ecol.* **2004**, *19*, 125–138. [CrossRef]

33. He, X.; Deng, Y.; Dong, A.; Lin, L. The relationship between acoustic indices, vegetation, and topographic characteristics is spatially dependent in a tropical forest in southwestern China. *Ecol. Indic.* **2022**, *142*, 109229. [CrossRef]

34. Scarpelli, M.D.A.; Ribeiro, M.C.; Teixeira, C.P. What does Atlantic Forest soundscapes can tell us about landscape? *Ecol. Indic.* **2021**, *121*, 107050. [CrossRef]

35. Sangermano, F. Acoustic diversity of forested landscapes: Relationships to habitat structure and anthropogenic pressure. *Landsc. Urban Plan.* **2022**, *226*, 104508. [CrossRef]

36. Mao, Q.; Sun, J.; Deng, Y.; Wu, Z.; Bai, H. Assessing Effects of Multi-Scale Landscape Pattern and Habitats Attributes on Taxonomic and Functional Diversity of Urban River Birds. *Diversity* **2023**, *15*, 486. [CrossRef]

37. Uebel, K.; Marselle, M.; Dean, A.J.; Rhodes, J.R.; Bonn, A. Urban green space soundscapes and their perceived restorativeness. *People Nat.* **2021**, *3*, 756–769. [CrossRef]

38. Hedblom, M.; Heyman, E.; Antonsson, H.; Gunnarsson, B. Bird song diversity influences young people's appreciation of urban landscapes. *Urban For. Urban Green.* **2014**, *13*, 469–474. [CrossRef]

39. Li, J.; Zhang, Z.; Jing, F.; Gao, J.; Ma, J.; Shao, G.; Noel, S. An evaluation of urban green space in Shanghai, China, using eye tracking. *Urban For. Urban Green.* **2020**, *56*, 126903. [CrossRef]

40. Liu, J.; Kang, J.; Luo, T.; Behm, H. Landscape effects on soundscape experience in city parks. *Sci. Total Environ.* **2013**, *454–455*, 474–481. [CrossRef] [PubMed]

41. Hasegawa, Y.; Lau, S.-K. Comprehensive audio-visual environmental effects on residential soundscapes and satisfaction: Partial least square structural equation modeling approach. *Landsc. Urban Plan.* **2022**, *220*, 104351. [CrossRef]

42. *GB50180-2018*; Standard for Urban Residential Area Planning and Design. Ministry of Housing and Urban-Rural Development, State Administration for Market Regulation: Beijing, China, 2018.

43. Chen, Z.; Zhu, T.-Y.; Liu, J.; Hong, X.-C. Before Becoming a World Heritage: Spatiotemporal Dynamics and Spatial Dependency of the Soundscapes in Kulangsu Scenic Area, China. *Forests* **2022**, *13*, 1526. [CrossRef]
44. Liu, J.; Xiong, Y.; Wang, Y.; Luo, T. Soundscape effects on visiting experience in city park: A case study in Fuzhou, China. *Urban For. Urban Green.* **2018**, *31*, 38–47. [CrossRef]
45. Li, W.; Zhai, J.; Zhu, M. Characteristics and perception evaluation of the soundscapes of public spaces on both sides of the elevated road: A case study in Suzhou, China. *Sustain. Cities Soc.* **2022**, *84*, 103996. [CrossRef]
46. Strohbach, M.W.; Lerman, S.B.; Warren, P.S. Are small greening areas enhancing bird diversity? Insights from community-driven greening projects in Boston. *Landsc. Urban Plan.* **2013**, *114*, 69–79. [CrossRef]
47. Wong, J.S.Y.; Soh, M.C.K.; Low, B.W.; Kenneth, B.H.E. Tropical bird communities benefit from regular-shaped and naturalised urban green spaces with water bodies. *Landsc. Urban Plan.* **2023**, *231*, 104644. [CrossRef]
48. Garizábal-Carmona, J.A.; Mancera-Rodríguez, N.J. Bird species richness across a Northern Andean city: Effects of size, shape, land cover, and vegetation of urban green spaces. *Urban For. Urban Green.* **2021**, *64*, 127243. [CrossRef]
49. Dzhambov, A.M.; Dimitrova, D.D. Urban green spaces' effectiveness as a psychological buffer for the negative health impact of noise pollution: A systematic review. *Noise Health* **2014**, *16*, 157. [CrossRef]
50. Hong, J.Y.; Lam, B.; Ong, Z.-T.; Ooi, K.; Gan, W.-S.; Kang, J.; Yeong, S.; Lee, I.; Tan, S.-T. Effects of contexts in urban residential areas on the pleasantness and appropriateness of natural sounds. *Sustain. Cities Soc.* **2020**, *63*, 102475. [CrossRef]
51. Liu, J.; Kang, J.; Behm, H. Birdsong as an element of the urban sound environment: A case study concerning the area of Warnemünde in Germany. *Acta Acust. United Acust.* **2014**, *100*, 458–466. [CrossRef]
52. Kurze, U.J.; Anderson, G.S.J.N.; Principles, V.C.E. Applications. Outdoor sound propagation. In *Noise and Vibration Control Engineering: Principles and Applications*, 2nd ed.; Wiley: Hoboken, NJ, USA, 2006; pp. 119–144. [CrossRef]
53. Han, X.; Huang, X.; Liang, H.; Ma, S.; Gong, J. Analysis of the relationships between environmental noise and urban morphology. *Environ. Pollut.* **2018**, *233*, 755–763. [CrossRef] [PubMed]
54. Guedes, I.C.M.; Bertoli, S.R.; Zannin, P.H.T. Influence of urban shapes on environmental noise: A case study in Aracaju—Brazil. *Sci. Total Environ.* **2011**, *412–413*, 66–76. [CrossRef]
55. Salomons, E.M.; Berghauser Pont, M. Urban traffic noise and the relation to urban density, form, and traffic elasticity. *Landsc. Urban Plan.* **2012**, *108*, 2–16. [CrossRef]
56. Amaya-Espinel, J.D.; Hostetler, M.; Henríquez, C.; Bonacic, C. The influence of building density on Neotropical bird communities found in small urban parks. *Landsc. Urban Plan.* **2019**, *190*, 103578. [CrossRef]
57. Benítez-López, A.; Alkemade, R.; Verweij, P.A. The impacts of roads and other infrastructure on mammal and bird populations: A meta-analysis. *Biol. Conserv.* **2010**, *143*, 1307–1316. [CrossRef]
58. Cooke, S.C.; Balmford, A.; Johnston, A.; Newson, S.E.; Donald, P.F. Variation in abundances of common bird species associated with roads. *J. Appl. Ecol.* **2020**, *57*, 1271–1282. [CrossRef]
59. Yu, W.L.; Kang, J. Relationship between traffic noise resistance and village form in China. *Landsc. Urban Plan.* **2017**, *163*, 44–55. [CrossRef]
60. Aletta, F.; Kang, J.; Fuda, S.; Astolfi, A. The effect of walking sounds from different walked-on materials on the soundscape of urban parks. *J. Environ. Eng. Landsc. Manag.* **2016**, *24*, 165–175. [CrossRef]
61. Payne, S.R.; Bruce, N. Exploring the Relationship between Urban Quiet Areas and Perceived Restorative Benefits. *Int. J. Environ. Res. Public Health* **2019**, *16*, 1611. [CrossRef] [PubMed]
62. Ren, Z.; Jiang, B.; Seipel, S. Capturing and characterizing human activities using building locations in America. *ISPRS Int. J. Geo-Inf.* **2019**, *8*, 200. [CrossRef]
63. Miller, S.A. Parents' attributions for their children's behavior. *Child Dev.* **1995**, *66*, 1557–1584. [CrossRef]
64. Hong, J.Y.; Jeon, J.Y. Relationship between spatiotemporal variability of soundscape and urban morphology in a multifunctional urban area: A case study in Seoul, Korea. *Build. Environ.* **2017**, *126*, 382–395. [CrossRef]
65. Sakieh, Y.; Jaafari, S.; Ahmadi, M.; Danekar, A. Green and calm: Modeling the relationships between noise pollution propagation and spatial patterns of urban structures and green covers. *Urban For. Urban Green.* **2017**, *24*, 195–211. [CrossRef]
66. Irvine, K.N.; Devine-Wright, P.; Payne, S.R.; Fuller, R.A.; Painter, B.; Gaston, K.J. Green space, soundscape and urban sustainability: An interdisciplinary, empirical study. *Local Environ.* **2009**, *14*, 155–172. [CrossRef]
67. Hewett, D. *Sound Propagation in an Urban Environment*; Oxford University: Oxford, UK, 2010.
68. Wu, J.; Zou, C.; He, S.; Sun, X.; Wang, X.; Yan, Q. Traffic noise exposure of high-rise residential buildings in urban area. *Environ. Sci. Pollut. Res.* **2019**, *26*, 8502–8515. [CrossRef] [PubMed]
69. Ow, L.F.; Ghosh, S. Urban cities and road traffic noise: Reduction through vegetation. *Appl. Acoust.* **2017**, *120*, 15–20. [CrossRef]
70. Coensel, B.D.; Vanwetswinkel, S.; Botteldooren, D. Effects of natural sounds on the perception of road traffic noise. *J. Acoust. Soc. Am.* **2011**, *129*, EL148–EL153. [CrossRef]
71. Wu, J.-G. *Landscape Ecology: Pattern, Process, Scale and Hierarchy*; Higher Education Press: Beijing, China, 2000.
72. Jaszczak, A.; Pochodyła, E.; Kristianova, K.; Małkowska, N.; Kazak, J.K. Redefinition of Park Design Criteria as a Result of Analysis of Well-Being and Soundscape: The Case Study of the Kortowo Park (Poland). *Int. J. Environ. Res. Public Health* **2021**, *18*, 2972. [CrossRef]

73. Hamilton, K.L.; Kaczynski, A.T.; Fair, M.L.; Lévesque, L. Examining the Relationship between Park Neighborhoods, Features, Cleanliness, and Condition with Observed Weekday Park Usage and Physical Activity: A Case Study. *J. Environ. Public Health* **2017**, *2017*, 7582402. [CrossRef]

74. Lauro, E.D.; Falanga, M.; Lalli, L.T. The soundscape of the Trevi fountain in COVID-19 silence. *Noise Mapp.* **2020**, *7*, 212–222. [CrossRef]

*Article*

# Harmonious Degree of Sound Sources Influencing Visiting Experience in Kulangsu Scenic Area, China

**Xuan Guo** [1,2], **Jiang Liu** [1,3,*], **Zhu Chen** [1,4] **and Xin-Chen Hong** [1,3]

1 School of Architecture and Urban-Rural Planning, Fuzhou University, Fuzhou 350108, China
2 Institute of Geography, Ruhr University Bochum, 44801 Bochum, Germany
3 Fujian Key Laboratory of Digital Technology for Territorial Space Analysis and Simulation, Fuzhou University, Fuzhou 350108, China
4 Institute of Environmental Planning, Leibniz University Hannover, Herrenhäuser Str. 2, 30419 Hanover, Germany
* Correspondence: jiang.liu@fzu.edu.cn; Tel.: +86-157-0595-8118

**Abstract:** Soundscapes are important resources and contribute to high-quality visiting experiences in scenic areas. Based on a public investigation of 195 interviewees in the Kulangsu scenic area, this study aimed to explore the relationships between the harmonious degree of sound sources (SHD) and visiting experience indicators, in terms of soundscape perception, as well as the satisfaction degree of visual landscape and comprehensive impression. The results suggested that the dominating positions of human sounds did not totally suppress the perception of natural sounds such as birdsong and sea waves in the scenic area. Natural sound sources also showed a higher harmonious degree than other artificial sounds. Significant relationships existed between the SHD of most sound sources and the visiting experience indicators. Natural sounds were closely related to pleasant and comfortable soundscape perception, while mechanical sound sources were mainly related to eventful and varied soundscapes. The close relationships between certain sound sources and the satisfaction degree of the visual landscape and comprehensive impression evaluation indicated the effectiveness of audio-visual and even multi-sensory approaches to enhance visiting experience. The structural equation model further revealed that (1) natural sound was the most influential sound source of soundscape and visual landscape perception; (2) human sounds and mechanical sounds all showed significant positive effects on soundscape perception; and (3) indirect relationships could exist in the SHD of sound sources with comprehensive impression evaluation. The results can facilitate targeted soundscape and landscape management and landsense creation with the aim of improving visiting experience.

**Keywords:** landscape; soundscape; sound perception; visiting experience; structural equation model; scenic area

**Citation:** Guo, X.; Liu, J.; Chen, Z.; Hong, X.-C. Harmonious Degree of Sound Sources Influencing Visiting Experience in Kulangsu Scenic Area, China. *Forests* **2023**, *14*, 138. https://doi.org/10.3390/f14010138

Academic Editor: Chi Yung Jim

Received: 1 December 2022
Revised: 5 January 2023
Accepted: 10 January 2023
Published: 12 January 2023

## 1. Introduction

As an important part of urban green infrastructure and urban ecosystems, scenic areas play the roles of meeting people's expectations or visions of the natural ecological and cultural environment and providing their needs in various aspects such as health-related, aesthetic, and cultural experiences [1–3]. In fact, the realization of these visions and needs is a complex process [4]. In practice, designers or managers usually endow or integrate one or more of their visions into a carrier through appropriate manifestation forms, so that others (including themselves) can graft these visions from this carrier and associated manifestation forms, and then satisfy the needs generated by their own visions [5]. This highlights the importance of the carrier, that is, the landsense element [6]. Effective management of the various landsense elements in the scenic areas will therefore help to optimize the planning and design of these areas and to achieve a resonance between the practices and the vision of people [7,8].

At the same time, it is worth noting that people's perception of these landsense elements is mainly achieved through the five senses, i.e., the interaction of landsense elements with human senses to produce landscape experiences (landsense effects) [5]. This is what landsenses ecology highlights—the interaction between human perception and landscape [6]. That is, the multidimensional sensory perception of the landscape is a central part of the overall visiting experience [9]. However, many of the previous researchs on the visiting experience have focused on the visual dimension, neglecting the role of the other senses [10–13], which limits our understanding of complex environments, as visualization only describes a fragment of a given landscape [14]. There is a growing interest in sensory perception other than vision, with new sensory perception research focusing on auditory perception, as it is the second most important way of perceiving the environment after the visual [15]. For example, Agapito et al. found that in the context of rural destinations, the frequency of auditory impressions reported by visitors regarding their sensory experiences (23%) was second only to visual elements (26%) [16]. The tourism industry, while prosperous, has also brought many problems, including the destruction of the acoustic environment by the noise generated during the visit, which affects people's perception of the unique sound sources in a scenic area, including natural and cultural sounds. Soundscape, as the acoustic environment perceived or experienced and/or understood by a person or people, in context [17], has been drawing increasing attention from researchers, also as an important resource in scenic areas. For example, it has been found that soundscapes can induce specific perceptions that cannot be experienced through visual stimuli, and can provide a unique set of emotional supports [18]. Soundscapes have a different impact on visitors' cognition and emotions than visual landscapes based on the cognitive–emotional model [19]. However, the coherence between soundscape and landscape has also been highlighted in several studies, both as a direct influence on overall visit satisfaction [19], and as a variable that mediates the impact of soundscape perception on the visiting experience [20].

In addition, the definition of soundscape differs from traditional acoustics in that it emphasizes the relationship between subjective human perception and the acoustic environment [17]. This is perfectly in line with the viewpoint emphasized by landsenses ecology [6]. In addition, what is highlighted by soundscape is precisely the interaction between the human auditory sense and the acoustic environment. It should be particularly noted that sound sources play key roles in this process, being important carriers or landsenses elements. Different sound sources, perceived by different individuals in different contexts, could form completely different soundscapes [21]. In this regard, some researchers have explored the perception characteristics of sound sources, such as perceived occurrences, perceived loudness, dominance and preference [1,22], in order to reveal how they contribute to soundscape perception, as well as visual perception, environmental satisfaction, and restorative benefits [23–26], etc. Therefore, it is essential to explore to what extent the existence of certain sound sources in a landscape corresponds to a person's preference, and how this correlation relates to the visiting experience. This will be an important guide for the management and protection of soundscapes in scenic areas and the enhancement of their quality. This is because, in concrete soundscape practice, people also often manage or change the sound source and the acoustic environment to achieve the ultimate soundscape creation [27]. In addition, this is a very important part of landsenses ecology, i.e., landsense creation. By integrating a vision with an existing carrier or a newly constructed one, people achieve the process of landsense creation [5]. This process is also reflected in the protection and management of soundscape resources in scenic areas [8].

Kulangsu was listed as a World Heritage Site by UNESCO on 8 July 2017, in acknowledgement of its outstanding value to humanity. Prior to this, it was already a famous scenic spot in China, with rich natural and cultural landscape and soundscape resources, attracting a large number of visitors. In addition, after receiving the World Heritage designation, the number of visitors to Kulangsu increased by 12.19% in the same month of the time of its listing on the World Heritage List, according to the "Monthly report on the completion of main economic indicators in Kulangsu" (Issued by: KulangSu Administrative Committee). This will undoubtedly disturb or even threaten the status of soundscapes on the island. Consequently, an in-depth exploration of the current situation and the role of the soundscapes in Kulangsu will provide theoretical guidance for the management and conservation of soundscapes resources in World Heritage sites under the pressure of tourism.

Therefore, in the framework of the landsenses ecology theory and based on a public questionnaire survey in the Kulangsu scenic area, this study aims to investigate the extent to which the objective presence of sound sources corresponds to the subjective preference, and how this status could influence visitors' visiting experience. By proposing the harmonious degree of sound sources (SHD) as a comprehensive indicator of the perceived occurrences, loudness, as well as preference for certain sound sources, we analyzed the impact of the SHD of different sound sources on the visiting experience from three aspects, including visual landscape, soundscape and comprehensive impression, in order to promote further understanding of this interactive process between landsense elements and human perception, and to achieve sustainable development of scenic areas.

## 2. Methods

### 2.1. Study Area

The study was conducted in Kulangsu Scenic Area in Xiamen city, China, a small island with an area of 1.88 km$^2$, a subtropical monsoon climate, as well as excellent light and heat conditions and an average temperature of 21.2 °C throughout the year (Figure 1). Due to the adequate weather conditions, there is rich flora, including over 40% of vegetation coverage, and more than 1000 species of trees, shrubs, vines, and ground cover plants. Kulangsu has been one of the most popular tourist resorts in China, especially after it was listed in the World Heritage List in the name of "Kulangsu: International Historic Community" in 2017. There are many famous scenic spots in Kulangsu, such as Sunlight Rock, Shuzhuang Garden, Haoyue Garden, Yu Garden, Kulangsu Stone, etc. In addition, Kulangsu has a unique historical and international culture. Based on the master tourism planning of the Kulangsu scenic area (2014), the island is divided into five functional zones, including the tourist service zone, musical zone, cultural and artistic zone, historical building zone, and natural landscape zone (Figure 1). This study was conducted in July 2019, two years after it was recognized as a world heritage site, as part of a series studied here [1]. Thus, it facilitates us to reveal what the soundscape status of the sonic area is after this change. Combining several field surveys and public investigation results, we identified the typical soundscape elements of Kulangsu as 12 sound sources in 3 sound categories (see Table 1).

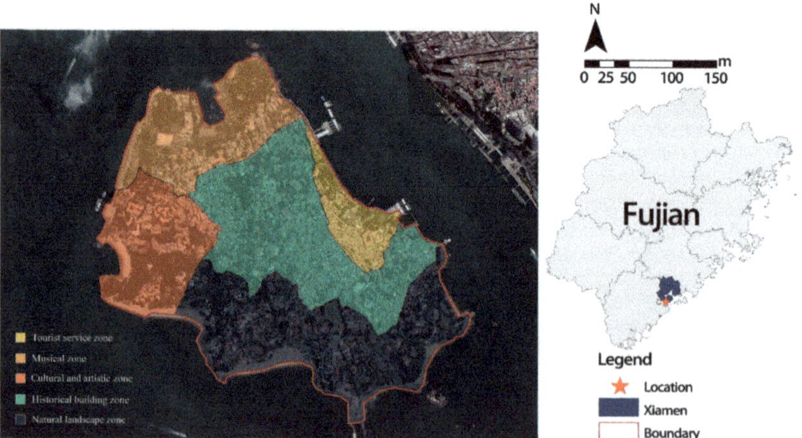


**Figure 1.** Aerial photo of Kulangsu scenic area (source: elaborated by the authors with Google earth).

**Table 1.** Identified sound sources in Kulangsu.

| Sound Category | Sound Source |
| --- | --- |
| Human sound | surrounding speech, playing children, sales calling, tour guide |
| Mechanical sound | broadcasting music, construction noise, traffic sound, |
| Natural sound | birdsong, insects, sea waves, tree rustling, water sound (fountain) |

*2.2. Questionnaire Design*

Questionnaires are an effective method supported by technical standards and previous research [28–30]. Data of this study were collected through a three-part questionnaire targeted to the visitors in Kulangsu.

The first part of the questionnaire was related to the basic personal information of the interviewee [22], including gender (male, female), age ($\leq$24, 25–30, 31–40, 41–50, 51–59, $\geq$60), educational background (primary, secondary, high school, university, postgraduate), occupation (student, enterprise and public institution, self-employed, retiree, other), residential status (Kulangsu resident, Xiamen resident, tourist, local merchant, foreign merchant, student), visit frequency (first time, second–third time, once a day, once a month, once a week, other), and length of residence (less than a week, less than a month, less than a year, less than five years, permanent residents).

The second part of the questionnaire was to evaluate each of the 12 sound sources according to their own perceptions [22,24], in terms of perceived occurrences (POS) (1—never, 2—occasionally, 3—normal, 4—often, 5—frequently), perceived loudness (PLS) (1—very weak, 2—weak, 3—normal, 4—strong, 5—very strong), and preference (PFS) (1—very dislike, 2—rather dislike, 3—normal, 4—rather like, 5— like very much).

The third part of the questionnaire was the evaluation of visiting experience from three aspects, including visual landscape, soundscape, and comprehensive impression [2,23]. Specifically, the satisfaction degree of the visual landscape (SVL) was evaluated in terms of natural scenery, architectural style, landscape design, sculpture (with other sketches), and pavement with a Likert 5-scale (1—very dissatisfied, 2—dissatisfied, 3—fair, 4—satisfied, 5—very satisfied) [2]. Six representative adjectives of soundscape perception (SSP) from previous studies were selected, including "harmonious", "pleasant", "vivid", "eventful", "comfortable", and "varied", and evaluated with a Likert 5-scale (1—strongly disagree, 2—disagree, 3—fair, 4—agree, 5—strongly agree) [23,31]. In addition, the comprehensive impression evaluation (CIE), as a comprehensive visiting experience evaluation and an overall impression of the scenic area to the visitors, was evaluated in terms of fascinating, interesting, harmonious, distinctive, and culturally profound with a Likert 5-scale (1—strongly disagree, 2—disagree, 3—fair, 4—agree, 5—strongly agree) [23].

The investigation was conducted on sunny days during July 2019, and 217 questionnaires were collected, including 195 valid questionnaires with an efficiency of 89.86%, with 35 to 43 in each functional zones. According to the requirements of partial lease squares structural equation modeling (PLS-SEM) as suggested by Hair, the sample size should be at least ten times the largest number of structural paths directed at a particular latent construct in the structural model [32]. In this study, the largest number of structural paths directed at a particular latent construct was 5, that is, the sample size was required to be more than 50. Therefore, the number of valid questionnaires collected in this study was able to meet the needs of the subsequent analysis.

The statistical results of the participants' personal information are shown in Table 2. In addition, the reliability and validity of the questionnaire were analyzed. After the reliability test, the Cronbach's alpha was 0.796 (>0.7), indicating a high reliability of the questionnaire. Validity analysis was carried out by KMO (Kaiser–Meyer–Olkin) and Bartlett's sphericity test, in which KMO = 0.813 (>0.6) and the significance value of Bartlett's sphericity test was 0.000 (<0.05), indicating a good validity of the questionnaire.

**Table 2.** Sample information of the questionnaire database, $N = 195$.

| Variable | Category | Sample Size | Proportion (%) |
|---|---|---|---|
| Gender | Male | 88 | 45.1 |
| | Female | 107 | 54.9 |
| Age | $\leq$24 | 75 | 38.5 |
| | 25–30 | 60 | 30.8 |
| | 31–40 | 33 | 16.9 |
| | 41–50 | 16 | 8.2 |
| | 51–59 | 8 | 4.1 |
| | $\geq$60 | 3 | 1.5 |
| Education background | Primary | 1 | 0.5 |
| | Secondary | 7 | 3.6 |
| | High school | 30 | 15.4 |
| | University | 142 | 72.8 |
| | Postgraduate | 15 | 7.7 |
| Occupation | Student | 55 | 28.2 |
| | Enterprise and public institution | 64 | 32.8 |
| | Self-employed | 27 | 13.8 |
| | Retiree | 7 | 3.6 |
| | Other | 42 | 21.5 |
| Residential status | Kulangsu resident | 7 | 3.6 |
| | Xiamen resident | 17 | 8.7 |
| | Tourist | 157 | 80.5 |
| | Local merchant | 4 | 2.1 |
| | Foreign merchant | 7 | 3.6 |
| | Student | 3 | 1.5 |
| Visit frequency | First time | 106 | 54.4 |
| | Second–third time | 32 | 16.4 |
| | Once a day | 21 | 10.8 |
| | Once a month | 1 | 0.5 |
| | Once a week | 9 | 4.6 |
| | Other | 26 | 13.3 |
| Length of residence | Less than a week | 169 | 86.7 |
| | Less than a month | 1 | 0.5 |
| | Less than a year | 3 | 1.5 |
| | Less than 5 years | 7 | 3.6 |
| | Permanent residents | 15 | 7.7 |

*2.3. Data Analysis*

2.3.1. Calculating the Harmonious Degree of Sound Sources

Soundscape perception is the result of the probability of cognitive stimulation and the perception of sound sources [33]. The usually used indicators from a single dimension, such as perceived occurrences and perceived loudness, as well as preference for certain sound sources, can provide useful soundscape information [24], but not in a comprehensive way reflecting the cognition process. Thus, we proposed a new harmonious degree indicator on the basis of the previous research [22], the harmonious degree of sound sources (SHD). It combines the three aforementioned sound source perception indicators, and indicates the degree to which the dominance of a sound source in the landscape matches the visitors' preference for the sound, which can reflect the harmonious status of the sound in the soundscape.

First of all, we conducted a process of formula manipulation to reduce the potential errors of data caused by subjective factors. The sound dominant degree (*SDD*), referring to the perception degree of a sound source, could then be acquired by Equation (1):

$$SDD_{ji} = POS_{ji} \times PLS_{ji} \tag{1}$$

where *POS* denotes perceived occurrences of individual sounds. *PLS* denotes the perceived loudness of individual sounds. Similarly, *j* represents the *j*th sample, and *i* represents the *i*th sound source.

In addition, we conducted the initial orientation of soundscape preference (*S*) as an indicator to distinguish the relative value of preference or dislike. The mean value of preference for individual sounds (*PFS*) was calculated from each sound source as a boundary value. If the preference for individual sounds of tourists is greater than the mean value of the preference for individual sounds, it means that tourists have a relative preference for the sound source; otherwise, it means they do not. The equation between *S* and *PFS* is as follows:

$$S_{ji} = \sum_{j=1}^{n} PFS_j/n - PFS_{ji} \tag{2}$$

where *S* denotes the initial orientation of soundscape preference. *PFS* denotes the preference for individual sounds. In addition, *n* represents the sample size. In this study, $j = 1, \ldots, n$, where $n = 195$ valid questionnaires.

Then, considering the extreme value of subjective data influencing the *PFS*, the *S* should be transformed. We adopted the final orientation of soundscape preference (*M*) as an indicator based on exponential function. If $M > 0$, the *M* value would represent like, and otherwise dislike. The equation between *M* and *S* is as follows:

$$M = (1/(e^{S_{ji}} + 1) - 0.5) \tag{3}$$

where *M* denotes the final orientation of soundscape preference, e represents an Euler, irrational and transcendental number.

Finally, the *SHD* combining *M* and *SDD* to express the orientation of sound dominant and preference degree, is built as follows:

$$SHD = M \times SDD \tag{4}$$

Moreover, this equation is equivalent to the equation as follows:

$$SHD_{ji} = (1/(e^{\sum_{j=1}^{n} PFS_{ji}/n - PFS_{ji}} + 1) - 0.5) \times POS_{ji} PLS_{ji} \tag{5}$$

These equations suggest that (1) *SHD* is determined by *SDD* and *PFS*; (2) due to using exponential function to relate the *PFS* and *S*, if a *PFS* value was higher than the mean value of *PFS*, high *SDD* would result in high *SHD*; (3) otherwise, if a *PFS* value was lower than the mean value of *PFS*, the *SDD* would reach a high value but the *SHD* would reach a low value. Ultimately, in combination with the study settings, the *SHD* value could range from $-12.5$ to $12.5$.

### 2.3.2. Modeling the Relationships among the SHD and Visiting Experience Indicators

Spearman's rho correlation analysis was carried out in SPSS 25.0 to detect the potential relationships between the SHD of different sounds and the visiting experience indicators. Furthermore, we conducted a structural equation model to explore (1) the effect of the SHD of different sound types on visiting experience, and (2) the significant variation of this effect in different functional zones. The procedure of the structural equation model was performed based on PLS-SEM (partial least squares SEM). The PLS-SEM has many advantages over CB-SEM (covariance-based SEM), including optimal consistency and target prediction [32]. The procedure of the structural equation model was carried out in Smart PLS 3.3.

## 3. Results

### *3.1. Characteristics of Sound Source Perception*

The mean values of the four sound source perception indicators, i.e., POS, PLS, PFS, and SHD are shown in Figure 2. The trend of POS and PLS are similar among different sound sources, reflecting relative dominating positions of human sounds represented by surrounding speech and sales calling, and natural sounds represented by sea waves and birdsong. In addition, the POS and PLS of traffic sound and construction noise were the lowest among all sound sources, indicating that these two sound sources were controlled effectively. In terms of the PFS, all natural sounds were favored by tourists, with sea waves showing the highest PFS, followed by birdsong. By contrast, construction noise showed the lowest PFS, followed by traffic sound. The SHD of natural sounds showed higher values than that of other sound sources, and water sound was the highest. The lowest SHD appeared mainly with human sounds, especially sales calling and surrounding speech, which were less preferred.

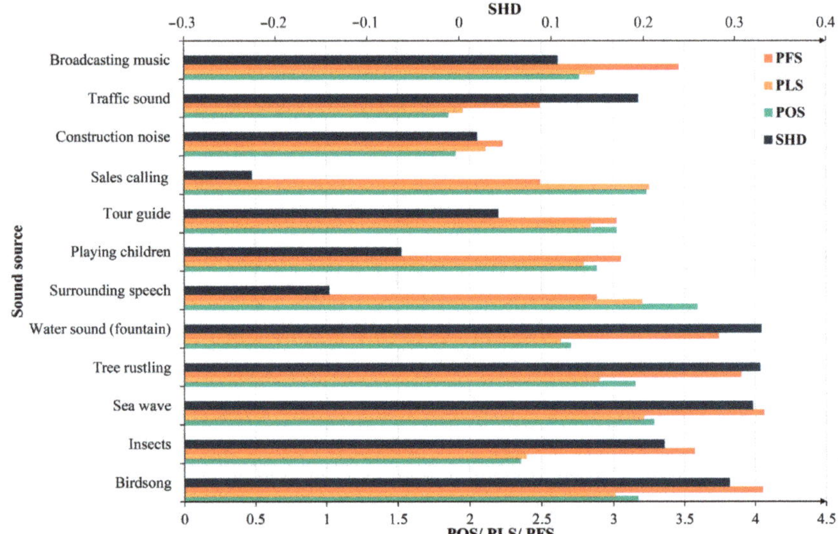


**Figure 2.** Mean values of sound source perception indicators, perceived occurrences (POS), perceived loudness (PLS), preference (PFS), harmonious degree of sound source (SHD).

Figure 3 shows the differences in the SHD of each sound source in different functional zones. It is obvious that the SHD values of the same type of sound source were different among different functional zones. Most of the natural sounds showed positive SHD values, with the highest one appearing with water sound in the musical zone, and limited negative values appearing in the natural landscape and historical building zones. More than half of the SHD values of human sounds were negative in different zones, with sales calling being the lowest one and playing children the highest, both appearing in the tourist service zone. In terms of mechanical sounds, they all showed relatively low SHD values, with a higher value of traffic sound in the cultural and artistic zone, and a lower value of broadcasting music in the historical building zone.

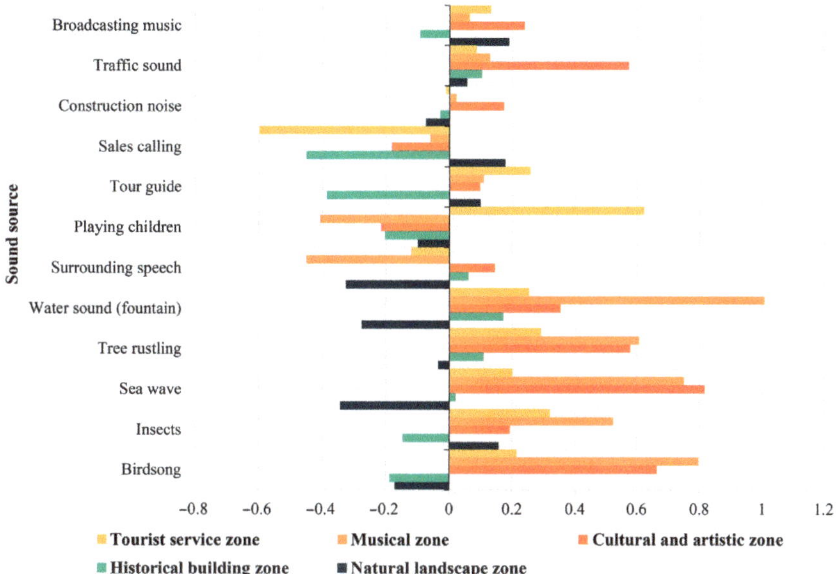



**Figure 3.** Mean values of harmonious degree of sound sources in different functional zones.

### 3.2. Relationships between the SHD and Visiting Experience Indicators

Spearman's rho correlation analysis results between the SHD of each of the 12 sound sources and visiting experience indicators are shown in Table 3. The harmonious status of nearly all the sound sources showed significant and positive relationships with at least one visiting experience indicator, except for surrounding speech. In terms of soundscape perception (SSP), pleasant soundscapes were more related to the SHD of 4 natural sounds, tour guide sound, and broadcasting music (6 out of 12), followed by varied soundscapes showing close relationships with all three mechanical sounds and insects sounds (4 out of 12). The SHD of broadcasting music showed the most significant asscociation with soundscape perception indicators (4 out of 6).

There are notable correlations between the SHD and the satisfaction degree of the visual landscape (SVL). The findings showed that the SHD of all natural sounds was correlated with the SVL, especially water sound (fountain) (all the 5 indicators) and birdsong (4 out of 5). Pavement conditions showed relatively more relationships with the SHD (5 out of 12), followed by natural scenery and architectural style (both 4).

In terms of the relationships between the SHD and comprehensive impression evaluation (CIE), the SHD of sea waves, water sound (fountain), and broadcasting music all showed significant relationships with 3 of the 5 indicators. The fascinating characteristic of the place was most related to the SHD (5 out of 12), followed by harmonious and distinctive (both 4).

In summary, it is clear that the SHD of nearly all the major sound sources in Kulangsu showed significant relationships with the visiting experience indicators. It is necessary to reveal how these indicators could interact with each other and contribute to the comprehensive visiting experience. Thus, in the next section, their relationships were further revealed by structural equation modeling.

**Table 3.** The correlation coefficients between the SHD of each of the sound sources and the visiting experience indicators, SSP: soundscape perception, SVL: satisfaction degree of visual landscape, CIE: comprehensive impression.

| Sound Category | SHD | SSP | SVL | CIE |
|---|---|---|---|---|
| Natural sound | Birdsong | Pleasant (0.146 *) | Natural scenery (0.159 *)<br>Architectural style (0.159 *)<br>Landscape design (0.196 **)<br>Pavement (0.155 *) | / |
| | Insects | Varied (0.164 *) | Pavement (0.151 *) | Fascinating (0.148 *) |
| | Sea waves | Pleasant (0.186 **)<br>Comfortable (0.165 *) | Natural scenery (0.195 **)<br>Architectural style (0.215 **)<br>Pavement (0.172 *) | Harmonious (0.197 **)<br>Distinctive (0.176 *)<br>Culturally profound (0.155 *) |
| | Tree rustling | Pleasant (0.176 *) | Architectural style (0.176 *) | Fascinating (0.164 *)<br>Harmonious (0.165 *) |
| | Water sound (fountain) | Pleasant (0.278 **)<br>Comfortable (0.196 **) | Natural scenery (0.228 **)<br>Architectural style (0.249 **)<br>Landscape design (0.227 **)<br>Sculpture (with other sketches) (0.144 *)<br>Pavement (0.175 *) | Fascinating (0.226 **)<br>Harmonious (0.227 **)<br>Distinctive (0.198 *) |
| Human sound | Surrounding speech | / | / | / |
| | Playing children | Harmonious (0.149 *) | / | / |
| | Tour guide | Pleasant (0.142 *)<br>Eventful (0.143 *) | Natural scenery (0.179 *)<br>Pavement (0.169 *) | Distinctive (0.153 *) |
| | Sales calling | Vivid (0.164 *) | / | Interesting (0.204 **) |
| Mechanical sound | Construction noise | Varied (0.154 *) | / | Interesting (0.173 *) |
| | Traffic sound | Eventful (0.216 **)<br>Varied (0.174 *) | Landscape design (0.143 *) | Fascinating (0.162 *)<br>Distinctive (0.141 *) |
| | Broadcasting music | Pleasant (0.168 *)<br>Comfortable (0.170 *)<br>Eventful (0.234 **)<br>Varied (0.198 **) | / | Fascinating (0.153 *)<br>Interesting (0.187 **)<br>Harmonious (0.169 *) |

Note: Significant correlations are marked with * ($p < 0.05$) and ** ($p < 0.01$).

### 3.3. Modeling the Effect of the SHD on Visiting Experience

#### 3.3.1. Measurement Model

The reliability and validity of the measurement models were assessed using individual item reliability, construct reliability, convergent validity, and discriminant validity, respectively. The results showed that the standardized factor loads of playing children (0.188) and sales calling (0.418) were less than 0.5 and failed to pass the significance test, which suggested that these two indicators should be removed. Then, we conducted a re-testing after removing these variables. As shown in Table 4, the standardized factor loads for all observed variables were significantly greater than 0.5 and all passed the significance test, indicating that these variables were acceptable. Meanwhile, the CR (construct reliability) of each latent variable ranged from 0.795 to 0.894, which was greater than 0.7, indicating that the latent variables had good construct reliability [34].

**Table 4.** Modified measurement model, SHD-NS: harmonious degree of natural sound, SHD-MS: harmonious degree of mechanical sound, SHD-HS: harmonious degree of human sound, SSP: soundscape perception, SVL: satisfaction degree of visual landscape, CIE: comprehensive impression evaluation.

| Latent Variables | Observed Variables | Standardized Factor Loading | CR | AVE |
| --- | --- | --- | --- | --- |
| SHD-NS | Insects | 0.681 *** | 0.883 | 0.602 |
| | Water sound (fountain) | 0.799 *** | | |
| | Sea waves | 0.813 *** | | |
| | Tree rustling | 0.797 *** | | |
| | Birdsong | 0.782 *** | | |
| SHD-MS | Traffic sound | 0.894 *** | 0.827 | 0.705 |
| | Construction noise | 0.782 *** | | |
| SHD-HS | Tour guide | 0.897 *** | 0.795 | 0.662 |
| | Surrounding speech | 0.721 *** | | |
| SSP | Pleasant | 0.804 *** | 0.894 | 0.585 |
| | Comfortable | 0.813 *** | | |
| | Harmonious | 0.755 *** | | |
| | Vivid | 0.796 *** | | |
| | Eventful | 0.754 *** | | |
| | Varied | 0.654 *** | | |
| SVL | Natural scenery | 0.724 *** | 0.886 | 0.609 |
| | Architectural style | 0.822 *** | | |
| | Landscape design | 0.833 *** | | |
| | Sculpture (with other sketches) | 0.780 *** | | |
| | Pavement | 0.736 *** | | |
| CIE | Fascinating | 0.809 *** | 0.878 | 0.591 |
| | Interesting | 0.840 *** | | |
| | Harmonious | 0.744 *** | | |
| | Distinctive | 0.739 *** | | |
| | Culturally profound | 0.702 *** | | |

Note: significant factors are marked with *** ($p < 0.001$)

The validity tests for the latent variables were further examined to include mainly convergent validity and discriminant validity. As shown in Table 4, the AVE (average variance extracted) values for all latent variables were between 0.585 to 0.705, which was greater than the threshold of 0.5 [35], indicating that the convergent validity of the latent variables was acceptable. While the discriminant validity was mainly tested by the Fornell–Larcker criterion and the Heterotrait–Monotrait ratio of correlations (HTMT) criterion [36]. The results showed that all latent variables could meet the Fornell–Larker criterion as the square root of AVE of each latent variable was higher than its correlation with other latent variables. Meanwhile, the HTMTs between the pairwise latent variables were all less than 0.9, indicating that there was good discriminant validity between each latent variable (See Table 5).

**Table 5.** The test of discrimination validity of the variables, SHD-NS: harmonious degree of natural sound, SHD-MS: harmonious degree of mechanical sound, SHD-HS: harmonious degree of human sound, SPE: soundscape perception, SVL: satisfaction degree of visual landscape, CIE: comprehensive impression evaluation.

| | Fornell–Larcker Criterion | | | | | |
| --- | --- | --- | --- | --- | --- | --- |
| | **SHD-NS** | **SHD-MS** | **SHD-HS** | **SSPE** | **SVL** | **CIE** |
| SHD-NS | **0.776** | | | | | |
| SHD-MS | −0.162 | **0.84** | | | | |
| SHD-HS | 0.058 | 0.305 | **0.813** | | | |
| SPE | 0.166 | 0.247 | 0.229 | **0.765** | | |
| SVL | 0.203 | 0.127 | 0.263 | 0.461 | **0.78** | |
| CIE | 0.195 | 0.11 | 0.142 | 0.578 | 0.478 | **0.768** |
| | HTMT Criterion | | | | | |
| | SHD-NS | SHD-MS | SHD-HS | SPE | SVL | CIE |
| SHD-NS | | | | | | |
| SHD-MS | 0.226 | | | | | |
| SHD-HS | 0.208 | 0.618 | | | | |
| SPE | 0.191 | 0.34 | 0.353 | | | |
| SVL | 0.247 | 0.165 | 0.37 | 0.52 | | |
| CIE | 0.229 | 0.173 | 0.221 | 0.682 | 0.563 | |

Note: values (bold) on the diagonal represent the square root of the AVE while the off-diagonals are correlations.

### 3.3.2. Conceptual Structural Equation Model

Based on the results of exploratory and confirmatory factor analysis and previous research [23,28], 5 main hypotheses and 12 sub-hypotheses were proposed as follows:

**H$_a$:** *SHD-NS has a significant effect on each of visiting experience indicators, with specific hypotheses including:* **H$_{a1}$:** *SHD-NS has a positive effect on SSP;* **H$_{a2}$:** *SHD-NS has a positive effect on SVL;* **H$_{a3}$:** *SHD-NS has a positive effect on CIE;*

**H$_b$:** *SHD-HS has a significant effect on each of the visiting experience indicators, with specific hypotheses including:* **H$_{b1}$:** *SHD-HS has a positive effect on SSP;* **H$_{b2}$:** *SHD-HS has a positive effect on SVL;* **H$_{b3}$:** *SHD-HS has a positive effect on CIE;*

**H$_c$:** *SHD-MS has a significant effect on each of visiting experience indicators, with specific hypotheses including:* **H$_{c1}$:** *SHD-MS has a positive effect on SSP;* **H$_{c2}$:** *SHD-MS has a positive effect on SVL;* **H$_{c3}$:** *SHD-MS has a positive effect on CIE;*

**H$_d$:** *SSP has a significant effect on SVL and CIE, with specific hypotheses including:* **H$_{d1}$:** *SSP has a positive effect on SVL;* **H$_{d2}$:** *SSP has a positive effect on CIE;*

**H$_e$:** *SVL has a significant effect on CIE, with specific hypotheses including:* **H$_{e1}$:** *SVL has a positive effect on CIE.*

A concept model of the SHD influencing visiting experience was proposed based on the hypotheses (Figure 4).

### 3.3.3. Evaluation of Structural Equation Model

We tested the validity of the structural equation model using bootstrap with 5000 replicate samples in SmartPLS 3.3 software [37], and the modified model is shown in Figure 5. As shown in Tables 6 and 7, the pathways passed the significance test, including H$_{a1}$, H$_{a2}$, H$_{b1}$, H$_{c1}$, H$_{d1}$, H$_{d2}$, and H$_{e1}$. All significant paths had a value of f$^2$ (effect size) greater than 0.02, which suggested that the measure of each path was statistically significant [38]. The effectiveness of the model was also verified through three indices, i.e., coefficient of determination, predict relevance, and goodness of fit [39], as shown in Table 7. The results indicated that the SHD of natural sounds ($\beta = 0.195$, $p < 0.01$), human sounds ($\beta = 0.147$, $p < 0.05$), and mechanical sounds ($\beta = 0.234$, $p < 0.01$) all showed significant positive effects on SSP. In addition, among the sound sources, only the SHD of natural sounds could directly affect the SVL ($\beta = 0.127$, $p < 0.05$). Among the three types of visiting experience indicators, both the SSP and SVL could positively affect the CIE, and the SSP could also contribute to the SVL.

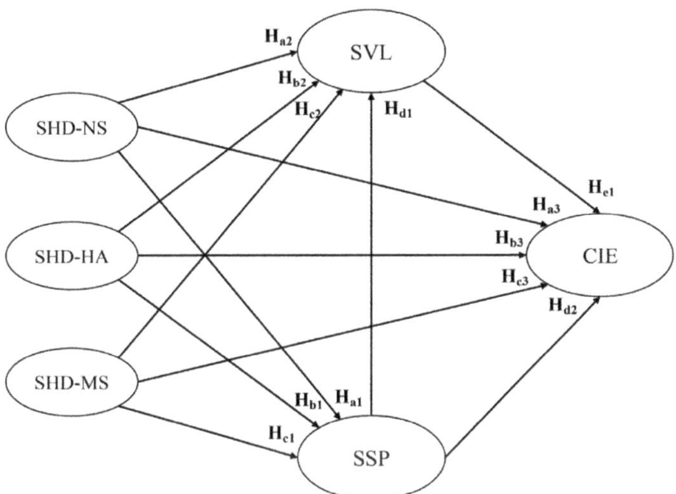

**Figure 4.** A conceptual model of the SHD influencing visiting experience in Kulangsu, SHD-NS: harmonious degree of natural sound, SHD-MS: harmonious degree of mechanical sound, SHD-HS: harmonious degree of human sound, SPE: soundscape perception, SVL: satisfaction degree of visual landscape, CIE: comprehensive impression evaluation.

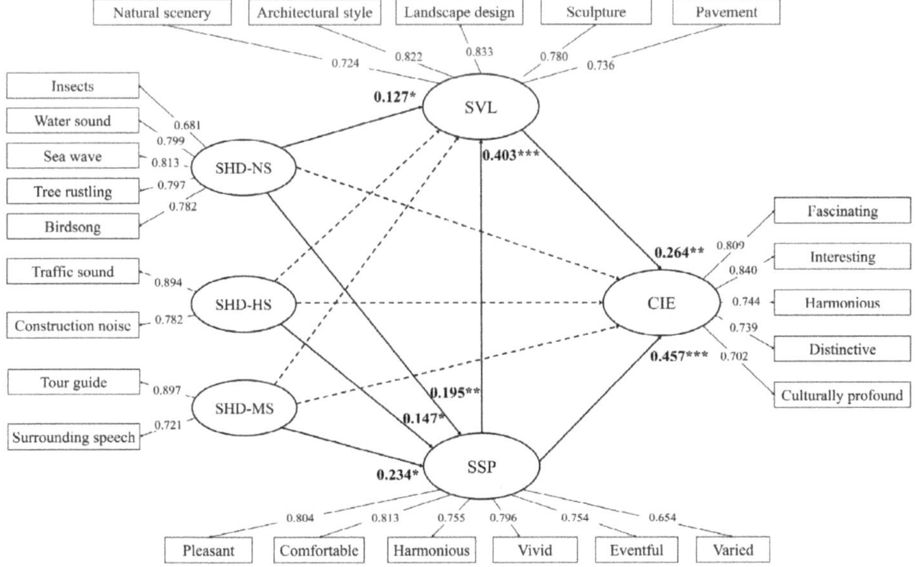

**Figure 5.** The modified model of the SHD influencing visiting experience in Kulangsu, SHD-NS: harmonious degree of natural sound, SHD-MS: harmonious degree of mechanical sound, SHD-HS: harmonious degree of human sound, SPE: soundscape perception, SVL: satisfaction degree of visual landscape, CIE: comprehensive impression evaluation, significant paths are marked with * ($p < 0.05$), ** ($p < 0.01$) or *** ($p < 0.001$).

**Table 6.** Testing of hypothesis paths in the structural equation model.

| Hypothesis Path | $\beta$ | Mean | S.E | T-Value | $p$-Value | 95% CI | | $f^2$ |
|---|---|---|---|---|---|---|---|---|
| | | | | | | 2.50% | 97.50% | |
| $H_{a1}$: SHD-NS → SPE | 0.195 | 0.206 | 0.069 | 2.816 | 0.005 ** | 0.024 | 0.311 | 0.042 |
| $H_{a2}$: SHD-NS → SVL | 0.127 | 0.131 | 0.055 | 2.283 | 0.022 * | 0.004 | 0.226 | 0.02 |
| $H_{a3}$: SHD-NS → CIE | 0.065 | 0.066 | 0.062 | 1.058 | 0.29 | −0.06 | 0.18 | 0.006 |
| $H_{b1}$:SHD-HS → SPE | 0.147 | 0.156 | 0.071 | 2.055 | 0.04 * | −0.012 | 0.269 | 0.022 |
| $H_{b2}$:SHD-HS → SVL | 0.164 | 0.158 | 0.098 | 1.664 | 0.096 | −0.027 | 0.36 | 0.032 |
| $H_{b3}$:SHD-HS → CIE | −0.031 | −0.029 | 0.069 | 0.448 | 0.655 | −0.174 | 0.096 | 0.001 |
| $H_{c1}$: SHD-MS → SSP | 0.234 | 0.223 | 0.099 | 2.36 | 0.018 * | 0.036 | 0.416 | 0.055 |
| $H_{c2}$: SHD-MS → SVL | −0.001 | 0.002 | 0.081 | 0.018 | 0.986 | −0.151 | 0.167 | 0 |
| $H_{c3}$: SHD-MS → CIE | −0.016 | −0.004 | 0.100 | 0.158 | 0.874 | −0.222 | 0.157 | 0 |
| $H_{d1}$: SSP → SVL | 0.403 | 0.412 | 0.092 | 4.388 | 0 *** | 0.213 | 0.568 | 0.191 |
| $H_{d2}$: SSP → CIE | 0.457 | 0.451 | 0.086 | 5.32 | 0 *** | 0.277 | 0.613 | 0.255 |
| $H_{e1}$: SVL → CIE | 0.264 | 0.276 | 0.086 | 3.074 | 0.002 ** | 0.094 | 0.418 | 0.086 |

Note: Significant paths are marked with * ($p< 0.05$), ** ($p< 0.01$) or *** ($p<0.001$).

**Table 7.** The validity of the structural equation model.

| Construct | SSP | SVL | CIE |
|---|---|---|---|
| Adjusted $R^2$ (coefficient of determination) | 0.110 | 0.238 | 0.381 |
| $Q^2$ (predict relevance) [a] | 0.059 | 0.125 | 0.222 |
| GoF (goodness of fit) [b] | | 0.38 | |

Note: a: $Q^2 > 0$ exhibiting predictive relevance; b: $0.1 \leq$ GoF small $< 0.25$, $0.25 \leq$ GoF medium $< 0.36$, GoF large $\geq 0.36$)

### 3.3.4. Comparison among Different Multi-Group Models

Based on the results of the previous part, we conducted a multi-group analysis across the functional zones' modeling (see Table 8). In terms of the SHD affecting visiting experience, the results suggested that only limited but different significant paths existed in different functional zones. Specifically, in the tourist service zone, the SHD showed the most significant effects on visiting experience, with natural sounds and human sounds showing opposite effects on the CIE, but the latter also showing positive effects on the SVL. In the cultural and artistic zone, only the SHD of natural and human sounds showed positive effects on SSP. The SHD of mechanical sounds affecting the SSP was the only significant path in the historical building zone, while the SHD of natural sounds affecting the CIE was also the only one in the natural landscape zone, and there was no significant path in the musical zone. In terms of the relationships among the three types of visiting experience indicators, audio-visual effects were more significant in the cultural and artistic zone and musical zone. The effects of the SSP on the CIE were significant in three different zones, including the musical zone, historical building zone, and natural landscape zone. In addition, the effects of the SVL on the CIE were only significant in the cultural and artistic zone and natural landscape zone.

Furthermore, we analyzed the differences in effectiveness of the same path between different functional zones. The results in Table 9 showed that the most differences were reflected in the effects of the SHD-NS on the CIE, followed by the audio-visual effects.

**Table 8.** Results of multiple-group analysis in different functional zones.

| Hypothesis path | Type I | Type II | Type III | Type IV | Type V |
|---|---|---|---|---|---|
| $H_{a1}$: SHD-NS → SSP | 0.028 | −0.277 | 0.443 ** | 0.306 | 0.15 |
| $H_{a2}$: SHD-NS → SVL | −0.163 | 0.071 | 0.207 | 0.276 | 0.282 |
| $H_{a3}$: SHD-NS → CIE | 0.433 * | 0.099 | −0.132 | −0.137 | 0.369 * |
| $H_{b1}$:SHD-HS → SSP | 0.129 | 0.177 | 0.362 * | 0.093 | −0.034 |
| $H_{b2}$:SHD-HS → SVL | 0.444 *** | 0.114 | 0.175 | 0.207 | −0.076 |
| $H_{b3}$:SHD-HS → CIE | −0.491 * | 0.125 | 0.104 | −0.037 | −0.132 |
| $H_{c1}$: SHD-MS → SSP | 0.328 | 0.179 | −0.065 | 0.297 * | 0.328 |
| $H_{c2}$: SHD-MS → SVL | 0.132 | −0.2 | −0.105 | −0.051 | −0.126 |
| $H_{c3}$: SHD-MS → CIE | −0.088 | 0.187 | 0.011 | 0.244 | 0.061 |
| $H_{d1}$: SSP → SVL | 0.178 | 0.789 *** | 0.423 * | 0.149 | 0.158 |
| $H_{d2}$: SSP → CIE | 0.549 | 0.566 ** | 0.253 | 0.576 *** | 0.387 * |
| $H_{e1}$: SVL → CIE | 0.341 | 0.131 | 0.656 *** | 0.328 | 0.352 * |

Note: β value of each path is shown in the table, Type I: tourist service zone, Type II: musical zone, Type III: cultural and artistic zone, Type IV: historical building zone, Type V: natural landscape zone, significant paths are marked with * ($p < 0.05$), ** ($p < 0.01$) or *** ($p < 0.001$).

**Table 9.** The coefficient difference between the same path in different functional zones.

| | Hypothesis Path | Path Coefficient-Difference | $p$ Value |
|---|---|---|---|
| Type I vs. Type II | SHD-HS → CIE | −0.615 | 0.034 |
| | SSP → SVL | −0.611 | 0.014 |
| Type I vs. Type III | SHD-HS → CIE | −0.595 | 0.018 |
| | SHD-NS → CIE | 0.565 | 0.02 |
| Type I vs. Type IV | SHD-NS → CIE | 0.57 | 0.032 |
| Type II vs. Type III | SHD-NS → SSP | −0.72 | 0.022 |
| | SVL → CIE | −0.526 | 0.021 |
| Type II vs. Type IV | SSP → SVL | 0.64 | 0.015 |
| Type II vs. Type V | SSP → SVL | 0.63 | 0.022 |
| Type III vs. Type V | SHD-NS → CIE | −0.501 | 0.013 |
| | SSP → SVL | −0.63 | 0.022 |
| Type IV vs. Type V | SHD-NS → CIE | −0.506 | 0.024 |

Note: Type I: tourist service zone, Type II: musical zone, Type III: cultural and artistic zone, Type IV: historical building zone, Type V: natural landscape zone.

## 4. Discussion

### 4.1. Perception Characteristics of Typical Sound Sources

The context, including location, landscape function and environmental characteristics of a landscape, contributes to a potential impact on the composition of sound sources in soundscapes (see Figure 3). In this case, certain human sounds (like surrounding speech) and natural sounds (like sea wave) occupied dominant positions than other sound sources, which was fitting with the environmental characteristics and functions of the Kulangsu scenic area. Mechanical sounds including construction sound and traffic sound showed a relatively weak dominating degree, which suggested effective noise control in Kulangsu. The preference values of all natural sounds were more than other sound sources [40,41], which suggested that tourists had a significantly higher preference for natural sounds that showed positive effects on visiting experience [41–43]. Furthermore, we found that broadcasting music also had a relatively high preference, which may be related to the personal information of the participants, such as cultural background, age, etc. [44].

In terms of the SHD, as a comprehensive sound source perception indicator, it can reflect the status of how could the dominating degree of certain sounds matches the visitors' preference for it. In this case, the dominating positions of natural sounds in Kulangsu matched the preference of the visitors, thus resulting their high SHDs. However, as a popular tourist resort that attracts a large number of tourists and resulted many human sound sources that are not always preferred simultaneously, the SHD of human sounds were the lowest ones. Especially, surrounding speech reflecting the crowd density, and sales calling reflecting extensive commercial promotion, could impair soundscape quality. Thus, in a scenic area (and a world heritage), it is necessary to control the daily amount of tourists and the volume of especially electronic equipment of the merchants for commercial promotion. In addition, it is noted in this study, the SHDs of each sound source in different functional zones were different. This is reasonable, as in different functional zones, the "context" for sound/soundscape perception is changing, and people could also have different expectations when they visit the different thematic zone.

### 4.2. The SHD Influencing Soundscape Perception

Visiting experience is influenced by various factors during the comprehensive experience in a scenic area, but from a perception perspective, visual and auditory perception characteristics are the most influential ones [22,23]. As sound perception is an essential process of soundscape perception, the SHD of all sound sources showed significant relationships with at least one soundscape perception indicator, except for surrounding speech. The contribution of natural sounds to positive soundscape perception was more obvious, especially to pleasant soundscapes. In Kulangsu, natural sounds including water sound (fountain), tree rustling, sea waves and birdsong were with the highest SHD values, and all significantly related to pleasant soundscape experience, which is similar to previous studies [45–47]. A number of studies have shown that natural sounds have more positive effects on people's physical and mental health, including physiological indicators and psychological feelings, than other sounds [48–52]. Thus, the preservation of the natural and ecological environment through thoughtful landscape planning and management is necessary for scenic areas, such as protecting the habitats of birds and insects, increasing berry fruit trees to attract birds, and building leisure trails near the coastal line [53].

Usually, artificial sounds are dominating sound sources in urbanized areas. The results suggest that all human sound sources except surrounding speech were related to soundscape perception indicators. Thus, it is necessary to control the amount of tourists to weaken the dominance of human sounds for a better soundscape experience. As the most preferred artificial sound sources, the SHD of broadcasting music showed the most significant relationships with soundscape perception indicator, especially pleasant and eventful. The results suggest that a potential match between the natural and cultural sound sources could contribute to higher soundscape quality, considering the rich musical resources in Kulangsu as a "Piano island". In addition, the SHD of mechanical sounds like construction noise and traffic sound were closely related to varied soundscape perception. This result confirms previous research that the dominance of mechanical sound such as traffic sound in the environment had a significant negative correlation with positive soundscape perception [28,54]. As the objective presence of such sounds in the environment increases in line with people's subjective preference, the people' positive soundscape perception can also increase. Therefore, there is a considerable need to control these sounds in the landscape, either by restricting relatively activities or by using vegetation or installing noise barriers to directly eliminate the presence of these sounds [55,56].

Furthermore, as indicated by the SEM in Figure 5, the SHD of all the three sound source types could positively affect the SSP. According to their preference characteristics, maintaining the dominance of natural sounds, rational controlling the dominance of human sounds and eliminating undesirable mechanical sounds such as construction sounds and traffic sounds, and properly introducing music are effective approaches to improve soundscape quality. Considering about the functional difference in different zones, only the SHD of natural and human sounds showed significant effects on the SSP in the cultural and artistic zone, and only mechanical sounds showed significant effects on the SSP in the historical building zone.

### 4.3. The SHD Influencing Visual Landscape Experience and Comprehensive Impression

The research results highlight the importance of natural sounds in visiting experience as reflected by the close relationships of their SHD and indicators of SVL as well as CIE [23,57]. This was further confirmed in the analysis of SEM, that only the SHD of natural sounds had significant and positive effects on the SVL, which verifies the existence of the audio–visual interaction in the scenic area [22,58,59]. In addition, the results also indicated the most effective landscape elements interacting with the SHD, such as pavement, natural scenery, and architecture. However, there is a relatively weak relationship between the SHD of both human and mechanical sounds and the SVL. Specific attention should be paid to tour guide sound and traffic sound in order to improve the SVL.

The CIE indicated by "fascinating" was most significantly related to the SHD of several sound sources, including insect, tree rustling, water sound (fountain), broadcasting music, and traffic sound. Although different sound sources could contribute to different comprehensive impression, sea waves, water (fountain), and broadcasting music together could be the most crucial sound sources in forming all the five comprehensive impressions, including fascinating, harmonious, distinctive, interesting, and culturally profound. In addition, as indicated by the SEM results, there was no direct effect of any sound source types on the CIE. However, they could indirectly affect it through the SSP which showed significant and even more effects than the SVL on the CIE [60,61].

In different functional zones, only the SHD of human sounds showed a significant and positive effect on the SVL in the tourist service zone, and it also showed a negative effect on the CIE in this zone. The SHD of natural sounds showed both significant and positive effects on the CIE in the tourist service zone and natural landscape zone. The relationships among the SSP, the SVL, and the CIE were changing in different functional zones as well. The results indicated that soundscape design or management strategies in different functional zones should be flexible, especially in targeting crucial sound sources and taking advantage of the audio–visual interaction, to contribute to a high-quality and comprehensive visiting experience [28,54,62].

### 4.4. Practical Implications

In this study, we found that the proposed indicators, harmonious degree of sound sources could better reflect the extent to which the dominance of sound sources in the environment matched the preferences of visitors and had a significant impact on the visiting experience. In practice, the research results could help designers, planners, managers to develop more detailed and effective management of soundscapes in the scenic area, and a better understanding of soundscape value and its role in visiting experience. Based on the findings of this study, we make the following proposals for the management of soundscapes in scenic areas.

(1)    Identify major negative sound sources

Until today, noise control has been the major focus of acoustic environment management. In this study, however, we can find that noise such as traffic sound and construction noise have been better controlled in Kulangsu. Instead, certain dominating human sounds have deviated significantly from the preferences of visitors. For example, surrounding speech and sales calling showed the lowest harmonious degree of sound sources. These sound sources, while reflecting the vitality of the scenic area to a certain extent, could actually blur or obscure the perception of other soundscapes such as natural soundscapes, so as to impair the soundscape quality in the scenic area. Therefore, effective control of certain human sounds is necessary in the Kulangsu scenic area, including fine-grained control of the daily number of visitors to the scenic area and some regulation of the volume and playing time of electronic devices by vendors, etc.

(2)     Emphasize the resource attributes of the positive soundscapes

The results of this study showed that the harmonious degree of sound sources could significantly enhance visiting experience, particularly that of natural sounds. Therefore, it's necessary to take steps to highlight the role of these soundscapes, thereby enhancing the attractiveness of the scenic area. For example, positive soundscapes could be labeled through a soundscape map and included in marketing materials such as brochures and tourist maps, so that visitors have sufficient information to know about these positive soundscapes and experience them better. In conjunction with the soundscape map, special routes such as recreational trails around the coastline can be created to guide visitors to experience these positive soundscapes. In addition, during this process, introductions by guides can also be used to enhance visitors' awareness of these positive soundscapes. It is also worth noting that the ability to make the most of these positive soundscapes to enhance the visiting experience and quality of the scenic area is based on the availability of adequate soundscape resources. This is reflected in the management of the soundscape on Kulangsu by actively creating more positive soundscape resources while protecting existing ones. For example, through thoughtful landscape planning and management to protect bird and insect habitats, adding berry species or trees with larger leaves, etc. In addition, the specific sound sources of Kulangsu can be increased through specific time and place events, such as the sound of various music sources, including pianos and live music, etc. These measures will help to increase soundscape resources to support an enhanced visiting experience.

(3)     Concern for the impact of context on soundscapes and visiting experience

This study has noted that different sound sources have different levels of harmony in different contexts and have different levels of impact on visiting experience. Therefore, when using soundscapes to stimulate positive emotions and promote visiting experience, attention needs to be paid to the impact of the context. For example, in the natural environment, it's necessary to pay more attention to the natural soundscapes, highlighting their dominant position and enhancing the visitor's perception of them. Whereas, in the human environment, more attention needs to be paid to cultural soundscapes. In certain contexts, there is also a potential for collaboration between different types of soundscapes. This suggests that soundscape management in scenic areas needs to be contextualised in order to develop appropriate solutions.

*4.5. Limitations and Future Research*

Although this study contributes to the understanding of the impact of soundscapes on visiting experience in scenic areas and how to conduct soundscape management accordingly, there are still some interesting questions for further research. Firstly, while the respondents in this study were all people with normal perceptual functions, sound is in fact the most important way for people with special needs, such as the blind, to perceive the environment. It is therefore essential to understand the impact of soundscape on their visiting experience in scenic areas. Secondly, we need more specific approaches and measures to achieve soundscape quality improvement in scenic areas. This type of research can be conducted through small-scale field experiments to modify the soundscapes of specific sites in a scenic area and to compare the visiting experience before and after the modification for validation, thus enabling evidence-based design and management.

## 5. Conclusions

The effective management of sound sources, as a crucial landsense element, is a reliable way of achieving soundscape quality control and landsense creation. However, this requires an in-depth understanding of how landsense elements interact with human senses, i.e., how sound source perception affects soundscape perception and other visiting experiences. In this study, based on a public investigation of 195 interviewees in the Kulangsu scenic area, we established a new sound source perception indicator, the harmonious degree of sound sources (SHD), integrating the perceived occurrences and loudness, and preference for sound sources. A statistical method was used to explore the relationships between the SHDs of different sound sources and visiting experience indicators, and a structural equation model was further constructed. The results indicate the following:

(1) Natural sounds had higher SHD values in the Kulangsu scenic area, with water sound (fountain) being the highest one, while human sounds, especially sales calling, surrounding speech, and playing children, showed lower SHD values. The SHD values of the same type of sound source were different among different functional zones. While most of the natural sounds showed positive SHD values in different zones, more than half of the SHD values of human sounds had negative values, and mechanical sounds normally had small but positive SHD values.

(2) The SHD, as a comprehensive sound perception indicator, is effective in building relationships with visiting experience. The harmonious status of nearly all the sound sources showed significant and positive relationships with at least one of the visiting experience indicators, except for surrounding speech. The SHD of natural sounds showed the most significant relationships with pleasant soundscape perception, while all three mechanical sounds were closely related to varied soundscapes, and human sounds showed the least but four different significant relationships with SSP indicators. Among all the sound sources, broadcasting music could be the most crucial sound source related to the SSP.

(3) The SHD of natural sounds also showed close relationships with the SVL, with water sound (fountain) and birdsong as the most prominent sounds, and pavement, natural scenery, and architecture as the most influential visual landscape elements. Although the SHD of both human and mechanical sounds showed relatively weak relationships with the SVL, certain sounds like tour guide sound and traffic sound could be influential to the SVL.

(4) Crucial sound sources related to the CIE were sea waves, water sound (fountain), and broadcasting music in Kulangsu. The SHD showed the most influence on the fascinating characteristic of the place, followed by harmonious and distinctive, but the effects could be indirectly through the SSP. In addition, audio-visual effects existed in the visiting experience in the scenic area, and the SSP showed more significant effects than the SVL on the CIE.

(5) The mechanism of the SHD affecting visiting experience was verified to be different according to the function of an area, reflected by different crucial sound sources, the significance of audio–visual interaction effects, as well as the contribution of the SSP and the SVL to the CIE. Thus, flexible soundscape design or management strategies should be adopted to promote a high-quality visiting experience in scenic areas.

**Author Contributions:** Conceptualization, J.L.; methodology, J.L. and X.G.; formal analysis, J.L. and X.G.; investigation, J.L., X.G., and Z.C.; writing—original draft preparation, J.L. and X.G.; writing—review and editing, J.L., X.-C.H. and Z.C.; supervision, J.L. All authors have read and agreed to the published version of the manuscript.

**Funding:** This research was funded by the National Key R&D Program of China, grant number 2022YFF1301303, National Natural Science Foundation of China, grant number 52208052 and the Program of Humanities and Social Science Research of Ministry of Education of China (21YJCZH038).

**Institutional Review Board Statement:** Ethical review and approval were waived for this study due to there is no direct or potential hazards of the questionnaire investigation to the subjects, and they were all informed of the purpose of this research and approved us to collect the data provided by them.

**Informed Consent Statement:** Informed consent was obtained from all subjects involved in the study.

**Data Availability Statement:** Data available on request due to restrictions, e.g., privacy or ethical.

**Acknowledgments:** The authors appreciate the valuable comments of editors.

**Conflicts of Interest:** The authors declare no conflict of interest.

## References


1. Chen, Z.; Zhu, T.-Y.; Liu, J.; Hong, X.-C. Before Becoming a World Heritage: Spatiotemporal Dynamics and Spatial Dependency of the Soundscapes in Kulangsu Scenic Area, China. *Forests* **2022**, *13*, 1526. [CrossRef]
2. Xiong, Y.; Yang, L.; Wang, X.; Liu, J. Mediating effect on landscape experience in scenic area: A case study in Gulangyu Island, Xiamen City. *Int. J. Sustain. Dev. World Ecol.* **2020**, *27*, 276–283. [CrossRef]
3. Hu, F.; Wang, Z.-J.; Sheng, G.-H.; Lia, X.; Chen, C.; Geng, D.-H.; Hong, X.-C.; Xu, N.; Zhu, Z.-P.; Zhang, Z.-F.; et al. Impacts of national park tourism sites: A perceptual analysis from residents of three spatial levels of local communities in Banff national park. *Environ. Dev. Sustain.* **2022**, *24*, 3126–3145. [CrossRef]
4. Dong, R.; Liu, X.; Liu, M.; Feng, Q.; Su, X.; Wu, G. Landsenses ecological planning for the Xianghe Segment of China's Grand Canal. *Int. J. Sustain. Dev. World Ecol.* **2016**, *23*, 298–304. [CrossRef]
5. Zhao, J.; Yan, Y.; Deng, H.; Liu, G.; Dai, L.; Tang, L.; Shi, L.; Shao, G. Remarks about landsenses ecology and ecosystem services. *Int. J. Sustain. Dev. World Ecol.* **2020**, *27*, 196–201. [CrossRef]
6. Zhao, J.; Liu, X.; Dong, R.; Shao, G. Landsenses ecology and ecological planning toward sustainable development. *Int. J. Sustain. Dev. World Ecol.* **2016**, *23*, 293–297. [CrossRef]
7. Zheng, T.; Yan, Y.; Lu, H.; Pan, Q.; Zhu, J.; Wang, C.; Zhang, W.; Rong, Y.; Zhan, Y. Visitors' perception based on five physical senses on ecosystem services of urban parks from the perspective of landsenses ecology. *Int. J. Sustain. Dev. World Ecol.* **2020**, *27*, 214–223. [CrossRef]
8. Dong, R.; Yu, T.; Ma, H.; Ren, Y. Soundscape planning for the Xianghe Segment of China's Grand Canal based on landsenses ecology. *Int. J. Sustain. Dev. World Ecol.* **2016**, *23*, 343–350. [CrossRef]
9. Buzova, D.; Sanz-Blas, S.; Cervera-Taulet, A. "Sensing" the destination: Development of the destination sensescape index. *Tour. Manag.* **2021**, *87*, 104362. [CrossRef]
10. Jiang, J. The role of natural soundscape in nature-based tourism experience: An extension of the stimulus–organism–response model. *Curr. Issues Tour.* **2020**, *25*, 707–726. [CrossRef]
11. An, W.; Alarcón, S. Exploring rural tourism experiences through subjective perceptions: A visual Q approach. *Span. J. Agric. Res.* **2020**, *18*, e0108. [CrossRef]
12. Fairweather, J.R.; Swaffield, R.S. Visitor Experiences of Kaikoura, New Zealand: An interpretative study using photographs of landscapes and Q method. *Tour. Manag.* **2001**, *22*, 219–228. [CrossRef]
13. Pérez, J.G. Ascertaining Landscape Perceptions and Preferences with Pair-wise Photographs: Planning rural tourism in Extremadura, Spain. *Landsc. Res.* **2002**, *27*, 297–308. [CrossRef]
14. Conniff, A.; Craig, T. A methodological approach to understanding the wellbeing and restorative benefits associated with greenspace. *Urban For. Urban Green.* **2016**, *19*, 103–109. [CrossRef]
15. Jo, H.I.; Jeon, J.Y. Overall environmental assessment in urban parks: Modelling audio-visual interaction with a structural equation model based on soundscape and landscape indices. *Build. Environ.* **2021**, *204*, 108166. [CrossRef]
16. Agapito, D.; Valle, P.; Mendes, J. The sensory dimension of tourist experiences: Capturing meaningful sensory-informed themes in Southwest Portugal. *Tour. Manag.* **2014**, *42*, 224–237. [CrossRef]
17. *ISO 12913-1:2014*; Acoustics—Soundscape—Part 1: Definition and Conceptual Framework. ISO: Geneva, Switzerland, 2014.
18. Watts, G.R.; Pheasant, R.J. Tranquillity in the Scottish Highlands and Dartmoor National Park–The importance of soundscapes and emotional factors. *Appl. Acoust.* **2015**, *89*, 297–305. [CrossRef]
19. Qiu, M.; Zhang, J.; Zhang, H.; Zheng, C. Is looking always more important than listening in tourist experience? *J. Travel Tour. Mark.* **2018**, *35*, 869–881. [CrossRef]
20. He, M.; Li, J.; Li, J.; Chen, H. A comparative study on the effect of soundscape and landscape on tourism experience. *Int. J. Tour. Res.* **2019**, *21*, 11–22. [CrossRef]
21. Kang, J.; Zhang, M. Semantic differential analysis of the soundscape in urban open public spaces. *Build. Environ.* **2010**, *45*, 150–157. [CrossRef]
22. Liu, J.; Xiong, Y.; Wang, Y.; Luo, T. Soundscape effects on visiting experience in city park: A case study in Fuzhou, China. *Urban For. Urban Green.* **2018**, *31*, 38–47. [CrossRef]
23. Liu, J.; Yang, L.; Xiong, Y.; Yang, Y. Effects of soundscape perception on visiting experience in a renovated historical block. *Build. Environ.* **2019**, *165*, 106375. [CrossRef]
24. Liu, J.; Wang, Y.; Zimmer, C.; Kang, J.; Yu, T. Factors associated with soundscape experiences in urban green spaces: A case study in Rostock, Germany. *Urban For. Urban Green.* **2019**, *37*, 135–146. [CrossRef]
25. Zhao, J.; Xu, W.; Ye, L. Effects of auditory-visual combinations on perceived restorative potential of urban green space. *Appl. Acoust.* **2018**, *141*, 169–177. [CrossRef]
26. Deng, L.; Luo, H.; Ma, J.; Huang, Z.; Sun, L.-X.; Jiang, M.-Y.; Zhu, C.-Y.; Li, X. Effects of integration between visual stimuli and auditory stimuli on restorative potential and aesthetic preference in urban green spaces. *Urban For. Urban Green.* **2020**, *53*, 126702. [CrossRef]
27. Chen, Z.; Hermes, J.; Liu, J.; von Haaren, C. How to integrate the soundscape resource into landscape planning? A perspective from ecosystem services. *Ecol. Indic.* **2022**, *141*, 109156. [CrossRef]

28. Hong, J.Y.; Jeon, J.Y. Comparing associations among sound sources, human behaviors, and soundscapes on central business and commercial streets in Seoul, Korea. *Build. Environ.* **2020**, *186*, 107327. [CrossRef]
29. Hong, X.-C.; Wang, G.-Y.; Liu, J.; Song, L.; Wu, E.T. Modeling the impact of soundscape drivers on perceived birdsongs in urban forests. *J. Clean. Prod.* **2021**, *292*, 125315. [CrossRef]
30. Guo, X.; Liu, J.; Albert, C.; Hong, X.-C. Audio-visual interaction and visitor characteristics affect perceived soundscape restorativeness: Case study in five parks in China. *Urban For. Urban Green.* **2022**, *77*, 127738. [CrossRef]
31. Hong, J.Y.; Jeon, J.Y. Influence of urban contexts on soundscape perceptions: A structural equation modeling approach. *Landsc. Urban Plan.* **2015**, *141*, 78–87. [CrossRef]
32. Hair, J.F.; Ringle, C.M.; Sarstedt, M. PLS-SEM: Indeed a Silver Bullet. *J. Mark. Theory Pract.* **2011**, *19*, 139–152. [CrossRef]
33. Jia, Y.; Ma, H.; Kang, J. Characteristics and evaluation of urban soundscapes worthy of preservation. *J. Environ. Manag.* **2020**, *253*, 109722. [CrossRef] [PubMed]
34. Wetzels, M.; Odekerken-Schröeder, G.; Van Oppen, C. Using PLS Path Modeling for Assessing Hierarchical Construct Models: Guidelines and Empirical Illustration. *MIS Q.* **2009**, *33*, 177–195. [CrossRef]
35. Bagozzi, R.P.; Yi, Y. On the evaluation of structural equation models. *J. Acad. Mark. Sci.* **1988**, *16*, 74–94. [CrossRef]
36. Henseler, J.; Ringle, C.M.; Sarstedt, M. A new criterion for assessing discriminant validity in variance-based structural equation modeling. *J. Acad. Mark. Sci.* **2015**, *43*, 115–135. [CrossRef]
37. Tenenhaus, M.; Vinzi, V.E.; Chatelin, Y.-M.; Lauro, C. PLS path modeling. *Comput. Stat. Data Anal.* **2005**, *48*, 159–205. [CrossRef]
38. Ketchen, D.J. A primer on partial least squares structural equation modeling. *Long Range Plan.* **2013**, *46*, 184–185. [CrossRef]
39. Shmueli, G.; Ray, S.; Estrada, J.M.V.; Chatla, S.B. The elephant in the room: Predictive performance of PLS models. *J. Bus. Res.* **2016**, *69*, 4552–4564. [CrossRef]
40. Yang, W.; Kang, J. Soundscape and Sound Preferences in Urban Squares: A Case Study in Sheffield. *J. Urban Des.* **2005**, *10*, 61–80. [CrossRef]
41. Sharifnejad, M.; Montazerolhodjah, M.; Montazerolhodjah, M. Soundscape preferences of tourists in historical urban open spaces. *Int. J. Tour. Cities* **2019**, *5*, 465–481.
42. Brown, A.; Kang, J.; Gjestland, T. Towards standardization in soundscape preference assessment. *Appl. Acoust.* **2011**, *72*, 387–392. [CrossRef]
43. Patón, D.; Delgado, P.; Galet, C.; Muriel, J.; Méndez-Suárez, M.; Hidalgo-Sánchez, M. Using acoustic perception to water sounds in the planning of urban gardens. *Build. Environ.* **2020**, *168*, 106510. [CrossRef]
44. Steele, D.; Tarlao, C.; Bild, E.; Rice, J.; Guastavino, C. How does activity affect soundscape assessments? Insights from an urban soundscape intervention with music. *J. Acoust. Soc. Am.* **2017**, *141*, 3622. [CrossRef]
45. Ma, K.W.; Mak, C.M.; Wong, H.M. Effects of environmental sound quality on soundscape preference in a public urban space. *Appl. Acoust.* **2021**, *171*, 107570. [CrossRef]
46. Liu, J.; Kang, J.; Behm, H. Birdsong as an Element of the Urban Sound Environment: A Case Study Concerning the Area of Warnemünde in Germany. *Acta Acust. United Acust.* **2014**, *100*, 458–466. [CrossRef]
47. Zhao, W.; Li, H.; Zhu, X.; Ge, T. Effect of Birdsong Soundscape on Perceived Restorativeness in an Urban Park. *Int. J. Environ. Res. Public Health* **2020**, *17*, 5659. [CrossRef]
48. Li, Z.; Kang, J. Sensitivity analysis of changes in human physiological indicators observed in soundscapes. *Landsc. Urban Plan.* **2019**, *190*, 103593. [CrossRef]
49. Alvarsson, J.J.; Wiens, S.; Nilsson, M.E. Stress Recovery during Exposure to Nature Sound and Environmental Noise. *Int. J. Environ. Res. Public Health* **2010**, *7*, 1036–1046. [CrossRef] [PubMed]
50. Krzywicka, P.; Byrka, K. Restorative Qualities of and Preference for Natural and Urban Soundscapes. *Front. Psychol.* **2017**, *8*, 1705. [CrossRef] [PubMed]
51. Li, Z.; Ba, M.; Kang, J. Physiological indicators and subjective restorativeness with audio-visual interactions in urban soundscapes. *Sustain. Cities Soc.* **2021**, *75*, 103360. [CrossRef]
52. Annerstedt, M.; Jönsson, P.; Wallergård, M.; Johansson, G.; Karlson, B.; Grahn, P.; Hansen, Å.M.; Währborg, P. Inducing physiological stress recovery with sounds of nature in a virtual reality forest—Results from a pilot study. *Physiol. Behav.* **2013**, *118*, 240–250. [CrossRef] [PubMed]
53. Jeon, J.Y.; Lee, P.J.; You, J.; Kang, J. Acoustical characteristics of water sounds for soundscape enhancement in urban open spaces. *J. Acoust. Soc. Am.* **2012**, *131*, 2101–2109. [CrossRef] [PubMed]
54. Hong, J.Y.; Jeon, J.Y. Relationship between spatiotemporal variability of soundscape and urban morphology in a multifunctional urban area: A case study in Seoul, Korea. *Build. Environ.* **2017**, *126*, 382–395. [CrossRef]
55. Li, W.; Zhai, J.; Zhu, M. Characteristics and perception evaluation of the soundscapes of public spaces on both sides of the elevated road: A case study in Suzhou, China. *Sustain. Cities Soc.* **2022**, *84*, 103996. [CrossRef]
56. Fan, Q.-D.; He, Y.-J.; Hu, L. Soundscape evaluation and construction strategy of park road. *Appl. Acoust.* **2021**, *174*, 107685. [CrossRef]
57. Jiang, J.; Zhang, J.; Zhang, H.; Yan, B. Natural soundscapes and tourist loyalty to nature-based tourism destinations: The mediating effect of tourist satisfaction. *J. Travel Tour. Mark.* **2018**, *35*, 218–230. [CrossRef]
58. Jeon, J.Y.; Hong, J.Y. Classification of urban park soundscapes through perceptions of the acoustical environments. *Landsc. Urban Plan.* **2015**, *141*, 100–111. [CrossRef]

59. Hong, J.Y.; Jeon, J.Y. Designing sound and visual components for enhancement of urban soundscapes. *J. Acoust. Soc. Am.* **2013**, *134*, 2026–2036. [CrossRef]
60. Kanngieser, A. A sonic geography of voice: Towards an affective politics. *Prog. Hum. Geogr.* **2012**, *36*, 336–353. [CrossRef]
61. Falahatkar, H.; Aminzadeh, B. The sense of place and its influence on place branding: A case study of Sanandaj natural landscape in Iran. *Landsc. Res.* **2020**, *45*, 123–136. [CrossRef]
62. Tan, J.K.A.; Hasegawa, Y.; Lau, S.-K.; Tang, S.-K. The effects of visual landscape and traffic type on soundscape perception in high-rise residential estates of an urban city. *Appl. Acoust.* **2022**, *189*, 108580. [CrossRef]

*Article*

# A Study on the Soundscape Preferences of the Elderly in the Urban Forest Parks of Underdeveloped Cities in China

Lei Luo [1,†], Qi Zhang [1,†], Yingming Mao [2], Yanyan Peng [3], Tao Wang [4] and Jian Xu [1,5,6,*]

1   Department of Tourism Management, South China University of Technology, Guangzhou 511442, China; luol19991124@163.com (L.L.); zhangqiqi3590@126.com (Q.Z.)
2   Graduate School of Horticulture, Faculty of Horticulture, Chiba University, Chiba 271-8510, Japan; skyisyui@gmail.com
3   Department of Urban Planning and Management, School of Public Administration and Policy, Renmin University of China, Beijing 100872, China; pengyanyan2018@ruc.edu.cn
4   Zhijiang College, Zhejiang University of Technology, Shaoxing 312000, China; wt@zzjc.edu.cn
5   Key Laboratory of Digital Village and Sustainable Development of Culture and Tourism, Guangzhou 510006, China
6   Tourism Strategy and Policy Research Center, Guangzhou 510006, China
*   Correspondence: jianxu@scut.edu.cn; Tel.: +86-181-4288-6885
†   These authors contributed equally to this work.

**Abstract:** Against the backdrop of the global aging trend, the proportion of the elderly population is severely increasing in the urban areas of underdeveloped regions. Despite evidence that urban forest parks are effective at enhancing the physical and mental well-being of the elderly, little has been done to investigate the connection between urban forest parks and the elderly in underdeveloped regions, and landscape studies in particular are lacking. This study attempted to address this gap, using a subjective evaluation method in which 725 elderly respondents were engaged in a questionnaire survey on their soundscape preferences in the urban forest parks of an underdeveloped city in China. The results revealed the elderly people's preferences for soundscapes, and a further analysis demonstrated the relationships between these preferences and landscape features. The effects of personal traits and living situations on soundscape preferences were determined by analyzing the impacts of living conditions, occupation, and education on soundscape preferences. By building a model with regression coefficients, the most powerful factors influencing soundscape choice were investigated. It was found that (1) the types of sound sources preferred by the elderly, in descending order, were natural sound, livestock sound, bird song, musical sound, other sounds. (2) The differences among education, occupation, and age all affected the participants' soundscape preferences, i.e., the mean values of the soundscape preferences among older adults varied with education, occupation, and age. The mean value of soundscape preference was higher among older adults who had received higher education, were government officials and business managers, and belonged to higher age groups. (3) Among the various factors influencing the soundscape preference of the elderly, the most influential factors were the length of time spent in the waterfront environment, the time spent in the forest park, and the importance of road signs. (4) The preference for soundscapes was strongly connected with happiness in life. (5) Wearing a mask significantly reduced soundscape perception scores under epidemic conditions, while vaccinated individuals were more tolerant of various noises. Recommendations for landscape design to improve the soundscape perception of elderly people are accordingly provided.

**Keywords:** soundscape preference; elderly; urban forest park; underdeveloped cities in China; subjective evaluation

**Citation:** Luo, L.; Zhang, Q.; Mao, Y.; Peng, Y.; Wang, T.; Xu, J. A Study on the Soundscape Preferences of the Elderly in the Urban Forest Parks of Underdeveloped Cities in China. *Forests* **2023**, *14*, 1266. https://doi.org/10.3390/f14061266

Academic Editor: Chi Yung Jim

Received: 18 May 2023
Revised: 5 June 2023
Accepted: 12 June 2023
Published: 19 June 2023

## 1. Introduction

As an important ecological service system for residential areas, urban forest parks are often referred to as the "heart of the city" due to their important role in maintaining the overall urban environment [1,2]. They harbor all urban trees, shrubs, lawns, pervious soils, roads, and other landscape facilities [3]. A quality urban forest park may enhance health, cleanse the air, and offer space for ecotourism, entertainment, and exercise, as well as aid in the prevention of obesity and the alleviation of chronic illnesses [4,5]. By concentrating on the relationships between individuals, sound perception, the acoustic environment, and society, soundscape studies regard the acoustic environment as a resource rather than as noise [6]. In addition, studies have shown that urban forest parks play an active role in providing mental relaxation via acoustic landscape design and implementation [7,8] Soundscape perception can be employed in the design and maintenance of forest parks to make them more appealing to park visitors. This not only serves to maximize visitor–park interactions but also helps to improve visitors' experience of the parks [9]. Recent years have witnessed a global concern for mental and physical well-being. Interestingly, research has shown that the proportion of people with depression is lowest near the equator and highest near the poles. Vulnerable groups, such as the elderly and patients with chronic illnesses, generally have a poorer understanding of weather-related risks [10], which further affects their mental and physical health [11]. Therefore, studying the activities of the elderly in high-latitude regions can contribute to understanding how to improve their well-being.

Meanwhile, urban forest parks are becoming increasingly significant to middle-aged and older people as the population continues to age and as their significance in residents' routines increases [9,12]. Urban forest parks are distinct from both urban parks and forest parks in that they are typically found in suburban or central metropolitan areas. The forest biological environment serves as a support system, while the human-made natural landscape serves as a complement [13]. Some of the benefits of urban forest parks are their outstanding acoustic environments, which can help with the elderly population's health issues.

In underdeveloped cities, the connection between the elderly and urban forest parks is even more prominent. Herein, underdeveloped cities refer to regions that have some economic strengths and potentialities but still lag behind developed regions, with uneven productivity development and underdeveloped technologies. Typical cases are the central, western, and northeastern regions of China [14]. Irregularities are not uncommon in the distribution of age groups in such residential areas. With younger age groups flooding into large, advanced, or prosperous developed metropolitan areas, the proportions of older age groups in less-developed urban areas are increasing to new highs. Due to a lack of recreational choices in these areas, parks are some of the primary locations for the aged groups to visit. It has been found that in these areas, older people spend 62.43% of their daytime hours in parks [15]. After the COVID-19 pandemic, a further decrease was observed in indoor activities among elder groups, and a further increase in the significance of urban forest parks was observed.

Compared to younger individuals, elderly people have lower levels of communication with the outside world and are less able to provide feedback in response to their surroundings [16]. According to demographic census data, the percentage of the population aged 60 and older nationwide had reached 18.7% by 2020, with the percentage of people aged 65 and older amounting to 13.5%. The aging problem is particularly salient in the Northeast, Sichuan, and Chongqing regions, with each surpassing 20% by 2020. Notably, these are also the least-developed regions in China [14]. Older adults may be more susceptible to experiencing higher levels of HA, which can significantly lower their quality of life and social adaptability. This is particularly true for countries such as Bulgaria, where there has been a steady increase in the proportion of older people at risk of poverty over the past decade [15].

In other words, it is impossible to overestimate the significance of the forest park as an area firmly connected to the elderly. The natural conditions in forest parks might vary, which may have an impact on the well-being of elderly individuals [17]. Soundscape factors in the natural environment in particular can have a direct impact on the mental health, cognitive function, and physical functioning of older adults [18]. Studies have revealed that elder adults are prone to many physical ailments, such as chronic obstructive pulmonary disease and cognitive decline [19], for which soundscape factors are crucial for an elderly person's physical recovery [20].

The word "soundscape" is used in the interdisciplinary field known as "soundscape studies," which was established by activist and composer R. Murray Schafer [21]. In the same way that visual pictures displayed in a specific location are considered landscapes in general, soundscapes may be thought of as audio landscapes [6]. Several studies on urban park soundscapes have been performed by academics in Asia, Oceania, Europe, and America [22], and it has been discovered that soundscape elements have an impact on the physical and mental health of children, the elderly, and those who are blind [23,24]. According to the study Soundscape Preference of Urban Peoples in China in the Post-Pandemic Era, soundscape preference is the preference of one or more persons for the sound environment of an area. Existing research shows that the elderly may suffer from a variety of medical conditions, including chronic obstructive pulmonary disease and cognitive decline [25,26], on which environmental soundscape elements can have a significant impact [27,28]. Additionally, urban forest parks can benefit elderly individuals' physical health and offer soundscapes [29,30]. It has been found that as society evolves, parks have become hideaways for older adults to escape the "urban disease" [31]. In the aftermath of the COVID-19 pandemic, the use of masks has increased greatly among the public. The elderly, with their organs becoming fragile, have a strong demand for soundscapes which is even more intense as wearing a mask greatly decreases the perceived level of a soundscape [32,33].

Nevertheless, in recent years, fewer researchers have concentrated on the findings of studies on audiovisual aspects conducted by Japanese researchers who combined forest-derived audiovisual stimuli to induce physiological and psychological relaxation, with the physiological relaxation effect being more pronounced under such conditions [34]. It is worth noting that no study has been carried out in this crucial research area on how the elderly perceive such environments in China's underdeveloped towns. With the advancement of society and technology, a variety of methods have emerged for surveying large numbers of respondents. Meanwhile, traditional data collection techniques, such as the subjective evaluation method and the questionnaire method, have been downplayed because of their inefficiencies in gathering big data [35].

Due to the limited knowledge of the elderly, the majority of previous studies on elderly populations have employed the more traditional subjective evaluation approach in combination with questionnaires to ensure that older adults can be maximally engaged in the studies. In addition, despite the heavy burden, the interaction with the elderly enables researchers to gain a deeper level of comprehension of the issue under examination [36].

The main objective of this study, which used Maoershan Forest Park as an example, was to investigate which acoustic elements of the park have an impact on the cognition of elderly individuals and their preferences for acoustic environments, as well as to provide suggestions for improving the mental and physical well-being of the elderly. A questionnaire was designed and administered to survey the preferred soundscape of the elderly in the underdeveloped city. A statistical analysis was then performed to detect the variables that affect soundscape preference. Finally, a regression coefficient model was developed to investigate the elderly participants' preferences for soundscapes. Answering the questions above helped us understand the soundscape preferences of elderly people in urban forest parks in China's underdeveloped urban areas, and our research team also offers some advice on how to create a soundscape.

## 2. Methodology

This research was based on the important theoretical foundations of ecology, acoustics, and landscape architecture, as well as the concept and academic background of urban public space soundscapes. The preferences of elderly people in urban forest parks in underdeveloped cities in China were investigated using a subjective evaluation method.

Specifically, a survey questionnaire was designed and administered to a group of frequent forest park visitors in the underdeveloped city of Yanji in Northeast China.

The methodological design of this study is demonstrated as follows (Figure 1).

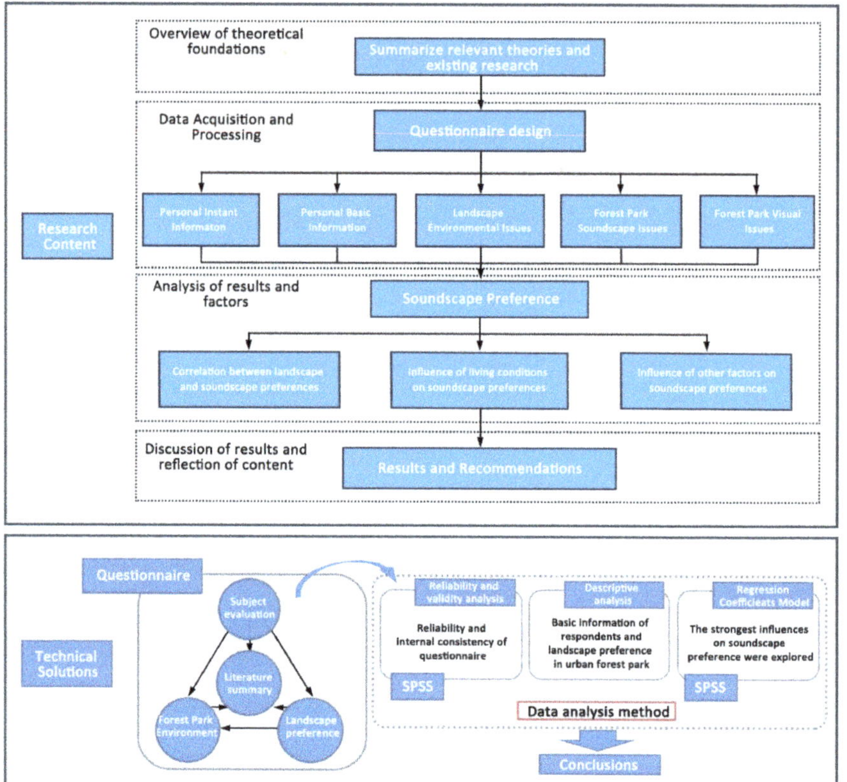


**Figure 1.** Research Contents and technical methods.

## 2.1. Research Context

Underdeveloped regions, due to limited healthcare conditions, inadequate economic development, and the mobility of younger populations, often have relatively high proportions of elderly populations. Research has shown that natural environments, particularly forest parks, have positive impacts on the physical and mental health of the elderly. Forest parks offer good air quality and atmospheres that are conducive to alleviating stress and anxiety among the elderly, promoting physical activity and social engagement, and enhancing happiness and quality of life. Therefore, studying the needs and utilization of forest parks among the elderly in underdeveloped regions can provide a better understanding of this specific population's demands and health conditions. In underdeveloped regions where healthcare resources may be limited, natural environments can become vital resources for the elderly to maintain their health, and studying the health benefits of forest parks among the elderly in such regions can provide scientific evidence for improving their quality of life.

China's population is the largest population in the world, and its elderly population is also the largest. Despite China's fast overall economic growth in recent decades, about 80% of the provinces and cities in China are still underdeveloped, with disproportionately large elderly populations and inefficient supplies of medical services. The city of Yanji in Jilin Province is a typical case in point. It has been rated as one of the key cities in northeast

China since China's reform and opening up. However, it is still in an underdeveloped state and faces many problems: there are fewer large projects, the supporting industries still rely on traditional industries, and the industrial structure is not balanced. Additionally, the city's transformation and upgrading still have a long way to go; the image of the capital city has not been fully manifested, the supporting functions of the city still need to be improved, the degree of refinement of urban management needs to be strengthened, the level of grass-roots social governance needs to be improved, and there is still much room for improving the quality of the city. Finally, the pressure on ecological environmental protection is extremely strong [37,38]. It is more worthwhile to investigate cities such as Yanji than provincial capitals such as Harbin, which is developing slowly and yet has a large population [7,39,40].

Since the central government's call to revive the northeast, Yanji, as a city near China's border with North Korea, is now facing the pressure of finding a clear path in its future development, partially due to its prolonged history of underdevelopment [41]. Realistic strategies are urgently needed to build up infrastructure that can ensure proper and efficient growth [42]. According to the data from the sixth population census, the proportion of the elderly population in Yanji City is approximately 22% [43]. The city's chronic underdevelopment has resulted in a large decline in the proportion of young people and an increase in the number of senior residents, making the problem of retirement all the more difficult for the local government authorities [44].

In spite of its underdevelopment, Yanji hosts several national forest parks, of which the Maoershan National Forest Park (its location is shown in Figure 2) is of significant value to the current research study as it is the primary urban forest park in Yanji and one of the largest urban forest parks in the neighboring provinces and cities. As noted above, forest parks can contribute to community development and economic growth. Underdeveloped regions often face economic challenges and social development issues. Studying the utilization of forest parks among the elderly in underdeveloped regions can help governments and community organizations understand the needs of the elderly, formulate relevant policies and plans, and promote community development and elderly participation. To put it simply, choosing the Maoershan National Forest Park in Yanji, Jilin, China, for the study of forest parks among the elderly in underdeveloped regions is not only helpful in gaining deeper insights into the needs and health conditions of this specific population but also in garnering scientific evidence for improving their quality of life and promoting community development.

In addition, the Maoershan National Forest Park has a natural environment that fits the purpose of the current study in that it has the characteristics of mountains, water, forests, fields, and cities and reflects the customs of Korean people. The park is home to a variety of pine trees, elm trees, poplar trees, and shrubs, wild animals such as pheasants and hares, and mushrooms. The climate is a temperate monsoon climate, with a dry and windy spring, a rainy season from June to August, and a cool autumn and cold winter. The average annual temperature is 2 °C–6 °C, with extreme minimum temperatures of −23 °C–34 °C and maximum temperatures of 34 °C–38 °C. The annual sunshine hours range from 2150–2480 h, and the average annual precipitation is 400 mm–650 mm. In addition, the Maoershan National Forest Park is easily accessible, most areas are free of charge, and the park has excellent forest ecological landscape resources and superior conditions for developing forest tourism [45]. The average daily visitor count at the forest park is approximately 10,000, and during holidays, it can reach a maximum of 60,000. The considerable number of visitors per day makes it possible to ensure ecological diversity for the current study.

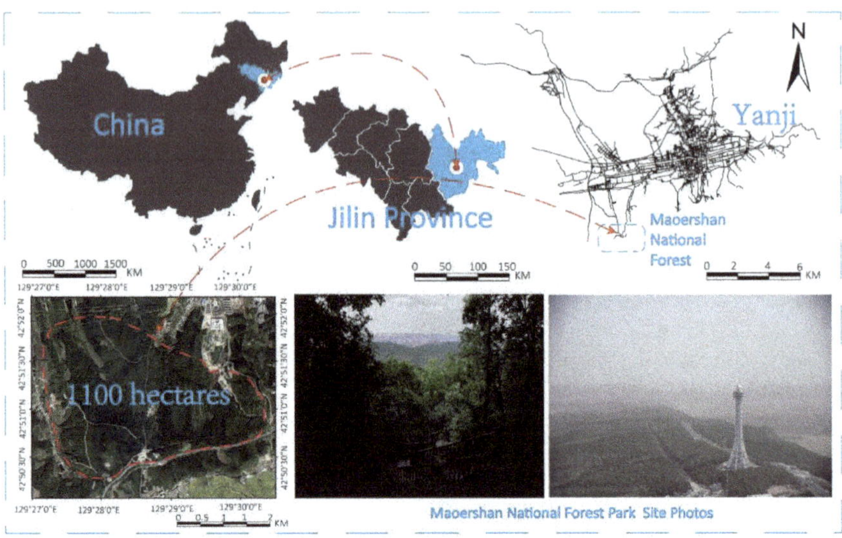

**Figure 2.** Location of the Maoershan National Forest Park in Yanji City.

Moreover, the Maoershan National Forest Park is geographically close to our unit, and our team was able to obtain information about what the elderly population cares about in the area and feedback from local residents regarding their daily lives. It can thus be used as an ideal example to explain the preferred soundscapes for elderly people living in undeveloped cities and to offer suggestions for the landscape planning and design of the forest park.

*2.2. Research Design*

The respondents were chosen at random from residential districts close to Maoershan National Forest Park in Yanji City to guarantee the questionnaire's thoroughness and to guarantee that the respondents were representative of the research. A pilot survey was first conducted in January 2022 in which the research team employed the simple random sampling method to select 68 elderly visitors to Maoershan National Forest Park. After cleaning the collected data, a total of 57 valid questionnaires were obtained, with a valid response rate of 83.8%. Among the 57 respondents, 25 were aged 60 and above, accounting for 30.2% of the total. The questionnaire was then adapted to avoid any misunderstanding among the elderly participants, and before conducting the formal survey, the surveyors received additional training that was aimed to facilitate the research process for all elderly individuals with normal hearing and to minimize the potential for misinterpretation. To identify survey participants, we randomly sampled adults who were able to subjectively assess public landscapes and soundscapes. Before the completion of the survey, a quick hearing test was performed to make sure that all chosen older adults had a normal degree of hearing. All respondents found to have a hearing problem or who were unable to understand the surveyor's instructions were disqualified.

The field investigation spanned from 10 February 2022 to 1 March 2022, and we visited the site for the survey on eighteen days with sunny weather within that stretch of time. Five volunteers were recruited to assist with the survey work. The combination of soundscapes with the routes and activity areas are displayed in Figure 3.

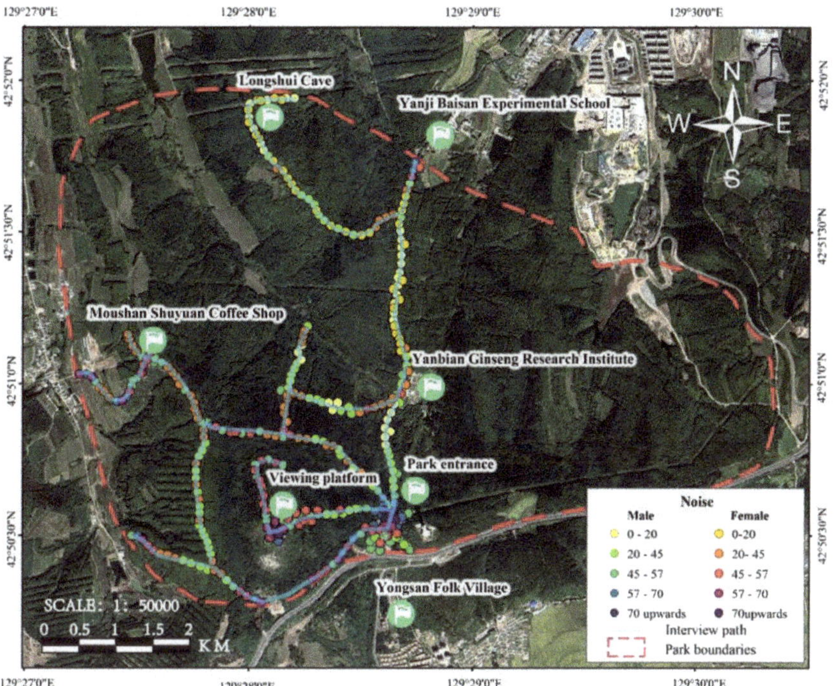


**Figure 3.** Research route map with surrounding sound environments.

To guarantee that the questionnaire would not take too long to complete, a pre-study was completed two days ahead of the survey. A small present was prepared to encourage the locals to participate in the survey. During the survey, a certain number of respondents were selected at random at several traffic intersections close to the park between 7 am and 10 am, between 3 pm and 6 pm, and between 7 pm and 9 pm. The total number of valid subjects was 725, accounting for approximately 10% of the daily visitor traffic. The respondents' key demographic information is displayed in Table 1.

The respondents were asked to complete the questionnaire by using the tablet PCs provided to them by the researchers. The questionnaire was delivered by "Questionnaire Star", a professional and authoritative questionnaire production and distribution platform which has published 154 million questionnaires and could completely fulfill the demands of this survey, including the quantity of questionnaires and question types. Special investigators were arranged to help the respondents with the questionnaire completion process. The researchers also prepared a moderate number of paper-based questionnaires for senior people who were unable to utilize electronic devices properly.

The questionnaire used in this study was designed in accordance with the existing literature and in keen consideration of the characteristics. The questionnaires used in previous related research studies were usually comprised of two general sections: a subjective section that aimed to collect the participants' personal responses toward the items under investigation and an objective section that pertained to environmental information, among other things [46]. It has been found that individual preferences are also influenced by immediate factors [4,47] such as visual features, especially the intricate coupling of auditory and visual elements [48]. In light of these findings, the questionnaire was designed to comprise 46 questions in total that were divided into five sections: time information, basic personal information, landscape environment, soundscape key, and initiatory environment.

**Table 1.** Demographic account of respondents.

| | Classification | Percentage |
|---|---|---|
| Gender | Female | 48.70% |
| | Male | 51.30% |
| Age | 60–64 | 43.30% |
| | 65–69 | 34.90% |
| | 70–74 | 12.00% |
| | 75–79 | 9.70% |
| | 80–84 | 0.10% |
| Education | Junior High School and Below | 48.00% |
| | High School or Vocational school | 42.90% |
| | Junior College or Undergraduate | 9.00% |
| | Master's degree and Above | 0.10% |
| Pension | EUR 0–100 | 26.10% |
| | EUR 101–300 | 73.70% |
| | EUR > 300 | 0.10% |
| Occupation | Experts, technicians and related workers | 4.40% |
| | Government officials and business managers | 4.60% |
| | Sales professionals | 17.00% |
| | Service professionals | 24.80% |
| | Agricultural, animal husbandry and forestry workers, fishermen and hunters | 20.40% |
| | Manufacturers and production-related workers, transportation equipment operators and workers | 18.80% |
| | Workers who cannot be classified by occupation | 10.10% |
| Physical Condition | Completely self-reliant | 89.80% |
| | In need of care | 10.20% |

The first section aimed to gather immediate data from the respondents as it has been found that the respondents' perceptions of the soundscape can be influenced to varying degrees by some external factors, such as the weather, i.e., how the climate felt that day, how they felt when they visited the forest that day, and how they felt on that day [49,50].

The second section focused on the respondents' personal information, including gender, age, education, occupation, pension, physical condition, visual condition, auditory condition, residence, the length of time it took to reach the forest park from where they lived, how often they visited the forest park, how they arrived at the forest park, why they went to the forest park, when they arrived at the forest park, how long they were inclined to stay at the forest park, what kind of signage they valued most in the forest park, whether the forest park signs could be obviously felt, the function of the forest park buildings, what kind of environment they liked best when staying in the forest park, and which buildings and services should be more important in the forest park [51,52].

The third section pertained to the question of landscape issues, specifically, the kind of water body that was preferred, the kind of tree environment that was preferred, whether pure green or color accents were preferred, and what kind of sky was preferred, as well as the visitors' understanding of noise, i.e., whether noise had an impact on their experience, whether they had complaints about noise, how to deal with noise pollution, how to reduce noise, and how to hear the location of the noise [53].

The fourth section aimed to collect information related to forest park soundscape issues. Respondents were expected to indicate the category and rate the volume of the vehicle sound, the bird song, the musical sound, the natural sound, and other sound in the forest park [24]. The last section concerned the visitors' perceptions of the visual aspects of the forest park, including their visual perceptions of the booths and icons established in the park. A 5-point Likert scale (strongly dislike (−2), dislike (−1), average (0), like (1), and like very much (2)) was adopted in the fourth and fifth sections to indicate their overall soundscape preferences. The pre-study administration of the questionnaire with 20 people showed that the average response time was 5 min and 19 s and that the questions were well designed as no respondents reported any doubts or objections.

### 2.3. Reliability and Validity Assessment

#### 2.3.1. Reliability Analysis

First, the Cronbach coefficient method was used to test the internal consistency of each dimension. The Cronbach coefficient takes values in the range of 0–1, and the higher the value of the coefficient, the better the reliability. In general, a coefficient of confidence below 0.6 is not credible, between 0.6 and 0.7 is credible, between 0.7 and 0.8 is relatively credible, between 0.8 and 0.9 is very credible, and between 0.9 and 1 is very credible. Based on the analysis results shown in Table 2, the reliability coefficient of vehicle sound in this analysis was 0.938, which fell within the range of 0.9–1, indicating that the vehicle sound dimension had a very good internal consistency and a very good reliability. The reliability coefficient of musical sound, which had the lowest value, is 0.870, within the range of 0.8–0.9, indicating that the reliability of the musical sound dimension was very credible. Taken together, all variables exhibited good reliability.

**Table 2.** Results of the reliability analysis of each variable.

| Variable | Abbreviations | Cronbach Alpha | Number of Items |
|---|---|---|---|
| Vehicle Sound | VS | 0.938 | 12 |
| Bird Song | BS | 0.928 | 8 |
| Livestock Sound | LS | 0.938 | 9 |
| Atmospheric Sound | AS | 0.885 | 5 |
| Musical Sound | MS | 0.870 | 3 |
| Natural Sound | NS | 0.874 | 4 |
| Other Sound | OS | 0.934 | 9 |
| Vision of Park Stands | VPS | 0.885 | 6 |
| Vision of Other Things | VOT | 0.872 | 6 |
| Vision of Signs | VOS | 0.834 | 2 |

#### 2.3.2. Validity Analysis

An exploratory factor analysis was used to test the structural validity of the study. In this study, there were 10 pre-defined dimensions in the scale section, and each dimension contained a certain number of measurement items. The results of the final factor categorization of the component matrix or the rotated component matrix were observed via exploratory factor analysis. If the categorization results are consistent with the predefined dimensions, the scale has good structural validity.

KMO values range from 0 to 1. The higher the coefficient value, the more suitable the data are for factor analysis. Generally, if the KMO value is less than 0.6, it is not suitable for factor analysis. As shown in Table 3, the KMO value was 0.922, which means that the dataset was suitable for factor analysis. In addition, the Bartlett test result ($p < 0.001$) rejected the original hypothesis, indicating that the data collected in this study were very suitable for factor analysis.

**Table 3.** KMO and Bartlett test.

| **KMO Values** | | 0.922 |
|---|---|---|
| | Approximate cardinality | 30,920.235 |
| Bartlett test | Degree of freedom | 2016 |
| | Significance | <0.001 |

2.3.3. Component Matrix after Transposition

The maximum variance method was used to extract the principal components according to the criteria of eigenvalues greater than 1. Finally, a total of 10 principal components were extracted, and the cumulative variance contribution rate was 67.02% indicating that the principal components extracted in this analysis could effectively replace the original data set (Table 4). The results of the commonality analysis showed that the commonality of each question item reached a standard of 0.4 or more, indicating that the results of categorizing the 10 principal component items were consistent with the preset dimensions of the questionnaire. Additionally, the factor loadings of each question were all greater than 0.5, basically above 0.7.

**Table 4.** Component matrix after transposition.

| | 1 | 2 | 3 | 4 | 5 | 6 | 7 | 8 | 9 | 10 |
|---|---|---|---|---|---|---|---|---|---|---|
| | VS1 | LS1 | OS1 | BS1 | VPS1 | VOT1 | AS1 | NS1 | MS1 | VOS1 |
| | VS2 | LS2 | OS2 | BS2 | VPS2 | VOT2 | AS2 | NS2 | MS2 | VOS2 |
| | VS3 | LS3 | OS3 | BS3 | VPS3 | VOT3 | AS3 | NS3 | MS3 | |
| | VS4 | LS4 | OS4 | BS4 | VPS4 | VOT4 | AS4 | NS4 | | |
| | VS5 | LS5 | OS5 | BS5 | VPS5 | VOT5 | AS5 | | | |
| Measure question items | VS6 | LS6 | OS6 | BS6 | VPS6 | VOT6 | | | | |
| | VS7 | LS7 | OS7 | BS7 | | | | | | |
| | VS8 | LS8 | OS8 | BS8 | | | | | | |
| | VS9 | LS9 | OS9 | | | | | | | |
| | VS10 | | | | | | | | | |
| | VS11 | | | | | | | | | |
| | VS12 | | | | | | | | | |
| Cumulative variance contribution rate | | | | | | 67.02% | | | | |

All abbreviations are shown in Table 2.

## 3. Results

### 3.1. Descriptive Statistics of the Elders' Preferences for Forest Park Soundscape

It can be seen in Table 5, that the elderly participants liked the sound made by nature in the forest park the most (mean = 3.66), followed by livestock sound (mean = 3.54), musical sound (mean = 3.45), and sound from birds of prey (mean = 3.50), and they disliked vehicle sound (mean = 2.35), atmospheric sound (mean = 2.34), and other sound (mean = 2.48) as well. In the category of natural sound, they liked the sound of rustling leaves the most (mean = 3.78) and the sound of falling stones the least (mean = 3.60); in the category of livestock sound, they liked the sound of cows the most (mean = 3.64) and the sound of goose the least (mean = 3.46); in the category of musical sound, they liked the sound of musical instruments the most (mean = 3.50) and the sound of electronic technology the least (mean = 3.50). In the category of musical sound, they liked the sound of musical instruments the most (mean = 3.50) and the sound of electronic technology products the least (mean = 3.42).

**Table 5.** Average description of the elders' overall soundscape preferences.

| Sound Category | Code | Sound Source | Average | Total Average |
|---|---|---|---|---|
| Vehicle Sound | 1 | Car | 2.22 | 2.35 |
| | 2 | Bus | 2.34 | |
| | 3 | Express train | 2.35 | |
| | 4 | Aircraft | 2.36 | |
| | 5 | Fighter | 2.34 | |
| | 6 | Motorcycle | 2.35 | |
| | 7 | Tractor | 2.36 | |
| | 8 | Bicycle | 2.38 | |
| | 9 | Truck | 2.34 | |
| | 10 | Police siren | 2.35 | |
| | 11 | Ambulance siren | 2.37 | |
| | 12 | Fire engine siren | 2.39 | |
| Bird Song | 13 | Pigeon | 3.61 | 3.50 |
| | 14 | Wild goose | 3.55 | |
| | 15 | Swallow | 3.48 | |
| | 16 | Eagle | 3.49 | |
| | 17 | Hawk | 3.49 | |
| | 18 | Swan | 3.50 | |
| | 19 | Egret | 3.45 | |
| | 20 | Sparrow | 3.46 | |
| Livestock Sound | 21 | Cattle | 3.64 | 3.54 |
| | 22 | Horse | 3.50 | |
| | 23 | Sheep/Goat | 3.50 | |
| | 24 | Chicken | 3.54 | |
| | 25 | Dog | 3.52 | |
| | 26 | Pig | 3.54 | |
| | 27 | Duck | 3.58 | |
| | 28 | Cat | 3.54 | |
| | 29 | Goose | 3.46 | |
| Atmospheric Sound | 30 | Rain | 2.24 | 2.34 |
| | 31 | Wind | 2.35 | |
| | 32 | Snow | 2.33 | |
| | 33 | Thunder | 2.38 | |
| | 34 | thunderstorm | 2.38 | |
| Musical Sound | 35 | Instrumental | 3.50 | 3.45 |
| | 36 | Vocal | 3.43 | |
| | 37 | Electronic | 3.42 | |
| Natural Sound | 38 | Leaves | 3.78 | 3.66 |
| | 39 | Falling stone | 3.60 | |
| | 40 | Flying dust | 3.61 | |
| | 41 | Flowing water | 3.64 | |
| Other Sound | 42 | Mechanical | 2.36 | 2.48 |
| | 43 | Construction site | 2.48 | |
| | 44 | Handwork | 2.51 | |
| | 45 | Human activities | 2.48 | |
| | 46 | Mobile ringtones | 2.50 | |
| | 47 | Children playing | 2.44 | |
| | 48 | Street performance | 2.54 | |
| | 49 | Sneezing | 2.51 | |
| | 50 | Nonlocal dialect | 2.52 | |

As shown in Figure 4, the specific sound can have particularly negative effects on elders' psychological perceptions.

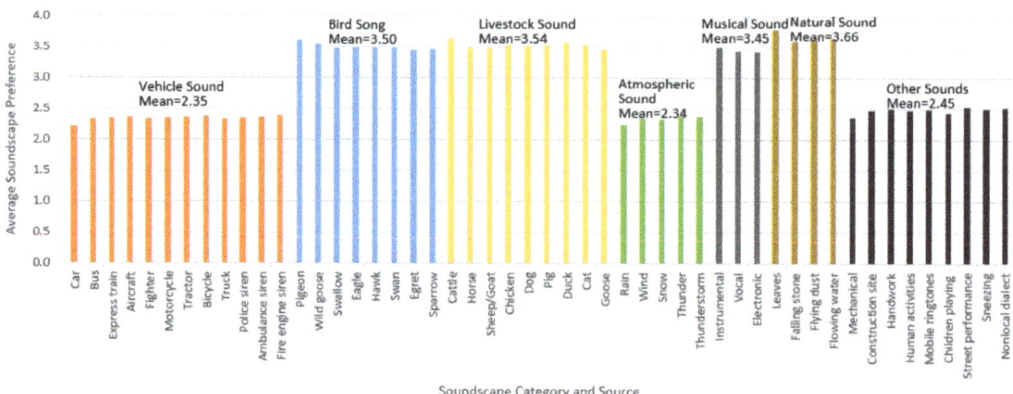

**Figure 4.** Columnar distribution of the mean value of soundscape preference (code is shown in Table 1).

### 3.2. Correlation between Landscape and Soundscape Preference

According to the correlation analysis (Table 6), the type of public landscape space and landscape structural design both influenced the subjective evaluation of the soundscape. In terms of overall landscape perception, those who preferred areas with sunshine demonstrated relatively higher tolerances for the sound of cars, buses, bicycles, fire trucks, ambulance sirens, apparatus, construction, handicrafts, human activities, stranger calls, children playing, street vendors, and local dialects. Those who preferred environments under the shade of trees had higher degrees of preference for the sound of planes, motorcycles, ambulance sirens, fire truck sirens, rainstorms, freezing rain, thunder, sound of instruments, children playing, etc. Those who preferred waterside environments had higher degrees of preference for the sound of passenger cars, ambulance sirens, heavy rain, wind, rain, and lightning. In the selection of landscape structural design, those who used walkways had relatively low preferences for the sound of police car sirens, while respondents who took the bus had relatively low preferences for the sound of leaves. Seniors who used private cars had relatively low preferences for the sound of the creek trickling. Elderly people who rode bicycles to the forest park had relatively low preferences for the sound of goose.

### 3.3. Influence of Living Conditions on Soundscape Preference

The results of the correlation analysis (Table 7) showed that the distance (Q12) to the green space, the frequency (Q13) of visiting the forest park, and the purpose (Q15) were related to the degree of soundscape preference. It can be inferred from the correlation results that (1) in terms of distance, the closer the elder was to the forest park, the higher their tolerance for sounds of birds, livestock, music, and nature; (2) in terms of the frequency of visiting the forest park, the higher the frequency of visiting green space, the more the elders liked the sounds of birds, livestock, music, and nature; (3) the seniors who visited the forest park for exercise preferred livestock sound, musical sound, and natural sound; (4) seniors who visited the Forest Park to walk their dogs and play chess preferred vehicle sound, atmospheric sound, and other sound; and (5) seniors who visited the forest park for square dancing preferred vehicle sound and atmospheric sound.

**Table 6.** Correlation between public landscape space preference and soundscape preference.

| Sound Category | Code | Sound Source | Q21: Which Environment Do You Prefer to Stay in the Forest Park? | | | | | | | | | | | Q14: How Do You Get to the Forest Park Now? | | | |
|---|---|---|---|---|---|---|---|---|---|---|---|---|---|---|---|---|---|
| | | | In Sunlit Areas | In the Shade of Trees | Near a Snack Stall | In Secluded Areas | On the Water-side | In Lounge Areas | In Open Pave-ments | On Roadside Seats | In Higher Places | In Park Build-ings | On the Open Lawn | On Foot | By Bus | By Car | By Bike |
| Vehicle Sound | 1 | Car | 0.100 ** | 0.110 ** | 0.072 | 0.134 ** | 0.064 | −0.061 | −0.003 | −0.073 | 0.071 | −0.004 | −0.008 | −0.061 | 0.080 * | 0.114 ** | 0.015 |
| | 2 | Bus | 0.087 * | 0.029 | 0.038 | 0.102 ** | 0.083 * | −0.011 | 0.018 | −0.038 | 0.102 ** | −0.001 | −0.026 | −0.055 | 0.058 | 0.129 ** | 0.004 |
| | 3 | Express train | 0.057 | 0.072 | 0.060 | 0.059 | 0.012 | −0.021 | −0.033 | −0.044 | 0.040 | −0.026 | 0.002 | −0.055 | 0.024 | 0.066 | 0.028 |
| | 4 | Aircraft | 0.053 | 0.088 * | 0.047 | 0.110 ** | 0.027 | −0.034 | 0.021 | −0.059 | 0.026 | −0.002 | −0.030 | −0.026 | 0.041 | 0.038 | 0.043 |
| | 5 | Fighter | 0.038 | 0.020 | 0.008 | 0.079 * | 0.046 | −0.076 * | −0.004 | −0.061 | 0.025 | 0.050 | 0.081 * | −0.009 | 0.030 | 0.060 | 0.011 |
| | 6 | Motorcycle | 0.064 | 0.092 * | 0.077 * | 0.049 | 0.033 | −0.041 | 0.021 | −0.052 | 0.040 | −0.014 | −0.026 | −0.041 | 0.052 | 0.085 * | −0.001 |
| | 7 | Tractor | 0.046 | 0.068 | 0.045 | 0.128 ** | 0.054 | −0.078 * | 0.012 | −0.041 | 0.040 | 0.024 | 0.023 | −0.005 | 0.030 | 0.086 * | 0.035 |
| | 8 | Bicycle | 0.086 * | 0.066 | 0.000 | 0.084 * | 0.016 | −0.023 | 0.053 | −0.019 | 0.023 | −0.016 | −0.011 | −0.033 | 0.042 | 0.064 | 0.055 |
| | 9 | Truck | 0.057 | −0.023 | 0.006 | 0.020 | 0.031 | 0.025 | 0.031 | −0.001 | 0.010 | −0.013 | −0.045 | 0.015 | 0.047 | 0.018 | 0.042 |
| | 10 | Police siren | 0.069 | 0.087 * | 0.067 | 0.125 ** | 0.044 | −0.039 | −0.032 | −0.074 * | 0.072 | −0.027 | 0.031 | −0.076 * | 0.075 * | 0.077 * | 0.045 |
| | 11 | Ambulance siren | 0.113 ** | 0.122 ** | 0.071 | 0.124 ** | 0.076 * | −0.098 ** | −0.046 | −0.126 ** | 0.073 | −0.003 | 0.007 | −0.075 * | 0.082 * | 0.115 ** | 0.036 |
| | 12 | Fire engine siren | 0.091 * | 0.132 ** | 0.097 ** | 0.114 ** | 0.021 | −0.027 | 0.010 | −0.100 ** | 0.022 | 0.034 | −0.036 | −0.071 | 0.051 | 0.110 ** | 0.049 |
| Bird Song | 13 | Pigeon | −0.118 ** | −0.191 ** | −0.191 ** | −0.207 ** | −0.178 ** | 0.127 ** | 0.007 | 0.133 ** | −0.037 | −0.057 | 0.081 * | 0.094 * | −0.104 ** | −0.099 ** | −0.099 ** |
| | 14 | Wild goose | −0.124 ** | −0.186 ** | −0.176 ** | −0.158 ** | −0.158 ** | 0.127 ** | −0.028 | 0.091 * | −0.017 | −0.050 | 0.064 | 0.115 ** | −0.090 * | −0.114 ** | −0.080 * |
| | 15 | Swallow | −0.141 ** | −0.140 ** | −0.157 ** | −0.152 ** | −0.146 ** | 0.145 ** | 0.024 | 0.113 ** | −0.027 | −0.068 | 0.077 * | 0.090 * | −0.066 | −0.088 * | −0.096 ** |
| | 16 | Eagle | −0.121 ** | −0.127 ** | −0.133 ** | −0.112 ** | −0.098 ** | 0.089 * | 0.029 | 0.081 * | −0.042 | −0.046 | 0.039 | 0.102 ** | −0.094 * | −0.079 * | −0.064 |
| | 17 | Hawk | −0.115 ** | −0.151 ** | −0.174 ** | −0.155 ** | −0.194 ** | 0.102 ** | 0.013 | 0.109 ** | −0.013 | −0.022 | 0.104 ** | 0.104 ** | −0.097 ** | −0.069 | −0.116 ** |
| | 18 | Swan | −0.167 ** | −0.166 ** | −0.112 ** | −0.141 ** | −0.116 ** | 0.092 * | 0.022 | 0.131 ** | 0.000 | −0.083 * | 0.073 * | 0.085 * | −0.058 | −0.068 | −0.093 * |
| | 19 | Egret | −0.092 * | −0.143 ** | −0.152 ** | −0.128 ** | −0.115 ** | 0.110 ** | 0.022 | 0.081 * | −0.011 | −0.043 | 0.017 | 0.047 | −0.103 ** | −0.044 | −0.060 |
| | 20 | Sparrow | −0.104 ** | −0.200 ** | −0.169 ** | −0.146 ** | −0.145 ** | 0.083 * | −0.039 | 0.115 ** | 0.002 | −0.068 | 0.088 * | 0.104 ** | −0.109 ** | −0.087 * | −0.082 * |
| Livestock Sound | 21 | Cattle | −0.075 * | −0.071 | −0.148 ** | −0.163 ** | −0.107 ** | 0.113 ** | 0.008 | 0.124 ** | −0.024 | −0.011 | 0.013 | 0.057 | −0.039 | −0.148 ** | −0.100 ** |
| | 22 | Horse | −0.058 | −0.057 | −0.086 * | −0.128 ** | −0.141 ** | 0.103 ** | 0.021 | 0.096 * | 0.000 | −0.024 | 0.016 | 0.070 | −0.017 | −0.118 ** | −0.079 * |
| | 23 | Sheep/Goat | −0.059 | −0.071 | −0.104 ** | −0.133 ** | −0.111 ** | 0.074 * | −0.006 | 0.145 ** | −0.087 * | −0.023 | 0.053 | 0.060 | −0.060 | −0.091 * | −0.072 |
| | 24 | Chicken | −0.062 | −0.048 | −0.102 ** | −0.194 ** | −0.124 ** | 0.101 ** | 0.052 | 0.156 ** | −0.017 | −0.039 | 0.001 | 0.062 | −0.037 | −0.102 ** | −0.108 ** |
| | 25 | Dog | −0.030 | −0.103 ** | −0.111 ** | −0.147 ** | −0.096 ** | 0.110 ** | 0.010 | 0.100 ** | −0.030 | −0.037 | 0.009 | 0.040 | −0.057 | −0.064 | −0.116 ** |
| | 26 | Pig | −0.069 | −0.044 | −0.065 | −0.136 ** | −0.108 ** | 0.080 * | 0.011 | 0.110 ** | 0.012 | −0.003 | −0.004 | 0.027 | 0.002 | −0.094 * | −0.049 |
| | 27 | Duck | −0.063 | −0.064 | −0.144 ** | −0.137 ** | −0.126 ** | 0.070 | 0.032 | 0.057 | −0.019 | −0.041 | 0.028 | 0.067 | −0.056 | −0.114 ** | −0.098 ** |
| | 28 | Cat | −0.008 | −0.064 | −0.115 ** | −0.175 ** | −0.121 ** | 0.056 | 0.003 | 0.091 * | −0.017 | 0.009 | 0.039 | 0.024 | −0.037 | −0.104 ** | −0.084 * |
| | 29 | Goose | −0.050 | −0.103 ** | −0.087 * | −0.151 ** | −0.079 * | 0.064 | 0.006 | 0.087 * | −0.041 | −0.057 | −0.002 | 0.057 | 0.007 | −0.104 ** | −0.137 ** |
| Atmospheric Sound | 30 | Rain | 0.089 * | 0.084 * | 0.100 ** | 0.108 ** | 0.090 * | −0.044 | −0.019 | −0.073 * | 0.035 | 0.093 * | −0.027 | −0.022 | 0.035 | 0.054 | 0.085 * |
| | 31 | Wind | 0.071 | 0.071 | 0.073 * | 0.070 | 0.068 | −0.037 | 0.036 | −0.063 | 0.010 | 0.038 | 0.001 | 0.001 | 0.026 | 0.034 | 0.060 |
| | 32 | Snow | 0.058 | 0.083 * | 0.031 | 0.068 | 0.043 | −0.035 | 0.016 | −0.051 | −0.005 | 0.064 | −0.019 | 0.002 | 0.005 | 0.082 * | 0.041 |
| | 33 | Thunder | 0.016 | 0.090 * | 0.078 * | 0.048 | 0.067 | −0.026 | −0.012 | −0.038 | −0.008 | 0.035 | −0.032 | 0.019 | 0.030 | 0.017 | 0.006 |
| | 34 | Thunderstorm | 0.029 | 0.027 | 0.040 | 0.086 * | 0.083 * | −0.016 | −0.036 | −0.024 | −0.008 | 0.060 | −0.003 | −0.019 | 0.026 | 0.080 * | 0.026 |
| Musical Sound | 35 | Instrumental | −0.086 * | −0.112 ** | −0.093 * | −0.135 ** | −0.145 ** | 0.120 ** | 0.029 | 0.041 | −0.014 | −0.058 | 0.075 * | 0.032 | −0.064 | −0.046 | −0.106 ** |
| | 36 | Vocal | −0.031 | −0.085 * | −0.109 ** | −0.102 ** | −0.116 ** | 0.094 * | 0.025 | 0.039 | −0.010 | 0.005 | 0.052 | −0.006 | −0.056 | −0.069 | −0.037 |
| | 37 | Electronic | −0.038 | −0.112 ** | −0.110 ** | −0.109 ** | −0.088 * | 0.081 * | 0.062 | 0.044 | 0.019 | −0.054 | 0.092 * | 0.051 | −0.053 | −0.049 | −0.081 * |
| Natural Sound | 38 | Leaves | −0.094 * | −0.111 ** | −0.201 ** | −0.244 ** | −0.158 ** | 0.047 | 0.046 | 0.054 | −0.035 | −0.009 | 0.095 * | 0.047 | −0.093 * | −0.161 ** | −0.102 ** |
| | 39 | Falling stone | −0.054 | −0.120 ** | −0.137 ** | −0.191 ** | −0.120 ** | 0.062 | −0.005 | 0.032 | −0.006 | −0.043 | 0.071 | 0.006 | −0.036 | −0.126 ** | −0.072 |
| | 40 | Flying dust | −0.082 * | −0.133 ** | −0.156 ** | −0.186 ** | −0.146 ** | 0.074 * | 0.051 | 0.056 | −0.007 | −0.079 * | 0.079 * | 0.062 | −0.049 | −0.111 ** | −0.101 ** |
| | 41 | Flowing water | −0.062 | −0.144 ** | −0.143 ** | −0.198 ** | −0.106 ** | 0.066 | 0.031 | 0.034 | −0.008 | −0.034 | 0.059 | 0.012 | −0.039 | −0.127 ** | −0.103 ** |

**Table 6.** *Cont.*

| Sound Category | Code | Sound Source | Q21: Which Environment Do You Prefer to Stay in the Forest Park? | | | | | | | | | | | Q14: How Do You Get to the Forest Park Now? | | | |
|---|---|---|---|---|---|---|---|---|---|---|---|---|---|---|---|---|---|
| | | | In Sunlit Areas | In the Shade of Trees | Near a Snack Stall | In Secluded Areas | On the Water-side | In Lounge Areas | In Open Pave-ments | On Roadside Seats | In Higher Places | In Park Build-ings | On the Open Lawn | On Foot | By Bus | By Car | By Bike |
| Other Sound | 42 | Mechanical | 0.086 * | 0.113 ** | 0.130 ** | 0.019 | −0.022 | −0.027 | −0.004 | −0.044 | 0.054 | −0.026 | −0.051 | −0.023 | 0.046 | 0.043 | 0.024 |
| | 43 | Construction site | 0.095 * | 0.087 * | 0.102 ** | 0.045 | −0.012 | −0.015 | −0.014 | −0.049 | 0.044 | −0.010 | −0.024 | −0.024 | 0.009 | 0.055 | 0.003 |
| | 44 | Handwork | 0.095 * | 0.080 * | 0.094 * | 0.013 | 0.028 | −0.021 | −0.021 | −0.024 | 0.057 | −0.024 | −0.037 | −0.023 | 0.053 | 0.019 | 0.026 |
| | 45 | Human activities | 0.084 * | 0.077 * | 0.097 ** | 0.030 | 0.036 | −0.075 * | −0.005 | −0.057 | 0.017 | −0.023 | −0.075 * | −0.027 | 0.117 ** | 0.050 | 0.010 |
| | 46 | Mobile ringtones | 0.122 ** | 0.055 | 0.133 ** | 0.040 | −0.005 | −0.028 | −0.006 | −0.042 | 0.046 | −0.031 | −0.036 | −0.030 | 0.051 | 0.027 | 0.030 |
| | 47 | Children playing | 0.076 * | 0.080 * | 0.105 ** | 0.066 | 0.025 | −0.067 | −0.019 | −0.053 | 0.013 | −0.027 | −0.041 | −0.051 | 0.064 | 0.056 | 0.053 |
| | 48 | Street performance | 0.109 ** | 0.088 * | 0.090 * | 0.047 | 0.043 | −0.045 | −0.002 | −0.029 | −0.001 | 0.046 | −0.047 | 0.014 | 0.010 | 0.042 | 0.005 |
| | 49 | Sneezing | 0.051 | 0.089 * | 0.089 * | 0.016 | −0.013 | −0.017 | −0.046 | −0.035 | 0.028 | 0.011 | −0.071 | −0.017 | 0.028 | 0.067 | 0.021 |
| | 50 | Nonlocal dialect | 0.100 ** | 0.110 ** | 0.072 | 0.134 ** | 0.064 | −0.061 | −0.003 | −0.073 | 0.071 | −0.004 | −0.008 | −0.061 | 0.080 * | 0.114 ** | 0.015 |

Note: *: $p \leq 0.05$, **: $p \leq 0.01$.

**Table 7.** Correlation between personal conditions and soundscape preference.

| Sound Category | Code | Sound Source | Q12 | Q13 | Question 15 | | | | |
|---|---|---|---|---|---|---|---|---|---|
| | | | | | Exercise | Dog walking | Playing Chess | Square Dancing | Socializing |
| Vehicle Sound | 1 | Car | 0.178 ** | 0.192 ** | −0.041 | 0.074 * | 0.092 * | 0.068 | −0.044 |
| | 2 | Bus | 0.154 ** | 0.158 ** | −0.037 | 0.047 | 0.064 | 0.052 | −0.013 |
| | 3 | Express train | 0.152 ** | 0.121 ** | 0.011 | 0.010 | 0.011 | 0.023 | −0.062 |
| | 4 | Aircraft | 0.132 ** | 0.153 ** | −0.048 | 0.025 | 0.057 | 0.043 | −0.037 |
| | 5 | Fighter | 0.144 ** | 0.150 ** | −0.009 | 0.001 | −0.013 | 0.050 | 0.016 |
| | 6 | Motorcycle | 0.028 | 0.102 ** | −0.019 | 0.006 | 0.013 | 0.016 | −0.036 |
| | 7 | Tractor | 0.173 ** | 0.147 ** | −0.057 | 0.075 * | 0.110 ** | 0.075 * | −0.053 |
| | 8 | Bicycle | 0.105 ** | 0.144 ** | −0.007 | 0.007 | 0.022 | 0.072 | −0.072 |
| | 9 | Truck | 0.091 * | 0.070 | −0.067 | 0.014 | 0.028 | 0.092 * | −0.050 |
| | 10 | Police siren | 0.122 ** | 0.183 ** | −0.029 | 0.038 | 0.062 | 0.030 | −0.032 |
| | 11 | Ambulance siren | 0.144 ** | 0.180 ** | −0.028 | 0.134 ** | 0.099 ** | 0.022 | −0.057 |
| | 12 | Fire engine siren | 0.117 ** | 0.125 ** | −0.008 | 0.058 | 0.099 ** | 0.023 | −0.048 |
| Bird Song | 13 | Pigeon | −0.211 ** | −0.303 ** | 0.049 | −0.155 ** | −0.182 ** | −0.102 ** | 0.035 |
| | 14 | Wild goose | −0.183 ** | −0.233 ** | 0.038 | −0.139 ** | −0.185 ** | −0.073 * | 0.015 |
| | 15 | Swallow | −0.101 ** | −0.255 ** | 0.009 | −0.159 ** | −0.157 ** | −0.054 | 0.042 |
| | 16 | Eagle | −0.162 ** | −0.235 ** | −0.011 | −0.123 ** | −0.148 ** | −0.060 | 0.013 |
| | 17 | Hawk | −0.170 ** | −0.224 ** | 0.024 | −0.175 ** | −0.198 ** | −0.034 | −0.015 |
| | 18 | Swan | −0.155 ** | −0.262 ** | 0.028 | −0.100 ** | −0.164 ** | −0.024 | 0.049 |
| | 19 | Egret | −0.183 ** | −0.251 ** | 0.012 | −0.103 ** | −0.148 ** | −0.043 | −0.004 |
| | 20 | Sparrow | −0.147 ** | −0.189 ** | 0.041 | −0.125 ** | −0.161 ** | −0.064 | 0.019 |
| Livestock Sound | 21 | Cattle | −0.254 ** | −0.219 ** | 0.111 ** | −0.224 ** | −0.127 ** | −0.056 | 0.028 |
| | 22 | Horse | −0.202 ** | −0.199 ** | 0.047 | −0.141 ** | −0.076 * | −0.047 | 0.031 |
| | 23 | Sheep/Goat | −0.202 ** | −0.199 ** | 0.087 * | −0.164 ** | −0.080 * | −0.073 * | 0.032 |
| | 24 | Chicken | −0.222 ** | −0.170 ** | 0.073 * | −0.143 ** | −0.088 * | −0.063 | 0.025 |
| | 25 | Dog | −0.228 ** | −0.151 ** | 0.110 ** | −0.185 ** | −0.100 ** | −0.068 | 0.014 |
| | 26 | Pig | −0.216 ** | −0.184 ** | 0.151 ** | −0.144 ** | −0.106 ** | −0.086 * | 0.001 |
| | 27 | Duck | −0.213 ** | −0.127 ** | 0.070 | −0.184 ** | −0.116 ** | −0.035 | 0.034 |
| | 28 | Cat | −0.251 ** | −0.189 ** | 0.065 | −0.162 ** | −0.084 * | −0.038 | −0.042 |
| | 29 | Goose | −0.202 ** | −0.136 ** | 0.071 | −0.156 ** | −0.094 * | −0.043 | 0.023 |
| Atmospheric Sound | 30 | Rain | 0.182 ** | 0.209 ** | −0.100 ** | 0.220 ** | 0.126 ** | 0.032 | 0.043 |
| | 31 | Wind | 0.110 ** | 0.133 ** | −0.126 ** | 0.172 ** | 0.095 * | 0.019 | 0.046 |
| | 32 | Snow | 0.151 ** | 0.104 ** | −0.081 * | 0.074 * | 0.052 | 0.076 * | 0.055 |
| | 33 | Thunder | 0.070 | 0.111 ** | −0.059 | 0.109 ** | 0.053 | 0.038 | 0.056 |
| | 34 | Thunderstorm | 0.131 ** | 0.195 ** | −0.049 | 0.109 ** | 0.066 | 0.015 | 0.020 |
| Musical Sound | 35 | Instrumental | −0.195 ** | −0.248 ** | 0.066 | −0.142 ** | −0.116 ** | −0.076 * | −0.022 |
| | 36 | Vocal | −0.133 ** | −0.219 ** | 0.073 * | −0.121 ** | −0.095 * | −0.046 | −0.026 |
| | 37 | Electronic | −0.124 ** | −0.193 ** | 0.085 * | −0.116 ** | −0.093 * | −0.057 | −0.041 |
| Natural Sound | 38 | Leaves | −0.287 ** | −0.306 ** | 0.083 * | −0.150 ** | −0.184 ** | −0.076 * | −0.039 |
| | 39 | Falling stone | −0.254 ** | −0.290 ** | 0.093 * | −0.141 ** | −0.179 ** | −0.110 ** | −0.009 |
| | 40 | Flying dust | −0.241 ** | −0.248 ** | 0.060 | −0.092 * | −0.158 ** | −0.047 | −0.051 |
| | 41 | Flowing water | −0.245 ** | −0.248 ** | 0.085 * | −0.162 ** | −0.198 ** | −0.070 | −0.016 |
| Other Sound | 42 | Mechanical | 0.139 ** | 0.173 ** | −0.034 | 0.077 * | 0.098 ** | 0.003 | 0.030 |
| | 43 | Machine noise | 0.127 ** | 0.167 ** | 0.003 | 0.099 ** | 0.076 * | 0.010 | 0.014 |
| | 44 | Construction noise | 0.051 | 0.137 ** | −0.008 | 0.123 ** | 0.118 ** | −0.044 | −0.019 |
| | 45 | Exercise sound | 0.077 * | 0.156 ** | 0.000 | 0.070 | 0.079 * | 0.014 | 0.021 |
| | 46 | Mobile ringtones | 0.105 ** | 0.123 ** | −0.075 * | 0.069 | 0.084 * | −0.007 | 0.014 |
| | 47 | Children playing | 0.144 ** | 0.148 ** | −0.075 * | 0.079 * | 0.115 ** | 0.024 | 0.049 |
| | 48 | Footstep | 0.105 ** | 0.192 ** | −0.063 | 0.115 ** | 0.119 ** | 0.028 | 0.019 |
| | 49 | Vehicle noise | 0.143 ** | 0.117 ** | −0.048 | 0.035 | 0.077 * | 0.047 | 0.021 |
| | 50 | Bus noise | 0.178 ** | 0.192 ** | 0.001 | −0.246 ** | −0.260 ** | −0.303 ** | −0.253 ** |

Note: *: $p \leq 0.05$, **: $p \leq 0.01$. Q12: How long does it take you to get to the Forest Park from where you live? Q13: How often do you go to the Forest Park? Q15: What's your purpose of going to the Forest Park?

### 3.4. Participants

#### 3.4.1. Gender

The results (as shown in Figure 5) showed that male and female older adults had the same average soundscape preferences. Among the four sound categories of vehicle sound, bird song, atmospheric sound, and other sound, men had a higher average soundscape preference than women.

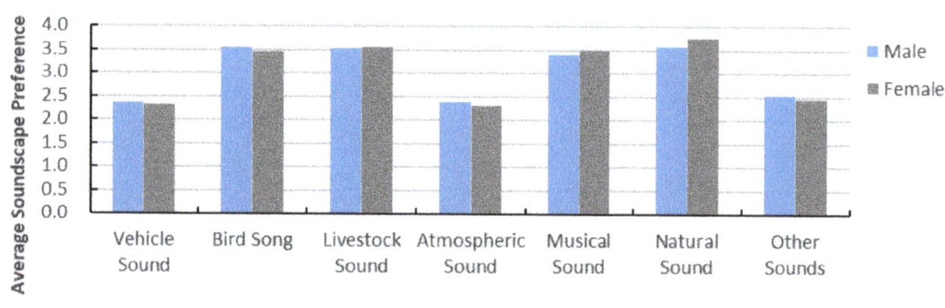


**Figure 5.** Average soundscape preferences of different genders.

### 3.4.2. Age

The results (as shown in Figure 6) showed that older adults of different ages had different preferences for different sound categories. Overall, the mean values of soundscape preference in all seven sound categories were higher for the individuals aged 80–84 years than for the other age groups. In contrast, the mean values of soundscape preference in the above seven sound categories were smaller for individuals aged 60–64 years than for other age groups.

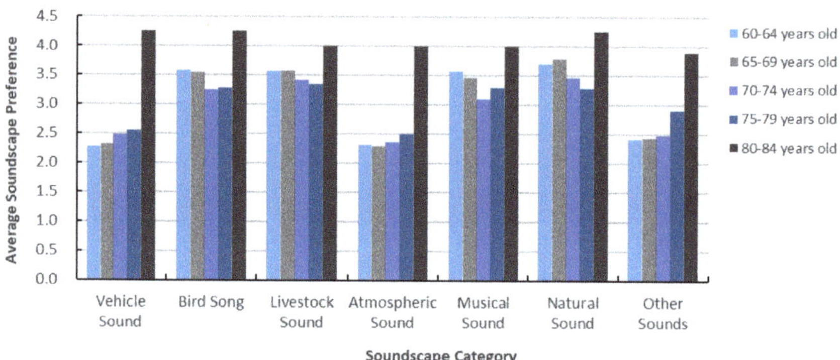


**Figure 6.** Average soundscape preference with different age groups.

### 3.4.3. Occupation

The results (as shown in Figure 7) show that older adults from different occupations had different preferences for different sound categories.

In the four sound categories of bird sound, livestock sound, musical sound, and natural sound, older adults from service worker occupations had higher soundscape preferences. In the three sound categories of vehicle sound, atmospheric sound, and other sound, older adults with the occupation of government officials and business managers had higher mean values of soundscape preference.

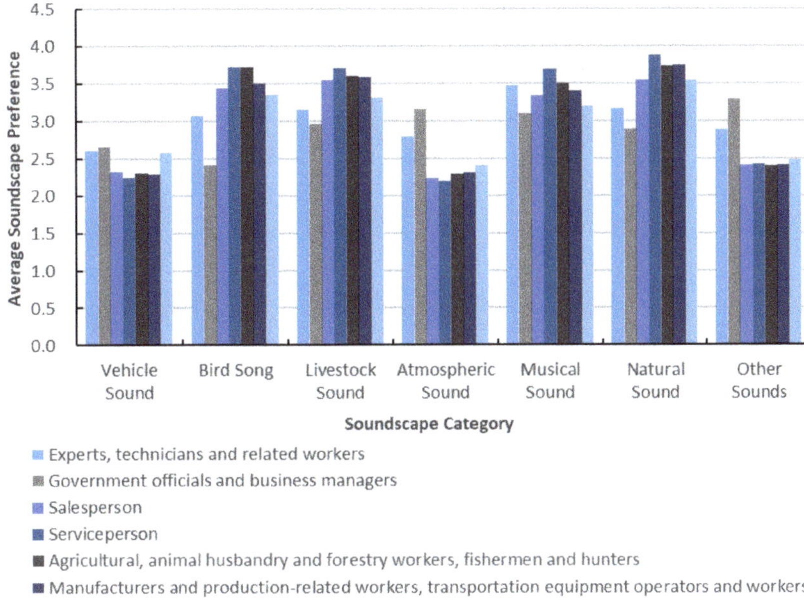



**Figure 7.** Average soundscape preferences with varying occupations.

### 3.4.4. Education Background

The results (as shown in Figure 8) showed that older adults with different educational backgrounds had different preferences for different sound categories. Figure 8 showed that across the seven sound categories, older adults with educations of a master's degree or higher had higher mean values of soundscape preference.

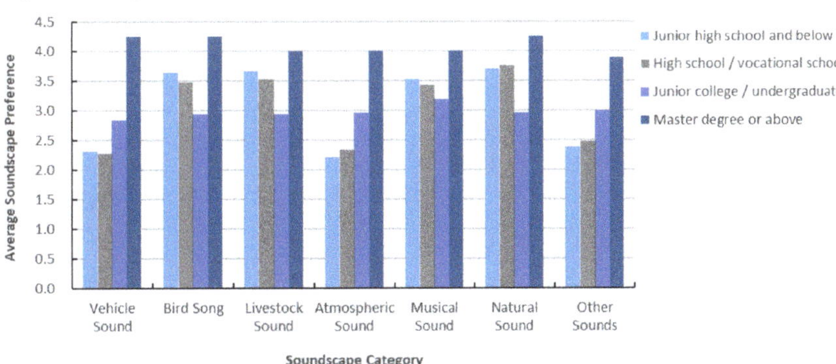



**Figure 8.** Average soundscape preferences with varying education backgrounds.

### 3.4.5. Living Conditions

The results (Figure 9) showed that older adults with different living conditions had different preferences for different sound categories.

Figure 9 also showed that in the four sound categories of bird song, livestock sound, musical sound, and natural sound, older adults living with a partner had higher mean values of soundscape preference. In the three sound categories of vehicle sound, atmospheric sound and other sounds, elderly people living in homes had higher mean values of soundscape preference.

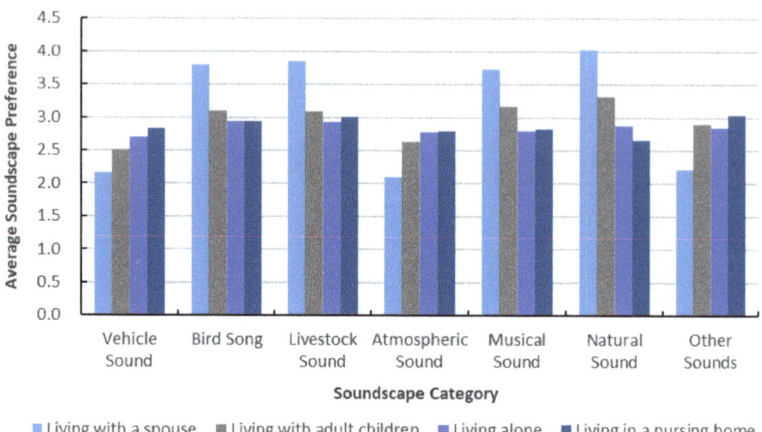


**Figure 9.** Average soundscape preferences with varying living conditions.

## 4. Regression Coefficient Model of Soundscape Preference

To further analyze the soundscape preferences, a regression coefficient model was applied with SPSS23.0, and the soundscape preference evaluation was divided into the target variables in Figure 10.

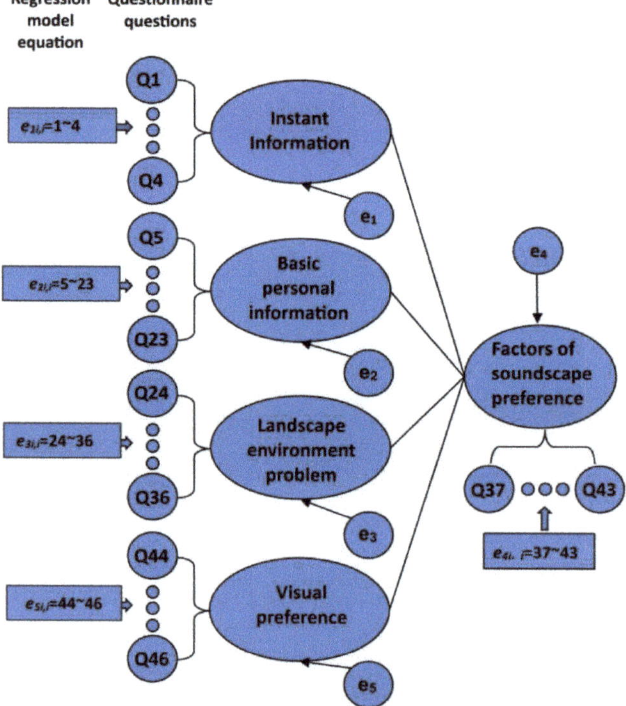


**Figure 10.** Structural equation model regression coefficients.

In this paper, the instant information, such as the weather condition, mood (Q1–Q4), basic personal information (Q5–Q23), landscape problem (Q24–Q36), and visual preference (Q44–Q46) on the day of visiting the forest park were used as independent variables, and soundscape preference was used as the dependent variable to construct a regression model. Considering the possible problem of covariance among the explanatory variables, the optimal model was constructed via stepwise regression, and the results of the stepwise regression are shown in Table 8.

**Table 8.** Results of the stepwise regression.

| Model | R | R Square | Adjusted R Square | Std. Error of the Estimate | Durbin–Watson |
|---|---|---|---|---|---|
| 1 | 0.272 a | 0.074 | 0.073 | 0.373 | |
| 2 | 0.322 b | 0.104 | 0.101 | 0.367 | |
| 3 | 0.345 c | 0.119 | 0.115 | 0.365 | |
| 4 | 0.363 d | 0.132 | 0.127 | 0.362 | |
| 5 | 0.377 e | 0.142 | 0.136 | 0.360 | |
| 6 | 0.389 f | 0.152 | 0.145 | 0.358 | |
| 7 | 0.399 g | 0.160 | 0.151 | 0.357 | |
| 8 | 0.406 h | 0.165 | 0.155 | 0.356 | |
| 9 | 0.412 i | 0.170 | 0.159 | 0.356 | 1.865 |

(a) Predictors: (constant); (b) predictors: (constant), Q28; (c) predictors: (constant), Q28, and Q16C; (d) predictors: (constant), Q28, Q16C, and Q13; (e) predictors: (constant), Q28, Q16C, Q13, and Q461; (f) predictors: (constant), Q28, Q16C, Q13, Q461, and Q21E; (g) predictors: (constant), Q28, Q16C, Q13, Q461, Q21E, and Q18A; (h) predictors: (constant), Q28, Q16C, Q13, Q461, Q21E, Q18A, and Q29; (i) predictors: (constant), Q28, Q16C, Q13, Q461, Q21E, Q18A, Q29, and Q20C.

Subsequently, an analysis of variance (ANOVA) was performed on the data, and the results obtained are shown in Table 9 below.

**Table 9.** Results of ANOVA.

| Model | | Sum of Squares | df | Mean Square | F | Sig. |
|---|---|---|---|---|---|---|
| 1 | Regression | 8.039 | 1 | 8.039 | 57.701 | 0.000 a |
| | Residual | 100.723 | 723 | 0.139 | | |
| | Total | 108.761 | 724 | | | |
| 2 | Regression | 11.257 | 2 | 5.628 | 41.678 | 0.000 b |
| | Residual | 97.504 | 722 | 0.135 | | |
| | Total | 108.761 | 724 | | | |
| 3 | Regression | 12.942 | 3 | 4.314 | 32.460 | 0.000 c |
| | Residual | 95.820 | 721 | 0.133 | | |
| | Total | 108.761 | 724 | | | |
| 4 | Regression | 14.314 | 4 | 3.579 | 27.280 | 0.000 d |
| | Residual | 94.447 | 720 | 0.131 | | |
| | Total | 108.761 | 724 | | | |
| 5 | Regression | 15.434 | 5 | 3.087 | 23.782 | 0.000 e |
| | Residual | 93.327 | 719 | 0.130 | | |
| | Total | 108.761 | 724 | | | |
| 6 | Regression | 16.499 | 6 | 2.750 | 21.400 | 0.000 f |
| | Residual | 92.262 | 718 | 0.128 | | |
| | Total | 108.761 | 724 | | | |
| 7 | Regression | 17.355 | 7 | 2.479 | 19.447 | 0.000 g |
| | Residual | 91.407 | 717 | 0.127 | | |
| | Total | 108.761 | 724 | | | |
| 8 | Regression | 17.914 | 8 | 2.239 | 17.649 | 0.000 h |
| | Residual | 90.847 | 716 | 0.127 | | |
| | Total | 108.761 | 724 | | | |
| 9 | Regression | 18.459 | 9 | 2.051 | 16.240 | 0.000 i |
| | Residual | 90.302 | 715 | 0.126 | | |
| | Total | 108.761 | 724 | | | |

(a) Predictors: (constant); (b) predictors: (constant), Q28; (c) predictors: (constant), Q28, and Q16C; (d) predictors: (constant), Q28, Q16C, and Q13; (e) predictors: (constant), Q28, Q16C, Q13, and Q461; (f) predictors: (constant), Q28, Q16C, Q13, Q461, and Q21E; (g) predictors: (constant), Q28, Q16C, Q13, Q461, Q21E, and Q18A; (h) predictors: (constant), Q28, Q16C, Q13, Q461, Q21E, Q18A, and Q29; (i) predictors: (constant), Q28, Q16C, Q13, Q461, Q21E, Q18A, Q29, and Q20C.

As it can be seen from Table 10, the stepwise regression showed that there were nine regression models to be constructed, among which the R-squared (0.170) and adjusted R-squared (0.159) values of model 9 had the largest values among the nine models and the standard error of estimation (0.35538) had the smallest value; therefore, model 9 was selected as the optimal regression model in this paper, and a subsequent analysis was conducted.

**Table 10.** Results of regression coefficient model.

| Model 9 | B | Std. Error | Beta | t | Sig. | Tolerance | VIF |
|---|---|---|---|---|---|---|---|
| (Constant) | 3.203 | 0.072 | | 44.752 | 0.000 | | |
| Q11 | −0.045 | 0.017 | −0.109 | −2.676 | 0.008 | 0.700 | 1.429 |
| Q28 | −0.039 | 0.019 | −0.083 | −2.051 | 0.041 | 0.709 | 1.410 |
| Q16C | −0.090 | 0.039 | −0.087 | −2.303 | 0.022 | 0.809 | 1.237 |
| Q13 | −0.044 | 0.014 | −0.117 | −3.082 | 0.002 | 0.810 | 1.235 |
| Q46_1 | 0.039 | 0.012 | 0.118 | 3.237 | 0.001 | 0.880 | 1.136 |
| Q21E | −0.174 | 0.068 | −0.095 | −2.578 | 0.010 | 0.851 | 1.174 |
| Q18A | 0.071 | 0.027 | 0.091 | 2.654 | 0.008 | 0.993 | 1.007 |
| Q29 | −0.050 | 0.024 | −0.074 | −2.116 | 0.035 | 0.948 | 1.055 |
| Q20C | −0.069 | 0.033 | −0.075 | −2.077 | 0.038 | 0.900 | 1.112 |

Dependent variable: soundscape preference.

Table 10 showed that the F-test statistic for model 9 was 16.240, with a significance value, $p$, of less than 0.001, indicating that there was a significant linear relationship between the independent variables.

As can be seen from Table 9, Model 9 contained a total of nine independent variables, namely Q11 (residence status), Q28 (what kind of sky you prefer to see), Q16C (coming to the forest park at noon), Q13 (frequency of visiting the forest park), Q46_1 (the forest street interface gives you a familiar feeling), Q21E (preferring to stay at the waterfront environment in the forest park), Q18A (placing more importance on road signs in forest parks), Q29 (hearing), and Q20C (placing more importance on the science education exhibition function of buildings in forest parks). The non-standardized regression coefficients of these nine independent variables were −0.045, −0.039, −0.090, −0.044, 0.039, −0.174, 0.071, −0.050, and −0.069, respectively. The regression coefficients of these nine independent variables were significant at the 5% level of significance, indicating a significant effect on soundscape preference for all of them. Among them, two independent variables, Q46_1 (forest street interface gives you a familiar feeling) and Q18A (more emphasis on road signs in forest parks), had positive effects on soundscape preference, while the rest of the factors had negative effects. Table 9 also shows that the tolerance values for each independent variable in model 9 were greater than 0.10 and the VIFs were less than 10, indicating that there was no multiple covariance problem.

The final regression model is summarized as:

$$y = 3.203 - 0.045*Q11 - 0.039*Q28 - 0.09*Q16C - 0.044*Q13 + 0.039*Q461 - 0.174*Q21E + 0.071*Q18A - 0.050*Q29 - 0.069*Q20C.$$

## 5. Discussion

In this study, an online questionnaire survey was conducted using a subjective evaluation method in the vicinity of Maoershan National Forest Park in Yanji, China. The data were analyzed and organized to understand the preferences of elderly people for forest soundscapes and the association between the soundscape preferences of elderly people and landscape characteristics. The investigation of soundscapes provides us with a new perspective for the development of urban forest parks [54]. The main findings are summarized as follows.

For the elderly, the preference for various sound sources in descending order is natural sound, animal sound, bird song, musical sound, vehicle sound, and atmospheric sound. In other words, compared with other sound sources, natural sound have an important influence on the elderly. Similarly, according to previous research, people prefer natural sound and most sounds associated with human activities [55]. However, they tend to dislike mechanical sound. As for the soundscape of forest parks, the main preferences of the elderly are the sound of leaves, the sound of falling stones, the sound of dust rising, the sound of tinkling brooks and the sound of birds.

In view of the preferences of the elderly for natural sound, the government can incorporate natural sound elements such as bird songs, wind sound, and water flowing sound into urban forest parks. Additionally, diverse soundscapes can be provided, such as music performance areas or musical fountains, to cater to the different preferences of the elderly. This can be achieved through well-planned and designed vegetation, water features, and landscapes in the park, creating a harmonious and pleasant acoustic environment.

Regarding the influences of various respondent characteristics on soundscape preference, the difference in gender was not statistically significant, and the average soundscape preferences of elderly men and elderly women are very similar; elderly people of different ages have different preferences for different sound categories; elderly people of different occupations have different preferences for different sound categories, and the those occupied as government officials and business managers have different preferences for different sound categories. The mean value of soundscape preference was higher for the elderly individuals with employment as government officials and business managers; the mean value of soundscape preference was higher for the elderly individuals with different educational backgrounds, and the mean value of soundscape preference was higher for the elderly individuals with master's degrees or above; the mean value of soundscape preference was higher for the elderly individuals with different living conditions, and the mean value of soundscape preference was higher for the elderly individuals living in homes.

Taking into consideration the preferences and needs of the various elderly individuals for soundscapes, policymakers can introduce interactive landscape elements in urban forest parks. This may include interactive musical installations, sound sculptures, or participatory music activities that have therapeutic qualities. These interactive landscapes can provide a sense of engagement and enjoyment for the elderly, enhancing their interaction with the sound environment and creating a positive acoustic environment and community atmosphere for them.

The regression equation model established in this study revealed that among the various factors influencing the soundscape preferences of the elderly, the top five most influential independent variables were whether they liked to stay in the waterfront environment in the forest park, whether they came to the forest park at noon, whether they valued the road signs in the forest park, whether they valued the science education exhibition function of the buildings in the forest park, and the auditory situation. In light of the elderly people's preferences for each sound source, more water-related natural environments such as artificial rivers should be built in the park.

### 6. Conclusions, Reflection, Limitations and Future Work

Although previous research has attempted to explore the relationship between the sensory perceptions and behavioral experiences of people in urban parks, few studies have approached the issue from the perspective of older adults, especially those in underdeveloped areas. In this study, we used a subjective evaluation method, namely, a questionnaire survey on Maoershan National Forest Park, to explore the relationships between older adults' preferences for urban forest park soundscapes in underdeveloped cities and made recommendations for urban forest park design in accordance with the findings of the survey and relevant theoretical foundations.

*6.1. Conclusions*

6.1.1. Landscape Design Recommendations to Enhance the Soundscape Experience

(1)    Overall Landscape Design

It was found that the subjective evaluation of a soundscape is closely related to landscape design. The results of the questionnaire showed that the elderly had higher preferences for the natural environment and the sounds of birds and animals. Older people of different genders, educational backgrounds, and physical qualities had different preferences for different soundscape elements. It is suggested that to cater to the preferences of each group, the integration of the soundscape should be fully considered in the overall landscape design to create a good layout. A harmonious environment can be created through the architectural design of the landscape structure, water features, plant design, etc., and a good audiovisual environment can be created by integrating soundscapes and different landscapes.

(2)   Green Environment Design

It was found that the degree of park greenery and the purpose and frequency of public green space use influenced the soundscape preferences of older adults. For example, as shown in Table 5, the mean values of the older adults' preferences for soundscapes were higher in environments with shade trees and open lawns. Previous studies have also found that the increased exposure of older adults to greenery can reduce mental stress and thus influence soundscape evaluations. Green landscaping on sidewalks and trails can produce sounds such as wind blowing in the leaves, thus stimulating resonance with natural sound. However, it is not suitable to set green landscapes outside of sports and leisure facilities, such as promenades, sports grounds, and benches, which mainly emphasize the soundscape of human activities.

Meanwhile, reasonable planning of plant shapes and colors, plant effects, plant distribution, plant types and terrace design can be used to divide areas for the elderly. In addition, different plant zones can be set up for elderly people with different plant preferences and physical health conditions, and water features should also be added near the plant zones to play with aesthetics and adjust the microclimate of the area.

### 6.1.2. Soundscape Design for the Elderly

It can be inferred from the questionnaire results that the elderly, as a group with a more complex situation, also have greater differences in their physical condition, and the elderly individuals with different physical qualities have different hobbies and different patterns of participating in urban forest parks. In view of this observation, the following landscape suggestions are proposed.

It was found that the elderly individuals in better physical condition preferred to visit places with mountains, rivers and forests, and these individuals had higher preferences for complex landscapes. On the contrary, for the elderly individuals who were in poor physical condition or had physical disabilities, the results of the questionnaire showed that they did not have a great preference for exercising in urban forest parks and even had a lower preference for some landscapes with high activity requirements. However, they still enjoyed hearing crowd activities, especially the rhythm and melody of square dancing. This finding implies that overly complex landscapes become a burden for such elderly people. Therefore, under the premise of protecting the safety of the elderly and controlling cost, designers should design landscapes that the elderly like. For example, the complex landscape can be set far away from the intersection, and an area near the entrance and exit can be set up to ensure the sound reception of such elderly people as much as possible by equipping speakers with appropriate volume to ensure that it is not noisy. For people with hearing impairments, visual cues such as text, lights, and guardrails will ensure their safety and improve their viewing experience as much as possible. Forest parks designed and built with such considerations will meet their soundscape preferences and entice them to spend time the space, which will be beneficial to their physical and mental health and can also satisfy the mobility requirements of the elderly individuals who are more physically active.

To ensure consistently favorable sound environments in urban forest parks, the government should undertake regular maintenance and management work. This includes monitoring and controlling sources of noise pollution, maintaining park facilities and sound equipment in good condition, and promptly addressing any issues that may affect the soundscape environment.

By implementing these policy recommendations, urban forest parks can provide an ideal soundscape environment that meets the preferences of the elderly for natural sounds and creates positive sound experiences and interactive opportunities for them. This will surely contribute to enhancing their quality of life and promoting their physical and mental well-being.

*6.2. Reflection*

This study adopted a subjective evaluation method to investigate the soundscape preferences of the older residents of underdeveloped cities in China for urban forest parks. Though the method proved reliable and valid, the data were only statistically analyzed from a correlation perspective, and other possible factors, such as social environment, epidemic context, level of urban management, and level of urban infrastructure, were not taken into account. It is suggested that such factors be incorporated into future research so as to provide a profound theoretical basis for the construction of soundscapes in urban forest park environments.

Overall, this study was successful in identifying the influence of different factors on the soundscape preferences of older adults in urban forest parks. These findings have rich implications for park designers and managers in developing relevant design and management strategies to enhance the evaluation and perception of elderly people.

*6.3. Limitations*

This study focused on social hotspots and analyzed the soundscape preferences of the elderly in urban forest parks in underdeveloped cities. The limitations and advantages of the research are presented and implications for future research as follows. Although studies on soundscape preference typically focus on people's perceptions and evaluations of sounds in natural environments rather than solely quantifying noise, soundscape preference research aims to understand people's preferences and evaluations of different sound environments as well as the impact of these sounds on their emotions, cognition, and behavior. This study dealt with the soundscape of a forest park, with a focus on elderly people's perceptions of natural sound such as bird songs, wind rustling, and water flow. However, for a more comprehensive understanding of the soundscape, future research should consider specific measurements of noise in urban forest parks.

As noted previously, soundscape preference was indirectly affected by the epidemic; in fact, all aspects of thought and life have been affected by the epidemic. Thus, it should be noted that the conclusions of this study might have been skewed due to the indirect potential impact of the epidemic on soundscape preference.

*6.4. Future Work*

This study constructed a comprehensive and in-depth questionnaire to capture the influence of many aspects on older individuals' soundscape preferences and arrived at convincing conclusions, lending support to the findings of earlier investigations. But there are three works are expected to be refined in the future.

(1) To maximize the understanding of the influences on the soundscape preferences of older adults, the relevant literature was comprehensively and profoundly reviewed, which generated rich implications for the design and administration of the questionnaire.

(2) To determine the key factors that influence the soundscape preference of the elderly, a regression coefficient model and an automatic linear model were established which effectively guaranteed the accuracy of the data analysis and interpretation.

(3) To bring light to the construction and maintenance of forest parks aimed at improving the well-being of elderly people, well-grounded recommendations were provided for landscape designers on how to cater to the soundscape preferences of different elderly groups.

**Author Contributions:** Conceptualization, J.X., L.L. and Q.Z.; data curation, L.L.; formal analysis, Q.Z. and Y.M.; investigation, J.X. and L.L.; methodology, Q.Z. and L.L.; resources, J.X.; software, Y.M. and Y.P.; validation, Y.M. and Y.P.; writing—original draft, Q.Z. and L.L.; writing—review and editing, T.W. All authors have read and agreed to the published version of the manuscript.

**Funding:** This research was supported by China youth fund of national natural science foundation projects (52208057), the Guangdong Provincial Natural Science Foundation General Project (2023A1515011191), the Ministry of Education humanities social sciences research project (20YJC760101), the Zhejiang cultural relics protection science and technology project (2021017), the Zhejiang Provincial Philosophy and Social Sciences Planning Project Youth Fund (21NDQN213YB), and the Major humanities and social sciences research program of colleges and universities in Zhejiang (2021QN053).

**Data Availability Statement:** The data in this study are available from the authors upon request.

**Conflicts of Interest:** The authors declare no conflict of interest.

# References

1. Cai, M.; Cui, C.; Lin, L.; Di, S.; Zhao, Z.; Wang, Y. Residents' Spatial Preference for Urban Forest Park Route during Physical Activities. *Int. J. Environ. Res. Public Health* **2021**, *18*, 11756. [CrossRef] [PubMed]
2. Nowak, D.J.; Hoehn, R.E.; Bodine, A.R.; Greenfield, E.J.; O'Neil-Dunne, J. Urban forest structure, ecosystem services and change in Syracuse, NY. *Urban Ecosyst.* **2016**, *19*, 1455–1477. [CrossRef]
3. Escobedo, F.J.; Kroeger, T.; Wagner, J.E. Urban forests and pollution mitigation: Analyzing ecosystem services and disservices. *Environ. Pollut.* **2011**, *159*, 2078–2087. [CrossRef] [PubMed]
4. Chen, W.Y.; Li, X. Urban forests' recreation and habitat potentials in China: A nationwide synthesis. *Urban For. Urban Green.* **2021**, *66*, 127376. [CrossRef]
5. Li, X.; Chen, C.; Wang, W.; Yang, J.; Innes, J.L.; Ferretti-Gallon, K.; Wang, G. The contribution of national parks to human health and well-being: Visitors' perceived benefits of Wuyishan National Park. *Int. J. Geoheritage Park.* **2021**, *9*, 1–12. [CrossRef]
6. Giannakopoulos, T.; Perantonis, S. Recognizing the quality of urban sound recordings using hand-crafted and deep audio features. In Proceedings of the 12th ACM International Conference on PErvasive Technologies Related to Assistive Environments, Rhodes, Greece, 5–7 June 2019; pp. 323–324.
7. He, M.; Wang, Y.; Wang, W.J.; Xie, Z. Therapeutic plant landscape design of urban forest parks based on the Five Senses Theory: A case study of Stanley Park in Canada. *Int. J. Geoheritage Park.* **2022**, *10*, 97–112. [CrossRef]
8. Liu, H.; Li, F.; Li, J.; Zhang, Y. The relationships between urban parks, residents' physical activity, and mental health benefits: A case study from Beijing, China. *J. Environ. Manag.* **2017**, *190*, 223–230. [CrossRef]
9. Chen, B.; Qi, X. Protest response and contingent valuation of an urban forest park in Fuzhou City, China. *Urban For. Urban Green.* **2018**, *29*, 68–76. [CrossRef]
10. Van De Vliert, E.; Rentfrow, P.J. Who is more prone to depression at higher latitudes? Islanders or mainlanders? *Curr. Res. Ecol. Soc. Psychol.* **2021**, *2*, 100012. [CrossRef]
11. Ratwatte, P.; Wehling, H.; Kovats, S.; Landeg, O.; Weston, D. Factors associated with older adults' perception of health risks of hot and cold weather event exposure: A scoping review. *Front. Public Health* **2022**, *10*, 939859. [CrossRef]
12. Dzhambov, A.M.; Dimitrova, D.D. Elderly visitors of an urban park, health anxiety and individual awareness of nature experiences. *Urban For. Urban Green.* **2014**, *13*, 806–813. [CrossRef]
13. Kumar, L.; Kahlon, N.; Jain, A.; Kaur, J.; Singh, M.; Pandey, A.K. Loss of smell and taste in COVID-19 infection in adolescents. *Int. J. Pediatr. Otorhinolaryngol.* **2021**, *142*, 110626. [CrossRef]
14. Han, F.; Lu, X.; Xiao, C.; Chang, M.; Huang, K. Estimation of Health Effects and Economic Losses from Ambient Air Pollution in Undeveloped Areas: Evidence from Guangxi, China. *Int. J. Environ. Res. Public Health* **2019**, *16*, 2707. [CrossRef]
15. Zhai, Y.; Li, D.; Wu, C.; Wu, H. Urban park facility use and intensity of seniors' physical activity—An examination combining accelerometer and GPS tracking. *Landsc. Urban Plan.* **2021**, *205*, 103950. [CrossRef]
16. Yu, W.; Sun, B.; Hu, H. Sustainable Development Research on the Spatial Differences in the Elderly Suitability of Shanghai Urban Parks. *Sustainability* **2019**, *11*, 6521. [CrossRef]
17. Yang, A.; Yang, J.; Yang, D.; Xu, R.; He, Y.; Aragon, A.; Qiu, H. Human Mobility to Parks under the COVID-19 Pandemic and Wildfire Seasons in the Western and Central United States. *GeoHealth* **2021**, *5*, e2021GH000494. [CrossRef]
18. Perry, M.; Cotes, L.; Horton, B.; Kunac, R.; Snell, I.; Taylor, B.; Wright, A.; Devan, H. "Enticing" but Not Necessarily a "Space Designed for Me": Experiences of Urban Park Use by Older Adults with Disability. *Int. J. Environ. Res. Public Health* **2021**, *18*, 552. [CrossRef]
19. Kaur, S.; Bhalla, A.; Kumari, S.; Singh, A. Assessment of functional status and daily life problems faced by elderly in a North Indian city. *Psychogeriatrics* **2019**, *19*, 419–425. [CrossRef]
20. Magnani, P.E.; Zanellato, N.F.G.; Genovez, M.B.; Alvarenga, I.C.; Faganello-Navega, F.R.; de Abreu, D.C.C. Usual and dual-task gait adaptations under visual stimulation in older adults at different ages. *Aging Clin. Exp. Res.* **2022**, *34*, 383–389. [CrossRef]
21. Fowler, M.D. Soundscape as a design strategy for landscape architectural praxis. *Des. Stud.* **2013**, *34*, 111–128. [CrossRef]

22. Haver, S.M.; Fournet, M.E.H.; Dziak, R.P.; Gabriele, C.; Gedamke, J.; Hatch, L.T.; Haxel, J.; Heppell, S.A.; McKenna, M.F.; Mellinger, D.K.; et al. Comparing the Underwater Soundscapes of Four U.S. National Parks and Marine Sanctuaries. *Front. Mar. Sci.* **2019**, *6*, 500. [CrossRef]
23. Shu, S.; Ma, H. Restorative effects of urban park soundscapes on children's psychophysiological stress. *Appl. Acoust.* **2020**, *164*, 107293. [CrossRef]
24. Zhao, W.; Li, H.; Zhu, X.; Ge, T. Effect of Birdsong Soundscape on Perceived Restorativeness in an Urban Park. *Int. J. Environ. Res. Public Health* **2020**, *17*, 5659. [CrossRef] [PubMed]
25. Nojiri, N.; Meng, Z.; Saho, K.; Duan, Y.; Uemura, K.; Aravinda, C.V.; Prabhu, G.A.; Shimakawa, H.; Meng, L. Apathy Classification Based on Doppler Radar Image for the Elderly Person. *Front. Bioeng. Biotechnol.* **2020**, *8*, 553847. [CrossRef] [PubMed]
26. Wen, C.; Albert, C.; Von Haaren, C. The elderly in green spaces: Exploring requirements and preferences concerning nature-based recreation. *Sustain. Cities Soc.* **2018**, *38*, 582–593. [CrossRef]
27. Ben Jemaa, A.; Irato, G.; Zanela, A.; Brescia, A.; Turki, M.; Jaïdane, M. Congruent auditory display and confusion in sound localization: Case of elderly drivers. *Transp. Res. Part F Traffic Psychol. Behav.* **2018**, *59*, 524–534. [CrossRef]
28. Tan, S.; Zhong, Y.; Yang, F.; Gong, X. The impact of Nanshan National Park concession policy on farmers' income in China. *Glob. Ecol. Conserv.* **2021**, *31*, e01804. [CrossRef]
29. Ali, M.J.; Rahaman, M.; Hossain, S.I. Urban green spaces for elderly human health: A planning model for healthy city living. *Land Use Policy* **2022**, *114*, 105970. [CrossRef]
30. Kweon, B.-S.; Sullivan, W.C.; Wiley, A.R. Green Common Spaces and the Social Integration of Inner-City Older Adults. *Environ. Behav.* **1998**, *30*, 832–858. [CrossRef]
31. Salgado, M.; Madureira, J.; Mendes, A.S.; Torres, A.; Teixeira, J.P.; Oliveira, M.D. Environmental determinants of population health in urban settings: A systematic review. *BMC Public Health* **2020**, *20*, 853. [CrossRef]
32. Liu, J.; Xu, J.; Wu, Z.; Cheng, Y.; Gou, Y.; Ridolfo, J. Soundscape Preference of Urban Residents in China in the Post-pandemic Era. *Front. Psychol.* **2021**, *12*, 750421. [CrossRef]
33. Xiao, Y.-H.; Liu, S.-R.; Tong, F.-C.; Kuang, Y.-W.; Chen, B.-F.; Guo, Y.-D. Characteristics and Sources of Metals in TSP and PM2.5 in an Urban Forest Park at Guangzhou. *Atmosphere* **2014**, *5*, 775–787. [CrossRef]
34. Song, C.; Ikei, H.; Miyazaki, Y. Effects of forest-derived visual, auditory, and combined stimuli. *Urban For. Urban Green.* **2021**, *64*, 127253. [CrossRef]
35. Lu, S.; Wu, F.; Wang, Z.; Cui, Y.; Chen, C.; Wei, Y. Evaluation system and application of plants in healing landscape for the elderly. *Urban For. Urban Green.* **2021**, *58*, 126969. [CrossRef]
36. Drieger, P. Semantic Network Analysis as a Method for Visual Text Analytics. *Procedia Soc. Behav. Sci.* **2013**, *79*, 4–17. [CrossRef]
37. Zeng, Z.; Kong, L.; Wang, J. The deformation and permeability of Yanji mudstone under cyclic loading and unloading. *J. Mt. Sci.* **2019**, *16*, 2907–2919. [CrossRef]
38. Giroud, N.; Hirsiger, S.; Muri, R.; Kegel, A.; Dillier, N.; Meyer, M. Neuroanatomical and resting state EEG power correlates of central hearing loss in older adults. *Brain Struct. Funct.* **2018**, *223*, 145–163. [CrossRef]
39. Hartig, T.; Kahn, P.H. Living in cities, naturally. *Science* **2016**, *352*, 938–940. [CrossRef]
40. Hedblom, M.; Gunnarsson, B.; Iravani, B.; Knez, I.; Schaefer, M.; Thorsson, P.; Lundström, J.N. Reduction of physiological stress by urban green space in a multisensory virtual experiment. *Sci. Rep.* **2019**, *9*, 10113. [CrossRef]
41. Leavell, M.A.; Leiferman, J.A.; Gascon, M.; Braddick, F.; Gonzalez, J.C.; Litt, J.S. Nature-Based Social Prescribing in Urban Settings to Improve Social Connectedness and Mental Well-Being: A Review. *Curr. Environ. Health Rep.* **2019**, *6*, 297–308. [CrossRef]
42. Levinger, P.; Cerin, E.; Milner, C.; Hill, K.D. Older people and nature: The benefits of outdoors, parks and nature in light of COVID-19 and beyond–where to from here? *Int. J. Environ. Health Res.* **2022**, *32*, 1329–1336. [CrossRef] [PubMed]
43. The Seventh National Population Census. Available online: https://www.gov.cn/guoqing/2021-05/13/content_5606149.htm (accessed on 1 May 2023).
44. Meneses-Barriviera, C.; Bazoni, J.; Doi, M.; Marchiori, L. Probable Association of Hearing Loss, Hypertension and Diabetes Mellitus in the Elderly. *Int. Arch. Otorhinolaryngol.* **2018**, *22*, 337–341. [CrossRef] [PubMed]
45. Peng, M.; Shah, S.S.; Wang, Q.; Meng, F. Response of Soil Bacterial Community Structure to Land-Use Conversion of Natural Forests in Maoershan National Forest Park, China. *Mol. Microbiol. Res.* **2017**, *7*. [CrossRef]
46. Song, C.; Ikei, H.; Miyazaki, Y. Physiological effects of forest-related visual, olfactory, and combined stimuli on humans: An additive combined effect. *Urban For. Urban Green.* **2019**, *44*, 126437. [CrossRef]
47. de Keijzer, C.; Bauwelinck, M.; Dadvand, P. Long-Term Exposure to Residential Greenspace and Healthy Ageing: A Systematic Review. *Curr. Environ. Health Rep.* **2020**, *7*, 65–88. [CrossRef]
48. Li, X.; Fan, L.; Leng, S.X. The Aging Tsunami and Senior Healthcare Development in China: Aging and Geriatrics Development in China. *J. Am. Geriatr. Soc.* **2018**, *66*, 1462–1468. [CrossRef]
49. Mediastika, C.E.; Sudarsono, A.S.; Kristanto, L.; Tanuwidjaja, G.; Sunaryo, R.G.; Damayanti, R. Appraising the sonic environment of urban parks using the soundscape dimension of visually impaired people. *Int. J. Urban Sci.* **2020**, *24*, 216–241. [CrossRef]
50. Li, A.; Zhao, P.; Haitao, H.; Mansourian, A.; Axhausen, K.W. How did micro-mobility change in response to COVID-19 pandemic? A case study based on spatial-temporal-semantic analytics. *Comput. Environ. Urban Syst.* **2021**, *90*, 101703. [CrossRef]
51. Liu, Y.; Hu, M.; Zhao, B. Audio-visual interactive evaluation of the forest landscape based on eye-tracking experiments. *Urban For. Urban Green.* **2019**, *46*, 126476. [CrossRef]

52. Lyu, Y.; Lu, S. Evaluation of Landscape Resources of Urban Forest Park Based on GIS—Take Beijing Tongzhou Grand Canal Forest Park as an Example. In Proceedings of the 2019 International Conference on Robots & Intelligent System (ICRIS), Haikou, China, 15–16 June 2019; pp. 418–424.
53. McCordic, J.A.; DeAngelis, A.I.; Kline, L.R.; McBride, C.; Rodgers, G.G.; Rowell, T.J.; Smith, J.; Stanley, J.A.; Stokoe, A.; Van Parijs, S.M. Biological Sound Sources Drive Soundscape Characteristics of Two Australian Marine Parks. *Front. Mar. Sci.* **2021**, *8*, 669412. [CrossRef]
54. Li, C.; Liu, Y.; Haklay, M. Participatory soundscape sensing. *Landsc. Urban Plan.* **2018**, *173*, 64–69. [CrossRef]
55. Fan, Q.; He, Y.; Hu, L. Soundscape evaluation and construction strategy of park road. *Appl. Acoust.* **2021**, *174*, 107685. [CrossRef]

*Article*

# Comparison of Soundscape Evaluation in Forest-Type and Urban-Type Han Chinese Buddhist Temples

Dongxu Zhang [1], Xueliu Liu [1] and Wei Mo [2,*]

1   College of Architecture and Urban Planning, Guangzhou University, Guangzhou 510000, China
2   School of Transportation, Civil Engineering and Architecture, Foshan University, Foshan 528000, China
*   Correspondence: mowei@fosu.edu.cn

**Abstract:** Soundscapes are one of the main means of creating a religious atmosphere in Han Chinese Buddhist temples, which are the most important religious sites in China. This paper selected several representative forest-type and urban-type Han Chinese Buddhist temples and employed a questionnaire and sound level measurement methods to conduct a comparative analysis of four aspects of acoustic environment evaluation, i.e., quietness, comfort, harmony, and sound preference, to identify and compare the characteristics of respondents' soundscape evaluation in these two types of temples. The results showed that compared with urban-type temples, respondents found the acoustic environment in forest-type temples to be quieter, more comfortable and more harmonious with the religious atmosphere. The sound level, measured with the questionnaire and respondents' social characteristics, such as age, occupation, level of education, purpose and frequency of visiting the temples, and attitude towards Buddhist thought, influenced their soundscape evaluation of urban-type and forest-type temples to different degrees. Among the various kinds of sounds in the temple, natural sounds, such as the sounds of flowing water, birds and insects, and rustling leaves, were preferred in forest-type temples, while Buddhism-related human-made sounds, including chanting and background music, were preferred in urban-type temples.

**Keywords:** forest-type temple; urban-type temple; Han Chinese Buddhism; soundscape evaluation; influencing factors

**Citation:** Zhang, D.; Liu, X.; Mo, W. Comparison of Soundscape Evaluation in Forest-Type and Urban-Type Han Chinese Buddhist Temples. *Forests* **2023**, *14*, 79. https://doi.org/10.3390/f14010079

Academic Editors: Jiang Liu, Xinhao Wang and Xin-Chen Hong

Received: 15 November 2022
Revised: 1 December 2022
Accepted: 2 December 2022
Published: 1 January 2023

## 1. Introduction

The term "soundscape" is defined as "the acoustic environment as perceived or experienced and/or understood by a person or people, in context" [1]. The notion of a soundscape was first proposed by Schafer and has continued to develop [2]. The soundscape of a place is considered to be a person's perceptual construct of the acoustic environment of that place [3,4]. In recent years, soundscapes have been one of the focuses of academic research. Many international journals have addressed the topic of soundscapes, and many interdisciplinary soundscape research organizations have been formed. The researchers come from the disciplines of acoustics, aesthetics, sociology, ecology, psychology, architecture, religious culture, environmental health, and urban studies. The research scope of soundscape places is constantly expanding, including parks [5], residential areas [6], historical buildings [7], historic towns [8], and religious architecture [9]. The research methods used include questionnaires [10], grounded theory approaches [11], soundwalk methods [12], laboratory experiments including binaural recordings [13], audio-visual interactions [14], predictive soundscape models, artificial neural networks [15], and structural equation modeling [16]. Unlike traditional acoustics, which focuses on the study of objective sound field characteristics, the soundscape approach enables the consideration of acoustic environments in positive terms, with soundscapes evaluated either positively or negatively [17]; therefore, the subjective evaluation of soundscapes is an important part of soundscape research.

Currently, many studies aim to evaluate soundscapes in forest parks or urban green spaces. Regarding the evaluation of the acoustic environment, some surveys have indicated that social/demographic/behavioral factors and visit motivations all showed significant relationships with individual sound perception in parks or green spaces [18–20]. Acoustic-related factors and park environment factors also influence the acoustic comfort evaluations of urban parks [21]. People's opinions of the meaning of tranquility may also influence the overall perceived quality of the soundscape during a park visit [22], and sounds caused by various human activities in parks play an important role in influencing the eventfulness of soundscape perceptions. LAeq is a useful indicator for the evaluation of environmental quietness [23]. Regarding the evaluation of sound preference in parks, some previous studies have shown that almost everyone likes sounds such as "songbirds" and "sparrows" in parks [24]. The presence of birds twittering, insects chirping, flowing water, light music, and ancient temple bells makes tourists feel more immersed [25], and running water and birdsong are the most commonly heard and most preferred sounds in national parks [16]. Recent literature also suggests that different types of birdsong exhibit different sound comforts in different seasons [26], and soundscapes with a rich array of perceived bird sounds and minimal perceived traffic noise offer the greatest perceived restorative value in parks [27]. There are many factors that affect sound preference. Age is one of the most influential dimensions in the perception of and preference for individual sounds in urban recreational forest parks [28]. In addition, perceptual responses to human sounds, birdsong, and water sounds differ significantly across cultural backgrounds [29].

There are also many studies on the soundscape evaluation of various urban spaces. Regarding the evaluation of the acoustic environment, results have shown that the perceived quality of the urban soundscape is very much an individually subjective experience [30]. For example, there are significant differences among different age groups in terms of acoustic comfort [4], and differences in the purpose of going to urban open spaces and education levels might lead to differences in the evaluation of acoustic comfort [31]. Interviewees' age, occupation, duration and purpose have a significant effect on their acoustic satisfaction in urban historical areas [32]. Visit frequency affects visitors' expectations of the general soundscape, and visitors' perceptions of loudness and satisfaction are associated with maximum sound levels [33]. Another study showed that acoustic comfort has a significant correlation with LAeq in public squares [34]. Regarding the evaluation of sound preference in urban space, relevant studies have revealed that "traffic" sounds and "birdsong" are critical factors that influence participants' initial perception of urban soundscape quality [35]. Birdsong plays an important and positive role in urban soundscape perception [36], and bird sounds are the most preferred among the natural sounds in urban streets [37]. Water sounds have been determined to be the best sounds to use for enhancing the urban soundscape [38], while traffic sounds are the dominant indicator that negatively affects pleasantness in urban residential areas [14]. The results of a questionnaire on the soundscape of a city square showed that the most unpleasant sounds were motorcycles, cars, and handcarts, while the most pleasant sound was water [39]. Some analyses of the influencing factors of sound preference have shown that demographic factors affect the evaluation of sound preference in urban open spaces; for example, with increasing age, people are generally more positive towards sounds related to nature, culture, or human activities [40]. Age and education level are two factors that universally influence sound preference, while gender and occupation generally do not significantly influence sound preference evaluation [41]. In brief, although natural sounds are perceived more favourably than urban sounds, an urban soundscape cannot be equated to noise, and its positive aspects should be more broadly acknowledged [42].

In recent years, scholars have analysed the relationship between the acoustic environments of religious spaces and human feelings from the perspective of soundscapes. Regarding the soundscape evaluation of temples, in contrast to an ordinary urban open space or simple natural landscape, natural sounds, cultural sounds, and historic sounds are widely appreciated in people's subjective feelings about Chinese Buddhist temples [43]. One author of this paper analysed the correlation between Chinese people's evaluations of Buddhist temple soundscapes and mental health [9] and studied sound preferences in Han

Buddhist temples [44]. In a Chinese Taoist temple, the soundscape evaluation was affected by the measured sound pressure level and the respondents' belief, type of activity, social factors, and spatial position [45]. Regarding the soundscape of churches, a previous study showed that 67% of observed visitors spent less than a minute in a chapel, yet 49% of the visitor comment cards mentioned the chapel or the chapel soundscape as their favourite part of the visit [46]. One study on the degree of acoustic comfort inside several churches in Sheffield suggested that there was no clear correlation between acoustic comfort and measured reverberation time [47]. Another study used a survey questionnaire to compare the soundscape around a Catholic church with the soundscape around a Buddhist temple in South Korea and proposed that sounds related to religious activities in the temple precincts are relatively more significant than those of cathedral precincts [48]. Regarding the soundscape of mosques, a study showed that the acoustic comfort conditions were perceived to be satisfactory in all case studies in historical and new mosques [49], and there was a correlation between the acoustic design of the mosque and the worshippers' comfort [50]. Another study proposed that the majority of respondents were in favour of a broadcast of music or prayer in both indoor and outdoor areas of a historical mosque [51].

In summary, the sites of most existing soundscape evaluations have mainly been common urban or forest areas. These studies noted the importance of studying the relationship between the acoustic environment and people's feelings from the perspective of the soundscape. Some studies have also analysed the acoustic environment of Christian churches or mosques from the perspective of traditional acoustic methods or soundscapes. However, there is currently relatively little research on the subjective evaluation of the soundscapes of Han Buddhist temples, and no research has focused on the respective characteristics of the soundscape evaluations of forest-type and urban-type temples, which are the two most important and most numerous types of religious architecture in China (approximately 47% of Han Chinese Buddhist temples are located in forests, 44% are urban-type temples, and the remaining 9% are rural-type temples [52]). A good acoustic environment both inside and outside the temples is the main means of facilitating a religious atmosphere. Especially for forest-type temples located in famous mountains and featuring beautiful scenery, all kinds of pleasant natural sounds dominate the acoustic environment of the temples, which can make a deep impression on people. In contrast, urban-type temples are located near city centres and are associated with more vehicles and pedestrians; accordingly, the sound environment of these temples is relatively noisy. There are obvious differences between the two types of temples. These differences undoubtedly affect people's perception of the acoustic environment of these two types of Buddhist temples. Therefore, we plan to analyse the respective characteristics of the soundscape evaluation of forest-type and urban-type temples and compare the differences among and factors influencing the soundscape evaluation of these two types of temples. Three specific research questions are addressed in this study.

(1)  What are the characteristics of respondents' evaluations of the acoustic environment of forest-type and urban-type temples? Urban-type temples have a certain function for public activities, while forest-type temples have the functions of leisure and relaxation, similar to parks. Are the characteristics of their soundscape evaluations different from ordinary public spaces?

(2)  To what extent do the objective measured sound level and the subjective sociological characteristics of the respondents affect the evaluation of the acoustic environment of the two types of temples?

(3)  Are there differences in the evaluation of sound preference between the respondents in the two types of temples? What are the influencing factors for these differences?

A large number of questionnaires were distributed in four typical Han Buddhist temples (including two forest types and two urban types), and sound pressure levels were synchronously measured. Subsequently, in accordance with statistics concerning the results of the questionnaire, differences in respondents' evaluations of the sound environment between these two types of temples were analysed and compared, as was the influence of objective factors (sound pressure level) and subjective factors, that is, respondents' sociological characteristics (including age, belief, occupation, purpose and frequency of visiting the temples, and level of education) on this difference. This study attempted to identify differences in the respondents' sound preferences between urban-type and forest-type temples. The research results are conducive to the better design of the soundscapes of the two types of temples and to the creation of a healthy and favourable religious acoustic environment for users.

## 2. Materials and Methods

### 2.1. Characteristics of Research Sites

In this study, four typical forest-type and urban-type Han Chinese Buddhist temples were selected for comparative analysis of the soundscape evaluation. Figure 1 shows the location and surrounding environment of each temple. With regard to the temples, the four research objects selected were all large-scale temples with a long history and many worshippers. Xiantong Temple and Longquan Temple were selected as representative forest-type temples. Xiantong Temple (hereinafter referred to as "FT1") is the largest and most historic Han Chinese Buddhist temple on Wutai Mountain. Longquan Temple ("FT2") is the largest Buddhist temple on Qianshan Mountain, Liaoning Province. Xiangguo Temple and Ci'en Temple were selected as representative urban-type temples. Xiangguo Temple ("UT1") is located in the centre of the city of Kaifeng. This temple is one of the ten most famous Han Chinese Buddhist temples. Ci'en Temple ("UT2") is one of the key temples of Chinese Buddhism in China and the largest existing Buddhist temple in the city of Shenyang. With respect to their geographical locations, "FT1" and "UT1" are located in central China, while "FT2" and "UT2" are located in northern China. The locations of the four temples represent different regions. Meanwhile, there are many similarities in culture, language and belief between the people in central China and people in northern China, and these would avoid the problem that the difference in social characteristics (especially the different attitudes towards Buddhist thought of the respondents in different regions) of the respondents in the temples could affect the results of the questionnaire. All four temples support monks' practice of Buddhism and are open to the public. They have an important position and extensive influence in the Chinese Buddhist circle; therefore, the acoustic environment of these temples is typical.

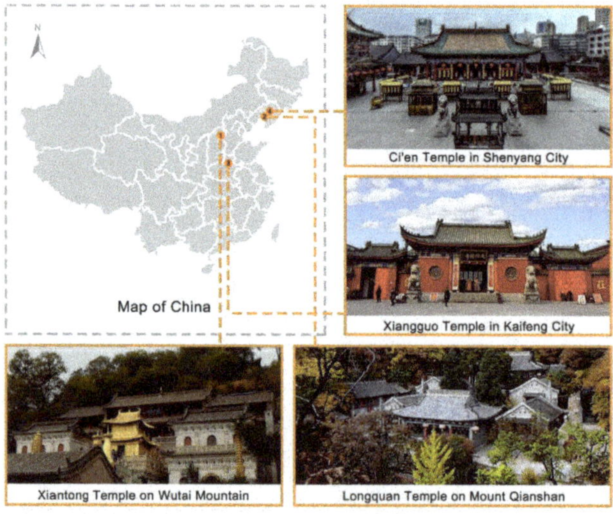


**Figure 1.** Location and surrounding environment of the four temples.

*2.2. Questionnaire Survey Method*

2.2.1. Design and Distribution of Questionnaire

(1) Before the formal investigation, this study first distributed 70 pilot questionnaires at Puji Temple in Putuo Mountain. Adjustments were made to the questionnaire items after the statistical results were obtained and analysed. Items that were invalid or lacked reliability and validity were revised, and the final version of the questionnaire was produced.

The formal questionnaire consisted of five parts. Previous studies have shown that the evaluation of soundscapes in various spaces is affected to varying degrees by objective acoustic environmental factors and human sociological characteristics. Therefore, the first part collected basic demographic information from the respondents, including gender, age, level of education, and occupation. In the questionnaire, people were divided into five groups according to age: younger than 18 years old (high school students and below), 18 to 30 (youth), 30 to 45 (middle-aged), 45 to 60 (middle-aged and elderly), and older than 60 (retiree). The aim was to analyse the differences between the soundscape evaluations of people of different age groups. However, the statistical results of this study showed that the soundscape evaluation results of respondents younger than 18 years old and those 18 to 30 years old were similar, and the results of respondents aged 45 to 60 years old and those older than 60 years old were similar. Therefore, in the current questionnaire analysis, the respondents were divided into three categories: younger than 30 years old, 30 to 45 years old and older than 45 years old. For the division of education level, the respondents were divided into four categories: primary school, middle school, university, and postgraduate.

The second part included items related to Buddhist beliefs, including attitudes towards Buddhist thought, the respondents' annual frequency of attending religious activities or visiting Buddhist temples and the respondents' purpose for visiting Buddhist temples. The questionnaire divided the respondents in accordance with their attitudes towards Buddhist thought into the categories of firm believers, partial believers (those who believe to a limited extent), and nonbelievers. The respondents' purpose was divided into four categories: visiting/tourism, worshiping the Buddha, exercising, and other purposes. In addition, in the pilot questionnaire, the respondents were asked to fill in the number of visits to the temple every year. The results showed that 70% of the respondents did not go to the temple more than 4 times a year. Therefore, the average number of annual visits to the temple was divided into less than once, 1 to 2 times, 3 to 4 times and more than 4 times in the formal questionnaire.

The third part was used to evaluate the respondents' attitudes towards the acoustic environment in Buddhist temples, including comfort, quietness, and harmony. Acoustic comfort has always been an important aspect of soundscape research [4,21], and the same is true for quietness [17,19,40]. Previous studies have suggested that religious precincts should be quiet and tranquil to allow people to engage in religious self-reflection [43,48,51,53]. This study chose the degree of acoustic harmony as an index in the soundscape evaluation of Buddhist temples, with reference to previous studies [18,54]. As Buddhist temples are the most important religious places in China, whether or not the acoustic environment is harmonious with the religious atmosphere is of great significance for facilitating a religious atmosphere.

The fourth part was used to evaluate the respondents' preferences for the typical sounds associated with Han Chinese Buddhist temples. Considering that the soundscape includes the entire acoustic environment resulting from natural and human-made sound sources [2,55], sound preference is an important aspect of soundscape research [56]. Previous study on the soundscape of religious places showed that the main components of sound elements were grouped into natural, social, and religious sounds [48]. In our pilot questionnaire survey and field observation, the natural sounds inside and outside the temple mainly included the sounds of flowing water, birds, insects, rustling leaves, and wind. Human-made sounds were divided into Buddhism-related sounds (including bells, chanting, various implements, drums, prayers, and background Buddhist music) and sounds unrelated to Buddhism (including footsteps, the voices of tour guides, tourist conversations, traffic sounds, and construction site noises). Buddhism-related human-made sounds often had the characteristics of the temples, while Buddhism-unrelated human-made sounds could occur in other places and did not have the unique place characteristics of Buddhist temples. The fifth part collected subjective suggestions concerning the acoustic environment of the temple.

Considering that there were few questions within the questionnaire, we did not use reverse-coded items. All items related to soundscape evaluation in the questionnaire were rated on a Likert five-level scale because the Likert scale is easy to design, its requirements for descriptive words are limited to no logical errors, it can be used to measure multidimensional complex concepts or attitudes [57], and it was suitable for the diverse population structure in this study. Level 1 in the questionnaire represented "quiet/comfortable/harmonious", and Level 5 represented "noisy/uncomfortable/inharmonious". Regarding the respondents' sound preferences, Level 1 indicated "liked the sound" and Level 5 indicated "disliked the sound". For specific questions and items, please see the questionnaire in the Appendix A.

(2) A total of 720 questionnaires were distributed at 4 temples, and 685 valid questionnaires were recovered: 177 from "FT1", 170 from "FT2", 160 from "UT1", and 178 from "UT2". The number of questionnaires distributed in a single temple was based on previous studies reporting that a range of 100 to 150 questionnaires could be considered representative in the context of an urban environment soundscape survey [2].

The questionnaires were distributed in the spring and summer, usually between 8:30 am and 17:30 pm, when the temple was open to the public. The questionnaires were distributed in the field at four temples by the researcher's team members (approximately 4 to 5 members for each temple). The specific locations at which the questionnaires were distributed were the courtyards of the four temples. The outdoor courtyard was chosen as the place for questionnaire distribution because quiet was generally required inside the hall in the temple and the indoor space of the hall allowed tourists to stay only for a short time, which made it difficult to complete the questionnaire. The target subjects of the questionnaire survey were randomly selected tourists and worshippers encountered in the temples. Because the four temples where the questionnaires were distributed were all tourist places with a large number of people, the genders and ages of the respondents should be balanced as much as possible to ensure the randomness and universality of the questionnaire survey. All questionnaires used anonymous methods (no record of the respondents' information, such as name and phone number). First, we asked whether the respondent was willing to participate in the survey. During the process of completing the questionnaire, the research team members provided consultations nearby. If the respondents provided answers quickly without carefully reading the items, the questionnaire was immediately marked as invalid after being returned.

In this study, the sound level measurements and the survey questionnaire were conducted at the same time; as the respondents answered the questionnaire, the on-site acoustic environment was simultaneously measured to ensure good correspondence between the psychological feelings expressed in the subjective survey and the objective measurement. During the measurement, the microphone of the sound level metre was positioned approximately 1 m away from any reflective surfaces and 1.5 m above the ground to reduce the effect of acoustic reflection [58]. The sound level corresponding to each questionnaire was measured more than 10 times. The interval between each measurement was five seconds, and the mean value was calculated. The instantaneous sound pressure level instead of LAeq was used in this research as the curiosity of other visitors may disturb and influence the measurement results.

### 2.2.2. Statistical Results of the Questionnaire

The correlation calculation conducted for this study mainly focused on the correlation coefficient between the acoustic environment evaluation and the sociological factors of the respondents in the two types of temples or the synchronous measured sound level. All results of the questionnaire survey were imported into SPSS software (version 26). Among these factors, the subjective acoustic quietness, acoustic comfort, acoustic harmony and sound preference, as dependent variables, were ordinal variables, while the independent variables, such as age, frequency of visiting, and attitude towards Buddhist thought were ordinal variables. Similarly, purpose and occupation were nominal variables, while gender and the types of temples were dichotomous nominal variables, and the measured sound level was a continuous variable. Due to the different types of dependent variables in question, the correlation calculation methods also differed between independent variables and dependent variables (as shown in Table 1).

**Table 1.** The calculation method for independent and dependent variables.

| Independent | Dependent Variables | SPSS Calculation Approach | Index | Variable Type |
|---|---|---|---|---|
| The measured sound level value synchronous with questionnaire | Acoustic comfort (harmony, quietness) evaluation | Bivariate correlation | Pearson | Continuous variable/ Ordinal variable |
| Gender, the types of temples | Acoustic comfort (harmony, quietness) evaluation, sound preference evaluation | Independent-samples *t*-test | Mean difference | Dichotomous (nominal) variable/ordinal variable |
| Purpose, occupation, different temples | Acoustic comfort (harmony, quietness) evaluation, sound preference evaluation | Crosstabs | Phi and Cramer's V | Nominal variable/ Ordinal variable |
| Age, frequency of visiting a temple, attitude towards Buddhist thought, education level | Acoustic comfort (harmony, quietness) evaluation, sound preference evaluation | Crosstabs | Gamma | Ordinal variable/ Ordinal variable |
| Acoustic comfort (harmony) evaluation | Acoustic quietness (harmony) evaluation | Crosstabs | Gamma | Ordinal variable/ Ordinal variable |
| Different demographic factors | Sound preference evaluation | Compare means | One-Way ANOVA | Nominal variable (or Ordinal variable)/ Ordinal variable |

Reliability and validity analyses of questionnaires are a necessary step before data analysis. The SPSS software's reliability analysis was used to perform a confidence test on the reliability. The calculation results showed that the Cronbach's $\alpha$ coefficient of the whole soundscape evaluation scale was 0.748, the coefficient of the acoustic environment evaluation was 0.727, and the coefficient of the sound preference was 0.708. All were within the acceptable range, indicating acceptable reliability of the data [59]. Then, factor analysis was used to verify the construct validity of the questionnaire. KMO = 0.778 for the acoustic environment evaluation. Accordingly, two factors were extracted with characteristic roots greater than 1, their cumulative contribution to all variables was 52.1%, and Bartlett's test of sphericity indicated $p < 0.001$. On the other hand, KMO = 0.778 for the sound preference evaluation. Four factors were extracted with characteristic roots greater than 1, their cumulative contribution to all variables was 51.6%, and Bartlett's test of sphericity indicated $p < 0.001$. These data showed that the results of the soundscape evaluation questionnaire met the requirement for construct validity [59,60].

In this paper, an independent-samples *t*-test was used to analyse whether the acoustic environment evaluation of respondents with the same characteristics between the two types of temples had significant differences. All assumptions for eligibility to perform the *t*-test were checked and passed before use, i.e., the samples were quantitative data, the two populations were normally distributed, and the two samples were random independent samples [60]. In this study, analysis of variance (ANOVA) was used to determine whether the mean value of two or more population samples with normal distribution was significantly different. Before performing ANOVA, the author checked that these datasets met the ANOVA assumptions, including that each population sample followed a normal distribution and the homogeneity of variance of each population and that each sample was independent and randomly selected [60].

Table 2 presents the distribution of the characteristics of 685 respondents in two types of temples, including gender, age, level of education, attitude towards Buddhist thought, and frequency and purpose of visiting temples. These distributions were the classification conditions when analysing the various evaluations of soundscapes in the temples. Next, the influence of subjective and objective factors on the soundscape evaluation in the two types of temples (including acoustic quietness, comfort, harmony and sound preference) was analysed, and the differences in these effects between the two types of temples were compared.

**Table 2.** Percentage of respondents' characteristics by the temple type in the questionnaire.

| Respondents' Characteristics | | Temple Type | |
|---|---|---|---|
| | | Forest-Type Temple | Urban-Type Temple |
| Gender | male | 48.70% | 46.40% |
| | female | 51.30% | 53.60% |
| Age | under 30 | 50.90% | 47.60% |
| | 30–45 | 30.50% | 32.50% |
| | above 45 | 18.60% | 19.90% |
| Frequency of visiting temple | less than one time | 62.80% | 47.00% |
| | 1 to 2 times | 10.70% | 9.80% |
| | 3 to 4 times | 7.80% | 7.70% |
| | more than 4 times | 18.70% | 35.50% |
| Education level | primary school or less | 7.90% | 8.30% |
| | middle school | 27.40% | 42.00% |
| | college or university | 56.20% | 41.30% |
| | postgraduate | 8.50% | 8.40% |
| Purpose | visiting tourism | 61.70% | 44.70% |
| | worshiping the Buddha | 22.70% | 35.80% |
| | exercising | 7.80% | 7.70% |
| | others | 7.80% | 11.80% |
| Attitude towards Buddhist thought | firmly believe | 30.00% | 41.10% |
| | partially believe | 59.00% | 48.80% |
| | do not believe | 11.00% | 10.10% |

It should be noted that our soundscape research referred to ISO 12193 (ISO Standard) [1,61,62], but considering the difference between the soundscape in Buddhist temples and that in other public spaces (the soundscape in religious buildings focuses more on the influence and inspiration of believers), the parameters and indices selected in the course of investigation and analysis were not completely consistent with those recommended in ISO 12193 (for example, the types of sound sources, performed effective quality, the assessment of surrounding sound environment, etc.).

## 3. Results and Discussion

### 3.1. Factors Influencing the Acoustic Environment Assessment

According to the statistical results of the questionnaire, the value of respondents' evaluation of the whole acoustic quietness in four temples was 2.13, that of acoustic comfort was 1.96 and that of the harmony between the acoustic environment and the religious atmosphere was 2.28. The evaluation value of the acoustic environment in the two types of temples is shown in Figure 2. The mean values of quietness, comfort, and harmony for forest-type temples (2.07, 1.87, and 2.17, respectively) were lower than those for urban-type temples (2.18, 2.06, and 2.39, respectively). The lower the value, the more quiet, comfortable, and harmonious the respondents considered the acoustic environment. The respondents' feelings about the overall acoustic environment were better for forest-type temples than for urban-type temples. The results of the correlation analysis showed that the values of the three acoustic environment evaluations were significantly correlated with each other, regardless of whether they were whole or divided into two types of temples. The correlation coefficient is shown in Table 3. The degree of correlation in forest temples was generally higher than that in urban temples. The results of the independent-samples *t*-test showed that there were significant differences in the evaluation of acoustic comfort or acoustic harmony between the two types of temples, but there was no significant difference in the evaluation of acoustic quietness. This study subsequently analysed the subjective and objective factors that affect the evaluation of the acoustic environment in the two types of temples.

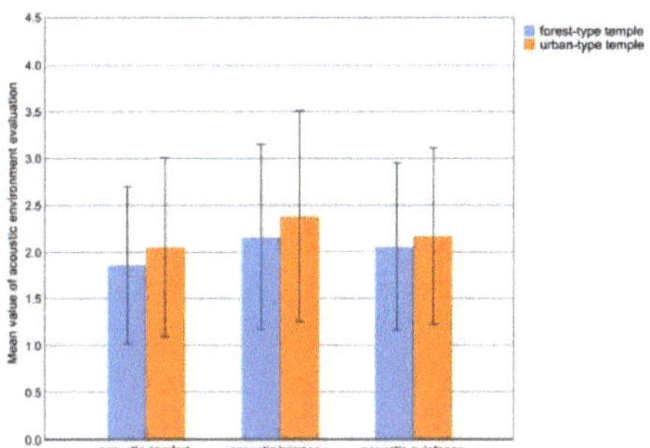


**Figure 2.** The evaluation value and standard deviation of the acoustic environment in forest-type and urban-type temples.

**Table 3.** Correlation between the subjective evaluation of the acoustic environment in the temple.

| Category | Overall Correlation Coefficient (Forest-Type Temple/Urban-Type Temple) | | |
|---|---|---|---|
| | Acoustic Comfort | Acoustic Harmony | Acoustic Quietness |
| Acoustic comfort | 1 | 0.67 ** (0.78 **/0.56 **) | 0.75 ** (0.78 **/0.71 **) |
| Acoustic harmony | | 1 | 0.59 ** (0.67 **/0.52 **) |
| Acoustic quietness | | | 1 |

Note: ** in the table indicates significance level that is $p < 0.01$.

### 3.1.1. Objective Factor

The mean value of the measured sound level synchronous with the questionnaire in the temple was 56.9 dBA (standard deviation of 7.3, below the same). The correlation between the measured sound level and the acoustic environment evaluation of the temple is shown in Table 4. The measured sound level was significantly correlated with acoustic comfort and quietness, with correlation coefficients of R = 0.118** and R = 0.195**, respectively, and had no correlation with acoustic harmony.

The mean value of the measured sound level synchronous with the questionnaire in urban-type temples was 58.0 dBA (8.2), and the maximum and minimum values were 76.8 and 45.0 dBA, respectively. The mean value of the measured sound level in forest-type temples was 55.9 dBA (6.3), and the maximum and minimum values were 72.6 and 36.4 dBA, respectively, which were all lower than those in urban-type temples, indicating that the sound field of forest-type temples was generally quieter than that of urban-type temples. In urban-type temples, the correlation coefficients between the measured sound level and the acoustic environment evaluation were 0.156** for acoustic comfort, 0.269** for quietness, and 0.006 ($p = 0.921$) for harmony. However, there was no significant correlation between the measured sound level and the three kinds of acoustic environment evaluations in the forest-type temples. The objective factor of the measured sound level synchronous with the questionnaire only affected acoustic comfort and quietness evaluations in urban-type temples.

**Table 4.** Correlation between objective or subjective factors and temple acoustic environment evaluation.

| Influence Factor | Correlation Coefficients/Significance Level | | |
|---|---|---|---|
| | **Acoustic Comfort** | **Acoustic Harmony** | **Acoustic Quietness** |
| Gender | 0.055/0.310 | 0.040/0.550 | −0.030/0.442 |
| Age | −0.138/0.013 (*) | −0.152/0.003 (**) | −0.268/0.000 (**) |
| Attitude towards Buddhist thought | 0.336/0.000 (**) | 0.340/0.000 (**) | 0.252/0.000 (**) |
| Purpose | −0.148/0.005 (**) | −0.174/0.001 (**) | −0.154/0.003 (**) |
| Occupation | 0.190/0.002 (**) | 0.182/0.009 (**) | 0.223/0.000 (**) |
| Average number of visits to the temple every year | −0.178/0.004 (**) | −0.134/0.026 (*) | −0.133/0.042 (*) |
| Education level | 0.156/0.007 (**) | 0.172/0.002 (**) | 0.156/0.004 (**) |
| The measured sound level synchronous with the questionnaire | 0.118/0.002 (**) | 0.195/0.000 (**) | 0.130/0.440 |

Note: * and ** in the table indicate significance level, * indicates $p < 0.05$, ** indicates $p < 0.01$.

### 3.1.2. Subjective Factor

The level of significance of the correlations between the respondents' sociological factors and the evaluation of the acoustic environment in the temples is shown in Table 4. The respondents' age, occupation, level of education, purpose and frequency of visiting temples and attitude towards Buddhist thought were significantly correlated with their evaluations of the acoustic environment (note: There was no correlation between gender and the three kinds of acoustic environment evaluations. This was consistent with previous studies that found no significant difference between males and females in acoustic environment evaluation [4,31], so no further analysis was conducted on this issue. Therefore, this study includes these sociological factors as independent variables to analyse the differences in evaluations of the acoustic environment (as dependent variables) between forest-type and urban-type temples.

(1)  Age

Figure 3a presents the mean values of acoustic comfort evaluation by respondents of different ages between the two types of temples (this item is given in Question 4 of the questionnaire in the Appendix A). With increased age, respondents tended to indicate more comfort in their evaluations of the acoustic environment of forest-type temples, while no such trend was observed with regard to evaluations of the acoustic environment in urban-type temples. The correlation coefficient between age and the acoustic comfort evaluation was calculated as −0.279** in forest-type temples and −0.018 ($p = 0.822$) in urban-type temples, indicating that age was correlated with the evaluation of acoustic comfort only in forest-type temples. The result of the independent-samples $t$-test showed that with regard to people younger than 30 years old or 30 to 45 years old, there were no significant differences in the acoustic comfort evaluation between the two types of temples. Among people older than 45 years old, the mean value of the acoustic comfort evaluation in forest-type temples was lower than that in urban-type temples by 0.47, and there was a significant difference between the two types of temples. This result suggested that for people older than 45 years, the type of temple affects their acoustic comfort.

The mean value of people's evaluations of acoustic harmony for the two types of temples across different ages is shown in Figure 3b (this item is given in Question 5 of the questionnaire in the Appendix A). These results are similar to those associated with acoustic comfort evaluations. With increasing age, respondents tended to indicate more harmony in their evaluation of the acoustic environment of forest-type temples, while no such trend was found for urban-type temples. The correlation coefficient between age and acoustic harmony evaluation was calculated as −0.227** for forest-type temples and −0.091 ($p = 0.187$) for urban-type temples, indicating that age was correlated with the evaluation of acoustic harmony only in forest-type temples. For people younger than 30 or those aged 30 to 45 years, no significant differences in acoustic harmony were found between the two types of temples. Among people older than 45 years, the mean value of acoustic harmony evaluation in forest-type temples was lower than that in urban-type temples by 0.45, and there was a significant difference in the evaluation of acoustic harmony between the two types of temples, which is similar to the findings concerning acoustic comfort evaluation.

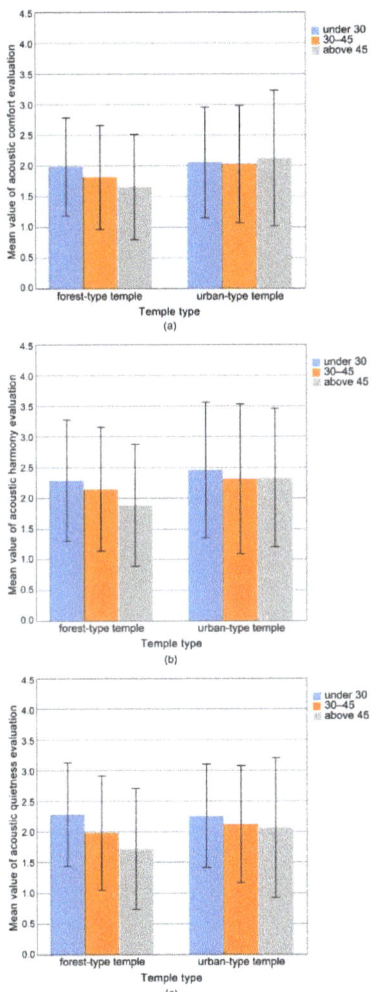

**Figure 3.** The evaluation value and standard deviation of acoustic environment by different respondents' age: (**a**) comfort; (**b**) harmony; (**c**) quietness.

Regarding the evaluation of quietness in the two types of temples (this item is given in Question 3 of the questionnaire in the Appendix A), Figure 3c shows that with increasing age, the respondents' evaluation of quietness in the two types of temples tended towards indications of quietness. The correlation coefficient between age and the evaluation of quietness was −0.351** in forest-type temples and −0.184* in urban-type temples. The results indicate that there were no significant differences in the evaluation of acoustic quietness reported by respondents of the same age group between the two types of temples. This finding differs from the results concerning evaluations of acoustic comfort and harmony.

(2)  Belief Factor

Figure 4a presents the relations between the respondents' attitudes towards Buddhist thought and the mean values of acoustic comfort evaluation for the two types of temples. (This item is given in Question 4 of the questionnaire in the Appendix A.) The same trends were found for both types of temples; that is, the more people believe in Buddhist thought, the more comfort they report in their evaluation of the acoustic environment. The correlation coefficient was 0.346** for forest-type temples and 0.360** for urban-type temples. The mean evaluation value of acoustic comfort by respondents who partially believed in Buddhist thought was 0.23 lower for forest-type temples than for urban-type temples (a lower value indicated that the respondents felt the acoustic environment to be more comfortable). The result of the independent-samples *t*-test indicated that between the two types of temples, only respondents who partially believed in Buddhist thought exhibited significant differences in their evaluation of acoustic comfort. For the other two groups of people, although the mean evaluation value of acoustic comfort by respondents in the forest-type temple was lower than that in the urban-type temple (0.21, firm believers; 0.34, nonbelievers), there were no significant differences in acoustic comfort evaluation between the two types of temples.

The relations between respondents' attitudes towards Buddhist thought and the mean values of acoustic harmony evaluation between the two types of temples are shown in Figure 4b. (This item is given in Question 5 of the questionnaire in the Appendix A). The trend is similar to that associated with the evaluations of acoustic comfort; that is, the more people believed in Buddhist thought, the more their evaluations tended to indicate harmony between the acoustic environment and the temple atmosphere. The correlation coefficient was 0.375** for forest-type temples and 0.336** for urban-type temples. Respondents who firmly believed or partially believed in Buddhist thought had a lower mean evaluation value of acoustic harmony for forest-type temples than for urban-type temples by 0.30 and 0.22, respectively. The result of the independent-samples *t*-test indicated the presence of significant differences in the evaluation values of acoustic harmony reported by people who firmly believed or partially believed in Buddhist thought between the two types of temples. Respondents deemed that the acoustic environment and temple atmosphere were more harmonious in forest-type temples. For those who did not believe in Buddhist thought, there was no significant difference in their evaluation of acoustic harmony between the two types of temples.

For the quietness evaluation, as shown in Figure 4c, the more the respondents believed in Buddhist thought, the more their evaluations of the temple acoustic environment tended towards quiet (this item is given in Question 3 of the questionnaire in the Appendix A). The correlation coefficient was 0.236** for forest-type temples and 0.291** for urban-type temples. These results indicated that between the two types of temples, only respondents who partially believed in Buddhist thought exhibited significant differences in terms of their evaluation of quietness, and those who partially believed in Buddhist thought in forest-type temples felt quieter (their mean evaluation value of acoustic quietness was 0.21 lower than that of urban-type temples). This is similar to the findings concerning the evaluation of acoustic comfort.

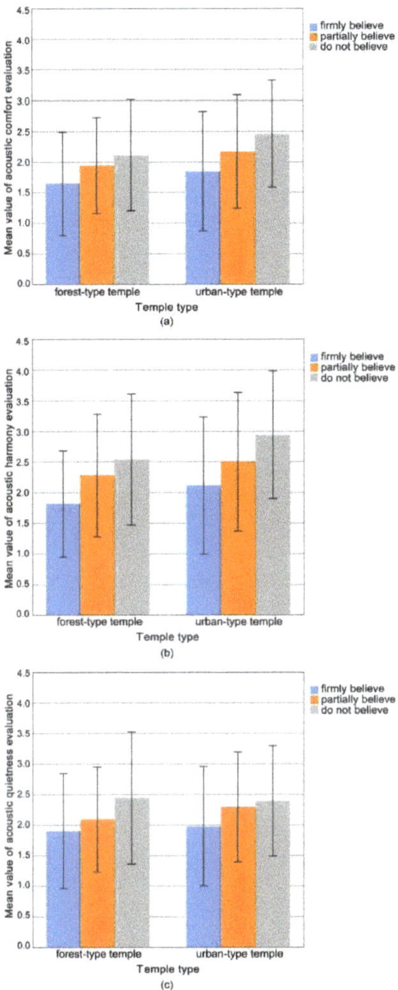

**Figure 4.** The evaluation value and standard deviation of acoustic environment by different respondents' attitude towards Buddhist thought: (**a**) comfort; (**b**) harmony; (**c**) quietness.

(3)  Purpose

Figure 5a presents the relations between the respondents' purpose and the mean value of acoustic comfort evaluation between the two types of temples (this item is given in Question 4 of the questionnaire in the Appendix A). Respondents who worshipped the Buddha in both types of temples believed that the acoustic environment of the temple was most comfortable. The correlation coefficient between the purpose of visiting the temple and acoustic comfort evaluation was $-0.200^*$ for forest-type temples and $-0.154^*$ for urban-type temples.

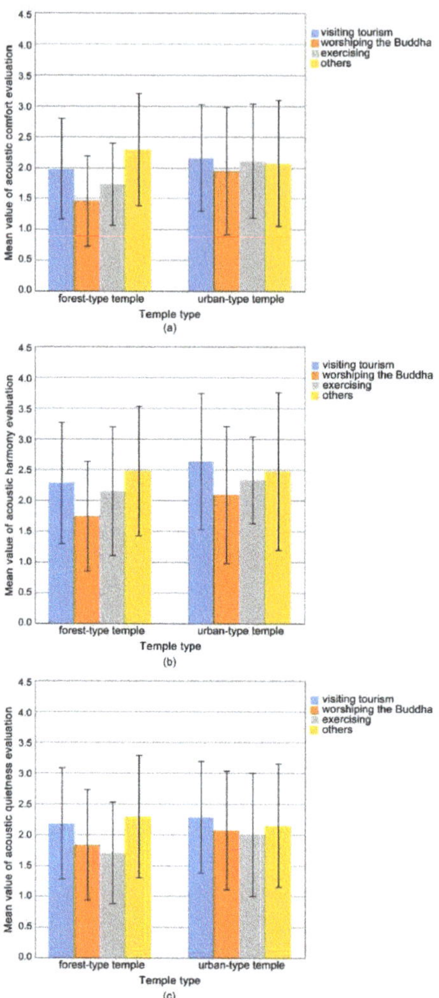

**Figure 5.** The evaluation value and standard deviation of acoustic environment by different respondents' purpose: (**a**) comfort; (**b**) harmony; (**c**) quietness.

The mean evaluation values of acoustic comfort by respondents who visited forest-type temples for tourism purposes or to worship the Buddha were lower than those for the same purpose in urban-type temples by 0.18 and 0.49, respectively. The result of the independent-samples *t*-test indicated that there were significant differences in the acoustic comfort evaluations between the two types of temples for tourists or worshipers; that is, respondents in forest-type temples for these two purposes felt that the acoustic environment was more comfortable. With respect to people exercising or other purposes, there were no significant differences in the evaluation of acoustic comfort between the two types of temples.

The relations between the respondents' purpose for visiting the temples and the mean value of acoustic harmony evaluation between the two types of temples are shown in Figure 5b. The correlation coefficient between the purpose of visiting the temples and the acoustic harmony evaluation was $-0.170*$ for forest-type temples and $-0.219**$ for urban-type temples. The mean values of acoustic harmony evaluation by respondents who visited forest-type temples for tourism purposes or to worship the Buddha were both lower than those for the same purpose in urban-type temples by 0.35 and 0.35, and the result of the *t*-test indicated that there were significant differences in the acoustic harmony evaluation values reported by people for tourism purposes or to worship the Buddha between the two types of temples. There were no significant differences in the evaluation of acoustic harmony between the two types of temples by people exercising or other purposes, similar to the findings concerning the evaluation of acoustic comfort.

Figure 5c shows that in the two types of temples, respondents who visited the temples to exercise reported that the acoustic environment of the temple was the quietest, followed by respondents who visited so to worship the Buddha. The correlation coefficient between their purpose in visiting the temple and the quietness evaluation value was $-0.184^*$ for forest-type temples and $-0.154^*$ for urban-type temples. The result of the *t*-test indicated that there were no significant differences between the two types of temples in the quietness evaluation of respondents who visited the temples for the same purpose. This finding is different from the results of the evaluation of acoustic comfort and harmony.

(4)  Frequency

We analysed the effect of frequency-related factors on the acoustic environment evaluation, as shown in Figure 6a. With an increase in the average number of annual visits to the temple, respondents tended to express more comfort in their evaluations of the acoustic environment. In general, this trend was especially evident in forest-type temples. The correlation coefficient between the average number of annual visits to the temple and the evaluation of acoustic comfort was $-0.267^{**}$ for forest-type temples and $-0.150^*$ for urban-type temples. The result of the *t*-test indicated that there were no significant differences in the evaluation of acoustic comfort between the two types of temples among people who visited the temples with the same frequency each year.

Figure 6b shows that respondents who visited the two types of temples 3 to 4 times per year evaluated the acoustic environment and religious atmosphere of the temples as the most inharmonious, while those who visited the temples more than 4 times a year thought that the temples' acoustic environment was the most harmonious. The correlation coefficient between frequency and acoustic harmony evaluation was $-0.160^*$ for forest-type temples and $-0.168^*$ for urban-type temples. The result of the *t*-test indicated that there were no significant differences in the evaluation of acoustic harmony between the two types of temples among people who visited the temples with the same frequency each year, which is similar to the findings concerning the evaluation of acoustic comfort.

Figure 6c presents the relationship between respondents' average number of visits to the temple per year and the mean value of the acoustic quietness evaluation between the two types of temples. The correlation coefficient was $-0.231^{**}$ for forest-type temples and $-0.025$ ($p = 0.769$) for urban-type temples, indicating that the average number of visits to the temple per year was correlated with the acoustic quietness evaluation only for forest-type temples. With an increase in the average number of visits to temples per year, respondents tended to report quiet in their evaluations of the acoustic environment of forest-type temples, while no such trend was observed in the context of urban-type temples. The results showed that among people who visited forest-type temples more than 4 times per year, the mean evaluation value of acoustic quietness was lower than that reported by those who visited urban-type temples by 0.31. The result of the *t*-test indicated that respondents who visited temples more than 4 times per year (and who are therefore more likely to be followers of Buddhism) exhibited significant differences in terms of their evaluation of acoustic quietness between the two types of temples. Respondents in forest-type temples felt that the environment was quieter. People who visited the temple with other frequencies exhibited no significant differences in quietness evaluations between the two types of temples. These findings are different from the findings concerning the evaluation of acoustic comfort and acoustic harmony.

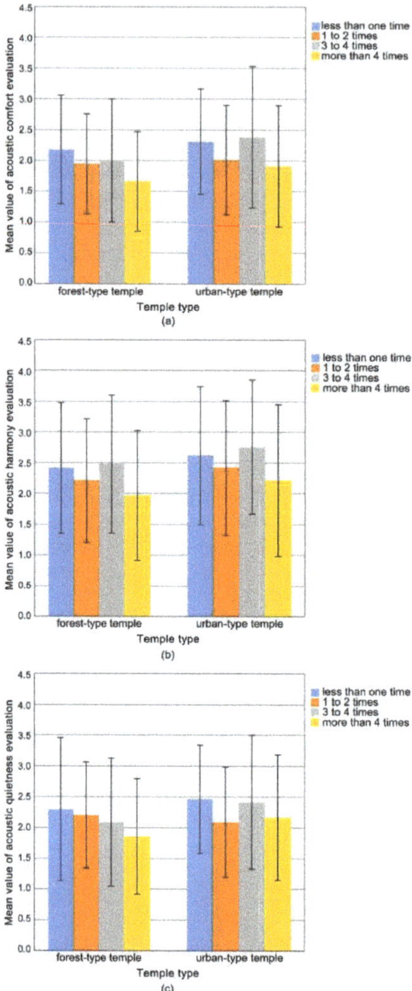

**Figure 6.** The evaluation value and standard deviation of acoustic environment by different respondents' average number of annual visits to the temple: (**a**) comfort; (**b**) harmony; (**c**) quietness.

(5)    Education and Occupation

Figure 7a presents the relations between the respondents' levels of education and the mean values of acoustic comfort evaluation regarding the two types of temples. The correlation coefficient between the level of education and the acoustic comfort evaluation was 0.341** for forest-type temples and 0.053 ($p = 0.506$) for urban-type temples, indicating that the level of education was correlated with acoustic comfort evaluation only in the context of forest-type temples. The mean evaluation values of acoustic comfort reported by people with primary school or middle school education were 1.00 and 0.42 lower for forest-type temples than for urban-type temples, respectively, and the results indicated that there were significant differences in the evaluation values of acoustic comfort reported by people with primary school education or middle school education between the two types of temples. For people with the same educational background, there was no significant difference between the two types of temples.

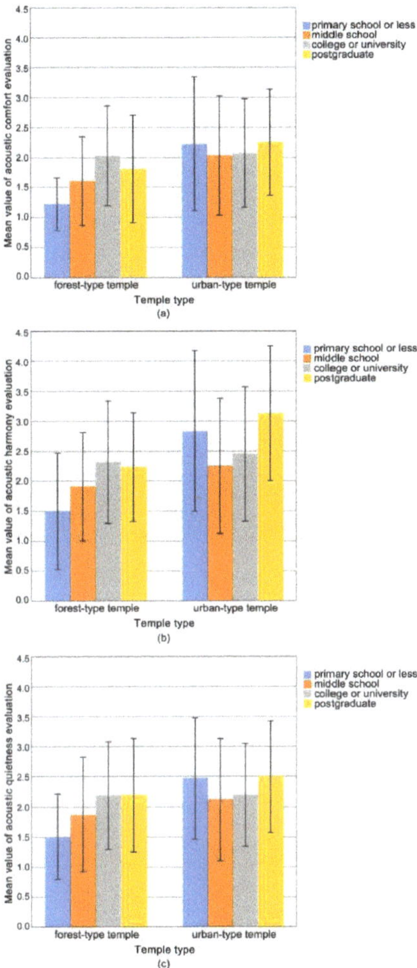

**Figure 7.** The evaluation value and standard deviation of the acoustic environment by different respondents' educational level: (**a**) comfort; (**b**) harmony; (**c**) quietness.

Figure 7b presents the relations between the respondents' levels of education and the mean value of acoustic harmony evaluation regarding the two types of temples. The correlation coefficient between these factors was 0.313** for forest-type temples and 0.105 ($p = 0.174$) for urban-type temples, indicating that the level of education was correlated with the acoustic harmony evaluation only in the context of forest-type temples. The mean evaluation values of acoustic harmony reported by people with primary school, middle school, and postgraduate education were lower for forest-type temples than for urban-type temples by 1.33, 0.34, and 0.90, respectively. The results of the $t$-test indicated that these three kinds of people with the same level of education exhibited significant differences in their evaluation of acoustic harmony between the two types of temples.

Figure 7c presents the relations between the respondents' level of education and the mean value of acoustic quietness evaluation regarding the two types of temples. The correlation coefficient was 0.267** for forest-type temples and 0.072 ($p = 0.356$) for urban-type temples, indicating that level of education was correlated with the quietness evaluation only in the context of forest-type temples, which is similar to the evaluation of acoustic comfort and acoustic harmony. The mean evaluation value of acoustic quietness reported by people with a primary school education was lower for forest-type temples than for urban-type temples by 0.97, and the result of the *t*-test indicated that only people with a primary school education exhibited significant differences in their evaluation of acoustic quietness between the two types of temples.

In addition, we analysed the influence of the respondents' occupations on their evaluation of the acoustic environment with regard to the two types of temples. To ensure the accuracy of the results, occupation types that referred to fewer than 15 people were removed from the survey numbers (the total number of remaining questionnaires was 660), and the results are shown in Figure 8a. The correlation coefficient between occupation and the acoustic comfort evaluation was 0.149** for forest-type temples and 0.132** for urban-type temples. A comparison of various occupations shows that the mean evaluation values of acoustic comfort reported by housewives and teachers were lower for forest-type temples than for urban-type temples by 1.02 and 0.55, respectively. The result of the *t*-test indicated that between the two types of temples, significant differences emerged in the evaluation values of acoustic comfort only in the case of housewives or teachers, while there were no significant differences for people who worked in other occupations.

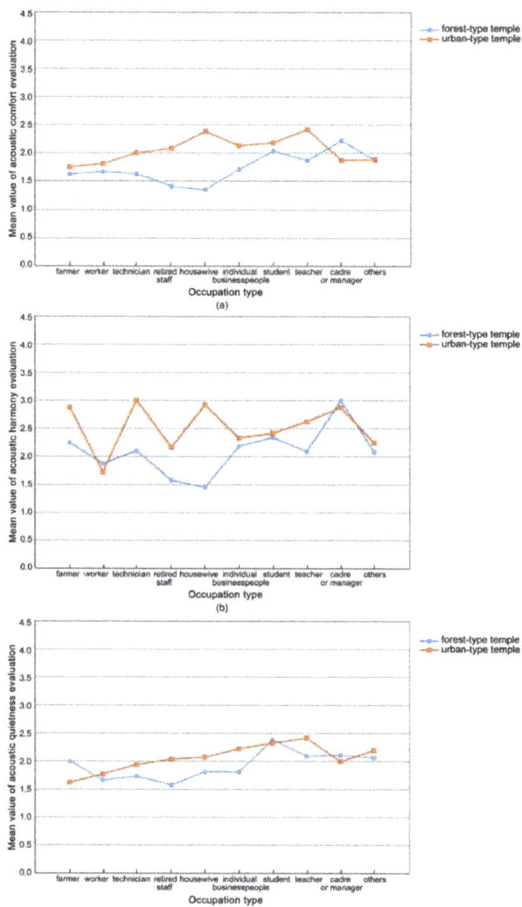

**Figure 8.** The evaluation value of the acoustic environment by different respondents' occupations: (**a**) comfort; (**b**) harmony; (**c**) quietness.

Figure 8b presents the relations between the respondents' occupations and the mean value of the acoustic harmony evaluation in the two types of temples. The correlation coefficient was 0.192** for forest-type temples and 0.143** for urban-type temples. A comparison of various occupations showed that the mean evaluation values of acoustic harmony reported by housewives and technicians were lower for forest-type temples than for urban-type temples by 1.47 and 0.89, respectively. The result of the *t*-test indicated significant differences in the evaluation of acoustic harmony between the two types of temples only in the case of housewives and technicians. There were no significant differences for people of other occupations.

Figure 8c presents the correlation between respondents' occupations and the mean value of the acoustic quietness evaluation between the two types of temples (this item is given in Question 3 of the questionnaire in the Appendix A). The correlation coefficient was 0.257** for forest-type temples and 0.218** for urban-type temples. The results of the *t*-test showed that there were no significant differences in the evaluation of acoustic quietness by people of the same occupation between the two types of temples.

### 3.2. Factors Influencing Sound Preference Evaluation

Various kinds of sounds occur in Han Chinese Buddhist temples, and respondents' preference evaluations of these different sounds might be affected by temple factors. The evaluation results concerning various sound preferences between the two types of temples are shown in Figure 9. With respect to the preference evaluation of natural sounds, Buddhism-related human-made sounds and Buddhism-unrelated human-made sounds, the *p* values of the independent-samples *t*-test between forest-type temples and urban-type temples were 0.000, 0.028 and 0.254, respectively. This indicated that there were significant differences in the evaluation of sound preference for natural sounds or Buddhism-related human-made sounds between the two types of temples, while there was no significant difference in the evaluation of Buddhism-unrelated human-made sounds. In general, the respondents in forest-type temples preferred natural sounds (1.40 for forest-type temples, 1.63 for urban-type temples), and respondents in urban-type temples preferred Buddhism-related human-made sounds (2.15 for forest-type temples, 2.00 for urban-type temples). For Buddhism-unrelated human-made sounds, the mean values of sound preference evaluation between the two temple types were similar (3.50 for forest-type temples, 3.55 for urban-type temples), indicating that people did not like Buddhism-unrelated human-made sounds in any environment and that this sound preference was unaffected by the factor of temple type.

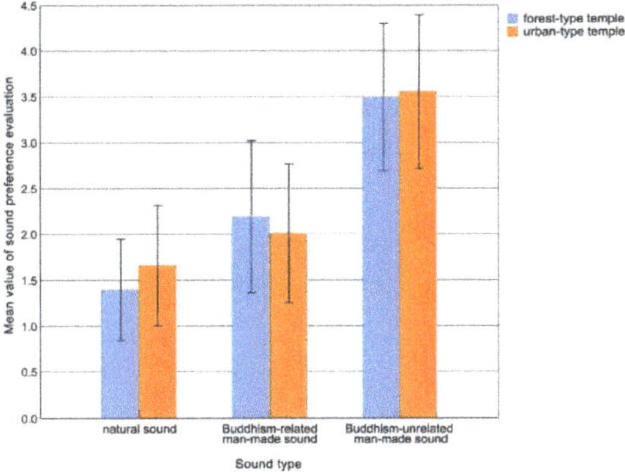


**Figure 9.** The evaluation value and standard deviation of preference for different sounds in the temple.

The preference for various sounds in the two types of temples was analysed. As shown in Figure 10, the mean values of the preference for the sounds of flowing water, birds and insects, and rustling leaves reported by respondents were lower for forest-type temples than for urban-type temples by 0.25, 0.21 and 0.27, respectively, while the mean values of the preferences for the sound of chanting and background music reported by respondents were lower for urban-type temples than for forest-type temples by 0.15 and 0.22, respectively. The results of the *t*-test indicated that there were significant differences in the evaluation values of respondents' preferences for these five kinds of sounds between the two types of temples. These results again showed that the environment of Buddhist temples had an impact on the evaluation of people's sound preferences. In forest-type temples, people preferred flowing water, birds and insects, and rustling leaves (all natural sounds), while in urban-type temples, people preferred chanting and background music, such as Buddhist odes (Buddhism-related human-made sounds).

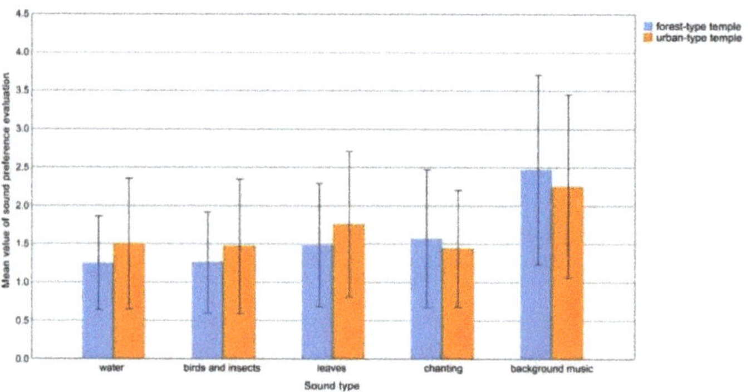


**Figure 10.** The evaluation value and standard deviation of sound preference for the five sounds with significant differences between the two types of temples.

Whether the preferences for the five sounds in the two types of temples were influenced by human sociological factors was analysed. In forest-type temples, the results showed that sound preferences for birds and insects and rustling leaves were affected by different educational levels (the correlation coefficients were −0.235* and −0.255**, respectively). The higher a person's education, the more he or she liked the sounds of birds and insects and rustling leaves. The preference for chanting in forest-type temples was affected by different ages, belief factors and purposes (−0.284**, 0.563**, 0.309**); the older people were, or the more they believed in Buddhist thought, the more they liked chanting. The preference for background music in forest-type temples was affected by belief factors (0.169*); the more people believed in Buddhist thought, the more they liked background music.

In urban-type temples, the preference for flowing water was affected by different purposes (the correlation coefficient was 0.327**). The preference for rustling leaves was affected by different ages (correlation coefficient was 0.159*); the older people were, the less they liked the sound of leaves. The preference for chanting was affected by different purposes, belief factors, and the frequency of visiting the temple (correlation coefficients were 0.321**, 0.489** and −0.311**, respectively); the more people believed in Buddhist thought or visited temples, the more they liked chanting. The preference for background music was affected by belief factors (correlation coefficient was 0.195**); the more people believed in Buddhist thought, the more they liked background music.

The above results show that in both urban-type and forest-type temples, the preference evaluation of Buddhism-related human-made sounds is obviously affected by belief factors, while the factors that affect the preference for natural sounds do not reflect a consistent rule (the preference for natural sounds in forest-type temples may be affected by different educational backgrounds, while in urban-type temples, it may be affected by different purposes and ages). In addition, the ANOVA results showed that among people with different belief factors, there were significant differences in their preference for background music or chanting, while there were no significant differences in their preference for other sounds. These results were consistent in urban and forest temples.

### 3.3. Discussion

#### 3.3.1. Comparison among Various Influencing Factors on Acoustic Environment Evaluation

(1) As shown in Figure 11, in forest-type temples, the three factors that had the most significant effect on the evaluation of acoustic comfort and harmony were attitudes towards Buddhist thought > education > age. The factors that had the most significant effect on the evaluation of acoustic quietness were education level > age > occupation. In urban-type temples, the three factors that had the most significant effect on the evaluation of acoustic comfort were attitudes towards Buddhist thought > frequency of visits to the temple per year > measured sound levels synchronous with the questionnaire; the factors for acoustic harmony were attitudes towards Buddhist thought > purpose > frequency of visits to the temple per year, and the factors for acoustic quietness were attitudes towards Buddhist thought > sound levels by synchronous measurement with the questionnaire > occupation. The above analysis shows that the most significant influencing factor of the evaluation of the acoustic environment was attitudes towards Buddhist thought in both forest-type and urban-type temples. In forest-type temples, the factors of education level and age were also important, while in urban-type temples, the frequency of visits to the temple per year and measured sound levels were also important.

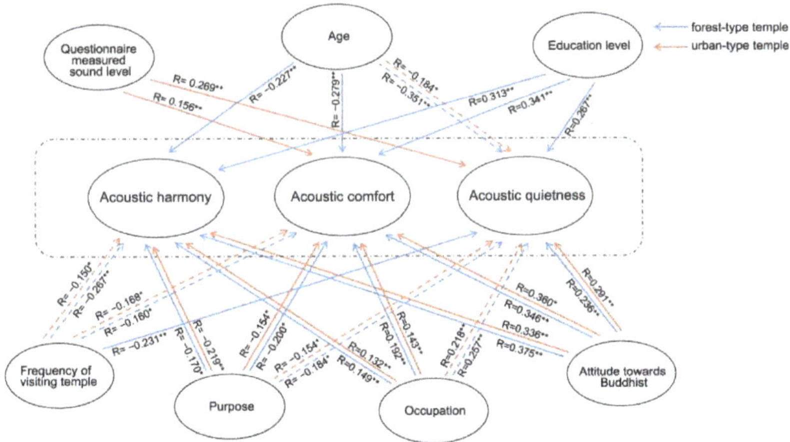


**Figure 11.** Schematic diagram of the influencing factors of three kinds of acoustic environment evaluations. Note: The double solid line indicates that certain people with the same characteristics have significant differences between the two types of temples, while the double dotted line indicates that people with the same characteristics have no significant differences between the two types of temples. In the figure, * and ** indicate significance level, * indicates $p < 0.05$, ** indicates $p < 0.01$.

Our research showed that in both types of temples, the influences of these factors usually differed between the three acoustic environment evaluations. This indicates that in research on special soundscapes such as religious buildings, it is not necessary to completely follow the standard assessment process of the surrounding sound environment recommended by ISO 12913, and different evaluation indices according to different architectural soundscape characteristics should be selected. These targeted soundscape evaluation indicators can enrich future editions of ISO 12913 (which should be under constant revision) by offering valuable approaches.

(2) In this study, the differences in the influencing factors on the evaluation of the acoustic environment between the two types of temples were analysed. The most significant factors were education level and attitudes towards Buddhist thought. For these two factors, there were significant differences in the three acoustic environment evaluations between the two types of temples. The measured sound levels were synchronous with the questionnaire, and there were significant differences in the evaluation of acoustic comfort and quietness between the two types of temples. Regarding the factors of purpose, occupation and age, there were significant differences in the evaluation of acoustic comfort and harmony between the two types of temples. The factor with the least difference was the frequency of visits to the temple per year; there was a significant difference only in the evaluation of acoustic quietness between the two types of temples.

### 3.3.2. Comparison with Previous Studies

Previous studies have not examined the difference in soundscapes between urban-type and forest-type religious sites. Considering that urban-type temples have the function of public activities while forest-type temples have the functions of leisure and relaxation, similar to parks, we compare the findings of this study with those of soundscapes in similar and single-place types. For example, the soundscape of forest-type temples was compared with the soundscape of parks, and the soundscape of urban-type temples was compared with the soundscape of urban public spaces.

(1) In terms of the relationship between the objective sound level and the acoustic environment evaluation, previous studies showed that when a pleasant sound such as music or water dominated the soundscape, the relationship between the acoustic comfort evaluation and the sound level was considerably weaker than that of other sound sources such as traffic and demolition sounds [4], and the overall noise annoyance had a significant relation with sound levels in an old town [6]. Our findings are consistent with these studies that showed that in urban-type temples (where the surrounding environment may be dominated by traffic sounds), there are significant correlations between sound levels and acoustic environment evaluations such as comfort and quietness, while in forest-type temples (where natural sounds usually dominate the soundscape), there is no correlation between sound levels and various kinds of acoustic environment evaluations.

(2) Among the subjective influencing factors of soundscape evaluation, in terms of age, Tarlao et al. found that older people rated the urban soundscape as less pleasant and less monotonous than younger people because they were more sensitive to noise [10], and the value of the acoustic satisfaction evaluation of retired or aged people was lower than that of other interviewees in urban historical areas [32]. Another study found the opposite: teenagers tended to be the most unsatisfied group, and older people were the most satisfied group in terms of acoustic comfort in urban open public spaces [4]. However, the research on urban temples in this paper showed that there was no relationship between age and the evaluation of acoustic comfort or harmony. This could indicate that the relationship between age and acoustic environment evaluation is complex in urban spaces and that there are no uniform results. In the condition of green spaces, a study conducted by Hedblom et al. showed that elderly individuals reported greater calmness when hearing bird songs and rustling leaves than younger and middle-aged individuals did [63]. This was similar to the results of the acoustic environment evaluation of forest-type temples in our paper; that is, with increasing age, the respondents in forest-type temples tended to evaluate the acoustic environment as more comfortable, quiet, and harmonious, indicating that older people have a higher tolerance to the acoustic environment than younger people in green space.

In terms of the relationship between soundscape evaluation and factors related to religious belief, Xie et al. showed that compared with laypeople, Taoist priests' acoustic comfort in a forest-type temple was more influenced by Taoist religious principles, emphasizing harmony between human beings and nature [45]. Similarly, our study indicated that people who completely believed in Buddhist thought felt that the acoustic environment in forest-type temples was quieter, more comfortable, and more harmonious with the religious atmosphere in the temples. As a result of studying urban-type Buddhist temples in Chengdu, Yi proposed that respondents who lacked religious response felt quiet or relatively quiet and comfortable or relatively comfortable and that the religious atmosphere was harmonious or relatively harmonious [64]. In contrast, we found that in urban-type temples, people who did not believe in Buddhist thought responded that the acoustic environment of temples was noisier or more uncomfortable and that the religious atmosphere was more disharmonious than people who firmly believed or partially believed in Buddhist thought. This difference may be because the two urban-type temples in our study were located in northern and central China, while the four temples investigated by Yi were located in southwestern China. Factors such as the personality of people in different regions (for example, different expectations for the acoustic environment of temples) might affect the soundscape evaluation of temples. Therefore, the influence of religious belief on the soundscape evaluation of Buddhist temples requires further analysis.

Regarding the relationship between the purpose of visiting temples and soundscape evaluation, Tse et al. found that with regard to the perception of park soundscapes, the motivation for visiting a park did not affect individuals' acoustic comfort evaluation [21]. However, some studies on the soundscapes of parks have suggested that different visit motivations could result in different soundscape experiences [18], and visit motivations have the most significant relationships with the perceived occurrences of individual sounds [20]. The last two studies were consistent with the findings of our research that people's purposes could affect their perceptions and evaluations of soundscapes in forest-type temples because the correlation coefficient between the evaluation of acoustic comfort and the purpose was $-0.200^*$ for forest-type temples. Respondents who came to worship the Buddha believed that the acoustic environment of the temple was more comfortable.

Regarding the frequency factor, previous studies have shown that people who came to the park more frequently showed more preference for natural sounds, especially wind blowing, tree rustling, raining, and water sounds [18], and visit frequency and length of stay were most frequently associated with the perception of individual sounds [20]. However, the research results of this paper did not show that the frequency of visits to the temple had an effect on sound preference. This may be because people who go to parks often may go for exercise, while people who go to temples often are more likely to be Buddhists. In terms of education level, a previous study revealed that educational background showed no significant relationship with the degree of harmony of any sound in the park [18]. However, another study on parks' soundscapes showed that higher education indicated lower tolerance towards sounds [28]. The results of our study supported the latter view; in forest-type temples, the higher the education level, the worse the evaluation of acoustic comfort, harmony and quietness (correlation coefficients of $0.341^{**}$, $0.313^{**}$, and $0.267^{**}$). Interestingly, there was no significant relationship between education level and the three kinds of acoustic environment evaluations in urban-type temples. We speculate that the possible reason is that people with higher education perceive the acoustic environment in urban-type temples to be affected by noisy surroundings and therefore have lower expectations of the acoustic environment.

In terms of occupation factors, several previous studies have shown that whether in green space or urban open space, occupation has no significant relationship with sound preference [20,40,41]. The findings of our research were similar to these studies; in both urban-type and forest-type temples, occupation had only a slight influence on sound preference.

(3) With regard to the sound preference for urban open space, a previous study showed that the sound of bells and the songs of birds were identified by passers-by as two of the most annoying sound sources in an old town [6]. However, the results of other studies differed: church bells were one of most preferred sounds in urban open public spaces [40], and one social survey of urban religious spaces that examined a cathedral and a Buddhist temple in Seoul suggested that bell sounds and sounds from religious ceremonies in the cathedral and temple were considered more important than other sounds, and human

sounds from religious activities, such as Buddhist chants, were perceived as pleasant by visitors [48]. Similar to the latter two studies, our research also found that Buddhism-related human-made sounds were preferred in urban-type temples.

For forest parks or green spaces, many studies on soundscapes have found similar results. In Jingci Temple (one Chinese forest-type temple in a famous West Lake scenic area), natural sounds were most widely praised [43]. In parks, the majority of visitors had a high preference for natural sounds [33], and birds, insects, and water sounds were the most preferred [16,20,28]. Natural sounds were used positively to reduce unpleasant feelings [19], and respondents' tolerance of natural sounds was high relative to that of traffic noise in nature parks [65]. Furthermore, natural sounds were associated with a much higher degree of favourability in natural and ecological areas than in modern commercial areas [66]. These results are consistent with our study; natural sounds were preferred in forest-type temples more than in urban-type temples.

## 4. Conclusions

The current study conducted comparative research to investigate subjective soundscape evaluations between typical forest-type and urban-type Han Chinese Buddhist temples. Based on an analysis of 685 valid soundscape questionnaires, it was concluded that there were differences in people's evaluation of the acoustic environment (including quietness, comfort and harmony) and sound preference between forest-type and urban-type temples. Objective factors, such as the measured sound levels synchronous with the questionnaire, and subjective factors, such as the respondents' sociological factors (including age, belief, the purpose of visiting the temple, the frequency of visits to the temple per year, education level, and occupation), had significant relationships with the evaluation of the acoustic environment of the temples, but these factors played different roles in the evaluation of the two types of temples, and some factors may not have contributed to the evaluation of the acoustic environment in certain types of temples. The respondents in the two types of temples had different sound preferences: respondents in forest-type temples preferred natural sounds, while respondents in urban-type temples preferred Buddhism-related human-made sounds. Overall, given the results of our quantitative questionnaire analysis, this study identifies the specific differences in the soundscape evaluation of the two types of temples so that the soundscapes of different types of temples can be designed to meet the needs of different people. These findings could be instructive in soundscape planning and designing processes in forest-type or urban-type temples. In the research process, the definition and data collection methods in ISO 12913 on soundscapes were referenced, and this paper also took into account the actual situation of soundscapes in religious buildings and referred to the research of other scholars [45,48]. These attempts might play a certain role in the development and subsequent revision of ISO 12913. In future work, more diversified and thorough research with the help of VR, eye-tracking technology, and physiological response monitoring should be adopted with the aim of identifying measures to effectively improve the acoustic environment of different types of temples and create a soundscape that is beneficial to people's mental health.

## 5. Research Limitations

To some extent, the aforementioned results elucidate the subjective soundscape evaluations of respondents with regard to forest-type and urban-type temples. The limitation of this study is that the survey was distributed in only four temples located in northern and central China, and more questionnaires and measurements focusing on more representative temples with a larger geographical area may be necessary. In addition, this study used univariate analysis in the correlation analysis without considering the interaction between the independent variables, that is, whether there was an influence of mediating variables or moderating variables. Only questionnaires and measured sound levels were used in the research without audio-visual interaction or observation of human physiological indicators.

Despite these limitations, the results of this study might provide effective and concrete guidelines for better acoustic environments in two types of Buddhist temple.

**Author Contributions:** Conceptualization, D.Z. and X.L.; methodology, D.Z.; software, D.Z. and X.L.; validation, D.Z. and X.L.; formal analysis, D.Z. and X.L.; investigation, D.Z. and W.M.; resources, D.Z. and W.M.; data curation, D.Z. and X.L.; writing—original draft preparation, X.L.; writing—review and editing, D.Z. and X.L. All authors have read and agreed to the published version of the manuscript.

**Funding:** This research was funded by the Program of Humanities and Social Science Research Program of Ministry of Education of China (22YJA760102) and Scientific Research Fund of Guangzhou University (PT252022006).

**Data Availability Statement:** The original contributions presented in the study are included in the article, and further inquiries can be directed to the corresponding authors.

**Acknowledgments:** The authors would like to express their thanks to all the respondents who volunteered to participate in the research.

**Conflicts of Interest:** The authors declare no conflict of interest.

## Appendix A

Questionnaire on the acoustic environment of Buddhist temples

Hello. This is a survey of the sound environment in Buddhist temples. This questionnaire is anonymous, and the results are only for this study, which will help to improve the quality of the sound environment in temples. Please fill out the questionnaire based on your feelings about the sound you heard inside and outside the temple. All questions are single-choice except where noted. Thank you for your support!

Time _________ Location (temple) _________ Weather ______ Temperature _________
Measured A sound level_______

---

**Gender** _________ (please mark √ under the selected items)
A. Male B. Female

**Age** _______
A. Younger than 18 years old B. 18 to 30 years old C. 30 to 45 years old
D. 45 to 60 years old E. Older than 60 years old

**Educational level** _________
A. Primary school B. Middle school C. College or university D. Postgraduate

**Occupation**: Farmer, worker, soldier, service personnel, technician, teacher, cadre, student, self-employed individual, managers, retired, unemployed, housewives, others ( … )

**The average times** that you would visit temple visits per year are____
A. Less than one time B. 1–2 times C. 3–4 times D. more than 4 times

---

1 You are here for__________ (mark √ under the selected items)
A. Visiting tourism B. Worshiping the Buddha C. Exercising D. others __________
2 What is your attitude towards Buddhist thought?
A. Firmly believe B. Partially believe C. Do not believe
3 What is your evaluation of the sound environment in your current location?
Quiet Somewhat quiet Neither quiet nor noisy Somewhat noisy Noisy
□ □ □ □ □
1 point 2 points 3 points 4 points 5 points
4 How do you feel about the sound environment among your current environment?
Comfortable Somewhat comfortable No feeling Somewhat uncomfortable Uncomfortable
□ □ □ □ □
1 points 2 points 3 points 4 points 5 points

5 Do you think the voices heard inside and outside the temple are in harmony with the overall religious atmosphere of the temple?

Harmonious Somewhat harmonious No feeling Somewhat inharmonious Inharmonious
□ □ □ □ □
1 points 2 points 3 points 4 points 5 points

6 In the temple, if you hear the following sounds, which ones do you like and which ones do you dislike? (please draw √)

| Sound heard | like | somewhat like | no feeling | somewhat dislike | dislike |
|---|---|---|---|---|---|
| | 1 points | 2 points | 3 points | 4 points | 5 points |
| a. Bell | | | | | |
| b. Tourist conversation | | | | | |
| c. Construction site noise | | | | | |
| d. Flowing water | | | | | |
| e. Tour-guide voice | | | | | |
| f. Wind | | | | | |
| g. Background music | | | | | |
| h. Traffic sound | | | | | |
| j. Drum | | | | | |
| k. Fountain | | | | | |
| l. Birds and insects | | | | | |
| m. Footsteps | | | | | |
| n. Rustling leaves | | | | | |
| o. Instrument | | | | | |
| p. Tourists' prayers | | | | | |
| q. Monks chanting | | | | | |

7 What advice do you have for sound environmrent improvement in Buddhist temples? _______________________________________________

_______________________________________________

_______________________________________________

## References


1. *ISO 12913–1:2014*; Acoustics-Soundscape—Part 1: Defifinition and Conceptual Framework. International Organization for Standardization (ISO): Geneva, Switzerland, 2014.
2. Schafer, R.M. *The Soundscape: Our Sonic Environment and The Tuning of The World*; Destiny Books: Rochester, NY, USA, 1993.
3. Porteous, J.D.; Mastin, J.F. Soundscape. *J. Archit. Plan. Res.* **1985**, *2*, 169–186.
4. Yang, W.; Kang, J. Acoustic comfort evaluation in urban open public spaces. *Appl. Acoust.* **2005**, *66*, 211–229. [CrossRef]
5. Zhang, Y.; Kang, J.; Kang, J. Effects of Soundscape on the Environmental Restoration in Urban Natural Environments. *Noise Health* **2017**, *19*, 65–72.
6. Gozalo, G.R.; Morillas, J.M.B. Perceptions and effects of the acoustic environment in quiet residential areas. *J. Acoust. Soc. Am.* **2017**, *141*, 2418–2429. [CrossRef]
7. Liu, J.; Yang, L.; Xiong, Y.; Yang, Y. Effects of soundscape perception on visiting experience in a renovated historical block. *Build. Environ.* **2019**, *165*, 106375. [CrossRef]
8. Garrioch, D. Sounds of the city: The soundscape of early modern European towns. *Urban Hist.* **2003**, *30*, 5–25. [CrossRef]
9. Zhang, D.; Kong, C.; Zhang, M.; Kang, J. Religious Belief-Related Factors Enhance the Impact of Soundscapes in Han Chinese Buddhist Temples on Mental Health. *Front. Psychol.* **2021**, *12*, 774689. [CrossRef]
10. Tarlao, C.; Steffens, J.; Guastavino, C. Investigating contextual influences on urban soundscape evaluations with structural equation modelling. *Build. Environ.* **2021**, *188*, 107490. [CrossRef]

11. Liu, F.; Kang, J. A grounded theory approach to the subjective understanding of urban soundscape in Sheffield. *Cities* **2016**, *50*, 28–39. [CrossRef]
12. Hwang, I.; Hong, J.; Jeon, J.Y. Investigation of tranquility in urban religious places. *J. Acoust. Soc. Am.* **2012**, *132*, 1926. [CrossRef]
13. Axelsson, O.; Nilsson, M.E.; Berglund, B. A principal components model of soundscape perception. *J. Acoust. Soc. Am.* **2010**, *128*, 2836–2846. [CrossRef] [PubMed]
14. Lu, Y.; Hasegawa, Y.; Tan, J.K.A.; Lau, S.-K. Effect of audio-visual interaction on soundscape in the urban residential context: A virtual reality experiment. *Appl. Acoust.* **2022**, *192*, 108717. [CrossRef]
15. Yu, L.; Kang, J. Modelling subjective evaluation of soundscape quality in urban open spaces: An artificial neural network approach. *J. Acoust. Soc. Am.* **2009**, *126*, 1163–1174. [CrossRef]
16. Wang, P.; Zhang, C.; Xie, H.; Yang, W.; He, Y. Perception of National Park Soundscape and Its Effects on Visual Aesthetics. *Int. J. Environ. Res. Public Health* **2022**, *19*, 5721. [CrossRef] [PubMed]
17. Payne, S.R.; Bruce, N. Exploring the Relationship between Urban Quiet Areas and Perceived Restorative Benefits. *Int. J. Environ. Res. Public Health* **2019**, *16*, 1611. [CrossRef] [PubMed]
18. Liu, J.; Xiong, Y.; Wang, Y.; Luo, T. Soundscape effects on visiting experience in city park: A case study in Fuzhou, China. *Urban For. Urban Green.* **2018**, *31*, 38–47. [CrossRef]
19. Jo, H.I.; Jeon, J.Y. The influence of human behavioural characteristics on soundscape perception in urban parks: Subjective and observational approaches. *Landsc. Urban Plan.* **2020**, *203*, 103890. [CrossRef]
20. Liu, J.; Wang, Y.; Zimmer, C.; Kang, J.; Yu, T. Factors associated with soundscape experiences in urban green spaces: A case study in Rostock, Germany. *Urban For. Urban Green.* **2019**, *37*, 135–146. [CrossRef]
21. Tse, M.S.; Chau, C.K.; Choy, Y.S.; Tsui, W.K.; Chan, C.N.; Tang, S.K. Perception of urban park soundscape. *J. Acoust. Soc. Am.* **2012**, *131*, 2762–2771. [CrossRef]
22. Filipan, K.; Boes, M.; De Coensel, B.; Lavandier, C.; Delaitre, P.; Domitrović, H.; Botteldooren, D. The Personal Viewpoint on the Meaning of Tranquility Affects the Appraisal of the Urban Park Soundscape. *Appl. Sci.* **2017**, *7*, 91. [CrossRef]
23. Jeon, J.Y.; Hong, J.Y. Classification of urban park soundscapes through perceptions of the acoustical environments. *Landsc. Urban Plan.* **2015**, *141*, 100–111. [CrossRef]
24. De Pessemier, T.; Van Renterghem, T.; Vanhecke, K.; All, A.; Filipan, K.; Sun, K.; De Coensel, B.; De Marez, L.; Martens, L.; Botteldooren, D.; et al. Enhancing the park experience by giving visitors control over the park's soundscape. *J. Ambient Intell. Smart Environ.* **2022**, *14*, 99–118. [CrossRef]
25. Liu, Y.; Hu, M.; Zhao, B. Audio-visual interactive evaluation of the forestlandscape based on eye-tracking experiments. *Urban For. Urban Green.* **2019**, *46*, 126476. [CrossRef]
26. Zhao, W.; Li, H.Y.; Zhu, X.; Ge, T.J. Effect of Birdsong Soundscape on Perceived Restorativeness in an Urban Park. *Int. J. Environ. Res. Public Health* **2020**, *17*, 5659. [CrossRef]
27. Uebel, K.; Marselle, M.; Dean, A.J.; Rhodes, J.R.; Bonn, A. Urban green space soundscapes and their perceived restorativeness. *People Nat.* **2021**, *3*, 756–769. [CrossRef]
28. Fang, X.; Gao, T.; Hedblom, M.; Xu, N.; Xiang, Y.; Hu, M.; Chen, Y.; Qiu, L. Soundscape Perceptions and Preferences for Different Groups of Users in Urban Recreational Forest Parks. *Forests* **2021**, *12*, 468. [CrossRef]
29. Jeon, J.Y.; Hong, J.Y.; Lavandier, C.; Lafon, J.; Axelsson, Ö.; Hurtig, M. A cross-national comparison in assessment of urban park soundscapes in France, Korea, and Sweden through laboratory experiments. *Appl. Acoust.* **2018**, *133*, 107–117. [CrossRef]
30. Hall, D.A.; Irwin, A.; Edmondson-Jones, M.; Phillips, S.; Poxon, J.E.W. An exploratory evaluation of perceptual, psychoacoustic and acoustical properties of urban soundscapes. *Appl. Acoust.* **2013**, *74*, 248–254. [CrossRef]
31. Meng, Q.; Kang, J. Effect of sound-related activities on human behaviours and acoustic comfort in urban open spaces. *Sci. Total Environ.* **2016**, *573*, 481–493. [CrossRef]
32. Zhou, Z.Y.; Kang, J.; Jin, H. Factors that influence soundscapes in historical areas. *Noise Control. Eng. J.* **2014**, *62*, 60–68. [CrossRef]
33. Ma, K.W.; Mak, C.M.; Wong, H.M. Effects of environmental sound quality on soundscape preference in a public urban space. *Appl. Acoust.* **2021**, *171*, 107570. [CrossRef]
34. Zhang, X.; Ba, M.; Kang, J.; Meng, Q. Effect of soundscape dimensions on acoustic comfort in urban open public spaces. *Appl. Acoust.* **2018**, *133*, 73–81. [CrossRef]
35. Jo, H.I.; Jeon, J.Y. Effect of the appropriateness of sound environment on urban soundscape assessment. *Build. Environ.* **2020**, *179*, 106975. [CrossRef]
36. Liu, J.; Kang, J.; Behm, H. Birdsong As an Element of the Urban Sound Environment: A Case Study Concerning the Area of Warnemünde in Germany. *Acta Acust. United Acust.* **2014**, *100*, 458–466. [CrossRef]
37. Hong, J.Y.; Jeon, J.Y. Designing sound and visual components for enhancement of urban soundscapes. *J. Acoust. Soc. Am.* **2013**, *134*, 2026–2036. [CrossRef]
38. Jeon, J.Y.; Lee, P.J.; You, J.; Kang, J. Perceptual assessment of quality of urban soundscapes with combined noise sources and water sounds. *J. Acoust. Soc. Am.* **2010**, *127*, 1357–1366. [CrossRef]
39. Ghalehnoee, M.; Ghaffari, A.; Haghighi, N.M. Soundscape quality assessment in Naghshe Jahan square. *Int. J. Architect. Eng. Urban Plan* **2018**, *28*, 215–225.
40. Kang, J.; Zhang, M. Semantic differential analysis of the soundscape in urban open public spaces. *Build. Environ.* **2010**, *45*, 150–157. [CrossRef]

41. Yu, L.; Kang, J. Factors influencing the sound preference in urban open spaces. *Appl. Acoust.* **2010**, *71*, 622–633. [CrossRef]
42. Krzywicka, P.; Byrka, K. Restorative Qualities of and Preference for Natural and Urban Soundscapes. *Front. Psychol* **2017**, *8*, 1705. [CrossRef]
43. Ge, J.; Guo, M.; Yue, M. Soundscape of the West Lake Scenic Area with profound cultural background—A case study of Evening Bell Ringing in Jingci Temple, China. *J. Zhejiang Univ. Sci. A* **2013**, *14*, 219–229. [CrossRef]
44. Zhang, D.; Zhang, M.; Liu, D.; Kang, J. Sounds and sound preferences in Han Buddhist temples. *Build. Environ.* **2018**, *142*, 58–69. [CrossRef]
45. Xie, H.; Peng, Z.; Kang, J.; Liu, C.; Wu, H. Soundscape Evaluation Outside a Taoist Temple: A Case Study of Laojundong Temple in Chongqing, China. *Int. J. Environ. Res. Public Health* **2022**, *19*, 4571. [CrossRef] [PubMed]
46. Brooks, J.; Kaufman, L.; Tyler-Jones, M. Soundscape Case Study: A Tudor Soundscape in a Historic House Chapel—Origins, Challenges and Evaluation. *Curator Mus. J.* **2019**, *62*, 439–451. [CrossRef]
47. Kang, J. Acoustic quality in nonacoustic public buildings. *Tech. Acoust.* **2006**, *25*, 513–522.
48. Jeon, J.Y.; Hwang, I.H.; Hong, J.Y. Soundscape evaluation in a Catholic cathedral and Buddhist temple precincts through social surveys and soundwalks. *J. Acoust. Soc. Am.* **2014**, *135*, 1863–1874. [CrossRef]
49. Kaymaz, E.; Sezer, F.S. Evaluation of the Environmental Comfort Conditions in Historical and New Mosques: A User Perception Survey. In Proceedings of the 4th International Conference the Importance of Place Cities and Cultural, Sarajevo, Bosnia and Herzegovin, 17 October 2017.
50. Othman, A.R.; Harith, C.M.; Ibrahim, N.; Ahmad, S.S. The Importance of Acoustic Design in the Mosques towards the Worshipers' Comfort. *Procedia—Soc. Behav. Sci.* **2016**, *234*, 45–54. [CrossRef]
51. Yilmazer, S.; Acun, V. A grounded theory approach to assess indoor soundscape in historic religious spaces of Anatolian culture: A case study on Hacı Bayram Mosque. *Build. Acoust.* **2018**, *25*, 137–150. [CrossRef]
52. Yuan, M. Studies on Chinese Contemporary Buddhist Architecture in the Han Area. Ph.D. Thesis, Tsinghua University, Beijing, China, 2008. (In Chinese)
53. Marin, L.D.; Newman, P.; Manning, R.; Vaske, J.J.; Stack, D. Motivation and Acceptability Norms of Human-Caused Sound in Muir Woods National Monument. *Leis. Sci.* **2011**, *33*, 147–161. [CrossRef]
54. Brown, A.L.; Kang, J.; Gjestland, T. Towards standardization in soundscape preference assessment. *Appl. Acoust.* **2011**, *72*, 387–392. [CrossRef]
55. Kull, R.C. Natural and urban soundscapes: The need for a multidisciplinary approach. *Acta Acust. United Acust.* **2006**, *92*, 898–902.
56. Brown, A.L. A review of progress in soundscapes and an approach to soundscape planning. *Int. J. Acoust. Vib.* **2012**, *17*, 2302. [CrossRef]
57. Johns, R. Likert items and scales. *Surv. Quest. Bank: Methods Fact Sheet* **2010**, *1*, 11–28.
58. Zahorik, P. Assessing auditory distance perception using virtual acoustics. *J. Acoust. Soc. Am.* **2002**, *111*, 1832–1846. [CrossRef]
59. Wu, M. *Questionnaire Statistical Analysis Practice–SPSS Operation and Application*; Chongqing University Press: Chongqing, China, 2010. (In Chinese)
60. Du, Z. *Sampling Survey and SPSS Applications*; Publishing House of Electronics Industry: Beijing, China, 2010. (In Chinese)
61. *ISO 12913-2:2018*; Acoustics-Soundscape—Part 2: Data Collection and Reporting Requirements. International Organization for Standardization (ISO): Geneva, Switzerland, 2018.
62. *ISO 12913-3:2019*; Acoustics-Soundscape—Part 3: Data Analysis. International Organization for Standardization (ISO): Geneva, Switzerland, 2019.
63. Hedblom, M.; Knez, I.; Sang, A.O.; Gunnarsson, B. Evaluation of natural sounds in urban greenery: Potential impact for urban nature preservation. *R. Soc. Open Sci.* **2017**, *4*, 170037. [CrossRef]
64. Yi, H.H. Soundscape Evaluation of Buddhist Temples in Chengdu Based on Hesitant Fuzzy Linguistic Analytic Hierarchical Process. Master's Thesis, Inner Mongolia University of Science and Technology, Baotou, China, 2021. (In Chinese)
65. Liu, F.; Kang, J.; Meng, Q. On the Influence Factors of Audio-Visual Comfort of Mountain Landscape Based on Field Surve. *J. Environ. Eng. Landsc. Manag.* **2020**, *28*, 48–61. [CrossRef]
66. Zhao, Z.; Wang, Y.; Hou, Y. Residents' Spatial Perceptions of Urban Gardens Based on Soundscape and Landscape Differences. *Sustainability* **2020**, *12*, 6809. [CrossRef]

*Article*

# Predicting and Visualizing Human Soundscape Perception in Large-Scale Urban Green Spaces: A Case Study of the Chengdu Outer Ring Ecological Zone

Yuting Yin [1], Yuhan Shao [1,*], Huilin Lu [1], Yiying Hao [2] and Like Jiang [3]

1    College of Architecture and Urban Planning, Tongji University, Shanghai 200092, China; yyin0326@tongji.edu.cn (Y.Y.)
2    Bureau Veritas, Atlantic House, Manchester M22 5PR, UK
3    Institute for Transport Studies, University of Leeds, Leeds LS2 9JT, UK
*    Correspondence: shaoyuhan@tongji.edu.cn

**Abstract:** Human soundscape perceptions exist through the perceived environment rather than the physical environment itself and are determined not only by the acoustic environment but also by the visual environment and their interaction. However, these relationships have mainly been established at the individual level, which may impede the efficient delivery of human-oriented considerations in improving the quality of large-scale urban spaces. Using the Chengdu Outer Ring Ecological Zone as an example, this study aims to develop an approach to predict human perceptions in large-scale urban green spaces. The site's visual attributes, i.e., landscape composition, were calculated using space syntax and the quantum geographic information system (QGIS); its aural attributes, i.e., the sound level, were measured on site using a multi-channel signal analyzer; and its functional attributes, i.e., vitality, were documented through on-site observations and mapping. This was performed whilst obtaining people's perceived soundscape through sound walks and a questionnaire-based on-site survey. The above environmental information was collected at micro-scale measurement spots selected within the site and then used together to formulate a model for predicting people's soundscape perceptions in the whole site. The prediction results suggested that people's perceived soundscape satisfaction increased as the distance from the ring road increased, and it gradually reached its highest level in the green spaces stretched outside the ring road. The prediction results of soundscape perception were then visualized using QGIS to develop planning and design implications, along with maps describing the site's visual, aural, and functional features. Planning and design implications were suggested, including setting green buffers between noise sources and vulnerable areas; identifying and preserving areas with special visual and acoustic characteristics; employing sound shields around traffic facilities; and using natural landscapes to distract people's attention from noise and to block their view of the source of noise. This study innovatively predicts individual-scale soundscape perception in large-scale UGSs based on environmental visual, aural, and functional characteristics through cross-level measurements, analyses, and model construction. By introducing a systematic perspective, the outcome of this study makes people's soundscape perceptions more applicable in the planning and design practices of large-scale urban settings.

**Keywords:** soundscape; human perception; prediction model; visual–aural attributes; urban green spaces

**Citation:** Yin, Y.; Shao, Y.; Lu, H.; Hao, Y.; Jiang, L. Predicting and Visualizing Human Soundscape Perception in Large-Scale Urban Green Spaces: A Case Study of the Chengdu Outer Ring Ecological Zone. *Forests* **2023**, *14*, 1946. https://doi.org/10.3390/f14101946

Academic Editor: Zhibin Ren

Received: 31 August 2023
Revised: 21 September 2023
Accepted: 22 September 2023
Published: 25 September 2023

## 1. Introduction

Soundscape is regarded as an important indicator of environmental quality and is essential in constituting the landscape of urban green spaces (UGSs). Differentiated from 'sound' and the 'acoustic/aural environment', soundscape emphasizes the perceptual construct of sounds and is defined as "the acoustic environment as perceived or experienced and/or understood by a person or people, in context" [1]. This definition clarifies that the soundscapes of UGSs exist through human perception rather than the physical environment itself and, thus, are determined not only by the acoustic environment but also by the visual environment and their interaction. However, its essence as an individual perception limits its practical application in providing efficient planning and design instructions for large-scale urban spaces. This study intends to develop an approach to predict people's soundscape perceptions in large-scale UGSs through cross-level measurements, analyses, and model construction, thus providing planning and design implications for improving the perceptual quality of large-scale urban spaces.

The perception of soundscape is formulated through a systematic process [2] and presented in multiple facets, among which perceived affective quality and physical acoustic information are the two major aspects. Perceived affective quality of soundscape was firstly proposed by Axelsson, Nilsson, and Berglund [3] as a two-dimensional model describing human perceived soundscape including four bipolar factors: pleasantness, eventfulness, calmness, and excitement. Many attempts have been made since to broaden the two-dimensional model. Excitement was replaced with vibrancy in a similar model with calmness as two underlying fundamental factors [4], while Aletta et al. [5] proposed that perceived affective quality could be measured by a three-dimensional model comprising pleasantness, eventfulness, and familiarity. A soundscape description tool including two axes, cacophony-hubbub and constant-temporal, which were developed to respectively present the numbers of sound sources and the levels of discordancy and variation within the soundscape perceived by people [6]. In addition to replacement, these basic factors have also been extended to incorporate more soundscape attributes. Axelsson, Nilsson, and Berglund enriched their two-dimensional model with eight unidirectional indicators: calm, pleasant, exciting, eventful, chaotic, annoying, monotonous, and uneventful [3]. The set of indicators was later included in the ISO 12913 [7] to unify soundscape perception descriptors and instruct soundscape evaluation. These perceived aural attributes of soundscapes are commonly evaluated using qualitative methods such as soundwalks, questionnaires, semantic analysis, and interviews [8]. Cao and Kang [9] used interviews and semantic analysis to explore how people perceived sound classification, sound appraisal (including sound features and psychological reactions), and soundscape preference in urban public spaces. Questionnaire-based surveys were employed in a study to evaluate the urban forest soundscape according to the eight perceptual attribute quality indicators outlined in ISO 12913 [7]. For similar studies, strict controls on experimental design and subject selection are required to ensure the accuracy of the obtained results, as these measurements are strongly influenced by individual perceptive differences [10] and the scope of settings [11].

In addition to perceptive attributes, physical parameters in relation to the acoustic environment, such as noise annoyance [12], sound level [13], and the composition and significance of sound sources [14], were also confirmed with effects on soundscape perceived by the individual. Noise annoyance was increased when the sound was regarded to have a negative impact on society [15] and was measured by loudness, sharpness, fluctuation strength, and impulsiveness in early research. After quantifiable acoustic instruments were invented, parameters (i.e., LAeq, LCeq, and $LA_{10}$) that could be recorded and analyzed to indicate sound pressure levels, fluctuation difference, and dB change at low frequencies were widely applied in relevant acoustic studies. Soundscape composition is normally used to objectively reflect the existence of different sound sources. It is another important parameter used in predicting people's soundscape perceptions, and it is often measured together with the loudness or significance of sound sources [16]. For studies evaluating soundscape perceptions through physical parameters, acoustic instruments such as digital acoustic recorders, spectrogram analyzers [17] and multi-channel signal analyzers [18] are commonly used to obtain and analyze the sound composition of the surrounding environment [14], the significance of sound sources [17], and the sound pressure level [13]. For example, recordings and EMuJoy software were used in a recent study to analyze the effect of sound sequence on soundscape emotions, focusing on three aspects of sound sources: the number of sound sources, changing trends in the number of sound sources (increment/decrement), and the category of the sound source [14]. The relationships between certain physical and psychoacoustic parameters (LAeq, LCeq, $LA_{10}$, and $LA_{90}$) and perceived soundscape composition parameters (perceived loudness of individual sound, perceived occurrences of individual sound, and soundscape diversity index) were investigated to develop a more effective way to design soundscapes in city parks [16]. Physical parameters were recorded with a Sony PCM-D50 sound recorder, while perceived soundscape composition parameters were obtained through sound walks. However, constrained by the accuracy and application scope of instruments, these relationships were primarily established at the individual level and are difficult to apply in predicting human perceptions in large-scale spaces. Some of them can even only be validated in laboratory settings.

In addition to aural aspects, environmental visual attributes can also influence soundscape perception [19]. The influences of visually perceived naturalness [20], waterscape [21], aesthetics [22], openness [23], layering and order [24], the visibility of noise sources [25], crowding [26], and maintenance level [27] have been proven in various contexts, such as urban forests [28], urban parks [29], urban public spaces, university campuses [30], and neighborhoods [31]. These indicators can generally be classified into perception type and physical type [32]. The perception type is normally measured and analyzed as qualitative indicators, while the physical type can be described both qualitatively and quantitatively. In a study of Chinese urban parks, the composition of landscape elements was calculated as percentages and considered an important factor in modeling soundscape perceptions [25]. Subjective soundscape attitudes were found to be directly influenced by visual indicators that physically described urban streets, including the street width, building height, and street width to building height in a study conducted with the help of 3D virtual reality [33]. The effects of visual attributes on soundscape perception have also been explored in studies investigating the interactions between visual and aural characteristics [34]. Scholars have found that a quiet environment with low sound pressure levels does not necessarily lead to high satisfaction with soundscape perception since visual information can supplement and enhance the meaning of the sound that people perceive [35,36], and vice versa [37]. High levels of environmental decibel values can significantly interfere with the process of perceiving visual environments, thus negatively affecting people's overall perceived soundscapes [38]. In addition, the consistency between visual and aural attributes plays an important role in constituting people's soundscape perceptions. When the two aspects are consistent, people are more likely to have better evaluations of environmental soundscapes [39]. Thus, it is essential to consider both the visual and aural attributes of an environment as well as their interaction in evaluating, modeling, and predicting individuals' soundscape perceptions.

Though efforts have been made, most relationships were constructed within perceptual qualities. People's perceived acoustic comfort was correlated with acceptability of the environment and preference to stay in an urban park context [6]. A study conducted in five different types of UGSs also found that perceived affective qualities of soundscape (PAQs) and perceived sensory dimensions of landscape (PSDs) could be used to measure overall soundscape satisfaction [6]. Very few studies have attempted to measure the soundscape perceptions of individuals based on the physical attributes of the environment, while relationships between subjective human experience and objective environmental indicators are actually key in delivering applicable design implications. Aletta, Kang, and Axelsson (2016) developed a conceptual framework for predictive soundscape models based on a literature review. Although it contained both physical and perceptive aural attributes, no visual indicator was considered. Also, these indicators were proposed to be measured with soundwalks, laboratory experiments, narrative interviews, and behavioral observations in the framework. Therefore, the application scope and practicality of the modeling results may be significantly constrained, and the results of applying the framework on a specific setting can rarely be used to provide insight for design improvements. This limitation was also observed in another model implementing perceptual aural attributes (e.g., 'calm' and 'pleasant') for a UK garden based on audio recordings of it [40]. Using physical, behavioral, social, demographical, and psychological factors as independent variables, an artificial neural network was employed to generate soundscape models accordingly. The outcome of this study still did not overcome the constraints on a research scale since the performance of the constructed models became worse as the number of sites increased.

Therefore, this study aims to establish a model to mathematically predict people's perceived soundscapes in large-scale UGSs based on environmental visual, aural, and functional characteristics. Taking the Chengdu Outer Ring Ecological Park as the research site, its visual attributes, specifically the landscape composition, were calculated using space syntax and the quantum geographic information system (QGIS); its aural attributes, specifically the sound level, were measured using a professional acoustic apparatus; and its functional attributes, namely vitality, were documented through on-site observations and mapping. To obtain people's perceived soundscapes, a questionnaire-based on-site survey was conducted during sound walks. Environmental data were collected at micro-scale measurement spots selected within the site and then used together to formulate a stochastic model. This was performed to investigate whether and how the model changed within different sound level ranges [41]. This model could not only disclose the correlation between environmental characteristics and human-perceived soundscapes but could also be used to predict people's perceived soundscapes in large-scale UGSs. Environmental data were then collected at the site level and input into the constructed model to predict people's perceived soundscapes in the site. The results were also visualized as a soundscape perception map to indicate where planning and design improvements should be introduced to enhance people's soundscape experiences.

## 2. Materials and Methods

### 2.1. Research Site

This study selected the Chengdu Outer Ring Ecological Zone (hereinafter referred as 'the Zone') as the research site, which is defined within 500 m on both sides of the Chengdu Outer Ring Road. The Zone covers an area of about 187.2 square km containing various land uses, such as green spaces, transportation, residences, commerce, and farmland. The planning objective of this project was to transform the Zone into a giant eco-park surrounding the inner city of Chengdu. To date, the southern part of the Zone has largely been built and widely praised by the public. However, it also suffers from noise issues. Traffic sounds from surrounding highways, railways, and Tianfu Airport, and mechanical sounds from nearby construction have seriously affected people's landscape experiences. High-level development requirements and current noise problems make the Chengdu Outer Ring Ecological Zone an appropriate research site for this study (see Figure 1).

Measurement spots were selected to obtain people's soundscape satisfaction evaluation results. At the spot scale, the influential indicators and mechanisms of people's soundscape satisfaction were determined so that a model to predict people's soundscape satisfaction for the whole site could be established. A total of 25 measurement spots were selected within the southern part of the site, which is built and open to the public. The selection was based on the following criteria: (1) the spots should present typical visual and acoustic features of the site and be distinct from one another; (2) an open view is required for each spot, along with enough space to allow participants to take free experiential walks without interfering with the acoustic recordings; and (3) the distance between measurement spots should be over 300 m to allow enough time for subjects to clear their heads before starting a new evaluation. Each measurement spot was delimited to a moderate size between 50 m × 50 m and 100 m × 100 m to ensure enough space for people to experience the environment without losing the sense of space [42] (see Figure 1).

**Figure 1.** Research site: Chengdu Outer Ring Ecological Park.

*2.2. Measurements*

Visual landscape composition, sound pressure level, and functional vitality were determined as environmental indicators to explore whether and how they can affect human soundscape perceptions. Given that the study aimed to predict human soundscape perceptions in large-scale UGSs, the visual, aural, and functional characteristics were first measured for the 25 measurement spots. These were then used to construct a prediction model for the Chengdu Outer Ring Ecological Park.

### 2.2.1. Visual Landscape Composition

Visual landscape attributes were calculated from people's perspectives rather than from the actual distribution of landscape elements. QGIS and space syntax were used to calculate the landscape composition people visually perceived at the site. According to the pilot study, five major landscape elements—woodland, buildings, grassland, waterscape, and pavement—existed within the site. Among which buildings and woodland were also taken as visual obstacles. Their boundaries were delimited using QGIS Maptiler and Google satellite images so that to identify areas that could be visually perceived by people. These visible areas were then analyzed in Depthmap over a 50 m × 50 m grid and imported into QGIS again to respectively calculate the areas of buildings (B), grassland (G), waterscape (Wa), woodland (Wo), and pavement (P) within each grid. Following this procedure, the visual landscape composition was first calculated for the whole site, and the composition of each measurement spot was then obtained from the overall calculation results based on its actual location and range (see Figure 2).

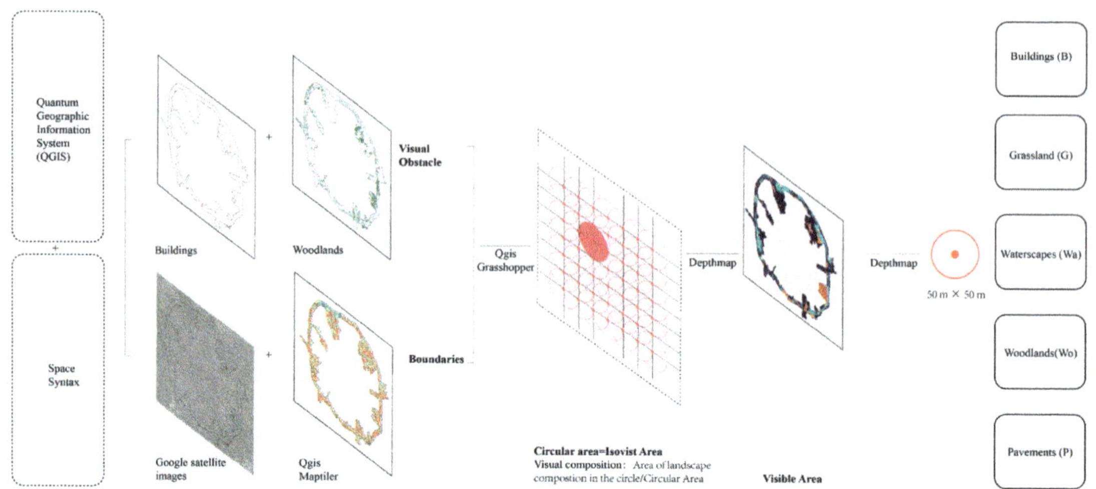


**Figure 2.** Measuring visual landscape calculations of the site using space syntax and QGIS.

### 2.2.2. Sound Pressure Level

The equivalent continuous A-weighted sound pressure level (LAeq) was selected to objectively describe environmental sound levels. It is a widely used parameter in existing studies to indicate the environmental decibel value and variations in sound intensity [42] and has been confirmed to have evident effects on people's soundscape perceptions [43]. Thus, sound pressure level was regarded as appropriate in objectively representing the aural attributes.

The sound level was first measured on a spot scale using a multi-channel signal analyzer (AWA6290L+). This acoustic instrument has proven to be effective in measuring and recording sound levels in outdoor environments owing to its multi-channel receptors and high sensitivity [44]. In the field measurements, it was used with a set of apparatus including two test microphones (AWA14423, frequency range 20–16,000 Hz, sensitivity about 40 mV/Pa), two preamplifiers (AWA14604, Integrated Circuits Piezoelectric, impedance conversion, frequency range 10–200 kHz, Gain 1), two 80 mm diameter wind balls, and two tripods (about 1.6 m high).

Considering the difficulty of instrumentally measuring the sound pressure level of the whole site, it was estimated based on the dominant sound source: traffic. The sound pressure level of traffic noise was calculated using the noise prediction model (NPL) [see http://resource.npl.co.uk/acoustics/techguides/crtn/ (accessed on 18 March 2022)]. It is an open-sourced model for calculating road traffic noise levels in non-complex situations [45]. Traffic volume data for the 30 urban roads that crossed the site and the penetration state of road pavement materials were retrieved from the Chengdu Communication Investment Company with their consent. Additionally, the proportion of heavy vehicles (Mean = 22%, SD = 2.87) in the traffic flow and active time period (18 h/day) was estimated through on-site investigation. This information was then entered into the NPL to calculate the decibel values for each road within a range of 10 m on both sides of its center line. The decibel values of the 30 roads generated by the NPL were then correlated with road hierarchy information and entered into QGIS to calculate the sound pressure level for the site's entire traffic network. Referring to the principle of sound attenuation, the OpenNoise plugin in QGIS was used to predict and visualize the noise distribution within the entire site (see Figure 3).

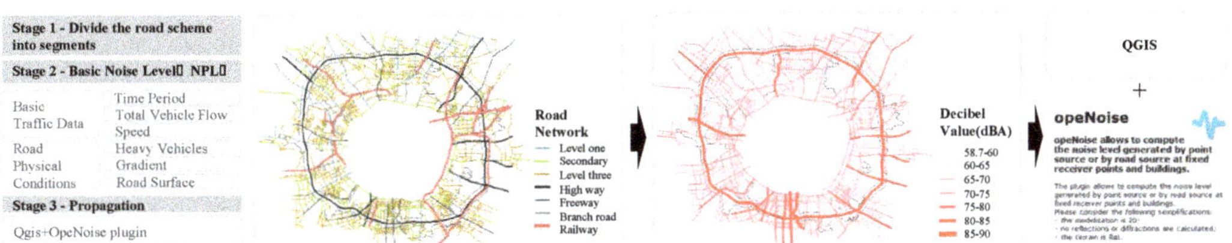

**Figure 3.** Sound pressure level calculated using NPL, QGIS, and the OpeNoise plugin.

### 2.2.3. Functional Vitality

Functional vitality (V) was adapted from the notion of urban vitality. It refers to people and their activities observable in a specific space throughout varied time schedules and is the product of the number and duration of various activities [46]. This study revised the term to "functional vitality" and defined it as the diversity and intensity of activities occurring within different areas of the site. The measurement of this indicator could be used to illustrate human behavior attributes in response to environmental characteristics.

Functional vitality was initially measured at the spot scale through on-site observations and mapping conducted by four groups of trained surveyors. During the site's open hours, surveyors ranked the diversity (total number of activity types) and intensity (total number of people) of each spot at two-hour intervals from 8:00 to 17:00 (10:00, 12:00, 14:00, and 16:00) using a five-level Likert scale (1–5, with 1 representing "not diverse/intensified at all" and 5 representing "strongly diverse/intensified"). The mean value of the four surveyors' ratings was then calculated to determine the functional vitality for each measurement spot.

Given that spatial vitality can be influenced by environmental functions [47] and landscape characteristics [48], the functional vitality obtained for each measurement spot was first correlated with its land use and visual landscape composition. The same value of functional vitality was then assigned to grids with similar land uses and visual landscape compositions using QGIS to ascertain the functional vitality of the entire site (see Figure 4).

### 2.2.4. Satisfaction with Perceived Soundscape

People's satisfaction with their perceived soundscape (S) was used to describe the overall soundscape perception in the site and was measured through a sound walk and questionnaire-based on-site survey conducted at each measurement spot. The questionnaire was designed as an online version in advance of the sound walk. It started with collecting subjects' background information, including gender and age, to assess whether individual perceptual differences existed, followed by a question to investigate whether people were satisfied with their perceived soundscape in each experienced measurement spot, also using a five-level Likert scale (1–5, with 1 representing "strongly unsatisfied" and 5 representing "strongly satisfied").

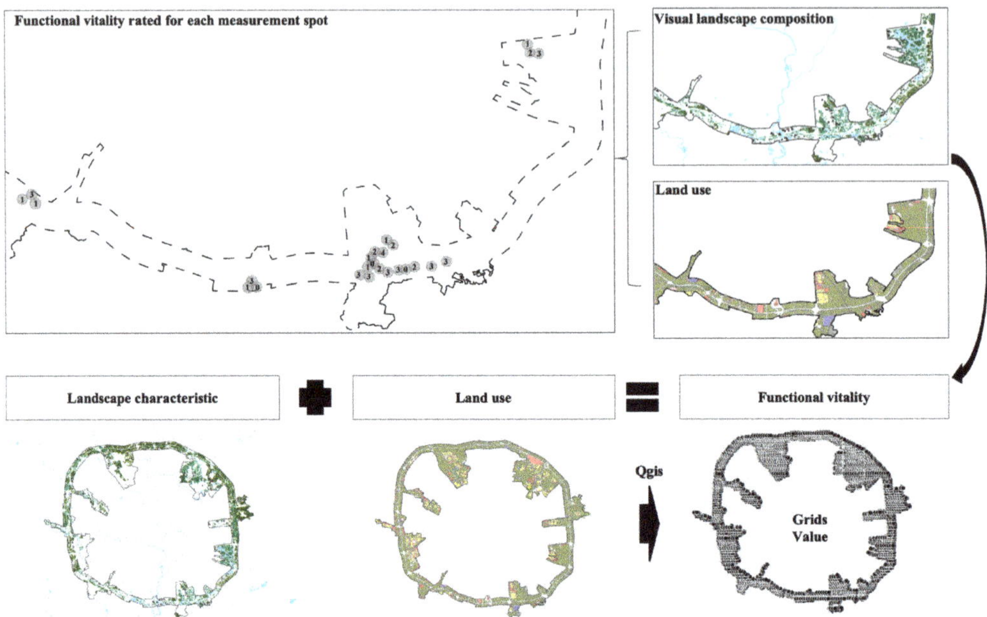


**Figure 4.** Functional vitality of the site calculated based on spot vitality ratings, land uses, and visual landscape characteristics.

### 2.3. Data Collection

Data collection mainly occurred at the spot level to gather information related to sound pressure level, functional vitality, and human-perceived soundscape satisfaction. Based on previous soundscape studies, the total number of valid responses needed for relevant studies should exceed 100. Considering the number of spots, 20 local university students aged between 20 and 30 years with normal hearing and vision abilities were recruited to participate in the sound walk and questionnaire-based survey. Before performing the sound walk and survey, all participants went through a training process and provided informed consent.

The field survey was conducted on three sunny days in July 2021, with an air quality index below 60 and wind speeds under 5 m/s. The 20 subjects were evenly split into four groups and each group was led by one researcher. The routes taken by the four groups to traverse the 25 measurement spots were different and determined before starting the survey. The subjects were instructed to take a sound walk and individually wander within each measurement spot for five minutes to experience their surrounding environment. Afterwards, they completed the online questionnaire through a link provided by the researcher. Subjects were asked repeatedly to complete the sound walk and questionnaire survey at each of the 25 measurement spots, yielding a total of 500 responses. Data deemed as "noise", such as incomplete questionnaires or those with identical ratings for every indicator, were manually removed. The final number of valid responses was 393 (approximately 16 per spot).

Another group of researchers recorded the environmental sound level at each spot using the multi-channel signal analyzer while the sound walk and questionnaire survey were being carried out. The set of equipment was positioned at the central point of each spot for recording. The sound pressure level of each spot was measured four times in accordance with the frequency of the sound walks conducted respectively by the four subject groups. For each measurement, the signal was firstly calibrated using a sound-level calibrator (AWA6021A, accuracy level 1) and then the formal recording was taken by the multi-channel signal analyzer (AWA6290L+) for a consecutive three minutes. Therefore, four sets of three-minute recordings (a total of 100 recordings) were collected for each spot and the sample frequency of each audio file was 96.0 kHz.

*2.4. Data Analysis*

Data analysis was conducted at the spot and site levels, while the prediction and visualization process was only undertaken at the site level.

At the spot level, metrics such as human-perceived soundscape satisfaction, sound pressure level, and functional vitality were collected. Human-perceived soundscape satisfaction was first checked for reliability and then presented as the mean value of the responses obtained at each spot. The sound data were analyzed using spectrum analysis incorporated in the SPL 1/3 octave bands analysis package of AWA6229 6.0 software. The analysis was performed on audio files to calculate the LAeq for each spot over the four sets of three-minute recordings. The average LAeq of each measurement spot represented the average of four sets of recordings and the average sound level recorded during three minutes. Regarding functional vitality, three steps were taken to determine the final value for each spot. First, the average of behavior diversity and intensity was rated by four surveyors at each measurement spot. Then, the average rating from the four surveyors was calculated. Given that observations and evaluations occurred in different time slots, the average functional vitality of the recorded slots was also determined. Therefore, the functional vitality represented the averages of the four surveyors' ratings, intensity and diversity, and all recorded time slots.

Before model construction, the turning point in the sound level was first determined. Correlational analysis was then employed to identify indicators relevant to people's satisfaction with the perceived soundscape. These relevant indicators were then used to formulate regression models to illustrate the relationships between people's soundscape satisfaction and its influencing visual and aural attributes.

Information about visual landscape composition, sound pressure level, and functional vitality was obtained and calculated at the site level. These data were then incorporated into the previously generated stochastic model to predict the overall satisfaction of people with their perceived soundscape for the site. The sound perception results were subsequently visualized using QGIS to develop planning and design implications.

## 3. Results
### 3.1. Manipulation Checks

The internal consistency of people's satisfaction responses was examined using Cronbach's alpha [49] in SPSS V26.0. The results suggested that the reliability of the evaluated soundscape perceptions was consistently greater than 0.6, indicating the effectiveness of the obtained data. An independent samples t-test was also conducted based on gender groups to investigate whether any perceptual differences existed. The Levene test for variance equality showed values greater than 0.05, indicating homogeneity of variance. Additionally, the $t$-test results for the mean equation were all greater than 0.05, demonstrating that gender differences had no impact on individual perceived soundscapes.

### 3.2. Descriptive Analysis

General descriptions of the visual landscape, sound pressure level, functional vitality, and soundscape perception results measured at the spot level are shown in Table 1.

**Table 1.** General descriptions of environmental attributes and human soundscape perceptions measured at the spot level.

| Measurement Spot (No.) | | Waterscape (Wa) | Woodland (Wo) | Grassland (G) | Buildings (B) | Pavement (P) | Sound Level (LAeq) | Functional Vitality (V) | Soundscape Satisfaction (S) | |
|---|---|---|---|---|---|---|---|---|---|---|
| 1 | Mean | 0.00% | 0.96% | 20.89% | 14.80% | 15.33% | 76.00 | 3.00 | Mean | 3.50 |
| | | | | | | | | | Std. Dev. | 0.52 |
| 2 | Mean | 0.00% | 0.00% | 88.27% | 0.00% | 11.73% | 78.00 | 1.00 | Mean | 4.00 |
| | | | | | | | | | Std. Dev. | 0.00 |
| 3 | Mean | 4.07% | 0.00% | 79.51% | 0.00% | 16.42% | 71.00 | 1.00 | Mean | 2.57 |
| | | | | | | | | | Std. Dev. | 0.51 |
| 4 | Mean | 6.02% | 0.29% | 74.85% | 1.70% | 8.96% | 76.00 | 1.00 | Mean | 4.00 |
| | | | | | | | | | Std. Dev. | 0.00 |
| 5 | Mean | 6.23% | 0.21% | 76.16% | 1.69% | 8.35% | 82.00 | 3.00 | Mean | 3.00 |
| | | | | | | | | | Std. Dev. | 0.00 |
| 6 | Mean | 6.10% | 0.12% | 74.56% | 1.64% | 8.57% | 79.00 | 0.00 | Mean | 4.00 |
| | | | | | | | | | Std. Dev. | 0.00 |
| 7 | Mean | 0.00% | 17.23% | 57.51% | 0.00% | 0.20% | 66.00 | 2.00 | Mean | 4.41 |
| | | | | | | | | | Std. Dev. | 0.51 |
| 8 | Mean | 27.71% | 16.26% | 48.92% | 0.07% | 1.54% | 84.00 | 1.00 | Mean | 4.00 |
| | | | | | | | | | Std. Dev. | 0.00 |
| 9 | Mean | 19.75% | 15.20% | 51.59% | 0.68% | 11.41% | 96.00 | 2.00 | Mean | 3.53 |
| | | | | | | | | | Std. Dev. | 0.52 |
| 10 | Mean | 15.75% | 13.84% | 53.60% | 0.54% | 11.95% | 76.00 | 1.00 | Mean | 3.53 |
| | | | | | | | | | Std. Dev. | 0.51 |
| 11 | Mean | 15.53% | 15.75% | 53.22% | 0.52% | 11.63% | 77.00 | 0.00 | Mean | 4.00 |
| | | | | | | | | | Std. Dev. | 0.00 |
| 12 | Mean | 0.00% | 22.64% | 52.63% | 2.60% | 14.16% | 69.00 | 4.00 | Mean | 3.63 |
| | | | | | | | | | Std. Dev. | 0.50 |
| 13 | Mean | 0.00% | 20.61% | 67.87% | 0.00% | 8.48% | 90.00 | 2.00 | Mean | 4.00 |
| | | | | | | | | | Std. Dev. | 0.00 |
| 14 | Mean | 21.66% | 12.15% | 46.64% | 0.77% | 14.96% | 100.00 | 1.00 | Mean | 3.58 |
| | | | | | | | | | Std. Dev. | 0.51 |
| 15 | Mean | 0.84% | 34.15% | 51.80% | 0.00% | 13.21% | 76.00 | 3.00 | Mean | 3.00 |
| | | | | | | | | | Std. Dev. | 0.00 |
| 16 | Mean | 2.12% | 17.25% | 64.50% | 0.00% | 16.12% | 76.00 | 3.00 | Mean | 3.00 |
| | | | | | | | | | Std. Dev. | 0.00 |
| 17 | Mean | 5.78% | 21.51% | 56.76% | 0.00% | 15.96% | 75.00 | 3.00 | Mean | 2.68 |
| | | | | | | | | | Std. Dev. | 0.48 |
| 18 | Mean | 13.84% | 28.27% | 45.41% | 0.39% | 12.10% | 71.00 | 3.00 | Mean | 3.50 |
| | | | | | | | | | Std. Dev. | 0.51 |
| 19 | Mean | 6.70% | 5.83% | 77.70% | 0.00% | 9.76% | 75.00 | 0.00 | Mean | 2.79 |
| | | | | | | | | | Std. Dev. | 0.42 |

**Table 1.** *Cont.*

| Measurement Spot (No.) | | Waterscape (Wa) | Woodland (Wo) | Grassland (G) | Buildings (B) | Pavement (P) | Sound Level (LAeq) | Functional Vitality (V) | Soundscape Satisfaction (S) | |
|---|---|---|---|---|---|---|---|---|---|---|
| 20 | Mean | 18.30% | 3.03% | 65.50% | 0.00% | 13.17% | 77.00 | 2.00 | Mean | 3.00 |
| | | | | | | | | | Std. Dev. | 0.00 |
| 21 | Mean | 16.68% | 17.53% | 54.22% | 0.12% | 11.45% | 79.00 | 3.00 | Mean | 3.00 |
| | | | | | | | | | Std. Dev. | 0.00 |
| 22 | Mean | 14.17% | 30.39% | 48.88% | 0.00% | 6.56% | 78.00 | 3.00 | Mean | 3.33 |
| | | | | | | | | | Std. Dev. | 0.48 |
| 23 | Mean | 52.33% | 2.25% | 30.56% | 1.95% | 1.66% | 76.00 | 1.00 | Mean | 3.42 |
| | | | | | | | | | Std. Dev. | 0.51 |
| 24 | Mean | 48.43% | 0.00% | 33.13% | 5.20% | 0.00% | 69.00 | 2.00 | Mean | 3.48 |
| | | | | | | | | | Std. Dev. | 0.51 |
| 25 | Mean | 48.60% | 0.03% | 33.49% | 4.59% | 0.00% | 72.00 | 3.00 | Mean | 3.88 |
| | | | | | | | | | Std. Dev. | 0.34 |
| Sum | Mean | 14.70% | 11.70% | 55.45% | 1.60% | 9.63% | 77.33 | 1.91 | Mean | 3.46 |
| | | | | | | | | | Std. Dev. | 0.62 |

The average satisfaction of people in terms of soundscape perception was rated as 3.46. People were most satisfied with their perceived soundscape in spot 7 (4.41/5.00) and least satisfied with spot 3 (2.57/5.00). Most measurement spots (21 out of 25) received ratings between 3.00 and 4.00. Three spots—spots 3 (2.57/5.00), 17 (2.65/5.00), and 19 (2.79/5.00)—were rated lower than 3.00. See Table 1.

3.2.1. Descriptive Analysis of Visual Landscape Composition

At the spot level, it was observed that the component of grassland (Mean = 55.45%) constituted the majority in almost all of the 25 measurement spots, followed by waterscape (Mean = 14.70%) and woodland (Mean = 11.70%). In most measurement spots, grassland accounted for over 50% of the landscape composition. The highest percentages of grassland appeared at spots 2 (88.27%), 3 (79.51%), 19 (77.70%), 5 (76.16%), 4 (74.85%), and 6 (74.56%). Only seven spots—8 (48.92%), 14 (46.64%), 18 (45.41%), 25 (33.49%), 24 (33.13%), 23 (30.56%), and 1 (20.89%)—had less than 50% grassland within their delimited areas. Among the 25 spots, spot 1 had the least component of grassland. In terms of another green landscape, woodland, it was negligible in spots 1 (0.96%), 2 (0.00%), 3 (0.00%), 4 (0.29%), 5 (0.21%), 6 (0.12%), 24 (0.00%), and 25 (0.03%), while spots 15 (34.15%), 22 (30.39%), 18 (28.27%), 12 (22.64%), 17 (21.51%), and 13 (20.61%) had relatively higher percentages of woodland. Waterscape was especially dominant in spots 23 (52.33%), 24 (48.43%), 25 (48.60%), 8 (27.71%), and 14 (21.66%). However, it was not a mandatory component for designing all spaces in UGSs; therefore, it was scarcely observed in spots 1 (0.00%), 2 (0.00%), 7 (0.00%), 12 (0.00%), 13 (0.00%), and 15 (0.84%). Pavement (Mean = 9.63%) and buildings (Mean = 1.60%) were the least dominant components in all of the measurement spots (see Table 1 and Figure 5).

In general, grassland was evenly distributed within the site but had some obvious aggregations in the stretched spaces located at the inner side of the ring road. The distribution of woodland, however, was mostly scattered and showed an opposite trend to that of buildings. Waterscape was primarily observed in the southern part of the site, where the construction process was more advanced. Pavement in the site was composed of roads and trails and was therefore distributed along the ring road and also within the parks on the site (see Figure 6).

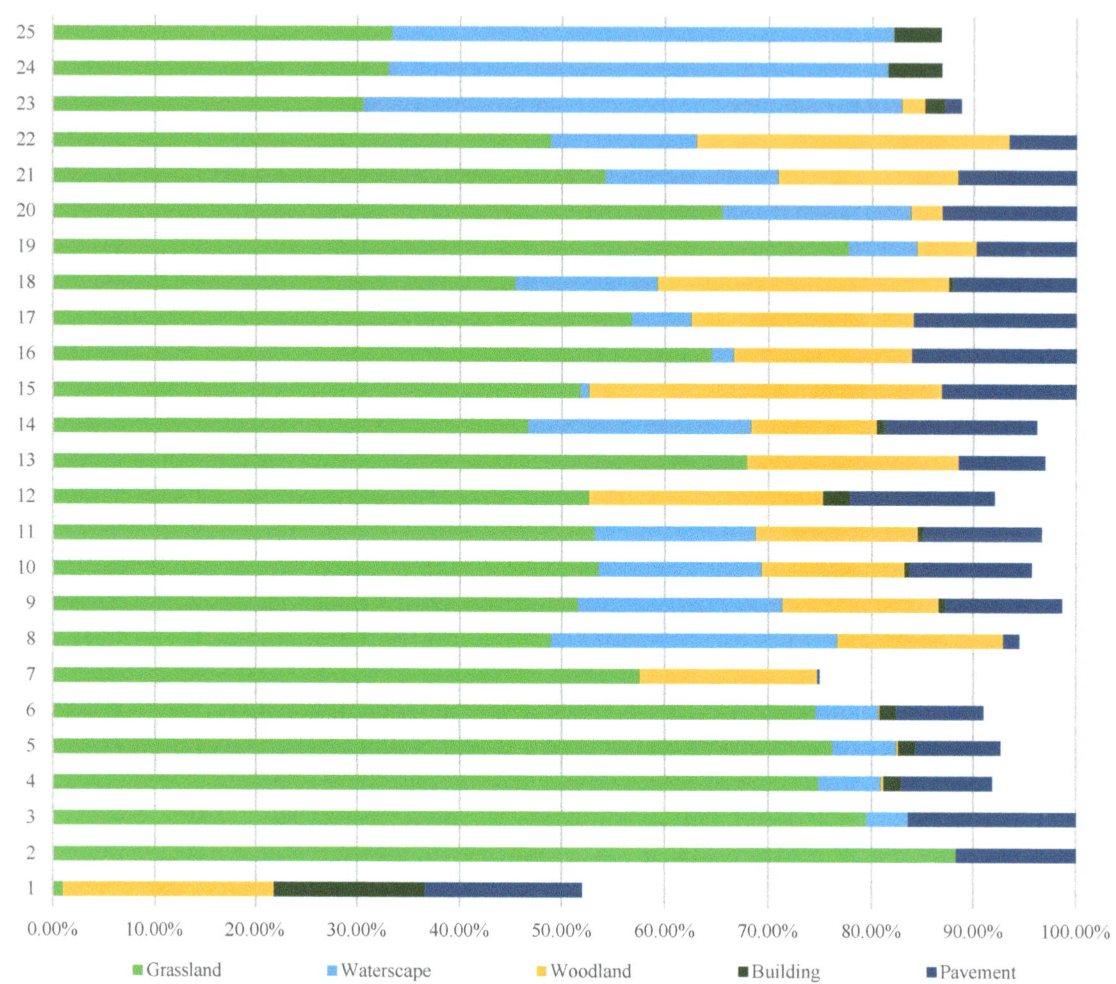



**Figure 5.** Visual landscape composition of each measurement spot.

### 3.2.2. Descriptive Analysis of Sound Pressure Levels

The average sound pressure level at the 25 measurement spots was 77.33 dBA. Twenty-three spots had decibel values exceeding the 70.00 dBA limit requested in the Standard of Sound Environment Quality (GB3096-2008). The sound levels at spots 9 (96.00 dBA) and 13 (90.00 dBA) surpassed 90.00 dBA and that at spot 14 even reached 100.00 dBA. Only three spots—spots 7 (66.00 dBA), 12 (69.00 dBA), and 24 (69.00 dBA)—had sound levels measured slightly lower than the 70 dBA limit. See Table 1 and Figure 7 for details.

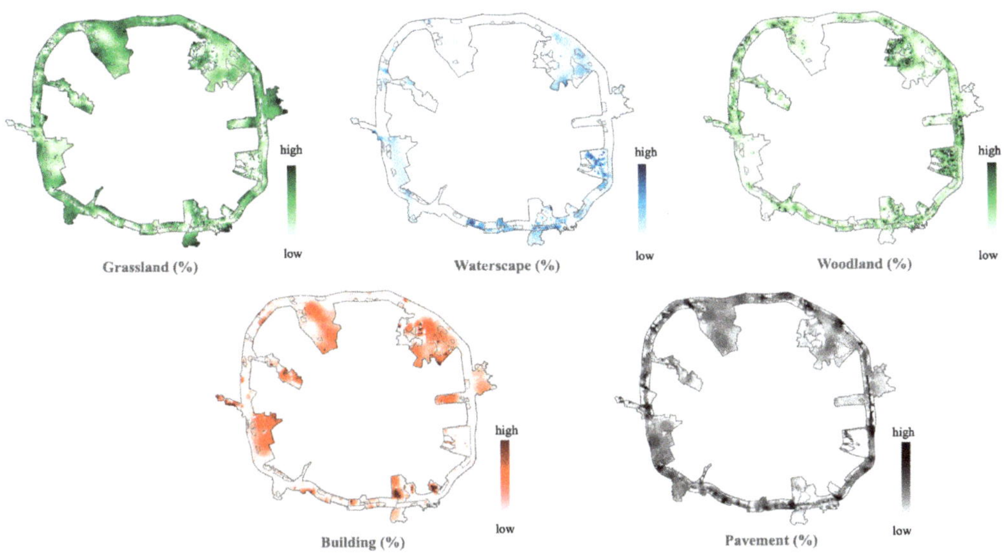


**Figure 6.** Distribution of major visual landscape elements within the site.

**Figure 7.** Sound pressure levels calculated for the site.

The modeling and visualization results suggested that the sound pressure level (above 80.00 dBA) was extremely high within a 150 m range on both sides of the ring road. For the ring road itself, the sound pressure level exceeded 95.00 dBA. Additionally, the sound map clearly showed that the environmental sound pressure level decreased evenly with increasing distance from both sides of the ring road (see Figure 7).

### 3.2.3. Descriptive Analysis of Functional Vitality

Functional vitality was relatively low at the spot level, with an average of 1.91 (SD = 1.14). Almost all spots were rated lower than 3.00, with the exception of spot 12 (4.00/5.00). The majority were evaluated at 3.00 (nine spots: spots 1, 5, 15–18, 21, 25), followed by 2.00 (five spots: spots 7, 9, 13, 20, 24), and 1.00 (seven spots: spots 2–4, 8, 10, 14, 23). Additionally, three spots—spots 6, 11, and 19—were rated at 0.00 in terms of functional vitality. See Table 1 and Figure 8b.

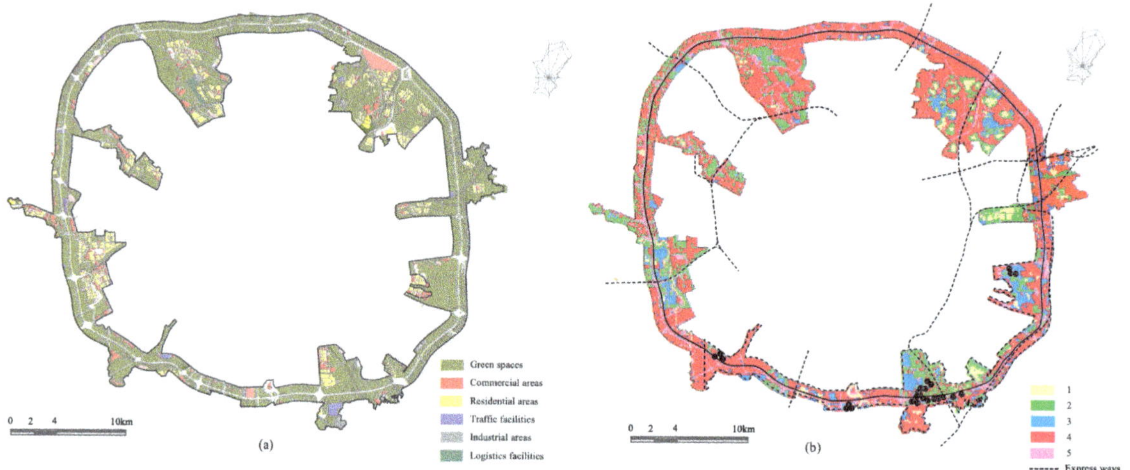


**Figure 8.** Land uses (**a**) and functional vitality (**b**) of the site.

The functional vitality predicted for the entire site is presented in Figure 8b. The road network was manually given a rating of 5.00 by the researcher to differentiate it from other spaces accessible to people. However, these highest-rated spaces will not be discussed further in the context of developing design and planning implications. Compared with the land uses shown in Figure 8a, functional vitality was relatively high in most green spaces and commercial areas, while lower levels of vitality were primarily observed around residential areas.

### 3.3. Satisfaction with People's Perceived Soundscape Predicted by Environmental Characteristics

### 3.3.1. The Influence of Sound Level Range on People's Soundscape Satisfaction

Previous evidence suggested that a certain range exists in the relationship between soundscape perception and sound level [50]; when the sound level exceeds that range, the relationship may vary accordingly [51]. Therefore, the sound pressure level was determined as an indicator and also the mediator in this study. To ensure the accuracy of the prediction model, this study first explored the sound level ranges of the site.

Correlational analysis was conducted first on all datasets to confirm that people's soundscape satisfaction was significantly influenced by sound level ($p < 0.01$). Afterwards, a scatter plot was used to visually present how people's soundscape satisfaction changed with sound level. It was quite evident that a negative relationship could be observed when the sound level was lower than 75 dBA. For levels above 75 dBA, there was first a slight increase between 75 and 84 dBA, and then stability was reached when the sound level exceeded 84 dBA (see Figure 9).

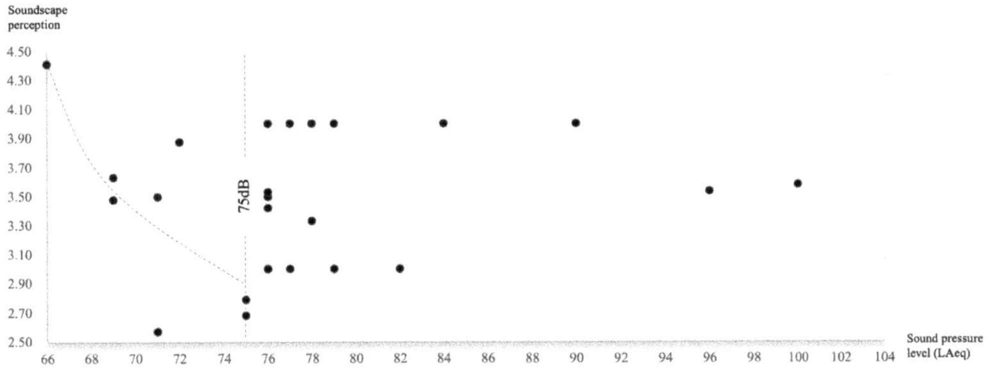

**Figure 9.** Scatter plot showing the relationship between overall soundscape perception and sound pressure level.

Correlation analyses were also performed to validate the turning point and found that when the sound level was lower than 75 dBA, a significant negative correlation between sound pressure level and soundscape satisfaction was evident ($R = -0.600$, $p < 0.01$). Additionally, the correlation coefficient decreased dramatically when the sound level exceeded this threshold ($R = 0.120$, $p = 0.062 > 0.05$). This indicated that when the sound level exceeded 75 dBA in this study, people's satisfaction with the surrounding soundscape became less predictable. Therefore, under and above 75 dBA were identified as two sound level ranges to carry out the following model construction.

### 3.3.2. Correlational Analysis between Soundscape Satisfaction and Visual and Aural Indicators

Data from the 25 measurement spots were divided into groups with sound pressure levels above and below 75 dBA. Pearson correlational analysis was then conducted separately for these two groups to identify visual, aural, and functional indicators related to people's soundscape perceptions in environments with sound pressure levels above and below 75 dBA.

The correlational analysis results indicated that for the dataset with sound pressure levels lower than 75 dBA, people's satisfaction with the soundscape could be influenced by the visual proportions of waterscape, grassland, buildings, and pavement, as well as the sound pressure level and functional vitality. Among them, the waterscape, buildings, and sound pressure level had positive effects, while the other three factors had negative influences. Fewer influential indicators were found for the group with sound pressure levels above 75 dBA; only grassland was positively related, while woodland, pavement, and functional vitality had negative effects. Only indicators with significant relationships with people's soundscape satisfaction were included in the regression analysis (see Table 2).

**Table 2.** Correlational analysis between soundscape satisfaction and environmental indicators.

| Groups | | Wo | Wa | G | B | P | LAeq | V |
|---|---|---|---|---|---|---|---|---|
| <75 dBA | Correlation (Pearson's r) | 0.145 | 0.192 * | −0.471 ** | 0.309 ** | −0.572 ** | 0.330 ** | −0.600 ** |
| N = 153 | Significant (p) | 0.073 | 0.017 | 0.000 | 0.000 | 0.000 | 0.000 | 0.000 |
| >75 dBA | Correlation (Pearson's r) | −0.211 ** | −0.083 | 0.202 ** | 0.004 | −0.180 ** | 0.120 | −0.548 ** |
| N = 240 | Significant (p) | 0.001 | 0.203 | 0.002 | 0.953 | 0.005 | 0.062 | 0.000 |

Indicator values with * and ** are those with obvious relationships with soundscape satisfaction.

### 3.3.3. Prediction Models of Soundscape Satisfaction

Regression analysis was also conducted, respectively, for the groups with sound levels above and below 75 dBA. Thus, two sets of linear regression models were established.

In the group with sound levels under 75 dBA, five environmental indicators, grassland ($p < 0.01$), buildings ($p < 0.01$), pavement ($p < 0.01$), sound pressure level ($p < 0.01$), and functional vitality ($p < 0.01$), were significantly related to people's satisfaction with their perceived soundscape. A model with an $R^2$ of 0.572 was found as follows:

$$S = 6.899 + 0.013 \times G - 0.079 \times B - 0.089 \times P - 0.059 \times LAeq + 0.368 \times V \qquad (1)$$

Correlational analysis results suggested that woodland ($p < 0.01$), grassland ($p < 0.01$), pavement ($p < 0.01$), and functional vitality ($p < 0.01$) were significantly influential indicators of people's satisfaction with their perceived soundscape. Thus, the model constructed for the group with sound levels above 75 dBA was ($R^2 = 0.312$):

$$S = 3.883 - 0.002 \times W_o + 0.003 \times G - 0.007 \times P - 0.253 \times V \qquad (2)$$

The estimated coefficient values for all indicators and model-fitting information are listed in Table 3.

**Table 3.** Model fit and estimation of B coefficients of the models.

| Group | Model Fit ($R^2$) | Attribute | Estimate B | Standard Error | t-Value | p-Value |
|---|---|---|---|---|---|---|
| <75 dBA (N = 153) | 0.572 | G | 0.013 | 0.008 | 1.699 | 0.091 |
| | | B | −0.079 | 0.034 | −2.319 | 0.022 |
| | | P | −0.089 | 0.014 | −6.513 | 0.000 |
| | | LAeq | −0.059 | 0.019 | −3.132 | 0.002 |
| | | V | 0.368 | 0.078 | 4.701 | 0.000 |
| >75 dBA (N = 240) | 0.312 | Wo | 0.002 | 0.003 | 0.624 | 0.624 |
| | | G | 0.003 | 0.002 | 1.618 | 1.618 |
| | | P | −0.007 | 0.007 | −1.110 | −1.110 |
| | | V | −0.253 | 0.030 | −8.333 | −8.333 |

Two sets of models were employed to visually represent people's satisfaction with the soundscape of the entire site (Figure 10). Compared with site mappings of visual landscape composition, sound pressure level, and functional vitality, the trend in soundscape satisfaction distribution was generally opposite to that of sound pressure level and functional vitality. The lowest level of soundscape satisfaction appeared alongside the ring road, where both the sound level and vitality were extremely high. Soundscape satisfaction increased as the distance from the ring road increased, and it gradually reached its highest level in the spaces stretched outside the ring road. Within the Chengdu Outer Ring Ecological Zone, spaces with higher ratings of soundscape perception also had high percentages of grassland, woodland, waterscape, and buildings (Figures 6–10).

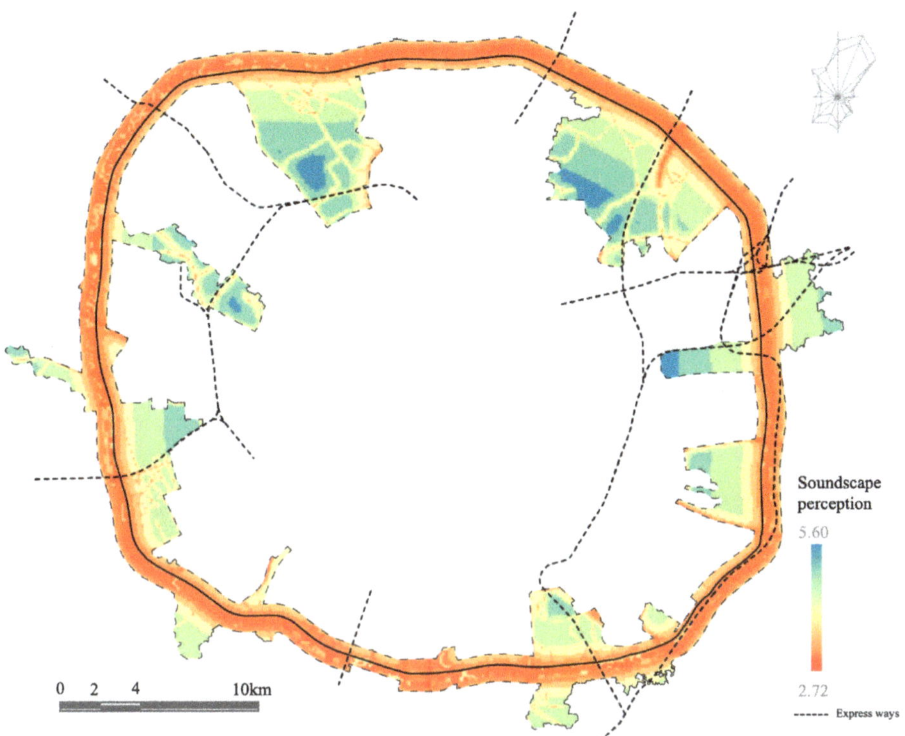



**Figure 10.** Soundscape perception map for the Chengdu Outer Ring Ecological Zone.

## 4. Discussion

### 4.1. Predicting Large-Scale Soundscape Perceptions Based on Small-Scale Measurements

This study innovatively predicted human soundscape perception in large-scale UGSs. In this study, the prediction of large-scale soundscape perceptions based on small-scale measurements was achieved through three major consecutive steps: (1) collecting visual, aural, and functional characteristics and people's soundscape satisfaction at the selected measurement spots; (2) building a stochastic model describing the relationships between human subjective perceptions and objective environmental attributes; and (3) collecting environmental information at the site level and inputting it into the stochastic model to predict people's soundscape satisfaction at the site, which was then visualized as a soundscape perception map.

This approach is expected to save time and effort in conducting sound walks and on-site evaluations. The model for predicting human satisfaction with the soundscape (dependent variable) simply requires quantifiable visual, aural, and functional attributes of the environment as independent variables. With the development of big data and modeling techniques, this environmental information becomes increasingly accessible to professionals, thereby increasing the accuracy and possibility of predicting human perceptions in large-scale urban spaces. A similar mindset was also found in a study assessing the visual comfort of streets at the city scale using artificial intelligence-based image analysis and perceptive evaluation methods [52]. These attempts have made individual perceptions more applicable in large-scale planning and design practices, although small-scale measurements are still needed at the initial model construction stage.

### 4.2. Applying Soundscape Perception Models and Maps in UGS Planning and Design

To provide cues for regulating planning and design, the soundscape satisfaction map should be considered in conjunction with sets of maps presenting the visual landscape, sound pressure levels, and functional vitality of the Chengdu Outer Ring Ecological Zone. The results suggested that strong planning and design interventions should be introduced within a 300 m range along both sides of the ring road. Tracing back to the distribution of visual, aural, and functional characteristics, a decrease in environmental sound level was particularly important for improving people's satisfaction. Given that the ring road serves essential traffic functions, controlling traffic flow is unrealistic. Thus, planning regulations should consider adjustments in land use. For example, residential land use and educational facilities should be planned as far away from the ring road as possible, and ecological green spaces that are inaccessible to people can serve as buffer zones [53] between them. Urban parks open to the public should not be planned too close to the ring road, if possible, while commercial and industrial lands can be prioritized in areas closer to the ring road. In addition to planning considerations, physical noise reduction methods such as acoustic shields [54] and noise-absorbing materials [55] should be encouraged along the ring road. In terms of the visual landscape, natural elements have long been confirmed to have positive effects on human soundscape satisfaction [56], which was also consistent with the distribution of grassland, waterscape, and woodland at the site. Spaces with high levels of satisfaction also featured a high percentage of buildings, possibly because high-rise buildings can obstruct people's view of noise sources [57]. Furthermore, a medium level of functional vitality should be encouraged to maintain people's satisfaction with soundscapes, which can also be modified through land use and visual landscape characteristics.

Therefore, it can be concluded that improving soundscape perception in large-scale UGSs requires an integrated approach from both planning and design perspectives. The application of this approach should also be considered in conjunction with spatial functions. Planning regulations for land use may include: (1) setting green buffers between noise sources and vulnerable areas (i.e., residential areas, educational facilities, and urban parks) that have low levels of sound tolerance; (2) placing land uses with high tolerance, such as traffic, industrial, and commercial functions, close to noise sources; and (3) identifying and preserving areas with special visual and acoustic characteristics. Design instructions may include: (1) using sound shields around traffic facilities; (2) utilizing natural landscapes to distract people's attention from noise and block their sight from the source of noise; and (3) introducing positive sounds (e.g., tree murmuring, bird chirping, and fish diving) [56,57].

### 4.3. Limitations

Although this study successfully predicted people's satisfaction with the soundscape across the entire site of the Chengdu Outer Ring Ecological Zone, it had limitations regarding the experimental design and data analysis. First, the measurement spots were primarily selected in the built southern part of the site, as the northern area was mostly under construction when the study was conducted. The model should be further refined with additional information from the northern side before implementation. Second, this study recruited only university students for the on-site soundscape perception survey to ensure all subjects had good vision and hearing and also to prevent potential perceptive differences induced by demographic variations. However, the unified background of the participants may have limited the comprehensiveness of the research results. In addition, participants' responses may also have been influenced by the high outdoor temperature (27 °C–30 °C), since the sound walk was carried out in early summer [58].

As for data analysis, the visual landscape composition of some spots was less than 100%. This was because some areas near the selected measurement spots were still under construction and they were largely mud lands protected with fence screens. These areas were eliminated in the visibility calculation since they were only temporary. In addition, the functional vitality and sound pressure levels of the site were not directly measured but were calculated based on other environmental information. Therefore, the predicted soundscape satisfaction was actually the result of two rounds of predictions and calculations, which might have weakened the accuracy of the established model and increased the complexity of applying this approach in practice. Also, the sound level of the site was estimated mainly based on traffic noise due to the realistic difficulty of on-site measurements. Although influences may exist, the dominancy of traffic sounds in constituting the overall acoustic environment supports this estimation being reasonable to some extent. Moreover, differences in the order of magnitude of the indicators may have slightly influenced the correlation and modeling results, even though standardization was applied.

Lastly, there may be limitations in the application of the research findings. The soundscape perception map provides no direct relationship between human satisfaction and physical planning and design features. Therefore, it needs to be used in conjunction with maps of sound pressure level, functional vitality, and visual landscape distribution as part of a comprehensive soundscape evaluation. Future studies should continue to explore ways to refine the collection and visualization of large-scale physical and perceptive data into a more systematic tool so as to increase its efficiency and broaden the contexts of its application.

### 5. Conclusions

Soundscape perception is essential in human-oriented multi-sensory landscape planning and design. However, it has obvious limitations in planning and design practices due to the high relevance of individual perceptions. This study aimed to predict human soundscape perception in large-scale UGSs based on environmental visual, aural, and functional characteristics through cross-level measurements, analyses, and model construction. Two innovative steps were carried out to achieve this aim. First, a stochastic model was constructed based on subjective soundscape perceptions and objective environmental visual and aural attributes. Although the data were collected at the measurement spot level, the relationships disclosed and the model established to present these relationships can be adopted in similar settings. The second innovation lies in utilizing the model for prediction by inputting objective environmental information collected at the site scale so that soundscape perception results can be estimated for the site. This approach not only manages to bridge subjective human perceptions and objective environmental characteristics, but it also offers insight for applying individual-scale landscape perceptions and experiences to instruct large-scale landscape planning and design. The research findings imply that both human soundscape perception and physical attributes should collectively inform the implementation of regulations and interventions to improve the soundscapes of UGSs. The perception and characteristic results are especially helpful when they are visually presented as a set of distribution maps. Nevertheless, future explorations are necessary to further clarify which planning regulations and design interventions should be introduced for each perception area. The outcome of this study, along with future relevant research, is expected to introduce a more human-oriented perspective into soundscape planning and design practices. It will also contribute to broadening the scope of human perceptual studies conducted in large-scale city spaces.

**Author Contributions:** Conceptualization, Y.Y., Y.H., L.J. and Y.S.; methodology, Y.Y., L.J. and Y.S.; software, H.L. and Y.Y.; validation, Y.Y. and Y.H.; formal analysis, Y.Y., H.L. and Y.S.; investigation, H.L. and Y.Y.; resources, Y.S. and Y.H.; data curation, Y.Y. and Y.H.; writing—original draft preparation, Y.Y.; writing—review and editing, Y.Y., L.J. and Y.S.; visualization, H.L.; supervision, Y.S.; project administration, Y.S.; funding acquisition, Y.Y. and Y.S. All authors have read and agreed to the published version of the manuscript.

**Funding:** This research was funded by the Ministry of Science and Technology, High-end Foreign Experts Introduction Plan, Restorative Urban Landscape Theory, Method and Practice (G2022133023L); the Restorative Urbanism Research Center (RURC), Joint Laboratory for International Cooperation on Eco-Urban Design, Tongji University (CAUP-UD-06); the Shanghai Key Laboratory of Urban

Design and Urban Science, NYU Shanghai Open Topic Grants (Grant No. 2022YTYin_LOUD); and the Shanghai Post-doctoral Excellence Program (2021357).

**Institutional Review Board Statement:** The study was conducted in accordance with the Declaration of Helsinki and approved by the Ethics Committee of Tongji University (protocol code 2020tjdx075 and date of approval 09/11/2020).

**Informed Consent Statement:** Informed consent was obtained from all subjects involved in the study.

**Data Availability Statement:** The data presented in this study are available upon request from the corresponding author. The data are not publicly available due to the confidentiality of participants' information.

**Conflicts of Interest:** The authors declare no conflict of interest.

# References


1. *12913-1*; I. Acoustics–Soundscape–Part 1: Definition and Conceptual Framework. Technical Report. International Organization for Standardization: Geneva, Switzerland, 2014.
2. Raimbault, M.; Dubois, D. Urban soundscapes: Experiences and knowledge. *Cities* **2005**, *22*, 339–350. [CrossRef]
3. Axelsson, Ö.; Nilsson, M.; Berglund, B. A principal components model of soundscape perception. *J. Acoust. Soc. Am.* **2010**, *128*, 2836–2846. [CrossRef]
4. Cain, R.; Jennings, P.; Poxon, J. The development and application of the emotional dimensions of a soundscape. *Appl. Acoust.* **2013**, *74*, 232–239. [CrossRef]
5. Aletta, F.; Axelsson, Ö.; Kang, J. Towards acoustic indicators for soundscape design. In Proceedings of the Forum Acusticum 2014 Conference, Kraków, Poland, 7–12 September 2014.
6. Davies, W.J.; Adams, M.D.; Bruce, N.S.; Cain, R.; Carlyle, A.; Cusack, P.; Hall, D.A.; Hume, K.I.; Irwin, A.; Jennings, P. Perception of soundscapes: An interdisciplinary approach. *Appl. Acoust.* **2013**, *74*, 224–231. [CrossRef]
7. *ISO/TS 12913-2*; 2018—Soundscape–Part 2: Data Collection and Reporting Requirements–What's It all About. International Organization for Standardization: Geneva, Switzerland, 2018; pp. 55–57.
8. Zhang, H.; Qiu, M.; Li, L.; Lu, Y.; Zhang, J. Exploring the dimensions of everyday soundscapes perception in spatiotemporal view: A qualitative approach. *Appl. Acoust.* **2021**, *181*, 108149. [CrossRef]
9. Cao, J.; Kang, J. A Perceptual Structure of Soundscapes in Urban Public Spaces Using Semantic Coding Based on the Grounded Theory. *Int. J. Environ. Res. Public Health* **2023**, *20*, 2932. [CrossRef] [PubMed]
10. Moreira, N.M.; Bryan, M. Noise annoyance susceptibility. *J. Sound Vib.* **1972**, *21*, 449–462. [CrossRef]
11. Li, Z.; Kang, J. Sensitivity analysis of changes in human physiological indicators observed in soundscapes. *Landsc. Urban Plan.* **2019**, *190*, 103593. [CrossRef]
12. Paiva, K.; Cardoso, M.-R.; Rodrigues, R. Noise pollution and annoyance: An urban soundscapes study. *Noise Health* **2015**, *17*, 125–133. [CrossRef]
13. Navickas, E. *Urban Soundscape: The Relationship between Sound Source Dominance and Perceptual Attributes along with Sound Pressure Levels*; University of Groningen: Groningen, The Netherlands, 2020.
14. Lavandier, C.; Defréville, B. The contribution of sound source characteristics in the assessment of urban soundscapes. *Acta Acust. United Acust.* **2006**, *92*, 912–921.
15. Schultz, T.J. Synthesis of social surveys on noise annoyance. *J. Acoust. Soc. Am.* **1978**, *64*, 377–405. [CrossRef] [PubMed]
16. Liu, J.; Kang, J. Soundscape design in city parks: Exploring the relationships between soundscape composition parameters and physical and psychoacoustic parameters. *J. Environ. Eng. Landsc. Manag.* **2015**, *23*, 102–112. [CrossRef]
17. Yilmazer, S.; Davies, P.; Yilmazer, C. A virtual reality tool to aid in soundscapes in the built environment (SiBE) through machine learning. In Proceedings of the INTER-NOISE and NOISE-CON Congress and Conference Proceedings, Glasgow, Scotland, 20–23 August 2023; pp. 737–747.
18. Sudarsono, A.S.; Lam, Y.W.; Davies, W.J. The validation of acoustic environment simulator to determine the relationship between sound objects and soundscape. *Acta Acust. United Acust.* **2017**, *103*, 657–667. [CrossRef]
19. Moors, A.; Ellsworth, P.C.; Scherer, K.R.; Frijda, N.H. Appraisal theories of emotion: State of the art and future development. *Emot. Rev.* **2013**, *5*, 119–124. [CrossRef]
20. Ode, Å.; Fry, G.; Tveit, M.S.; Messager, P.; Miller, D. Indicators of perceived naturalness as drivers of landscape preference. *J. Environ. Manag.* **2009**, *90*, 375–383. [CrossRef]
21. Luo, J.; Zhao, T.; Cao, L.; Biljecki, F. Water View Imagery: Perception and evaluation of urban waterscapes worldwide. *Ecol. Indic.* **2022**, *145*, 109615. [CrossRef]
22. Palmer, S.E.; Schloss, K.B.; Sammartino, J. Visual aesthetics and human preference. *Annu. Rev. Psychol.* **2013**, *64*, 77–107. [CrossRef]
23. Weitkamp, G.; Bregt, A.; Van Lammeren, R. Measuring visible space to assess landscape openness. *Landsc. Res.* **2011**, *36*, 127–150. [CrossRef]

24. Grahn, P.; Stigsdotter, U.K. The relation between perceived sensory dimensions of urban green space and stress restoration. *Landsc. Urban Plan.* **2010**, *94*, 264–275. [CrossRef]
25. Yin, Y.; Shao, Y.; Meng, Y.; Hao, Y. The effects of the natural visual-aural attributes of urban green spaces on human behavior and emotional response. *Front. Psychol.* **2023**, *14*, 1186806. [CrossRef] [PubMed]
26. Kim, S.-O.; Shelby, B. Effects of soundscapes on perceived crowding and encounter norms. *Environ. Manag.* **2011**, *48*, 89–97. [CrossRef] [PubMed]
27. Herranz-Pascual, K.; García, I.; Diez, I.; Santander, A.; Aspuru, I. Analysis of field data to describe the effect of context (Acoustic and Non-Acoustic Factors) on urban soundscapes. *Appl. Sci.* **2017**, *7*, 173. [CrossRef]
28. Liu, Y.; Hu, M.; Zhao, B. Audio-visual interactive evaluation of the forest landscape based on eye-tracking experiments. *Landsc Urban Plan* **2019**, *46*, 126476. [CrossRef]
29. Peschardt, K.K.; Stigsdotter, U.K.; Schipperrijn, J. Identifying features of pocket parks that may be related to health promoting use. *Landsc. Res.* **2016**, *41*, 79–94. [CrossRef]
30. Wang, R.; Jiang, W.; Lu, T. Landscape characteristics of university campus in relation to aesthetic quality and recreational preference. *Urban For. Urban Green.* **2021**, *66*, 127389. [CrossRef]
31. Khachatryan, H.; Rihn, A.; Hansen, G.; Clem, T. Landscape aesthetics and maintenance perceptions: Assessing the relationship between homeowners' visual attention and landscape care knowledge. *Land Use Policy* **2020**, *95*, 104645. [CrossRef]
32. Zube, E.H.; Sell, J.L.; Taylor, J.G. Landscape perception: Research, application and theory. *Landsc. Plan.* **1982**, *9*, 1–33. [CrossRef]
33. Liu, F.; Kang, J. Relationship between street scale and subjective assessment of audio-visual environment comfort based on 3D virtual reality and dual-channel acoustic tests. *Build. Environ.* **2018**, *129*, 35–45. [CrossRef]
34. Wang, X. Research on spatial acoustic landscape optimization of green spaces in street parks. *Build. Environ.* **2014**, *169*, 106544. [CrossRef]
35. Brambilla, G.; Gallo, V.; Asdrubali, F.; D'Alessandro, F. The perceived quality of soundscape in three urban parks in Rome. *J. Acoust. Soc. Am.* **2013**, *134*, 832–839. [CrossRef]
36. Liu, J.; Kang, J.; Luo, T.; Behm, H. Landscape effects on soundscape experience in city parks. *Sci. Total Environ.* **2013**, *454*, 474–481. [CrossRef] [PubMed]
37. Liu, J.; Yang, L.; Xiong, Y.; Yang, Y. Effects of soundscape perception on visiting experience in a renovated historical block. *Build. Environ.* **2019**, *165*, 106375. [CrossRef]
38. Gidlöf-Gunnarsson, A.; Öhrström, E. Noise and well-being in urban residential environments: The potential role of perceived availability to nearby green areas. *Landsc. Urban Plan.* **2007**, *83*, 115–126. [CrossRef]
39. Jo, H.I.; Jeon, J.Y. Effect of the appropriateness of sound environment on urban soundscape assessment. *Build. Environ.* **2020**, *179*, 106975. [CrossRef]
40. Yu, L.; Kang, J. Modeling subjective evaluation of soundscape quality in urban open spaces: An artificial neural network approach. *J. Acoust. Soc. Am.* **2009**, *126*, 1163–1174. [CrossRef]
41. Schirpke, U.; Tappeiner, G.; Tasser, E.; Tappeiner, U. Using conjoint analysis to gain deeper insights into aesthetic landscape preferences. *Ecol. Indic.* **2019**, *96*, 202–212. [CrossRef]
42. Lindeberg, T. Scale-space theory: A basic tool for analyzing structures at different scales. *J. Appl. Stat.* **1994**, *21*, 225–270. [CrossRef]
43. Nilsson, M.; Botteldooren, D.; De Coensel, B. Acoustic indicators of soundscape quality and noise annoyance in outdoor urban areas. In Proceedings of the 19th International Congress on Acoustics, Madrid, Spain, 2–7 September 2007.
44. Song, X.; Lv, X.; Yu, D.; Wu, Q. Spatial-temporal change analysis of plant soundscapes and their design methods. *Urban For. Urban Green.* **2018**, *29*, 96–105. [CrossRef]
45. Gulliver, J.; Morley, D.; Vienneau, D.; Fabbri, F.; Bell, M.; Goodman, P.; Beevers, S.; Dajnak, D.; Kelly, F.J.; Fecht, D. Development of an open-source road traffic noise model for exposure assessment. *Environ. Model. Softw.* **2015**, *74*, 183–193. [CrossRef]
46. Gehl, J. A changing street life in a changing society. *Places* **1989**, *6*, 8–17.
47. Li, Q.; Cui, C.; Liu, F.; Wu, Q.; Run, Y.; Han, Z. Multidimensional urban vitality on streets: Spatial patterns and influence factor identification using multisource urban data. *ISPRS Int. J. Geo-Inf.* **2021**, *11*, 2. [CrossRef]
48. Meng, Y.; Xing, H. Exploring the relationship between landscape characteristics and urban vibrancy: A case study using morphology and review data. *Cities* **2019**, *95*, 102389. [CrossRef]
49. Yang, W.; Kang, J. Acoustic comfort evaluation in urban open public spaces. *Appl. Acoust.* **2005**, *66*, 211–229. [CrossRef]
50. Wang, X. *Research on Optimization of Sound Landscape of Green Space in Street Park*; Tianjin University: Tianjin China, 2014.
51. Shao, Y.; Yin, Y.; Xue, Z.; Ma, D. Assessing and Comparing the Visual Comfort of Streets across Four Chinese Megacities Using AI-Based Image Analysis and the Perceptive Evaluation Method. *Land* **2023**, *12*, 834. [CrossRef]
52. Wen, X.; Lu, G.; Lv, K.; Jin, M.; Shi, X.; Lu, F.; Zhao, D. Impacts of traffic noise on roadside secondary schools in a prototype large Chinese city. *Appl. Acoust.* **2019**, *151*, 153–163. [CrossRef]
53. Van Renterghem, T.; Botteldooren, D. On the choice between walls and berms for road traffic noise shielding including wind effects. *Landsc. Urban Plan.* **2012**, *105*, 199–210. [CrossRef]
54. Bendtsen, H. *Noise Barrier Design: Danish and Some European Examples*; University of California Pavement Research Center: Davis, CA, USA, 2010.
55. Jo, H.I.; Jeon, J.Y. Overall environmental assessment in urban parks: Modelling audio-visual interaction with a structural equation model based on soundscape and landscape indices. *Build. Environ.* **2021**, *204*, 108166. [CrossRef]

56. Tan, J.K.A.; Hasegawa, Y.; Lau, S.-K.; Tang, S.-K. The effects of visual landscape and traffic type on soundscape perception in high-rise residential estates of an urban city. *Appl. Acoust.* **2022**, *189*, 108580. [CrossRef]

57. Brown, A.L. Soundscape planning as a complement to environmental noise management. In Proceedings of the Inter-Noise and NOISE-CON Congress and Conference Proceedings, Melbourne, Australia, 16–19 November 2014; pp. 5894–5903.

58. Guan, H.; Hu, S.; Lu, M.; He, M.; Mao, Z.; Liu, G. People's subjective and physiological responses to the combined thermal-acoustic environments. *Build. Environ.* **2020**, *172*, 106709. [CrossRef]

*Article*

# Forest Visitors' Multisensory Perception and Restoration Effects: A Study of China's National Forest Parks by Introducing Generative Large Language Model

Yu Wei [1,2,*] and Yueyuan Hou [3]

1 College of Economics and Management, Nanjing Forestry University, Nanjing 210037, China
2 NJFU Academy of Chinese Ecological Progress and Forestry Development Studies, Nanjing 210037, China
3 Swarma Research Center, Beijing 102300, China; houyy.cm@gmail.com
* Correspondence: weiyutour@njfu.edu.com

**Abstract:** Sensory perception of forests is closely related to human health and well-being. Based on attention recovery theory and stress relief theory, this paper investigates the influence of sensory perception of forests on visitors' restoration effects from a multidimensional and multisensory perspective, integrating the use of a generative large language model, regression analysis, and semantic analysis. The results of the study show that (1) the application of a generative large language model provides new ideas and methods to solve the dilemma caused by the traditional self-report scale measurement and provides a possible way to explore a new research paradigm in the context of the rapid development of generative artificial intelligence; (2) the effects of each sensory quantity differed, with the sensory quantities of sight, hearing, touch, and taste having a significant positive effect on visitors' restoration effects, and the sense of smell having a significant negative effect on visitors' restoration effects; (3) sensory psychological distance partially had a significant effect on visitors' restoration effects, both proximal psychological distance and distal psychological distance were significantly correlated with visitors' restoration effects, and intermediate psychological distance had a negative effect on visitors' restoration effects, but the effect was not significant; (4) the sensory dimension has a significant positive effect on visitors' restoration effects, the integration and synergistic effect of the senses are enhanced, and multidimensional sensory cross-perception has a positive effect on visitors' restoration effects at the social health level; and (5) the sensory elements of National Forest Parks that influence visitors' restoration effects are mainly natural attributes, and the elements related to "people" also play an important role in visitors' restoration effects. This study provides a useful complement to the study of forest sensory perception, and at the same time has an important reference value for exploring the management of forest recreation experience and sensory marketing practices.

**Keywords:** multisensory perception; restoration effects; forest recreation; generative large language model; National Forest Parks

**Citation:** Wei, Y.; Hou, Y. Forest Visitors' Multisensory Perception and Restoration Effects: A Study of China's National Forest Parks by Introducing Generative Large Language Model. *Forests* **2023**, *14*, 2412. https://doi.org/10.3390/f14122412

Academic Editor: Cate Macinnis-Ng

Received: 9 October 2023
Revised: 2 December 2023
Accepted: 7 December 2023
Published: 11 December 2023

## 1. Introduction

Currently, in the context of global climate change, biodiversity destruction, and pollution issues, health is increasingly becoming a major public health issue and a prominent social problem worldwide [1]. The uncertainty after the COVID-19 pandemic has intensified the potential health threats to human society, and physical and mental health has become a highly prioritized global issue. Studies show that global rates of anxiety and depression increased by more than 25% in the first year of the COVID-19 pandemic [2]. Previous research demonstrates that being in a natural environment can effectively enhance physical, mental, and emotional well-being [3–7]. Among various natural settings, forests have been extensively shown to offer advantages for psychological, physiological, and social well-being. Various cultural ecosystem services are provided by forests around the world, including forest therapy, forest bathing, and shinrin-yoku [8].

Tourism and recreational activities carried out in forests are often in multisensory stimulating environments [9,10]. Sensory perception in forests is usually associated with attention and emotional regulation. Restorative properties are typically experienced through sensory perception, and the quality of sensory perception and experience can also impact restoration effects. Research indicates that engaging all senses enhances the relaxation and restorative potential of an environment [11], while different sensory stimuli may produce different restorative experiences.

However, current research has limitations. Firstly, studies predominantly concentrate on single sensory perception. In real forest settings, individuals' sensory perception is often all-inclusive and even too complex to discern entirely. The restorative effects of a single sensory input experienced by visitors have primarily been examined through a controlled laboratory environment [8], which is somewhat different from the multidimensional and multisensory perception of the real environment. Secondly, numerous studies have examined the relationship between forest sensory perception and visitors' restoration effects. However, many of these studies only scratch the surface of the relationship without delving into the specific sensory elements. Moreover, exploring only a certain type or a few types of sensory perception in a generalized manner still holds limited practical significance. Once again, due to the high cost of collecting actual environmental data, studies are generally restricted to particular forests or regions, and large-scale, national-level, region-wide, and long-term studies remain infrequent. Finally, self-reported scales are often used to collect data in studies of forest sensory perception and visitors' restoration effects, which have many limitations in terms of data reliability and consistency, individual subject preferences, memory bias, and the influence of social expectations [12,13], which affect the reliability of the studies to a certain extent.

Benefiting from the rapid improvement of current computer performance, the generative model that learns the probability density of observable samples and generates new samples randomly has become a prominent topic [14]. This type of model can not only realize the mining of shallow semantic features but can also achieve the learning of abstract deep semantic features, and the model performance is superior. The generative AI known as ChatGPT is a standard large language model with minimal data annotation needs, the ability to generate cross-modal content, and robust logic and organization skills [15–17]. Recently, there has been a growing interest in utilizing generative large language models within the field of social sciences [18], with some even considering it a new paradigm [19].

The emergence of generative large language models offers potential ideas and approaches to overcome constraints in the present research area of forest sensory perception and the restorative effects on visitors. Initially, conduct multi-dimensional sensory studies in actual surroundings. The generative large language model utilizes genuine user-generated data from the environment, such as travel notes, to overcome the limitations of laboratory data, which can be divorced from the real environment and target a single sensory perception to some extent. As a result, it better reflects individual behavior in the actual forest environment. Additionally, there is extensive, inter-temporal, and massive data batch processing. The generative, large language model facilitates batch processing of travelogues, spanning various research groups and temporal-spatial dimensions. This enables large-scale, territory-wide, and time-spanning research with enhanced data representativeness and higher data processing efficiency. Furthermore, it mitigates the limitations of self-reported scale measurements. The large generative language model can produce relatively consistent data when operating under identical input conditions, thus eliminating potential bias caused by individual variations and subjectivity in self-reported questionnaires. Additionally, this approach sidesteps the issues arising from social expectations, self-presentation, and recall delay that often plague manual completion. Finally, this paves the way for expanding research horizons and areas of investigation. The use of a generative large language model, based on rich corpus training, can offer a more extensive analysis of the scale. For instance, it can provide a comprehensive assessment of the various dimensions of the perceptual restorative scale concerning environmental elements, sensory perception, and comfort. This, in turn, enables researchers to broaden their horizons and gain potential insights.

Therefore, the research objectives of this paper are threefold: (1) to expand the application of generative large language models in the field of forest recreation and tourism, and to provide possible ideas and methods for solving the dilemmas posed with traditional self-reported scale measurements; (2) to explore the influence relationship between multisensory perception and visitors' restoration effects of forest tourists; and (3) to clarify the deep-level association between various sensory elements and visitors' restoration effects, and to construct a multidimensional and multi-level spectrum of forest sensory perception elements. The research results provide a theoretical basis and practical guidance for the users of the forest environment and the improvement of the health and well-being of tourists.

## 2. Theoretical Analysis and Research Hypotheses

### 2.1. Attention Restoration Theory and Stress Recovery Theory

In exploring the multidimensional and multisensory experience and restoration effects of forests, attention restoration theory and stress recovery theory provide theoretical ideas for our study. The attention restoration theory was proposed by the Kaplans in 1995 and argues that human attention can be divided into two types, voluntary attention and involuntary attention, and that voluntary attention causes inhibition of nerve centers and overuse of function, resulting in fatigue, difficulty in concentrating, and ease of agitation, which may lead to errors in work [11,20]. On the contrary, involuntary attention occurs spontaneously without the need to invest a lot of energy in paying attention, so it is conducive to the recovery of intentional attention [21]. Attention restoration theory suggests that human beings inherently favor the natural environment so that when they are in nature, they mainly focus on involuntary attention, which leads to the restoration of voluntary attention [22]; this kind of environment is called the restorative environment [23], and forests, grasslands, and urban green spaces are typical restorative environments.

Stress recovery theory was developed by Ulrich in 1991. The theory states that the natural environment has a positive effect on the physical and emotional recovery of human beings [24]. Specifically, there are three elements that play a role in relieving stress in people: non-threatening landscapes, greenery, and specific natural landscapes [25]. Ulrich argues that this ability to respond to stress relief in specific natural environments comes as a result of human evolution [26]. Humans respond to the three types of natural environments mentioned above instantly and unconsciously, requiring minimal cognitive resources to process responses, and are thus able to quickly recover their strength from stress [20].

### 2.2. Sensory Quantity and Visitors' Restoration Effects

The term 'sensory quantity' pertains to the amount of different senses that individuals encounter when engaging in recreational activities in forested areas. According to the attention recovery theory, which discusses the relationship between the experience of natural environments and involuntary attention, the increase in sensory stimulation in forest environments usually occurs without the need to actively devote attention and thus reduces the brain's tension in conscious tasks and helps the brain to achieve rest and recovery. For instance, forests are typically associated with attractive natural landscapes and peaceful surroundings [27]. Additionally, the sounds of birds chirping, water flowing, and wind blowing within forests are often experienced as enjoyable [28]. Furthermore, physical contact with trees may cultivate a stronger sense of connectedness between individuals and natural environments. The scents of trees and flowers promote physical and mental relaxation [29,30] while indulging in a picnic or dining outdoors in the forest enhances one's positive emotions. According to this theory, more sensory stimulation may better facilitate involuntary attention, thus helping to restore resources for voluntary attention. An increase in sensory experience involves not only an increase in mere quantity but also an enhanced diversity of sensory stimuli, which in turn stimulates different feelings and emotions and increases the levels and dimensions of attentional recovery, thus positively affecting visitors' restoration effects. Therefore, we propose the following hypotheses:

**H1a:** *The number of visual senses has a significant positive effect on the visitors' restoration effects.*

**H1b:** *The number of hearing senses has a significant positive effect on the visitors' restoration effects.*

**H1c:** *The number of smell senses has a significant positive effect on the visitors' restoration effects.*

**H1d:** *The number of touch senses has a significant positive effect on the visitors' restoration effects.*

**H1e:** *The number of taste senses has a significant positive effect on the visitors' restoration effects.*

### 2.3. Sensory Psychological Distance and Visitors' Restoration Effects

Psychological distance can be defined as the subjective experience that something is very close or very far from the self, here and now [31,32]. Various dimensions determine the way in which objects are dispersed in psychological space, resulting in different types of psychological distance. According to the level of interpretation theory, when one's psychological distance is close, greater focus is placed on concrete details, utilizing fewer abstract mental representations, whereas when one's psychological distance is more distant, intensified attention is paid to points of principle and deeper meanings, disregarding both concrete content and details, thereby utilizing more abstract mental representations [31,33,34].

For the psychological distance of the senses, Ryan argues that the maximum distance perceived by the senses varies in different sensory experiences [35]. Specifically, things can only be tasted or touched when they are held in the mouth or touched by hand, so touch and taste belong to the sensory experience of proximal psychological distance; through light reflection and sound waves, things can be seen or heard even at a longer distance, so vision and hearing belong to the sensory experience of distal psychological distance. The sense of smell relies on the diffusion of molecules and can be perceived by individuals within a certain physical distance, so the sense of smell is between the proximal and distal end of the psychological distance sensory experience [36].

Different tourism and recreation activities exhibit varying psychological distances for tourists, which align with different levels of interpretation. For example, relaxing tourism activities possess a closer psychological distance, corresponding with a lower level of interpretation. In contrast, tourism activities that provide challenges tend to be associated with farther psychological distances, indicating higher levels of interpretation. Ensuring alignment between various psychological distances and levels of interpretation can result in stimulating positive psychological states [37]. Tourism and recreation activities in forests are typically relaxing tourism activities, and thus, the proximal psychological distance senses of taste and touch will have a more positive impact on the visitors' restoration effects. Further, from the perspective of the methods and properties of perception, proximal psychic distance sensory experiences like touch and taste, which are typically experienced through physical contact or ingestion, are more likely to create profound interactions and emotional connections and foster positive restorative effects in visitors. Meanwhile, distal psychic distance sensory experiences such as visual and auditory experiences possess passive features and might be adversely impacted by excessive stimuli or noise pollution. Therefore, we contend that all sensory experiences in forest environments are impacted by psychological distance. The senses' restorative effects are more pronounced the closer the distance, and conversely, less favorable when greater. Thus, we formulate the hypothesis accordingly.

**H2a:** *The proximal psychological distance sensory has a significant positive effect on the visitors' restoration effects.*

**H2b:** *The medial psychological distance sensory has a significant negative effect on the visitors' restoration effects.*

**H2c:** *The distal psychological distance sensory has a significant negative effect on the visitors' restoration effects.*

*2.4. Sensory Dimensions and Visitors' Restoration Effects*

In the real world, the human brain can combine cues from various sensory channels to perceive external objects and events [38,39]. This process effectively merges information from different cues within the same sensory channel and cues from different sensory channels into a unified, coherent, and robust perception, known as perceptual integration of cues from multiple sensory channels, or multisensory integration for short [40]. Multiple fields, including neurophysiology, electrophysiology, and psychology, have conducted studies on multisensory integration. Scholars in recreation and tourism research have emphasized the synesthesia effect as an interconnected, cross-sensory experience [41]. Certain scholars examine the impact of multisensory interactions on the identity of tourism destinations, with sight and smell deemed as the most effective forms of synesthesia [42]. The research indicates that the way in which natural environments' soundscapes are characterized, either directly or through the mediation of visual landscapes, contributes significantly to the tourists' attentional recovery and quality of life [43]. Compared to the rise in the quantity of senses in terms of total count, sensory dimensions predominantly explore the impact of sensory integration, crossover, and complementarity among multiple dimensions on visitors' restoration effects. Augmenting the number of senses does not necessarily lead to an increase in sensory dimensions, whereas enhancing sensory dimensions can potentially heighten the depth and richness of the experience. Therefore, we posit the following hypothesis:

**H3:** *Sensory dimension has a significant positive impact on visitors' restoration effects.*

### 3. Methodology

With the continuous development of artificial intelligence, computational intelligence and perceptual intelligence are gradually developing toward the level of cognitive intelligence [44]. The further improvement of computing power and the explosive growth of massive data have led to the application of deep learning methods in natural language processing. Since 2018, the concept of transfer learning has been introduced, and large language models with a huge training corpus scale and parameters have emerged. Representative language models include ELMo, ULMFiT, BERT, PaLM, BLOOM, and OpenAI GPT.

In recent years, there has been growing skepticism toward the use of self-reported scale measurements in psychology due to the high cost of questionnaires, the limited representativeness and generalization ability of data, the difficulty in guaranteeing the reliability and consistency of data, and the interference of subjects' individual preferences, memory bias, and social expectations [12]. The rapid development of generative large language models provides possible ideas and methods for generating scale questionnaires to solve the above dilemmas to a certain extent by using travelogue texts. First of all, semantic analysis of texts of different types and genres through machine learning and deep learning models, and then understanding the subjective emotions and mental states of the authors, has achieved some results in previous studies [45–47]. Texts targeting the travelogue aspect of tourists have been similarly applied to the study of travelers' emotions, subjective preferences, and mental states [48,49]. Secondly, travelogue texts serve as a factual account of tourists' travel activities and corresponding emotions. They capture vivid sensory experiences and can demonstrate the sensory, emotional, and psychological states of tourists [50]. This aligns with the primary focus of psychological scale questions and lays the groundwork for developing scale questionnaires from travelogue texts. Finally, the deep learning-based generative language model exhibits a remarkable performance in reading comprehension tasks, enabling intricate text mining and profound semantic analysis. Moreover, it presents the opportunity to construct extensive questionnaires by means of travelogue texts at a methodological level.

For this reason, this study tries to introduce a generative large language model, on the basis of which the research methodology and analysis route of this paper is formed (Figure 1), which includes a total of five modules: data collection, data preprocessing, generative large language model construction, regression model construction, and semantic analysis.

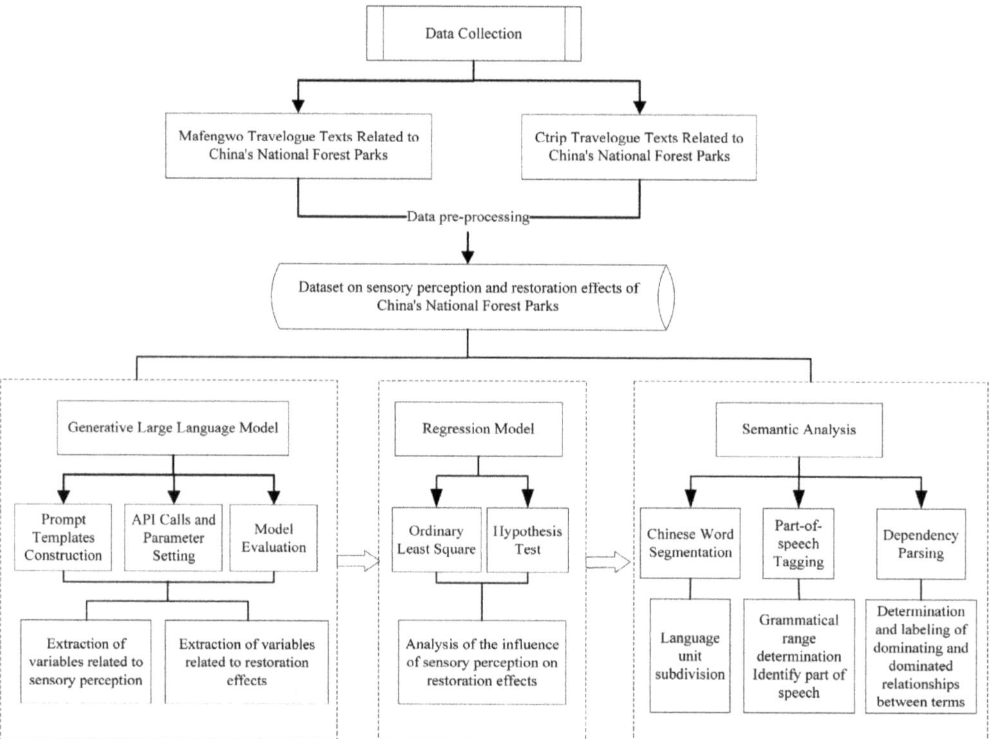


**Figure 1.** Research methodology and analysis route.

### 3.1. Study Area

This study takes China's National Forest Parks as the research area (Figure 2). These parks were established by China's National Forestry and Grassland Administration to rationally utilize forest landscape resources and promote forest tourism. The parks are classified into three tiers, of which the national tier is the pinnacle. It encompasses forest parks with stunning forest views, concentrated cultural landscapes, high aesthetic value, as well as scientific and cultural importance. The parks also occupy significant geographical locations, have a certain degree of regional representativeness, and provide a complete range of tourism services, making them highly popular. The attributes and characteristics of National Forest Parks are representative and typical for this paper to study the sensory experience and visitors' restoration effects of forests. The study selects 897 of China's National Forest Parks.

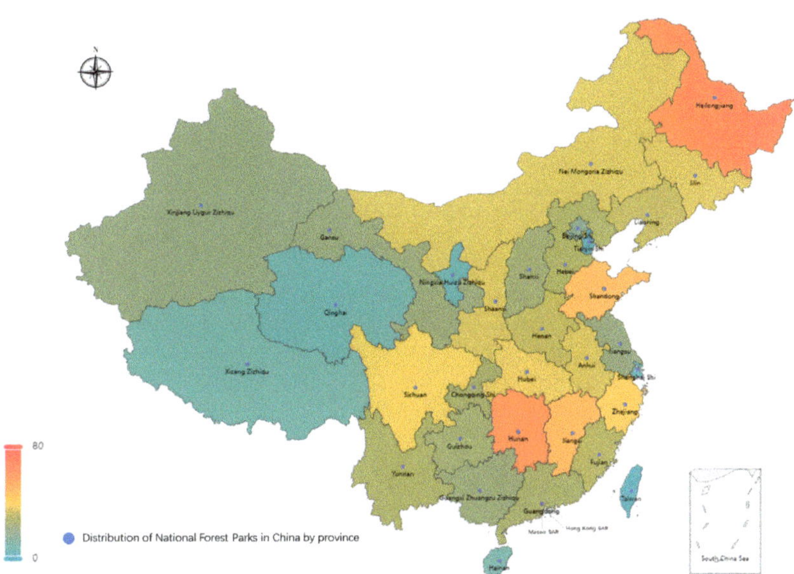



**Figure 2.** Distribution of China's National Forest Parks.

### 3.2. Data Collection

In this study, we collected travelogue text data to conduct a study on forest tourists' multisensory perception and recreational effects. User-generated social media data have certain advantages in conducting such studies: first, the data scale is larger and sample-rich. The large scale of social media data contains rich information from a large number of users, allowing researchers to observe large and diverse sensory experiences. Second, authenticity and real time. Social media data are generated in real time and can capture users' sensory experiences in real life with timeliness. Third, lower cost. Compared to professional guidance, sensory perception research based on user-generated social media data has lower time and labor costs.

In accordance with the selection criteria of being manageable, reliable, and representative, combined with the characteristics of the research object of this paper, Mafengwo and Ctrip were selected as the text collection platforms of National Forest Parks.

Mafengwo is the largest UGC tourism website in China, with more than 100 million users [51,52]; Ctrip is one of the largest comprehensive tourism service websites in China, which has built a more comprehensive UGC content service system, and many scholars have used its data to conduct tourism-related research [53–55]. The study takes the name of the National Forest Park as the keyword, and searches in the platform of Mafengwo and Ctrip, using the collector crawler software (version 4.0.1) to obtain 1557 Mafengwo travelogues and 292 Ctrip travelogues, totaling 1849 travelogues and 3,272,687 words.

### 3.3. Data Preprocessing

Before analyzing the data, it is imperative to carry out essential preprocessing steps such as cleaning and filtering to enhance the quality of the data. The data preprocessing in this study comprises two components: data cleaning and data selection. The primary objective of the data cleaning process is to remove unnecessary HTML tags, nonsensical emojis, and other symbols from the data. On this basis, data selection is carried out, which means that for the characteristics of the research object, a number of rules applicable to the National Forest Park are proposed in a targeted manner, and the noisy data with less relevance to it are eliminated to ensure the quality of model generation. The principles are as follows: firstly, exclude travelogues that do not mention the name of the National Forest

Park or only contain relevant pictures. Secondly, exclude travelogues that do mention the National Forest Park but lack substantive content or only use the park's name as a caption for pictures. Third, when dealing with travel notes that touch on National Forest Parks but are not the main focus of the text, it is necessary to locate the paragraph containing the park's name and the one preceding and following it. Once identified, the text should be deduplicated. Fourth, due to the complexity of the visitors' restoration effects, if the visitors' restoration effect-related items in the model generation results are extracted as "none" and the results are correct after manual review, such travelogues will be deleted.

### 3.4. Research Methods

This study utilizes a combination of a large language model, regression analysis, and semantic analysis. First, the independent and dependent variables of forest visitors' sensory perception were generated by a generative large language model. Second, the relationship between sensory quantity, sensory distance and sensory dimensions, and the visitors' restoration effects was empirically analyzed using the ordinary least squares regression model. Finally, relying on the semantic analysis method of natural language processing, we use Chinese word segmentation, part-of-speech tagging, and dependency parsing to identify the specific sensory elements that affect the visitors' restoration effects, and construct the sensory element spectrum.

## 4. Generative Large Language Model

The application of generative large language models mainly includes three modules: construction of prompt templates, API calls and parameter settings, and model evaluation (Figure 3).

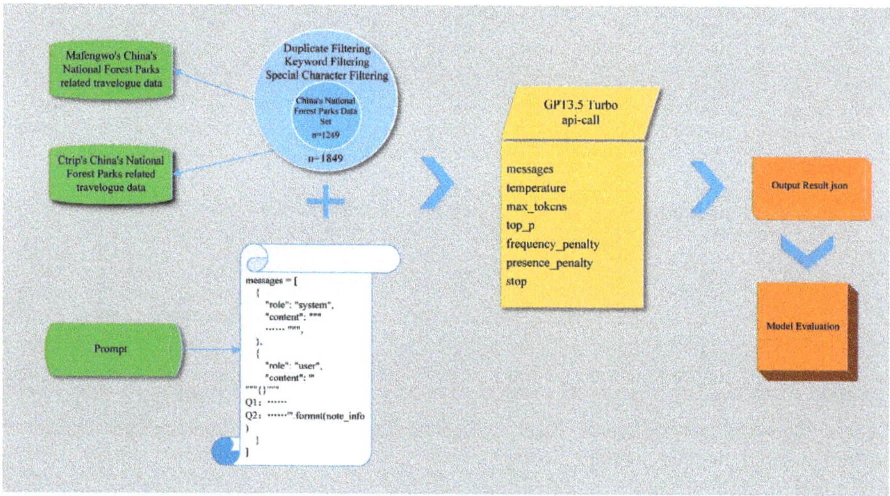


**Figure 3.** Generative large language model calling.

### 4.1. Prompt Templates Construction

Prior to large-scale experiments, we conducted experiments in a sandbox environment. First, the bias problem. Due to the possible bias of the training data and the algorithm itself, the content generated by generative big language models may be biased toward a certain group of people or produce discriminatory results, and the general bias can be classified into social bias and linguistic bias [56]. Because the travelogue data in this study contains less content in the areas of politics, religion, gender, etc., the social bias is more obvious; coupled with the fact that the model uses the existing travelogue data to generate the training results, the input data is relatively more controllable and deterministic, and thus the social bias has less impact on the study. Another type of bias is language bias, but as Chinese data were used as one of the ChatGPT training data sources, language bias has less impact. However, taking into account that the English training data compared to the Chinese data is much larger and richer, this study still tries to use English for testing; that is, the prompt template and the travelogue text are translated into English, and it is found that affected by the translation effect, the output effect of the model is not significantly improved compared to that of Chinese. Second, the problem of AI hallucinations. AI hallucinations refer to the content generated by the model that is not based on any real data but is a product of the imagination of the large language model [57,58]. Because the input data of this study is relatively controllable and certain, and the use of the model is 'something out of something' rather than 'something out of nothing', the illusion problem has less impact on the study.

On this basis, the prompting strategy was optimized for the characteristics of this study. First, clarity. Formulate questions as clearly as possible for sensory experience and visitors' restoration effects; for example, appropriately use separators to distinguish different questions and expressions in the perceptual restoration measure to eliminate ambiguity in the prompts. Second, stepwise. The extraction and generation of variables, such as sensory perception and restoration effect, require a methodical approach to prevent mistakes during the model's inference process. Therefore, we adopt a step-by-step methodology. Thirdly, providing examples. We supply examples to establish consistency in the model output's style, format, and content details. This assists in gathering statistics related to variables including sensory perception and visitors' restoration effects. Finally, providing reasoning time. There is a large number of travelogue texts and a common issue is that a single travelogue can be lengthy. To prevent the model from making mistakes when extracting content from corresponding variables, subsequent queries are used in the prompts. This ensures that the model does not miss relevant content and obtains better performance.

### 4.2. API Calls and Parameter Setting

Access the GPT-3.5 model through the OpenAI API and set the model parameters [59], including temperature, max tokens, top P, presence penalty, and frequency penalty. Temperature controls the randomness and determinism of the model's output, with values ranging from 0 to 2. The lower the value, the more centralized and deterministic the model output results, and the higher the value, the more random and diverse. The presence penalty determines the probability of introducing a new subject in the output, with a range of values from −2.0 to 2.0. The frequency penalty controls the likelihood that the output results in the model repeating the same line verbatim, taking values between −2.0 and 2.0. Positive values penalize new tokens based on their existing frequency in the text, thus reducing the likelihood of the model repeating the same line word by word. The presence penalty is a one-time addition that applies to all tokens that have been sampled at least once, while the frequency penalty is proportional to the frequency with which a particular token has been sampled. The model parameters for this study were determined through iterative debugging with small samples (Table 1).

**Table 1.** Parameter setting.

| Name | Setting |
| --- | --- |
| temperature | 1 |
| max_tokens | 5000 |
| top_p | 0.95 |
| presence penalty | 0 |
| frequency penalty | 0 |

### 4.3. Model Evaluation

The large language model evaluation in this study can be regarded as a class of classification model performance evaluation. In order to evaluate the model results, we extracted the text of the travelogues from the experimental sample travelogues for manual annotation. The model performance is examined by calculating the *Precision* and *Recall*, and the balance between the two is quantified using the *F1 Score*. The formula is shown below:

$$Precision = \frac{TP}{TP + FP} \tag{1}$$

$$Recall = \frac{TP}{TP + FN} \tag{2}$$

$$F1\ Score = 2 * \frac{Precision\ *\ Recall}{Precision + Recal} \tag{3}$$

where *TP* denotes the number of true positives, *FP* denotes the number of false positives, and *FN* denotes the number of false negatives.

## 5. Results

### 5.1. Generative Large Language Model Evaluation

The generative large language model evaluation is mainly divided into two parts. The first is the classification effect evaluation for the five sensory categories of sight, hearing, smell, touch, and taste (Table 2, Figure 4); the second is the scoring effect evaluation for the restoration effect and its various subitems (Tables 3 and 4, Figure 4). In this study, the model performance was mainly evaluated using the three indicators of *Precision*, *Recall*, and *F1 Score*. Overall, we paid more attention to *Precision* than *Recall*; this is because we hope that the model classifies the travelogue text correctly according to the senses or scores it correctly according to a scale of 1–5 better than it omits the related travelogue text, in this way ensuring the accuracy of the model's output results.

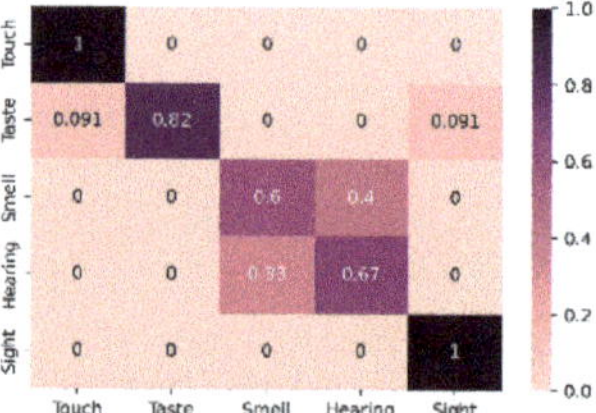

**(a)** Sensory categorization confusion matrix

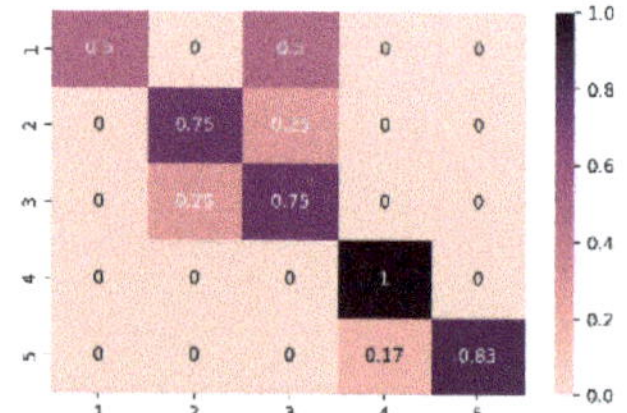

**(b)** Restoration effects confusion matrix

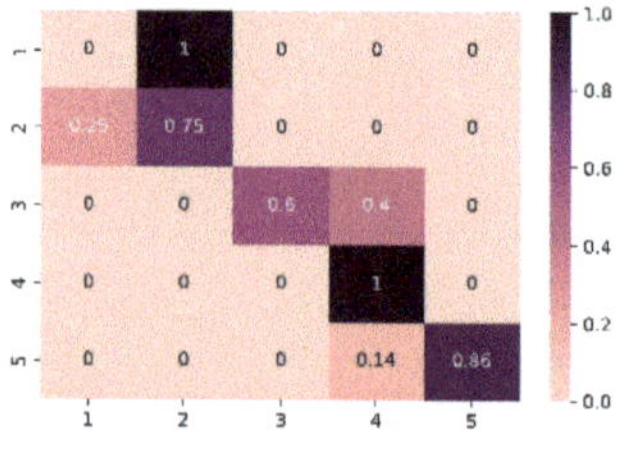

**(c)** Q1 Confusion Matrix

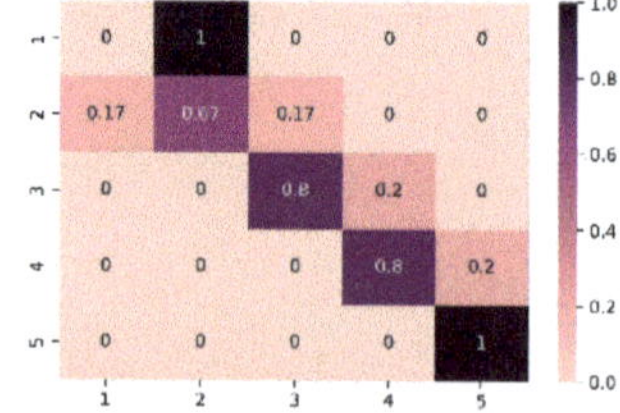

**(d)** Q2 Confusion Matrix

**Figure 4.** *Cont.*

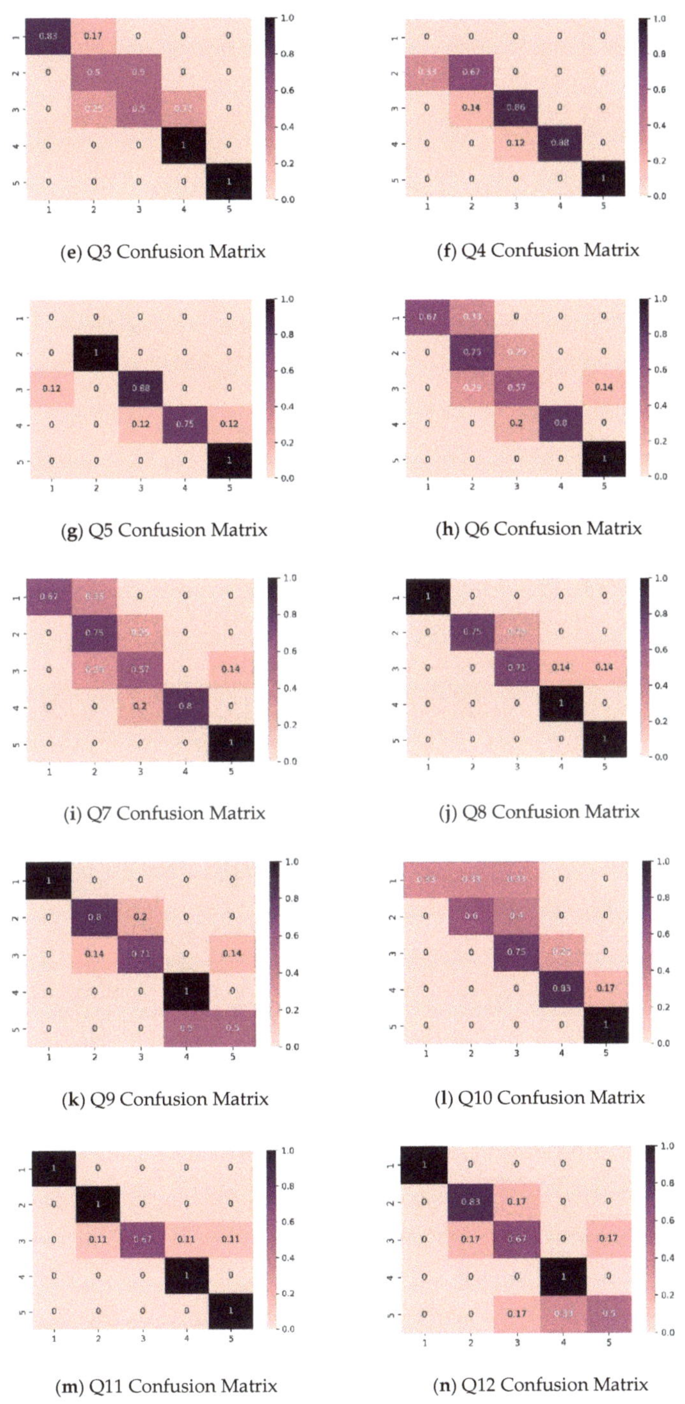

**Figure 4.** Confusion matrix for sensory categorization, visitors' restoration effects, and their subtopic items. (**a**) is the evaluation results of sensory classification, (**b**) is the evaluation results of the overall score of the restoration effect, and (**c–n**) is the subtopic of the restoration effect.

**Table 2.** Model evaluation results of sensory categorization.

| Term | Precision | Recall | F1 Score |
| --- | --- | --- | --- |
| Hearing | 0.67 | 1.00 | 0.80 |
| Sight | 1.00 | 0.82 | 0.90 |
| Smell | 0.75 | 0.60 | 0.67 |
| Taste | 0.50 | 0.67 | 0.57 |
| Touch | 0.50 | 1.00 | 0.67 |
| Accuracy | | | 0.77 |
| Macro avg | 0.68 | 0.82 | 0.72 |
| Weighted avg | 0.82 | 0.77 | 0.78 |

**Table 3.** Model evaluation results of visitors' restoration effects.

| Term | Precision | Recall | F1 Score |
| --- | --- | --- | --- |
| 1 | 1.00 | 0.50 | 0.67 |
| 2 | 0.80 | 0.80 | 0.80 |
| 3 | 0.60 | 0.75 | 0.67 |
| 4 | 0.86 | 1.00 | 0.92 |
| 5 | 1.00 | 0.83 | 0.91 |
| accuracy | | | 0.83 |
| macro avg | 0.85 | 0.78 | 0.79 |
| weighted avg | 0.85 | 0.83 | 0.83 |

For the effect of sensory classification, the overall *Precision* reaches 77%, and the macro average and weight average are 72% and 78%, respectively. Among them, in terms of *Precision*, sight, smell, and hearing had higher classification *Precision*, and in terms of *F1 Score*, sight and hearing had the best classification effect, especially sight perception, which exhibits minimal classification bias and performs with an *F1 Score* of 90%. The following in classification performance are smell and touch. Taste, however, exhibits the poorest score due to the intermingling of smell perception, such as food aroma and smell, and taste perception, such as food flavor and taste, during the classification process.

Regarding the classification effect of the visitors' restoration effects, the overall accuracy reaches 83%, and the macro average and weight average are 83% and 79%, respectively. Among them, from the perspective of *Precision*, the classification *Precision* scores of 5, 1, and 4 are higher. From the *F1 Score*, the evaluation results of 5 and 4 are better, reaching, respectively, 91% and 92%, and 1 and 3 points are more unstable. Judging from the 12 subitems, the overall accuracy rate is between 73% and 86%. In terms of the scoring results of each item, on the whole, the results with scores of 5, 4, and 1 are more stable and have a higher *Precision*, while the results with scores of 3 and 2 fluctuate more. Regarding specific sub-topics, those that are more associated with tourists' personal feelings or experiences, such as "I am mesmerized by this place", "This place suits my personality", and "This place meets my expectations", generally produce less stable results. However, topics that are more precise or closely related to physical space tend to have better accuracy and *F1 Score*. This highlights differences in the impact of model training on subjective questions versus objective descriptions and implies that the model's outputs could be influenced by subjective elements within the training data. Nonetheless, it is important to acknowledge that in practical evaluations, individual variation in perceptions of emotions and experiences can also contribute to fluctuations in evaluation outcomes. Therefore, it is crucial to include the impact of this type of volatility on the results of the study at the data level by expanding the sample size or utilizing big data as much as possible.

**Table 4.** Model evaluation results of the 12 subtopic items of visitors' restoration effects.

| Term | | Precision | Recall | F1 Score | Term | | Precision | Recall | F1 Score |
|---|---|---|---|---|---|---|---|---|---|
| Q1 The material arrangements here are well organized | 1 | 0.00 | 0.00 | 0.00 | Q7 This place is different from where I usually live | 1 | 1.00 | 0.50 | 0.67 |
| | 2 | 0.75 | 0.75 | 0.75 | | 2 | 0.80 | 0.67 | 0.73 |
| | 3 | 1.00 | 0.60 | 0.75 | | 3 | 0.40 | 0.50 | 0.44 |
| | 4 | 0.62 | 1.00 | 0.77 | | 4 | 0.80 | 0.80 | 0.80 |
| | 5 | 1.00 | 0.86 | 0.92 | | 5 | 0.83 | 1.00 | 0.91 |
| | accuracy | | | 0.77 | | accuracy | | | 0.73 |
| | macro avg | 0.68 | 0.64 | 0.64 | | macro avg | 0.77 | 0.69 | 0.71 |
| | weighted avg | 0.82 | 0.77 | 0.78 | | weighted avg | 0.75 | 0.73 | 0.73 |
| Q2 Everything here matches the overall environment | 1 | 1.00 | 0.50 | 0.67 | Q8 This place makes me feel rested | 1 | 1.00 | 1.00 | 1.00 |
| | 2 | 0.80 | 0.80 | 0.80 | | 2 | 1.00 | 0.75 | 0.86 |
| | 3 | 0.60 | 0.75 | 0.67 | | 3 | 0.83 | 0.71 | 0.77 |
| | 4 | 0.86 | 1.00 | 0.92 | | 4 | 0.83 | 1.00 | 0.91 |
| | 5 | 1.00 | 0.83 | 0.91 | | 5 | 0.80 | 1.00 | 0.89 |
| | accuracy | | | 0.83 | | accuracy | | | 0.86 |
| | macro avg | 0.85 | 0.78 | 0.79 | | macro avg | 0.89 | 0.89 | 0.88 |
| | weighted avg | 0.85 | 0.83 | 0.83 | | weighted avg | 0.87 | 0.86 | 0.86 |
| Q3 This place seems infinite and allows exploration | 1 | 1.00 | 0.83 | 0.91 | Q9 This place is a haven for me | 1 | 1.00 | 1.00 | 1.00 |
| | 2 | 0.50 | 0.50 | 0.50 | | 2 | 0.80 | 0.80 | 0.80 |
| | 3 | 0.50 | 0.50 | 0.50 | | 3 | 0.83 | 0.71 | 0.77 |
| | 4 | 0.75 | 1.00 | 0.86 | | 4 | 0.60 | 1.00 | 0.75 |
| | 5 | 1.00 | 1.00 | 1.00 | | 5 | 0.67 | 0.50 | 0.57 |
| | accuracy | | | 0.77 | | accuracy | | | 0.77 |
| | macro avg | 0.75 | 0.77 | 0.75 | | macro avg | 0.78 | 0.80 | 0.78 |
| | weighted avg | 0.78 | 0.77 | 0.77 | | weighted avg | 0.79 | 0.77 | 0.77 |
| Q4 This place makes me curious about things | 1 | 0.00 | 0.00 | 0.00 | Q10 This place suits my personality | 1 | 1.00 | 0.33 | 0.50 |
| | 2 | 0.67 | 0.67 | 0.67 | | 2 | 0.75 | 0.60 | 0.67 |
| | 3 | 0.86 | 0.86 | 0.86 | | 3 | 0.50 | 0.75 | 0.60 |
| | 4 | 1.00 | 0.88 | 0.93 | | 4 | 0.83 | 0.83 | 0.83 |
| | 5 | 1.00 | 1.00 | 1.00 | | 5 | 0.80 | 1.00 | 0.89 |
| | accuracy | | | 0.86 | | accuracy | | | 0.73 |
| | macro avg | 0.70 | 0.68 | 0.69 | | macro avg | 0.78 | 0.70 | 0.70 |
| | weighted avg | 0.91 | 0.86 | 0.88 | | weighted avg | 0.77 | 0.73 | 0.72 |
| Q5 This place has a lot of interesting things that catch my attention | 1 | 0.00 | 0.00 | 0.00 | Q11 There are few boundaries here that restrict my movement | 1 | 1.00 | 1.00 | 1.00 |
| | 2 | 1.00 | 1.00 | 1.00 | | 2 | 0.80 | 1.00 | 0.89 |
| | 3 | 0.88 | 0.88 | 0.88 | | 3 | 1.00 | 0.67 | 0.80 |
| | 4 | 1.00 | 0.75 | 0.86 | | 4 | 0.80 | 1.00 | 0.89 |
| | 5 | 0.75 | 1.00 | 0.86 | | 5 | 0.80 | 1.00 | 0.89 |
| | accuracy | | | 0.86 | | accuracy | | | 0.86 |
| | macro avg | 0.72 | 0.72 | 0.72 | | macro avg | 0.88 | 0.93 | 0.89 |
| | weighted avg | 0.92 | 0.86 | 0.88 | | weighted avg | 0.89 | 0.86 | 0.86 |
| Q6 I am mesmerized by this place | 1 | 1.00 | 0.67 | 0.80 | Q12 This place meets my expectations | 1 | 1.00 | 1.00 | 1.00 |
| | 2 | 0.50 | 0.75 | 0.60 | | 2 | 0.83 | 0.83 | 0.83 |
| | 3 | 0.67 | 0.57 | 0.62 | | 3 | 0.67 | 0.67 | 0.67 |
| | 4 | 1.00 | 0.80 | 0.89 | | 4 | 0.60 | 1.00 | 0.75 |
| | 5 | 0.75 | 1.00 | 0.86 | | 5 | 0.75 | 0.50 | 0.60 |
| | accuracy | | | 0.73 | | accuracy | | | 0.73 |
| | macro avg | 0.78 | 0.76 | 0.75 | | macro avg | 0.77 | 0.80 | 0.77 |
| | weighted avg | 0.77 | 0.73 | 0.73 | | weighted avg | 0.74 | 0.73 | 0.72 |

## 5.2. Regression Analysis and Hypothesis Test

The regression analysis (Table 5) clearly shows that the quantity of different senses, except in the case of smell, has a significant positive correlation with the visitors' restoration effects. H1a, H1b, H1d, and H1e are confirmed, but H1c is not confirmed. Specifically, the quantity of the senses of sight, taste, hearing, and touch all had significant positive effects on the visitors' restoration effects, with regression coefficients of 5.66, 2.182, 0.712, and 0.384, respectively. Of all sensory perceptions, sight tends to dominate as the primary mode of acquiring information. Correspondingly, in the National Forest Park, sight had the most significant impact on visitors' restoration effects. The sensitivity of sight perception is further enhanced by the characteristics of various types of recreational behaviors and activities in the National Forest Park. The sense of touch has the least influence on the visitors' restoration effects, which is due to the fact that compared with the touch experience, National Forest Parks tend to be more visually and auditorily stimulating; in addition, the sense of touch perception requires more initiative than other sensory perceptions, and it often requires people to actively touch the trees, leaves, bodies of water, or other landscapes, flora and fauna in order to obtain it, which leads to the fact that the sense of touch experience in the forests is usually relatively more consistent and smoother. The sense of smell plays a significant negative influence on the visitors' restoration effects, indicating that the smell perception in the National Forest Park can inhibit the visitors' restoration effects to a certain extent. For example, air pollution, environmental hygiene problems, and some irritating odors can bring about adverse sensory perceptions and directly inhibit visitors' restoration effects. The following semantic analysis further proved this point.

**Table 5.** Hypothesis Test.

| Hypothesis | Coef | $p$ | Hypothesis Result |
|---|---|---|---|
| H1a: Sight sensory number → Visitors' restoration effects | 5.66 | 0.000 *** | True |
| H1b: Hearing sensory number → Visitors' restoration effects | 0.712 | 0.018 ** | True |
| H1c: Smell sensory number → Visitors' restoration effects | −1.297 | 0.004 *** | False |
| H1d: Touch sensory number → Visitors' restoration effects | 0.384 | 0.001 *** | True |
| H1e: Taste sensory number → Visitors' restoration effects | 2.182 | 0.000 *** | True |
| H2a: Proximal psychological distance sensory → Visitors' restoration effects | 1.764 | 0.000 *** | True |
| H2b: Medial psychological distance sensory → Visitors' restoration effects | −1.558 | 0.281 | False |
| H2c: Distal psychological distance sensory → Visitors' restoration effects | −7.165 | 0.000 *** | True |
| H3: Sensory dimension → Visitors' restoration effects | 0.119 | 0.058 * | True |

*** $p < 0.01$; ** $p < 0.05$; * $p < 0.1$.

Among the sensory psychological distances, both proximal and distal psychological distances were significantly correlated with the visitors' restoration effects, and H2a and H2c were both confirmed. Specifically, the regression coefficient of proximal psychological distance is 1.7648, which has a significant positive impact on the visitors' restoration effects, and the regression coefficient of distal psychological distance is −7.165, which has a significant negative impact on the visitors' restoration effects. Medial psychological distance has a negative impact on visitors' restoration effects, but the impact is not significant, and hypothesis H2b was not confirmed. This may be due to the fact that as the subjective psychological distance increases, the external stimuli and disturbances perceived by visitors' senses increase, leading to information processing overload and making it more difficult to bring about positive visitors' restoration effects.

Sensory dimensions had a significant positive effect on the visitors' restoration effects, and H3 was confirmed. Compared with single-dimensional senses, multidimensional sensory perception increases the sensory richness and hierarchy of the recreational experience in the National Forest Park, synthesizes the advantages of sensory perception in terms of quantity and psychological distance, and has a beneficial impact on visitors' restoration effects.

### 5.3. Semantic Analysis

Use Chinese word segmentation, part-of-speech tagging, and dependency parsing in natural language processing to identify sensory elements and conduct further semantic analysis.

First, Chinese word segmentation is used to segment the target sentence output by the large language model into word sequences, forming a basic module for natural language processing analysis. The original sentences are segmented through sequence annotation to obtain word segmentation results, as shown in Figure 5. In general, there are three sequential positions of a character in a word: beginning of word (B), end of word (E), and single word (S) [60].

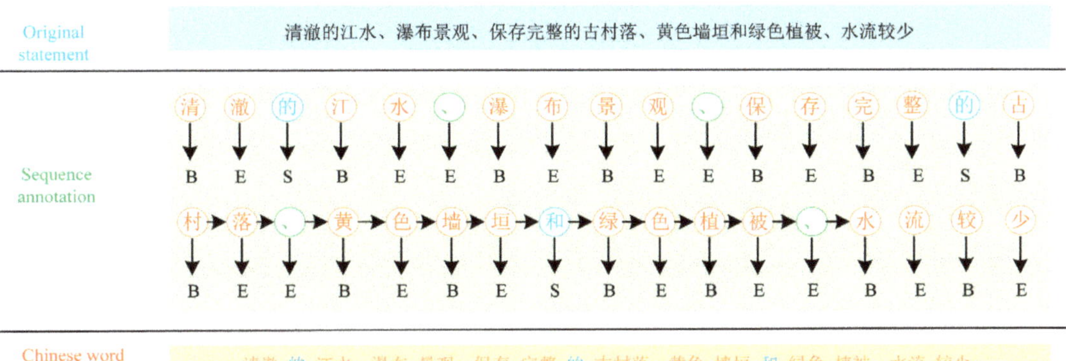

Figure 5. Example of Chinese word segmentation sequence annotation. In Figure 5, "清澈的江水、瀑布景观、保存完整的古村落、黄色墙垣和绿色植被、水流较少" is a Chinese text, which is the semantic text corresponding to the senses extracted by the model from the travelogue. Orange is multiple words, blue is single words, and green is punctuation.

Secondly, part-of-speech tagging is utilized to analyze and annotate the results obtained from word segmentation. According to the semantic characteristics of the sensory elements, terms other than the part-of-speech "NN" are filtered out (Figure 6).

Figure 6. Example of part-of-speech tagging. In Figure 6, "清澈的江水、瀑布景观、保存完整的古村落、黄色墙垣和绿色植被、水流较少" is a Chinese text, which is the semantic text corresponding to the senses extracted by the model from the travelogue.

Finally, dependency parsing is implemented to recognize and mark the dominant and subordinate connections between each expression (Figure 7). Through interpreting the syntactic relations labeled by the dependency arcs of sensory elements and their related semantic layers, we filter the semantic layers with the dependency relations "compound:nn", "amod", "nmod", and at the same time, filter the lexically labeled "NN" entries in the semantic layer of sensory elements, thus completing the recognition and extraction of sensory elements.

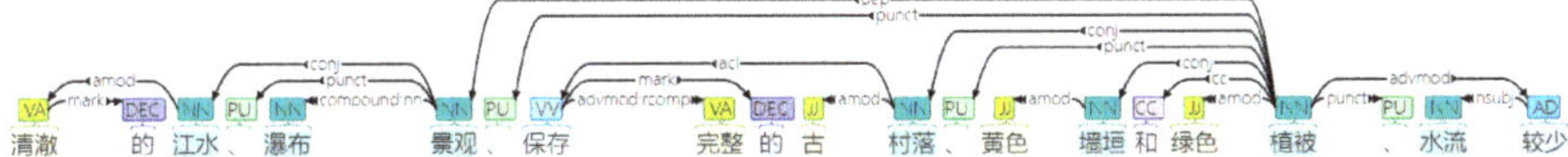

Figure 7. Example of dependency parsing. In Figure 7, "清澈的江水、瀑布景观、保存完整的古村落、黄色墙垣和绿色植被、水流较少" is a Chinese text, which is the semantic text corresponding to the senses extracted by the model from the travelogue.

The analysis considers the percentage of each type of element, visualizing and analyzing the top 10 elements. As shown in Figure 8, among the sight sensory elements, natural landscape elements and architectural elements such as vegetation (13.55%), water (10.49%), mountains (10.23%), and ancient architecture (3.07%) ranked highly, which is a direct reflection of the sight sensory level of China's National Forest Parks, combining ecological, protective, recreational and leisure functions. Forest vegetation is the primary focus of visitors to National Forest Parks, with attention given to the color, form, and seasonal variations of various types of forests and trees. These elements combine to create a unique sight experience in the form of group or combined landscapes.

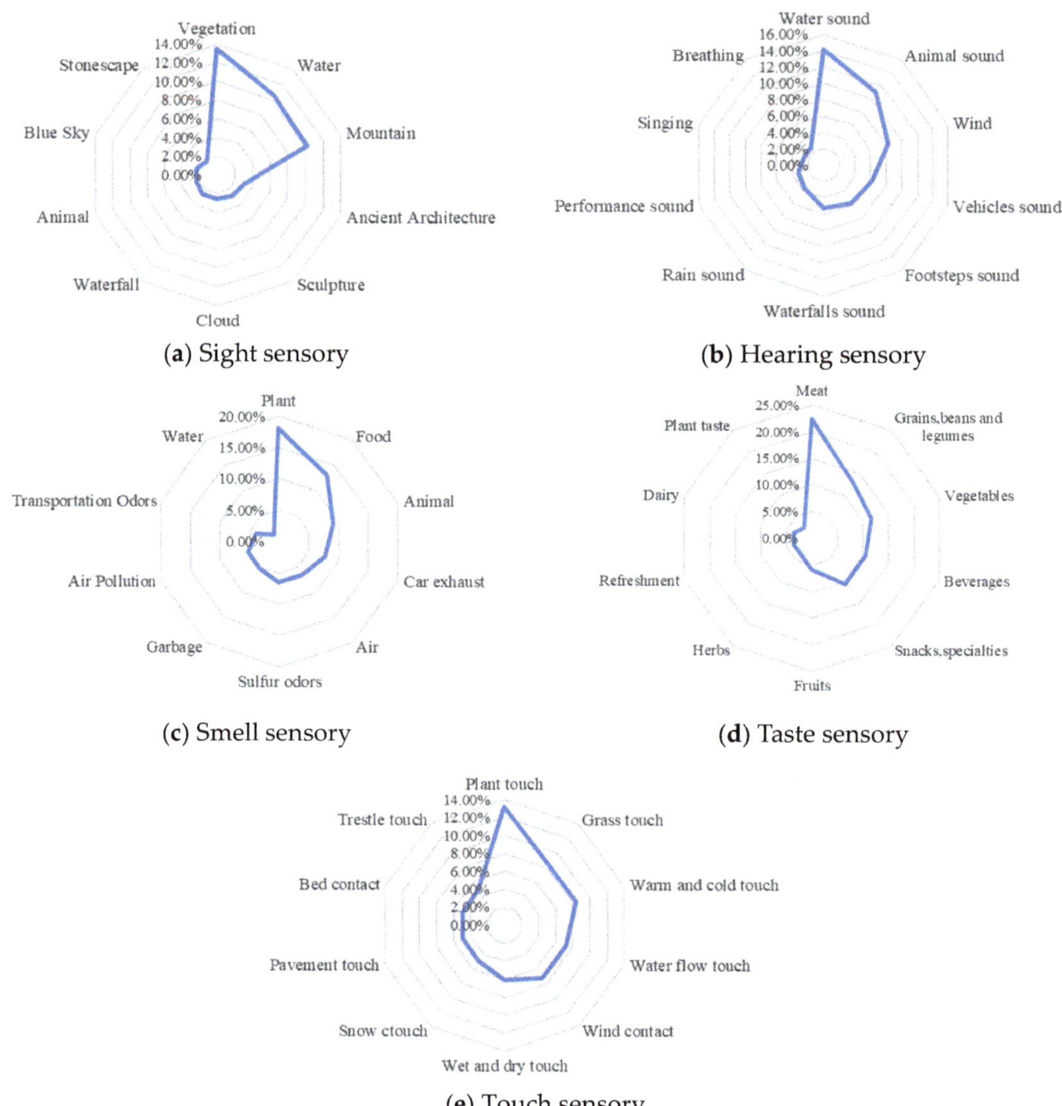



**Figure 8.** Percentage of major elements for each sensory category.

Among the hearing sensory elements, elements such as water sound (14.14%), animal sound (10.99%), wind (8.38%), vehicle sound (6.28%), and footstep sounds (5.76%) are ranked highly. Forested landscapes are often closely associated with water features such as lakes, streams, and rivers. The sounds of "gurgling water", "singing birds" and "wind blowing leaves" are typical hearing sensory perceptions of forest restoration. In addition to natural sound elements, various human activities in forest parks also become the main source of hearing experience, including the sound of traffic, footsteps of visitors, etc. It is important to point out that these sounds do not always have a positive effect on the visitors' restoration effects.

Among the smell sensory elements, plants (18.18%) and food (12.99%) accounted for a relatively high proportion; the fresh smell of plants, the aroma of food, and fresh air are all smell sensory perceptions with a positive effect, but there has to be concern about the large presence of negative experiences in the smell senses, such as the relatively high proportion of odors relating to animals (9.09%), car exhaust (7.79%), garbage (5.19%), and air pollution (5.19%), odors in transportation (3.90%), and water odors (1.30%), etc. Rotting plants, animal excrement, and mold smell brought by the humid environment in the forest environment, as well as garbage and pollutants produced by human activities, are all negative factors that cannot be ignored in the sensory perception of the forest, and the negative impact of this type of negative sensory perception on the visitors' restoration effects is also obvious.

Taste sensations are well-represented among sensory categories in some of China's National Forest Parks, which also function as tourist attractions where food and beverage activities are an integral part of forest recreation. Delicious or novel dining experiences are an important component of the taste senses and positively influence visitors' restoration effects. Meat (22.35%); grains, beans, and legumes (12.94%); vegetables (11.76%); and refreshing mouthfeel (3.53%) emerged as highly ranked elements of the taste experience.

The top-ranked touch sensory elements can be categorized into two groups. One is contact with the natural forest environment, such as contact with plants through touch (13.25%), grass through touch (8.43%), water flow through touch (7.23%), and wind contact (7.23%), especially contact with grassland, woodland, soft land, and contact with rivers and streams in the forest; the second comprises cold and warm touch (8.43%) and dry and wet touch (6.02%) related to the weather environment. As a typical proximal psychological distance sense, touch is one of the main sensory ways for tourists to have direct contact with the forest environment. It is more likely to cause immediate physical and emotional responses and plays a positive role in the visitors' restoration effects.

After the identification of sensory elements, sensory categories are formed after merging and thematic refinement to identify the intrinsic connections between the elements at the conceptual and content levels. This process aligns with the construction process of constructivist theory. Through multiple rounds of coding operations, all sensory elements with similar meanings are aggregated into sensory categories until all sensory elements are coded. Finally, a spectrum of sensory elements consisting of five sensory categories, 21 subcategories, and a number of single elements was formed (Figure 9). Among them, the sight sensory category includes five subcategories of natural landscape, architecture, climate, facilities, and people; the hearing sensory category includes four subcategories of climate sounds, plant and animal sounds, human voice, and human activity sounds; the smell sensory category includes five subcategories of food, natural smells, environmental odors, irritating odors, and others; the touch sensory category includes four subcategories of skin irritation sensation, landscape touch, facility touch, and nature touch; and the taste sensory category includes three subcategories of vegetative flavors, food flavors, and mouthfeel.

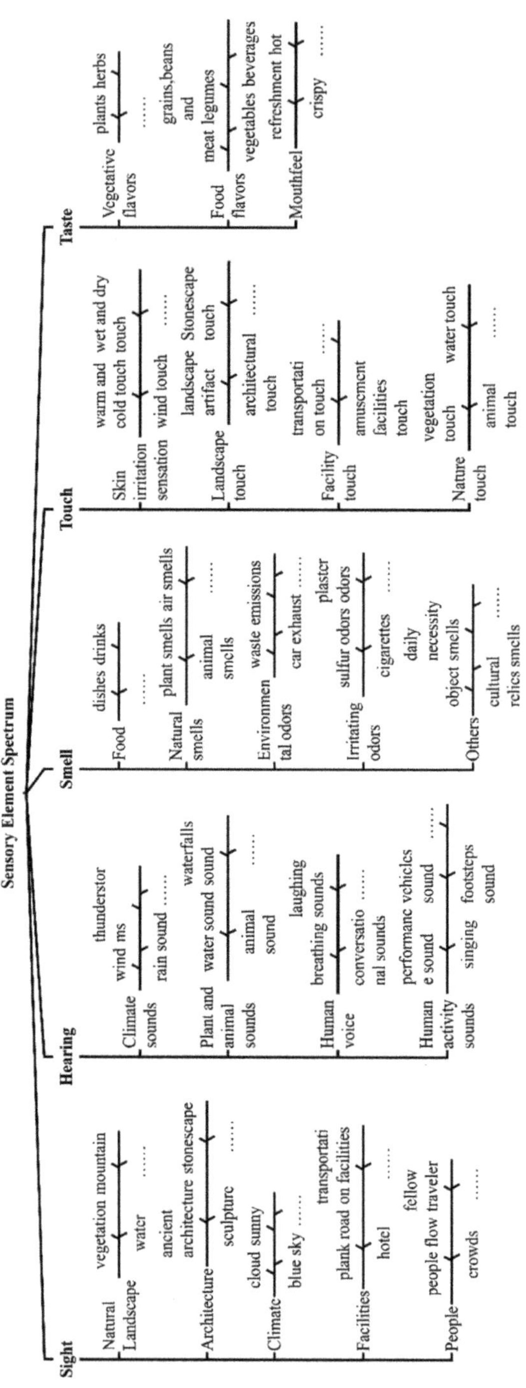


**Figure 9.** Sensory element spectrum.

## 6. Discussion

This study provides possible ideas and methods for generating scale questionnaires through travelogue texts to solve the dilemma of self-reported scale measurement to a certain extent through the application of a generative large language model. In terms of data quality, it improves the problems of self-reported scale measurement in data reliability and consistency, subjects' individual preferences, memory bias, and the influence of social expectations to a certain extent, and better characterizes the behaviors of recreationists and tourists in the real environment of National Forest Parks. In terms of data scale, it covers a wider range of research subjects and research time and space, providing a more convenient database for conducting large-scale, cross-time, and multi-group research. The results support the benefits of the extensive language model in both data generation and analysis, consistent with Shen et al.'s research [19].

In addition, our research also broadens the application of large language models in forest parks, recreation, restorative experiences, and other related fields. Specifically, the overall accuracy of the model training results for the senses reached 77%, and the overall accuracy of the model training results for the visitors' restoration effects reached 83%. The model performance is relatively ideal. Optimizing prompts has a significant impact on training results, which is consistent with Tiffany's research conclusion [61]. By trying models such as GPT-3.5 and text-davinci, it is shown that different models have an impact on the training results to a certain extent [62]. The impact of model parameters is not as significant as in the research of Kristy, Rodolfo, and others; temperature, frequency penalty, and presence penalty exert a more evident effect on training results. In addition, the limitations of large language models such as bias, discrimination, and AI hallucinations have been mentioned frequently, but they are not notably significant in our study.

There was a significant positive effect of the quantity of each sense, except smell, on visitors' restoration effects, with sight having the greatest impact, followed by taste, hearing, and touch, respectively. Sight is the most dominant sense in human perception and understanding of the environment. Some studies have shown that more than 80% of the information received by human beings is acquired through sight [63], which is also confirmed in our study, and the number of sight senses is at the top of the number of all senses. In National Forest Parks, the development of various recreational activities, such as viewing nature and recognizing attractions and trails, relies heavily on the sight senses, and a variety of visual perceptions enriches the forest recreational experience, which in turn positively affects the visitors' restoration effects. The second is the sense of taste. The significance of taste experience in forest recreation tends to be undervalued, which is also a common issue in previous studies regarding recreation and tourism. The rise of gastronomic tourism has brought forth renewed interest in taste among scholars of tourism and recreation [64]. Furthermore, a substantial number of China's National Forest Parks double as tourist attractions, thereby elevating the value of taste experiences within these parks. Additionally, the role of taste in physical and mental rejuvenation is becoming an increasingly vital area of study [65]. Tasting food, experiencing various flavors and cultures, and sharing meals with other travelers to enhance social interaction all positively impact visitors' restorative effects. The third is the sense of hearing, which is one of the important senses for participating in recreational activities in National Forest Parks; natural sounds in the forest such as birdsong, running water, wind, and the sound of falling leaves all have a positive effect on the restoration effect, which is consistent with the findings of Deng's study [66]. However, our study also found that sounds of human activity, such as footsteps on leaves, conversations, and music, were also crucial to promoting positive effects of restoration among visitors. These hearing experiences help to enhance emotional bonds with nature and people, leading to improved social connections and contributing to physical and mental recovery. Fourthly, touch offers microscopic and direct sensory perceptions, such as touching the bark of a tree, walking on grass, stepping into a stream, and interacting with animals in the forest. It is considered one of the most effective ways to enhance one's connection with nature. Several studies have demonstrated that touching natural elements can reduce heart rate, blood pressure, stress and anxiety [67], findings which are supported by our study.

From the perspective of the direction of influence, our study found that the sense of smell plays a significant negative role in the visitors' restoration effects. This is a point that is often overlooked in past research [68]. Some studies have shown that forests do not always play a positive role in the impact of visitors' restoration effects. This was confirmed in our research on China's National Forest Parks. The smell of decaying leaves, animal excrement, environmental garbage, and car exhaust in the forest have all become widely mentioned adverse smell experiences. We believe that the significant negative impact of the sense of smell is mainly influenced by two aspects: one is the characteristics of the sense of smell itself. The sense of smell is the only sense that has been shown to be directly related to memory, and memory is emotional [69], making it easier for such negative sensory perceptions to leave a deep impression on tourists and users, thereby affecting the recovery effect. At the same time, smell is generally considered to be one of the senses closely linked to emotional processing [70]. As emotions have a higher priority in cognitive processes, when the sense of smell produces negative stimuli, people may be more susceptible to its influence, which in turn affects the overall assessment. In addition, in the long process of human evolution, the sense of smell has always been one of the senses closely linked to survival [71]. Humans are usually very sensitive to odors. Therefore, rotting plants, animal excrement, and humid environments in the forest bring mold odors, and odors such as garbage and pollutants generated by human activities are more likely to produce negative recreational effects. The second is the characteristics of the study area. A considerable number of the National Forest Parks in China, where this article is studied, have the functions of tourism, recreation, and ecological protection. Therefore, in such National Forest Parks, the number of tourists is larger, and recreational activities are more frequent and diverse. In addition, the concentrated travel brought about by China's holiday system results in a considerable number of National Forest Parks having a large number of people and a high density at the same time, which is certain to lead to more negative sensory experiences, such as garbage and pollutants generated by human activities, which in turn have negative recreational effects.

Sensory psychological distance partially had a significant effect on the visitors' restoration effects. Both proximal psychological distance and distal psychological distance have a significant impact on the visitors' restoration effects, while medial psychological distance has a negative impact on the visitors' restoration effects, but the impact is not significant. Recreational activities in National Forest Parks are typical relaxing tourism activities [72]. These activities align with the proximal psychological distance senses, resulting in significant positive restorative effects for visitors. In contrast, the distal psychological distance senses have a significant negative impact. The effects of the medial psychological distance senses fall between the other two but are not significant. Moreover, proximal psychological distance senses, such as taste and touch, often align with low interpretation levels [36], are more vivid and concrete, and are more conducive to attracting tourists' attention, improving concentration, and thus promoting visitors' restoration effects. At the same time, proximal psychological distance senses are more likely to produce emotional connections so that travelers can experience more emotional closeness, which means that they may be more likely to produce positive emotions conducive to visitors' restoration effects, such as contentment and calmness. In contrast to the significant positive effects of proximal psychological distance senses on visitors' restoration effects, distal psychological distance senses are not compatible with relaxing tourism activities such as forest recreation, and more importantly, distal psychological distance senses such as sight and hearing are consistent with high levels of interpretation [36], which tend to be associated with complex and deeper meanings, leading to increased cognitive load. In addition, it was found that the sight and hearing senses, which tend to be subject to passive external stimulation and interference, have a more pronounced negative impact on recovery. Therefore, although from the perspective of the quantity of senses, an increase in the number of sight, hearing, and other types of senses enhances positive visitors' restoration effects, from the perspective of psychological distance, the senses at the proximal end of the psychological distance are more likely to positively influence visitors' restoration effects, in comparison.

Sensory dimensions have a significant positive effect on visitors' restoration effects. Our study of China's National Forest Parks further supports the notion that multidimensional sensory engagement aligns with human cognition of the environment and surrounding objects [73,74]. Expanding sensory dimensions, rather than the mere addition of individual senses, amplifies the integration and synergistic effects of sensory inputs. This enhances the depth and structure of the forest recreation experience. This has been described by visitors, who reported: "Climbing Yushan Mountain, enjoying the red maple, and drinking the refreshing Yushan green tea. When the maple leaves fall on the ground, step on the creaking sound, everything is so perfect". At the same time, multidimensional senses also play a complementary effect, and the stimulation of different senses can complement each other to provide a more comprehensive forest recreation experience, as one tourist said: "Canyons in the mist, falling waterfalls, the sound of cascading water in our ears, nature has fully endowed us with the most beautiful scenery!" Furthermore, our research has revealed that interactions between humans and nature through multidimensional senses can also promote interpersonal interactions, leading to cross-perceptual effects. For instance, in the context of forest environments as opposed to urban environments, human conversations and footsteps cease to be a source of jarring noise and instead emerge as a means to augment social connectivity and social integration. This, in turn, fosters correlated positive effects on visitors' restorative experiences in relation to social health.

The sensory elements of National Forest Parks that affect the restoration effect have predominantly natural attributes while revealing the interaction between people and people, and between people and nature. First, natural attributes are predominant, which is basically consistent with most of the existing studies that take the forest environment as the observation object. Natural elements are involved to varying degrees in the five categories of senses, among which, the vast majority of sight, hearing, and touch are natural sensory elements, including plants, animals, mountains, water, and natural climate. These elements are closely related to the environmental features found in the National Forest Park, which is consistent with prior research on forest restoration [75]. Second, sensory elements reveal the interaction between people and between people and nature. In addition to traditional natural sensory elements, our research also observed that elements related to "people" also play a vital role in the impact of visitors' restoration effects, such as architecture and facilities in sight sensory elements, and people in auditory sensory elements. For instance, architectural designs and amenities impact the sight sensory experience while the sound of conversations and footsteps, as well as the rustling of leaves, affect the hearing sensation. The smell and taste senses are stimulated by the aroma and flavor of food whereas the touch sensory experience involves physical interactions with people and surroundings. On the one hand, this is related to the fact that China's National Forest Parks have the attributes of both nature reserves and tourist attractions, especially for some of the National Forest Parks that are A-level scenic spots, and that more frequent human-to-human and human-to-nature interactions constitute an important part of the impact on visitors' restoration effects. On the other hand, it is imperative to emphasize the significance of forest areas in encouraging social interaction, fostering emotional connections, and promoting interpersonal contentment. This is as crucial as maintaining physical and mental well-being and constitutes an essential factor in achieving positive restoration effects for visitors. Empirical studies on urban green spaces like city parks and urban green fields have verified this proposition [76–78].

## 7. Conclusions

### 7.1. Research Conclusions

Based on the attention restoration theory and stress recovery theory, this paper explores the influence of sensory perception of the forest on visitors' restoration effects from a multidimensional and multisensory perspective, comprehensively utilizing methods such as the generative large language model, regression analysis, and semantic analysis. The main conclusions are as follows:

First, by introducing the generative large language model, this study expands its application in fields such as tourism and recreation and provides new ideas and methods for solving the dilemma posed by the traditional self-reported scale measurement. Moreover, it potentially paves the way for exploring novel research paradigms in light of the burgeoning development of generative artificial intelligence. In this study, we aim to use UGC data to extract independent variables such as sensory quantity, sensory psychological distance, and sensory dimension, alongside the dependent variable of visitors' restoration effects, via the use of a generative large language model. Evaluation of the model's output demonstrates a high level of feasibility for this method.

Second, in the study of the quantity of senses and visitors' restoration effects, compared to previous studies that only examined some specific senses and less often compared the differential effects of the five categories of senses with different dimensions, our study found that the effects of the quantity of the five categories of senses differed, with the quantity of sight, hearing, touch, and taste senses having a significant positive effect on the visitors' restoration effects, and the sense of smell having a significant negative effect. This is due to the fact that smell is closely related to memory, directly affects the emotional center, and is closely related to survival instincts, as well as the fact that the study area of this paper, the National Forest Parks of China, combines the functions of tourism, recreation, and ecological protection. The results of this study provide new insights that are different from most previous studies and give us more inspiration and space to further explore the effects of smell perception on forest visitors' restoration effects.

Third, the sensory psychological distance and visitors' restoration effects in the study allowed us to step out of the perspective of simply examining the increase in the quantity of senses, and to reconceptualize the senses in terms of the level of interpretation and psychological distance, which proved to be a useful attempt. Our research shows that forest recreation and tourism activities with lower levels of interpretation are better suited to proximal psychological distance senses, resulting in better recovery. However, as psychological distance increases, the mismatch in levels of interpretation also increases, causing the visitors' restoration effects to gradually shift from positive to negative. Therefore, it is important to consider psychological distance when designing nature-based activities for tourists. Indeed, as mental distance increases, external stimuli and distractions to the sight and hearing senses also increase, and to some extent, cognitive load is likely to increase. Although an increase in the quantity of a single sense, including sight and hearing, has a significant positive effect on restoration in most cases, whether individuals still use involuntary attention as referred to in the attention restoration theory at this time, and whether involuntary attention and voluntary attention are mutually exclusive, are worthy of further exploration [20]. Thus, some studies have suggested that restoration of voluntary attention does not necessarily require the use of involuntary attention and that there may be other mechanisms, such as exposure to natural environments that can activate an individual's positive emotions, which can have a positive effect on the restoration of voluntary attention [20].

Fourth, the sensory dimension has a significant positive impact on the visitors' restoration effects. The expansion of sensory dimensions heightens the amalgamation and synchronicity of the senses, offering a supplementary impact that enriches the depth and hierarchy of the forest recreational experience. Furthermore, the multidimensional senses advance human interaction with nature, producing a perceptual interaction that positively influences restoration outcomes concerning social health.

Fifth, through the semantic analysis method in natural language processing, we further identified the specific sensory elements that affect the visitors' restoration effects and constructed a multidimensional and multi-level sensory element spectrum. The study results indicate that the visitors' restoration effects in National Forest Parks are mainly influenced by the natural attributes of the sensory elements. Additionally, elements related to people also have a significant impact on the visitors' restoration effects.

*7.2. Research Limitations and Future Research*

This study still has some limitations and aspects to be explored in depth in future research:

First, in terms of the use of generative large language models, the model training effect needs to be evaluated from multiple perspectives such as different model selections and different parameter settings. Especially in terms of model parameters, it is still necessary to conduct a comparative analysis of the training effects of different parameter settings based on the characteristics of the training objectives, training tasks, and training data [62]. The second aspect is to focus on the cross-modal interaction of sensory perception. For example, sight perception in the forest may be affected by hearing or touch-related features, there may also be a cross-modal interaction effect between senses with different psychological distances. Third, it is important to consider the characteristics of olfactory perception and how it affects the recovery of tourists. The function of odor and how it interacts with other senses can be assessed from three perspectives when examining forest tourists: smell has a strong link to memory, plays a direct role in affecting the emotional center, and has close ties to the instinct for survival. Fourth, the multisensory perception and restoration effects of forest tourism among tourists from different cultural and geographic backgrounds should be studied. Differences in the multisensory perception of forest tourists may arise due to varying cultural backgrounds, and cultural concepts may shape the therapeutic influence of forest surroundings. Cognitive variances may also impact people's desire to seek psychological healing and relaxation in natural environments. Meanwhile, unique environmental conditions, like varying climates, landscapes, biodiversity, and regional cultures, influence not only the forest's natural characteristics but also the tourists' multisensory experiences within the forest. Fifth, the similarities and differences in the perception of the five senses stimuli based on the SNS data and the professional guidance from different points of view can be compared.

**Author Contributions:** Conceptualization, Y.W.; data curation, Y.W. and Y.H.; formal analysis, Y.W.; funding acquisition, Y.W.; methodology, Y.H.; software, Y.H.; visualization, Y.H.; writing—original draft, Y.W. All authors have read and agreed to the published version of the manuscript.

**Funding:** This research is funded by the Nanjing Forestry University Metasequoia Research Fund, grant number 163060200.

**Data Availability Statement:** The data presented in this study are available on request from the corresponding author.

**Conflicts of Interest:** The authors declare no conflict of interest.

# References

1.  UNDP. *Human Development Report 2021–2022*; UNDP: New York, NY, USA, 2022.
2.  World Health Organization. COVID-19 Pandemic Prompts 25% Increase in Global Anxiety and Depression Prevalence. Available online: https://www.who.int/zh/news/item/02-03-2022-covid-19-pandemic-triggers-25-increase-in-prevalence-of-anxiety-and-depression-worldwide (accessed on 2 March 2022).
3.  Wang, J.; Zhang, C. The effect of perceived environmental aesthetic qualities on tourists' positive emotions: The multiple mediation model. *Tour. Trib* **2022**, *37*, 80–94.
4.  Mayer, F.S.; Frantz, C.M.; Bruehlman-Senecal, E.; Dolliver, K. Why is nature beneficial? The role of connectedness to nature. *Environ. Behav.* **2009**, *41*, 607–643. [CrossRef]
5.  Xie, B.; Lu, Y.; Zheng, Y.J.U.F.; Greening, U. Casual evaluation of the effects of a large-scale greenway intervention on physical and mental health: A natural experimental study in China. *Urban For. Urban Gree.* **2022**, *67*, 127419. [CrossRef]

6. Larson, L.R.; Mullenbach, L.E.; Browning, M.H.; Rigolon, A.; Thomsen, J.; Metcalf, E.C.; Reigner, N.P.; Sharaievska, I.; McAnirlin, O.; D'Antonio, A. Greenspace and park use associated with less emotional distress among college students in the United States during the COVID-19 pandemic. *Environ. Res.* **2022**, *204*, 112367. [CrossRef] [PubMed]

7. Brancato, G.; Van Hedger, K.; Berman, M.G.; Van Hedger, S.C. Simulated nature walks improve psychological well-being along a natural to urban continuum. *J. Environ. Psychol.* **2022**, *81*, 101779. [CrossRef]

8. Nilsson, K.; Sangster, M.; Konijnendijk, C.C. *Forests, Trees and Human Health and Well-Being: Introduction*; Springer: Berlin/Heidelberg, Germany, 2011.

9. Li, H.; Li, M.; Zou, H.; Zhang, Y.; Cao, J. Urban sensory map: How do tourists "sense" a destination spatially? *Tour. Manag.* **2023**, *97*, 104723. [CrossRef]

10. Li, H.; Wang, C.R.; Meng, F.; Zhang, Z. Making restaurant reviews useful and/or enjoyable? The impacts of temporal, explanatory, and sensory cues. *Int. J. Hosp. Manag.* **2019**, *83*, 257–265. [CrossRef]

11. Kaplan, S. The restorative benefits of nature: Toward an integrative framework. *J. Environ. Psychol.* **1995**, *15*, 169–182. [CrossRef]

12. Chan, D. So why ask me? Are self-report data really that bad? In *Statistical and Methodological Myths and Urban Legends*; Routledge: London, UK, 2010; pp. 329–356.

13. Friedman, A.; Katz, B.A.; Cohen, I.H.; Yovel, I. Expanding the scope of implicit personality assessment: An examination of the questionnaire-based implicit association test (qIAT). *J. Pers. Assess.* **2021**, *103*, 380–391. [CrossRef]

14. Hu, M.; Zuo, X.; Liu, J. Survey on Deep Generative Mode. *Acta Autom. Sin.* **2022**, *48*, 40–74.

15. Sihong, Z.; Jian, Z. Impact and countermeasures of generative AI represented by ChatGPT on the telecom industry. *Telecommun. Sci.* **2023**, *39*, 67–75.

16. Luo, Y.; Zhang, X. Focusing on ChatGPT: Development, impact, and issues. *Nat. Mag.* **2023**, *2*, 106–108.

17. Gui, C.; LI, Y.; Jiang, W. Development and challenge of ar tificial intelligence: Taking ChatGPT as an example. *Telecom Eng. Technol. Stand.* **2023**, *3*, 24–28.

18. Aher, G.V.; Arriaga, R.I.; Kalai, A.T. Using large language models to simulate multiple humans and replicate human subject studies. In Proceedings of the International Conference on Machine Learning, Honolulu, HI, USA, 23–29 July 2023; pp. 337–371.

19. Shen, H.; Li, T.; Li, T.J.-J.; Park, J.S.; Yang, D. Shaping the Emerging Norms of Using Large Language Models in Social Computing Research. *arXiv* **2023**, arXiv:2307.04280.

20. Chen, X.; Wang, B.; Zhang, B. Far from the "madding crowd": The positive effects of nature, theories and applications. *Adv. Psychol. Sci.* **2016**, *24*, 270. [CrossRef]

21. Faber Taylor, A.; Kuo, F.E. Children with attention deficits concentrate better after walk in the park. *J. Atten. Disord.* **2009**, *12*, 402–409. [CrossRef] [PubMed]

22. Kaplan, S.; Talbot, J.F. Psychological benefits of a wilderness experience. In *Behavior and the Natural Environment*; Springer: Berlin/Heidelberg, Germany, 1983; pp. 163–203.

23. Yongrui, G.; Jie, Z.; Shaojing, L.; Yuling, Z.; Sifeng, N.; Bingjin, Y. The Difference and Structural Model of Tourist's Perceived Restorative Environment. *Tour. Trib.* **2014**, *29*, 93–102.

24. Gaekwad, J.S.; Moslehian, A.S.; Roös, P.B. A meta-analysis of physiological stress responses to natural environments: Biophilia and Stress Recovery Theory perspectives. *J. Environ. Psychol.* **2023**, *90*, 102085. [CrossRef]

25. Ulrich, R.S. Aesthetic and affective response to natural environment. In *Behavior and the Natural Environment*; Springer: Berlin/Heidelberg, Germany, 1983; pp. 85–125.

26. Tan, S.; Li, J. Restoration and Stress Relief Benefits of Urban Park and Green Space. *Chin. Landsc. Arch.* **2009**, *25*, 79–82.

27. Mundher, R.; Abu Bakar, S.; Al-Helli, M.; Gao, H.; Al-Sharaa, A.; Mohd Yusof, M.J.; Maulan, S.; Aziz, A. Visual Aesthetic Quality Assessment of Urban Forests: A Conceptual Framework. *Urban Sci.* **2022**, *6*, 79. [CrossRef]

28. Yamada, Y. Soundscape-based forest planning for recreational and therapeutic activities. *Urban For. Urban Green.* **2006**, *5*, 131–139. [CrossRef]

29. Henshaw, V. *Urban Smellscapes: Understanding and Designing City Smell Environments*; Routledge: London, UK, 2013.

30. Dann, G.; Jacobsen, J.K.S. Tourism smellscapes. *Tour. Geogr.* **2003**, *5*, 3–25. [CrossRef]

31. Trope, Y.; Liberman, N. Construal-level theory of psychological distance. *Psychol. Rev.* **2010**, *117*, 440. [CrossRef] [PubMed]

32. Chen, H.; He, G. The effect of psychological distance on intertemporal choice and risky choice. *Acta Psychol. Sin.* **2014**, *46*, 677–690. [CrossRef]

33. Kim, J.; Cui, Y.G.; Choi, C.; Lee, S.J.; Marshall, R. The influence of preciseness of price information on the travel option choice. *Tour. Manag.* **2020**, *79*, 104012. [CrossRef]

34. Liberman, N.; Trope, Y. The role of feasibility and desirability considerations in near and distant future decisions: A test of temporal construal theory. *J. Pers. Soc. Psychol.* **1998**, *75*, 5. [CrossRef]

35. Elder, R.S.; Schlosser, A.E.; Poor, M.; Xu, L. So close I can almost sense it: The interplay between sensory imagery and psychological distance. *J. Consum. Res.* **2017**, *44*, 877–894. [CrossRef]

36. Tingting, G.; Baoku, L. Visual or Touchable?—The Imagery Experience Effect Induced by Sensory Clues of Online Reviews. *Collect. Essays Financ. Econ.* **2019**, *250*, 82.

37. Chen, G.; Chen, Z.; Lin, M.; Zhou, X.; Tang, C.; Zhang, H.; Jiang, Y.; Lei, Z.; Xie, J. Series of Writing Talks on "Frontier Theory of Social Science and its Application inTourism Research". *Tour. Forum* **2021**, *14*, 1–20.

38. Li, Y.-C.; Zhou, T.-R.; Zhou, X. Theoretical Models of Multisensory Cues Integration...... WEN Xiao-Hui, LIU Qiang, SUN Hong-Jin; et al.(666) Construal Level Theory: From Temporal Distance to Psychological Distanc. *Adv. Psychol. Sci.* **2009**, *17*, 667.
39. Wen, X.-H.; Li, G.-Q.; Liu, Q. Processing of Audiovisual Integration and Its Neural Mechanism. *Adv. Psychol. Sci.* **2011**, *19*, 976.
40. Ernst, M.O.; Bülthoff, H.H. Merging the senses into a robust percept. *Trends Cogn. Sci.* **2004**, *8*, 162–169. [CrossRef] [PubMed]
41. Dong, Y.; Qu, Y. How Does Sensory Stimulation Stimulate Emotions? The Continued Influence Mechanism of Multi-sensory Experience on Destination Attachment. *Tour. Sci.* **2022**, *36*, 101–121.
42. Kah, J.A.; Shin, H.J.; Lee, S.-H. Traveler sensoryscape experiences and the formation of destination identity. *Tour. Geogr.* **2022**, *24*, 475–494. [CrossRef]
43. Qiu, M.; Jin, X.; Scott, N. Sensescapes and attention restoration in nature-based tourism: Evidence from China and Australia. *Tour. Manag. Perspect.* **2021**, *39*, 100855. [CrossRef]
44. Feng, Z. *A Concise Tutorial on Natural Language Processing*; Shanghai Foreign Language Education Press: Shanghai, China, 2012.
45. Kim, D.-M.; Lee, S.-H.; Cheong, Y.-G. Predicting emotion in movie scripts using deep learning. In Proceedings of the 2018 IEEE International Conference on Big Data and Smart Computing (BigComp), Shanghai, China, 15–17 January 2018; pp. 530–532.
46. Ahmad, S.; Asghar, M.Z.; Alotaibi, F.M.; Khan, S. Classification of poetry text into the emotional states using deep learning technique. *IEEE Access* **2020**, *8*, 73865–73878. [CrossRef]
47. Fourati, M.; Jedidi, A.; Gargouri, F. A deep learning-based classification for topic detection of audiovisual documents. *Appl. Intell.* **2023**, *53*, 8776–8798. [CrossRef]
48. Zhang, Y.; Prayag, G.; Song, H. Attribution theory and negative emotions in tourism experiences. *Tour. Manag. Perspect.* **2021**, *40*, 100904. [CrossRef]
49. Pan, S.; Ryan, C. Tourism sense-making: The role of the senses and travel journalism. *J. Travel Tour. Mark.* **2009**, *26*, 625–639. [CrossRef]
50. Wong, S.F.; Kler, B.K.; KPD Balakrishnan, B. Sense of Place: Narrating Emotional Experiences of Malaysian Borneo through Western Travel Blogs. *Tour. Hosp.* **2022**, *3*, 666–684. [CrossRef]
51. Li, M.; Xu, W.; Chen, Y. Young children's vacation experience: Through the eyes of parents. *Tour. Manag. Perspect.* **2020**, *33*, 100586. [CrossRef]
52. Wang, W.; Yi, L.; Wu, M.-Y.; Pearce, P.L.; Huang, S.S. Examining Chinese adult children's motivations for traveling with their parents. *Tour. Manag.* **2018**, *69*, 422–433. [CrossRef]
53. Song, Z.; Zhang, G.; Li, S. Is dark tourism all darkness? A group comparison study of tourist experience based on UGC. *Tour. Trib.* **2019**, *34*, 90–104.
54. Yao, C.; Peixue, L.; Jianxin ZHANG, C.X.; Lu, T. Distant home: A study on the accommodation choice and experience of Chinese tourists in sharing accommodation. *World Reg. Stud.* **2020**, *29*, 181.
55. Yaobin, W.; Ling, Y.; Chuanling, S. Comparative research on travel sharing of typical travel website based on text mining-taking Gansu Province as an example. *Resour. Dev. Mark.* **2017**, *33*, 100–104.
56. Schramowski, P.; Turan, C.; Andersen, N.; Rothkopf, C.A.; Kersting, K. Large pre-trained language models contain human-like biases of what is right and wrong to do. *Nat. Mach. Intell.* **2022**, *4*, 258–268. [CrossRef]
57. Leiser, F.; Eckhardt, S.; Knaeble, M.; Maedche, A.; Schwabe, G.; Sunyaev, A. From ChatGPT to FactGPT: A Participatory Design Study to Mitigate the Effects of Large Language Model Hallucinations on Users. In *Proceedings of the Mensch und Computer 2023*; Association for Computing Machinery: New York, NY, USA, 2023; pp. 81–90.
58. Ji, Z.; Lee, N.; Frieske, R.; Yu, T.; Su, D.; Xu, Y.; Ishii, E.; Bang, Y.J.; Madotto, A.; Fung, P. Survey of hallucination in natural language generation. *ACM Comput. Surv.* **2023**, *55*, 1–38. [CrossRef]
59. Brown, T.; Mann, B.; Ryder, N.; Subbiah, M.; Kaplan, J.D.; Dhariwal, P.; Neelakantan, A.; Shyam, P.; Sastry, G.; Askell, A. Language models are few-shot learners. *Adv. Neur.* **2020**, *33*, 1877–1901.
60. Deng, L.P.; Luo, Z.Y. Domain Adaptation Of Chinese Word Segmentation on Semi-Supervised Conditional Random Fields. *J. Chin. Inf. Process.* **2017**, *31*, 9–19.
61. Kung, T.H.; Cheatham, M.; Medenilla, A.; Sillos, C.; De Leon, L.; Elepaño, C.; Madriaga, M.; Aggabao, R.; Diaz-Candido, G.; Maningo, J. Performance of ChatGPT on USMLE: Potential for AI-assisted medical education using large language models. *PLoS Digit. Health* **2023**, *2*, e0000198. [CrossRef]
62. Carpenter, K.A.; Altman, R.B. Using GPT-3 to build a lexicon of drugs of abuse synonyms for social media pharmacovigilance. *Biomolecules* **2023**, *13*, 387. [CrossRef] [PubMed]
63. Zhang, T.; Han, B. Psychological effect of red: Phenomenon and mechanism review. *Adv. Psychol. Sci.* **2013**, *21*, 398. [CrossRef]
64. Hillel, D.; Belhassen, Y.; Shani, A. What makes a gastronomic destination attractive? Evidence from the Israeli Negev. *Tour. Manag.* **2013**, *36*, 200–209. [CrossRef]
65. Zhang, Q.; Zhang, H.; Xu, H. Health tourism destinations as therapeutic landscapes: Understanding the health perceptions of senior seasonal migrants. *Soc. Sci. Med.* **2021**, *279*, 113951. [CrossRef] [PubMed]
66. Deng, L.; Luo, H.; Ma, J.; Huang, Z.; Sun, L.-X.; Jiang, M.-Y.; Zhu, C.-Y.; Li, X. Effects of integration between visual stimuli and auditory stimuli on restorative potential and aesthetic preference in urban green spaces. *Urban For. Urban Green.* **2020**, *53*, 126702. [CrossRef]
67. Townsend, M. Feel blue? Touch green! Participation in forest/woodland management as a treatment for depression. *Urban For. Urban Green.* **2006**, *5*, 111–120. [CrossRef]

68. Grilli, G.; Sacchelli, S. Health benefits derived from forest: A review. *Int. J. Environ. Res. Public Health* **2020**, *17*, 6125. [CrossRef] [PubMed]
69. Lv, X.; McCabe, S. Expanding theory of tourists' destination loyalty: The role of sensory impressions. *Tour. Manag.* **2020**, *77*, 104026. [CrossRef]
70. Krusemark, E.A.; Novak, L.R.; Gitelman, D.R.; Li, W. When the sense of smell meets emotion: Anxiety-state-dependent olfactory processing and neural circuitry adaptation. *J. Neurosci.* **2013**, *33*, 15324–15332. [CrossRef]
71. Engen, T. The sense of smell. *Annu. Rev. Psychol.* **1973**, *24*, 187–206. [CrossRef]
72. Su, L.; Tang, B.; Nawijn, J. How tourism activity shapes travel experience sharing: Tourist well-being and social context. *Ann. Tour. Res.* **2021**, *91*, 103316. [CrossRef]
73. Tang, X.; Wu, J.; Shen, Y. The interactions of multisensory integration with endogenous and exogenous attention. *Neurosci. Biobehav. Rev.* **2016**, *61*, 208–224. [CrossRef] [PubMed]
74. Krishna, A.; Schwarz, N. Sensory marketing, embodiment, and grounded cognition: A review and introduction. *J. Consum. Psychol.* **2014**, *24*, 159–168. [CrossRef]
75. Takayama, N.; Korpela, K.; Lee, J.; Morikawa, T.; Tsunetsugu, Y.; Park, B.-J.; Li, Q.; Tyrväinen, L.; Miyazaki, Y.; Kagawa, T. Emotional, restorative and vitalizing effects of forest and urban environments at four sites in Japan. *Int. J. Environ. Res. Public Health* **2014**, *11*, 7207–7230. [CrossRef]
76. Jennings, V.; Bamkole, O. The relationship between social cohesion and urban green space: An avenue for health promotion. *Int. J. Environ. Res. Public Health* **2019**, *16*, 452. [CrossRef]
77. Wang, Y.-C.; Liu, C.-R.; Huang, W.-S.; Chen, S.-P. Destination fascination and destination loyalty: Subjective well-being and destination attachment as mediators. *J. Travel Res.* **2020**, *59*, 496–511. [CrossRef]
78. Jennings, V.; Baptiste, A.K.; Osborne Jelks, N.T.; Skeete, R. Urban green space and the pursuit of health equity in parts of the United States. *Int. J. Environ. Res. Public Health* **2017**, *14*, 1432. [CrossRef]

*Article*

# The Effects of Spatial Characteristics and Visual and Smell Environments on the Soundscape of Waterfront Space in Mountainous Cities


Bingzhi Zhong [1], Hui Xie [2,3,*], Tian Gao [1], Ling Qiu [1], Heng Li [2] and Zhengkai Zhang [1]

[1]  College of Landscape Architecture and Art, Northwest A&F University, Yangling 712199, China
[2]  Faculty of Architecture and Urban Planning, Chongqing University, Chongqing 400044, China
[3]  Key Laboratory of New Technology for Construction of Cities in Mountain Area, Ministry of Education, Chongqing University, Chongqing 400044, China
*   Correspondence: xh@cqu.edu.cn



**Abstract:** The soundscape of waterfront space in mountainous cities (WSMC) can affect people's physical and mental health. Taking seven WSMCs in Chongqing, China, as the study area, this study aimed to investigate the soundscape and explore the influence of spatial characteristics and visual and smell environments on the soundscape of WSMCs through a sensewalking approach. The results show that the soundscape evaluations of WSMCs are of poor quality, and traffic sounds are dominant (33%). Among spatial characteristics, the position relative to the road (including vertical and horizontal distances) had a greater impact than other spatial indicators on soundscape evaluations. Elevation was positively correlated with the A-weighted equivalent sound level ($L_{Aeq}$) and negatively correlated with the soundscape comfort degree (SCD). In terms of visual elements, the proportions of paved ground, pedestrians, and buildings had negative effects on the soundscape, while those of the sky, water, and natural terrain had positive effects. High visual and smell environment quality can enhance soundscape evaluations, although the smell environment had a greater impact on the SCD than the visual environment in WSMCs. Finally, this study summarizes the recommended values of spatial characteristics and visual and smell environment indicators to put forward references for the soundscape design of WSMCs.


**Keywords:** soundscape; waterfront space in mountainous cities; spatial elements; multisensory interaction


**Citation:** Zhong, B.; Xie, H.; Gao, T.; Qiu, L.; Li, H.; Zhang, Z. The Effects of Spatial Characteristics and Visual and Smell Environments on the Soundscape of Waterfront Space in Mountainous Cities. *Forests* **2023**, *14*, 10. https://doi.org/10.3390/f14010010

Academic Editors: Jiang Liu, Xinhao Wang and Xin-Chen Hong

Received: 19 November 2022
Revised: 16 December 2022
Accepted: 19 December 2022
Published: 21 December 2022

## 1. Introduction

Urban waterfront space is a general term for a certain area connected by land and water in the city and is generally formed by water areas, water boundaries, and land areas [1]. Water spaces not only constitute a natural ecological transition between water and land to enrich urban landscapes but also foster a close connection between nature and people [2]. Waterfront space in mountainous cities (WSMC) has the characteristics of both urban waterfront space and mountainous topography, which creates landscape diversity and uniqueness in the residential environment [3]. With the spread of COVID-19, lockdowns and a decrease in outdoor activities led to an increase in both psychological stress and mortality by suicide [4]. There has been a growing demand for relaxation and entertainment by the public. As natural places, WSMCs can provide entertainment and perceived restoration to the public, with important health, ecological, and economic value.

Soundscapes have multiple impacts on environmental health. The World Health Organization (WHO) notes that high sound pressure levels (SPLs) can increase cardiovascular disease risk, sleep disturbance, and annoyance, which may reduce productivity and increase accident rates [5]. In contrast, studies have shown that there is a significant association between a positive soundscape and health-related effects, including increased restoration and reduced stress-inducing mechanisms [6,7]. The soundscape is vital to building a distinctive cultural atmosphere, attracting visitors, increasing the district's vitality, and enhancing the landscape evaluation of an area [8]. Moreover, soundscape assessment can be used to monitor ecological conditions and reflect human disturbances to the ecosystem [9,10]. Some acoustic indices, such as acoustic complexity, acoustic evenness, and the normalized soundscape difference index (NDSI), are significantly correlated with biodiversity [11,12]. Therefore, the soundscape quality of WSMC can have important impacts on public well-being and environmental health. However, as most studies of WSMCs have focused on urban planning and landscape design based on visual features, the soundscape and the factors that influence soundscape evaluations have not yet been studied.

WSMCs have a variety of spatial elements that influence soundscape evaluations. As a natural spatial element, water bodies have special characteristics and important effects on the urban soundscape [13]. Watts et al. [14] found that the presence of water can create a sense of tranquility. The sound of water can not only mask noise but also increase the positive perception of urban green space [15]. However, urban waterfront space has different characteristics according to the different types and nature of adjacent water bodies. For example, in littoral areas, ocean visibility improves soundscape evaluation [16]. In areas close to streams or waterfalls, soundscape evaluation remains positive even with a high A-weighted equivalent sound level ($L_{Aeq}$) [17]. In addition, the soundscape of mountainous cities is closely related to the mountainous topography. Some studies have shown that, compared with flat ground, mountainous terrain may not only lead to an increase in the SPL but also limit people's activities, giving the soundscape specific spatial–temporal variations [18–20]. In addition, due to the dense urban roads and undulating terrain, the traffic noise in mountainous cities is often more complex than that in plains cities [21,22]. Scholars have achieved some understanding of the impact of single spatial characteristics. However, few studies have discussed the impact of various spatial elements on the soundscape with a background of multiple spatial characteristics, which is essential for WSMCs.

Multiple senses can convey more profound and comprehensive information than a single sense [23]. As important media for perceiving the environment, both vision and olfaction can interact with auditory perception [24,25]. Aural preferences and visual elements are intrinsically linked. The proportion of visual elements (such as buildings, vegetation, and sky), visual factors (such as distance and color), and visual perception indicators can all significantly affect soundscape evaluation [26–29]. Although studies on the interaction between auditory perception and olfaction have been limited to date, some studies have shown that odor can affect the response time to auditory stimuli, and the evaluation of sound and odor shows analogous trends of sensory comfort and preference [30,31]. Adams and Askins [32] found that sensewalking, as a varied method, can provide an effective way to study the urban environment from sensory perspectives, but existing studies on sensory interactions are mostly limited to the laboratory environment and have focused more on the influence of single sensory factors, resulting in limited external validity [33]. To date, few studies have considered both the perceptual characteristics and potential effects of visual and smell environments on the soundscapes of WSMCs in the field.

In order to bridge these gaps, this study investigated the soundscapes, spatial characteristics, and visual and smell environments of seven typical WSMCs in Chongqing, China, using the sensewalking approach. The aims of this study were divided into four parts. The first part was to investigate the current soundscape quality of WSMCs. The second part was to explore the influence of spatial characteristics on the soundscape of WSMCs. The third part was to explore visual environment–soundscape and smell environment–soundscape interactions. The fourth was to construct models to summarize the recommended variable values to achieve positive soundscape evaluations. Ensuring sufficient sampling sites and sensewalking participants, as well as organizing measurements under the influence of the epidemic, posed certain challenges in this study. Overall, the results of this study can provide data support and references for the soundscape design of WSMCs while providing better-perceived restoration sites for the public and improving environmental health and people's happiness.

## 2. Materials and Methods

### 2.1. Studied Areas

Chongqing is a typical mountainous city in Southwest China, where 90% of the total area is mountains and hills. Located on both sides of the Yangtze and Jialing Rivers, seven WSMCs in Chongqing were randomly selected as the study areas, namely, Jiangbeizui (JB), Shacixiang (SC), Chaotianmen Square (CT), CBD Riverside Park (CB), Liziba Park (LZ), Jiulongpo Park (JL), and Nanbin Park (NB). Figure 1 shows their plans and the locations of walking points in Chongqing (JB: 6 points; SC: 9 points; CT: 9 points; CB: 10 points; LZ: 9 points; JL: 10 points; and NB: 10 points). The walking routes align with the main touring route in each WSMC. Table 1 provides basic information about the study areas.

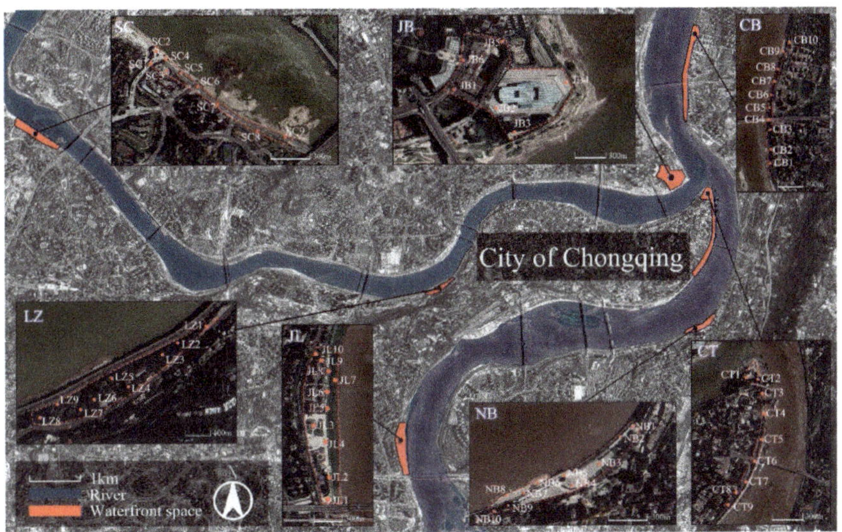


**Figure 1.** Plans of the seven studied WSMCs and the locations of walking points in Chongqing. WSMC = waterfront space in mountainous cities.

**Table 1.** Basic information on the seven WSMCs.

|  | JB | SC | CT | CB | LZ | JL | NB |
|---|---|---|---|---|---|---|---|
| Area (km$^2$) | 110.7 | 48.1 | 112.1 | 117.1 | 32.2 | 163.7 | 51.0 |
| Year of completion | 2009 | 2018 | 1998 | 2005 | 2010 | 2012 | 2005 |
| Affiliated urban district | Jiangbei | Shapingba | Yuzhong | Nan'an | Yuzhong | Jiulongpo | Nan'an |
| Function | Civic square | Business district | Business district | Park | Park | Civic square | Civic square |

### 2.2. The Sensewalking Approach and Questionnaire Design

Sensewalking is a common way to study one or more aspects of the sensory environment and usually involves a researcher walking alone or with one or more participants [34]. In this study, the subjective evaluations of the soundscape, as well as the visual and smell environments, were obtained using the sensewalking method. The participants comprised 172 architectural students (78 males and 94 females, with a mean age of 21 years old) from Chongqing University, with normal hearing and a basic knowledge of soundscapes and landscapes, who voluntarily participated in sensewalking. In total, 23–26 participants took part in sensewalking in each WSMC. They were chosen to understand future designers' perspectives of the urban soundscape. Sensewalking was undertaken between 10 a.m. and 12 p.m. on summer weekdays. The weather conditions were stable, with a light breeze, no rain, and a temperature ranging from 25 to 32 °C (Figure 2). Each participant spent 5 min at each of the walking points to evaluate the soundscape quality and fill out the questionnaire. All participants underwent a pre-investigation in another local park before performing the formal sensewalk. The pre-investigation involved (a) familiarity with the survey process of sensewalking, (b) the rapid identification of sound sources and odors, and (c) the determination of appropriate subjective evaluation indicators of the soundscape and visual and smell environments.

**Figure 2.** Sensewalking survey and mountainous topography status, taking JL as an example.

This study referred to the existing research and feedback on the pre-investigation design questionnaire. The questionnaire consists of three parts, covering the soundscape, visual environment, and smell environment. The subjective evaluation in this study was measured using a 5-point Likert scale.

In terms of the soundscape, in order to investigate the composition of the sound sources and the overall soundscape experience of the site, the questionnaire referred to the suggestions in ISO/TS 12913-2: 2018 and related research [35–37]. First, participants were asked to list all of the sound sources they noticed while listening at each walking point. The sound sources were classified as traffic sounds (e.g., cars, buses, trains, and airplanes), human sounds (e.g., conversation, laughter, children at play, and footsteps), natural sounds (e.g., biological and geophysical sounds), and mechanical sounds (e.g., sirens, construction, and industrial sounds). Second, the overall soundscape comfort was evaluated by the soundscape comfort degree (SCD), from 1 = "uncomfortable" to 5 = "comfortable".

In terms of the visual environment, previous studies have shown that comfort, complexity, and naturalness are valid for evaluating the visual landscape [38]. The subjective evaluation of the visual environment was obtained through the visual environment comfort degree (VECD), from 1 = "uncomfortable" to 5 = "comfortable"; the visual environment natural degree (VEND), from 1 = "artificial" to 5 = "natural"; and the visual environment diversity degree (VEDD), from 1 = "simple" to 5 = "complex".

In terms of the smell environment, few studies have explored criteria for subjective smell environment evaluations [39]. Therefore, this study used the smell environment comfort degree (SECD) as a subjective evaluation indicator to compare it with visual and soundscape comfort (1 = "uncomfortable" to 5 = "comfortable"). In addition, the human sense of smell is currently the most sensitive tool available for assessing the smell environment [40]. Therefore, to identify the odor composition, participants were also asked to name the main odors at each walking point.

The details of the questionnaire are shown in Appendix A. Finally, 1544 valid questionnaires were obtained. With a KMO index of 0.806 and a Cronbach's Alpha of 0.791, the questionnaire had high validity and reliability.

*2.3. Objective Soundscape Evaluation Parameter Measurement*

In order to objectively evaluate the physical and ecological characteristics of the acoustic environments of the WSMCs, the $L_{Aeq}$ and NDSI were selected as the objective soundscape evaluation parameters in this study. The NDSI can be used to indicate human disturbances to biodiversity (such as the richness of bird species), ranging from $-1$ to 1 [10]. The higher the proportion of the artificial sound, the smaller the value. The formula is as follows, where $\alpha$ represents anthrophony (1–2 kHz), and $\beta$ represents biophony (2–11 kHz):

$$NDSI = \frac{(\beta - \alpha)}{(\beta + \alpha)}. \tag{1}$$

The $L_{Aeq}$ was calculated every 5 min ($L_{Aeq\_5min}$), and the audio (using a binaural method) was recorded at each walking point ($N = 63$) during the sensewalk. The $L_{Aeq\_5min}$ was measured using an AWA 6228$^+$ sound level meter (Class 1, Aihua Instruments Co., Ltd., Hangzhou, China). Audio samples were recorded using a PCM-M10 audio recorder (Sony Corporation, Tokyo, Japan). All equipment was located 1.2 m above the ground and more than 2 m from nearby buildings. ISO1996/2-2017 was followed throughout the field measurements [41]. RStudio (Version 1.1.463, RStudio, Inc., Boston, MA, USA) was used to analyze the NDSI from audio files. Detailed specifications of all measurement equipment and associated data processing software in this study are shown in Appendix B.

*2.4. Spatial Indicator Measurements*

Previous studies have shown that indicators of space, such as elevation, have significant impacts on soundscape evaluations in mountainous cities [42]. Based on the spatial characteristics of WSMCs, five indicators—elevation, vertical distance from the shoreline (VDS), horizontal distance from the shoreline (HDS), vertical distance from the road (VDR), and horizontal distance from the road (HDR)—were selected as spatial indicators in this study (Table 2) and measured using a YILI X28 altimeter (Appendix B).

**Table 2.** The ranges of spatial indicators of seven WSMCs. VDS = vertical distance from the shoreline. HDS = horizontal distance from the shoreline. VDR = vertical distance from the road. HDR = horizontal distance from the road.

|  | JB | SC | CT | CB | LZ | JL | NB |
|---|---|---|---|---|---|---|---|
| Elevation (m) | 244–276 | 232–247 | 271–278 | 189–210 | 196–229 | 192–218 | 200–213 |
| VDS (m) | 16.9–48.9 | 11.5–26.2 | 27.4–32.7 | 4.2–25.5 | 3.0–35.9 | 1.7–27.1 | 5.4–19.2 |
| HDS (m) | 109.0–410.2 | 9.0–98.5 | 65.7–74.7 | 9.4–84.8 | 30.6–104.0 | 29.5–151.7 | 18.7–80.6 |
| VDR (m) | 6.4–238.9 | 9.6–187.7 | 8.1–116.8 | 29.9–74.3 | 15.1–46.2 | 35.2–184.9 | 17.3–76.1 |
| HDR (m) | $-25.6$–6.4 | 9.2–5.5 | $-3.7$–5.5 | $-21.5$––0.2 | $-32.6$–0.3 | $-22.7$–2.7 | $-11$–2.8 |

*2.5. Identification of the Proportion of Visual Elements*

In addition, to obtain the visual elements experienced by participants during the sensewalk, panoramic street-view images were taken using smartphones (Appendix B) at each walking point. A fully connected network (FCN) model (GUC. HPSCIL, University of Geo-sciences, China) was used to identify different visual elements in street-view images (Figure 3) [43]. Coupled with the calculation of the per-pixel loss, the FCN produces the area ratio of each visual element in the image by counting the number of pixels in each segmentation mask. The FCN model can extract 150 visual elements and achieve a pixel-wise accuracy of 81% for training data and 67% for actual data (Appendix B). Seven visual elements that occupy a relatively large proportion of WSMCs were identified in this study, namely, paved ground, buildings, vegetation, sky, water, natural terrain, and pedestrians and animals.

**(a)**

**(b)**

**Figure 3.** Visual element recognition in street-view images: (**a**) raw image and (**b**) the same image segmented with the FCN model (fully connected network).

*2.6. Data Analysis*

Based on the one-sample Kolmogorov–Smirnov test, data were not normally distributed. Spearman's rho correlation analysis was performed to identify the relationships between spatial indicators (elevation, VDS, HDS, VDR, and HDR), the proportions of visual elements, subjective evaluations of the visual and smell environments (VECD, VEND, VEDD, and SECD), and soundscape evaluation parameters ($L_{Aeq\_5min}$, NDSI, and SCD). The Spearman's correlation coefficients r and $p$ were used to find correlations between variables and soundscape evaluation parameters; a $p$-value of less than 0.05 was considered statistically significant.

Previous studies have shown that some important variables for soundscape evaluations, such as distance and environment subjective evaluations, behave linearly [44,45]. The final aim of this study was to summarize the recommended values of the variables and provide references for the design of WSMCs. Therefore, in order to further explore the relationship between variables and soundscape evaluation parameters, multiple linear regression analyses were used to model spatial indicators, the proportions of visual elements, and subjective evaluations of visual and smell environments (dependent variables) with soundscape evaluation parameters (independent variables). Adjusted $R^2$ and β coefficients were used to assess the quality of the obtained models. Variables with $p < 0.05$ and VIF < 2 were retained in the model using the stepwise method.

All statistical analyses were performed using Statistical Package for Social Scientists (SPSS) software version 22.0 (IBM SPSS, Inc., Chicago, IL, USA). The minimum hardware required for data analysis was a personal computer with a 2.5 GHz Intel Core i5 processor and at least 4 GB of RAM.

## 3. Results

### 3.1. Soundscape Evaluations of Waterfront Spaces in Mountainous Cities (WSMCs)

The sound sources listed by participants were classified as traffic sounds, human sounds, natural sounds, and mechanical sounds according to ISO/TS 12913-2: 2018 [35], and the proportions of sound sources in seven studied WSMCs were calculated (Table 3). It is known that the high $L_{Aeq}$ of traffic noise can affect soundscape comfort, while natural sounds can significantly improve soundscape comfort [46]. However, the results showed that, on average, the proportion of traffic sounds was 33%, higher than in urban parks in mountainous cities (from 4.9% to 9.2%), while the proportion of natural sounds was only 27%, lower than in urban parks in mountainous cities (from 31.4% to 53.3%) [19]. This indicates that the soundscape components of WCMCs need to be improved by controlling the interference of traffic noise and improving the proportion of natural sounds.

**Table 3.** Proportions of traffic sounds, human sounds, natural sounds, and mechanical sounds in seven WSMCs.

|  | JB | SC | CT | CB | LZ | JL | NB | Mean |
|---|---|---|---|---|---|---|---|---|
| Traffic sounds | 34% | 20% | 42% | 36% | 37% | 27% | 37% | 33% |
| Human sounds | 23% | 21% | 28% | 15% | 20% | 27% | 18% | 22% |
| Natural sounds | 25% | 28% | 14% | 37% | 28% | 32% | 23% | 27% |
| Mechanical sounds | 18% | 32% | 16% | 11% | 15% | 14% | 22% | 18% |

Figure 4 shows the mean values of the soundscape evaluation parameters ($L_{Aeq\_5min}$, NDSI, and SCD) at walking points in seven WSMCs. It can be seen that the overall $L_{Aeq\text{-}5min}$ of WSMCs was high (Figure 4a). In fact, the $L_{Aeq\text{-}5min}$ values of the six WSMCs and 79% of the walking points were higher than 55 dBA, exceeding the national recommended value (daytime, city park, and green space) [47]. All NDSIs of WSMCs were negative, ranging from $-0.425$ to $-0.004$ (Figure 4b). This indicates that human disturbance was dominant in the WSMCs. The NDSI was the highest ($-0.004$) in JL and the lowest ($-0.425$) in NB. This might be related to the fact that JL is located in the old town and the pedestrian flow was low, while NB is located in a popular scenic spot and the pedestrian flow was high. The SCD of the seven WSMCs was in the range of 2.4–3.0, between "a little uncomfortable" and "moderate" (Figure 4c). Specifically, only 31.7% of the walking points were positively evaluated in terms of the SCD. In general, the soundscape evaluations of WSMCs were of poor quality.

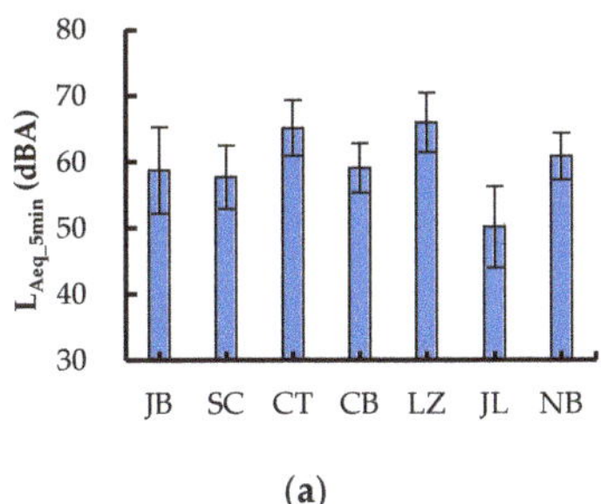

(a)

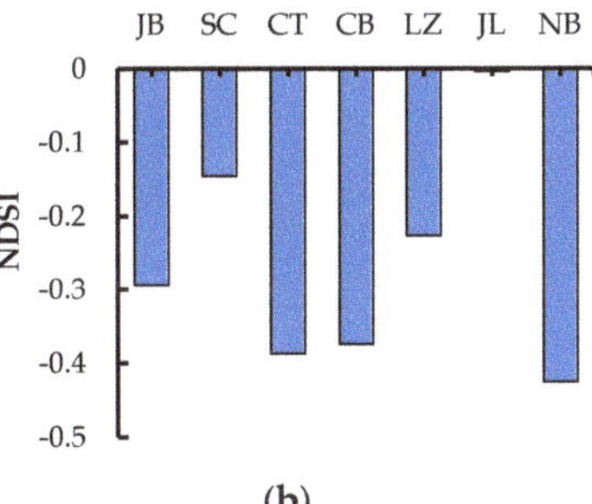

(b)

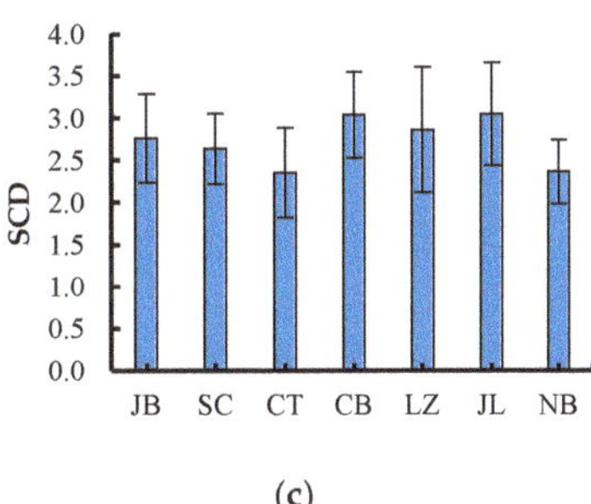

(c)

**Figure 4.** The mean values of soundscape evaluation parameters at walking points in seven WSMCs. (a) $L_{Aeq\_5min}$, (b) NDSI, and (c) SCD. $L_{Aeq\_5min}$ = A-weighted equivalent sound level calculated every 5 min; NDSI = normalized soundscape difference index; SCD = soundscape comfort degree. All soundscape evaluation parameters are the average values for the area.

### 3.2. The Influence of Spatial Characteristics on the Soundscape

Combined with the spatial characteristics of WSMCs, spatial indicators, including the elevation, VDS, HDS, VDR, and HDR, were extracted for correlation analysis using soundscape evaluation parameters (Table 4). The results show that both elevation and VDS had significantly positive correlations with the $L_{Aeq\_5min}$ (r = 0.323 and 0.344, respectively, $p$ < 0.01) and significantly negative correlations with the SCD (r = −0.375 and −0.344, respectively, $p$ < 0.01). This was due to the significant autocorrelation between elevation and VDS (r = 0.734, $p$ < 0.01). In terms of road-related indicators, the VDR had a significantly negative correlation with the SCD (r = −0.450, $p$ < 0.01). The HDR had a significantly negative correlation with the $L_{Aeq\_5min}$ (r = −0.635, $p$ < 0.01) and significantly positive correlations with the NDSI and SCD (r = 0.306 and 0.402, respectively, $p$ < 0.01). This indicates that, regardless of whether in the vertical or horizontal direction, the closer the distance to the road, the lower the comfort of the soundscape. The correlation coefficient is a statistical tool used to measure the extent of the relationship between variables. Since the HDR and VDR are the spatial indicators with the highest correlation coefficients with the $L_{Aeq\_5min}$ (r = −0.635) and SCD (r = −0.450), respectively, the HDR may have a greater influence on the $L_{Aeq\_5min}$, and the VDR may have a greater influence on the SCD.

**Table 4.** Spearman's rho correlation coefficients for the relationship between spatial indicators and soundscape evaluation parameters.

|  | Elevation | VDS | HDS | VDR | HDR |
|---|---|---|---|---|---|
| $L_{Aeq\_5min}$ | 0.323 ** | 0.344 ** | 0.049 | 0.112 | −0.635 ** |
| NDSI | −0.103 | −0.143 | −0.079 | −0.145 | 0.306 ** |
| SCD | −0.375 ** | −0.344 ** | −0.087 | −0.450 ** | 0.402 ** |

Notes: ** significance at the 0.01 level (2-tailed).

In contrast to Spearman's correlation coefficients, multiple linear regression found that the VDS was left out in the $L_{Aeq\_5min}$ model, and elevation and VDS were left out in the SCD model (Table 5). A possible reason for this might be the multicollinearity of the VDS with the $L_{Aeq\_5min}$ (VIF = 2.875), and the correlations of elevation and the VDS with the SCD were relatively low. Furthermore, the spatial indicators accounted for 42.5%, 21.7%, and 35.7% of the variability in the $L_{Aeq\_5min}$, NDSI, and SCD, respectively (adjusted $R^2$ = 0.425, 0.217, and 0.357, respectively). Referring to similar soundscape studies, an adjusted $R^2$ over 0.3 provides sufficient reliability for the linear model [27,48]. It can be seen that the performance of the linear regression model of the NDSI was relatively low. This may mean that although the linear regression model was significant, the impact of the HDR on the NDSI was limited. The most influential variables on the $L_{Aeq\_5min}$, NDSI, and SCD were the HDR ($\beta$ = −0.578), HDR ($\beta$ = 0.479), and VDR ($\beta$ = 0.503), respectively.

**Table 5.** Results of multiple linear regression analyses of spatial indicators and soundscape evaluation parameters.

| Dependent Variables | Factors | Adjusted $R^2$ | $\beta$ | SE | *t*-Value | $p$ | VIF |
|---|---|---|---|---|---|---|---|
| $L_{Aeq\_5min}$ | HDR | 0.425 | −0.578 | 0.011 | −5.967 | 0.000 | 1.013 |
|  | Elevation |  | −0.271 | 0.023 | 2.791 | 0.007 | 1.013 |
| NDSI | HDR | 0.217 | 0.479 | 0.001 | 4.264 | 0.000 | 1.000 |
| SCD | VDR | 0.357 | −0.503 | 0.008 | −4.727 | 0.000 | 1.094 |
|  | HDR |  | 0.236 | 0.001 | 2.213 | 0.031 | 1.094 |

Linear regression curves between spatial indicators and soundscape evaluation parameters are shown in Figure 5. It can be seen that when the HDR was above 90 m, the $L_{Aeq\_5min}$ could drop below 55 dBA; when the elevation was below approximately 220 m, the $L_{Aeq\_5min}$ of WSMCs could drop below 55 dBA (Figure 5a). In terms of the NDSI, when the HDR was more than 90 m, the NDSI became positive (Figure 5b). This indicates that, at this distance, there was a higher proportion of biological sounds, and a better ecological status could be achieved. In terms of the SCD, this study found that when the HDR was more than 70 m, or the VDR was less than −10 m, the evaluation of the SCD was higher than moderate (evaluation = 3). Small changes in the vertical distance could result in large variances in the SCD (Figure 5c).

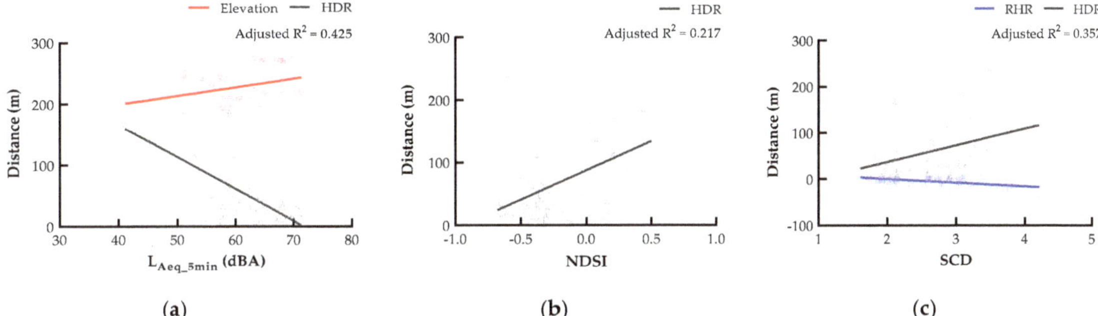


**Figure 5.** Linear regression curves between spatial indicators and soundscape evaluation parameters: (**a**) spatial indicators and L_Aeq_5min; (**b**) spatial indicators and NDSI; (**c**) spatial indicators and SCD.

### 3.3. The Influence of Visual and Smell Environments on the Soundscape

3.3.1. Visual Environment of WSMCs

Table 6 shows the average proportions of visual elements in seven WSMCs. Paved ground and buildings can form an artificially closed spatial enclosure. It was found that the average proportion of paved ground and buildings were 14.6% and 20.8%, respectively, and their total proportion reached 41%. The average proportion of vegetation was 20.8%. In terms of the sky, except for the proportion of sky in LZ, which was only 4.1%, little difference was found between the other six WSMCs, which were within the range of 10.0%–25.4%. LZ is located in a historic district where the trees are leafy, covering the sky. The waterfront is an important feature of WSMCs. However, water only accounted for 0.1%–7.2% of the seven WSMCs, with an average of only 2.6%. In fact, most of the waterfront spaces in the studied WSMCs were designed to be open for viewing water. The reason for the small proportion of water was that there were few waterfront spaces in the planning of the WSMCs, and many walking points were sheltered by vegetation and buildings. Other visual elements, such as natural terrain and pedestrians and animals, made up a relatively small proportion of WSMCs and were within the range of 0.2%–1.9%.

**Table 6.** Average proportions of nine visual elements in seven WSMCs.

|  | JB | SC | CT | CB | LZ | JL | NB | Average |
|---|---|---|---|---|---|---|---|---|
| Paved ground | 33.0% | 30.7% | 26.9% | 21.1% | 32.7% | 16.5% | 24.1% | 26.4% |
| Buildings | 20.1% | 17% | 23.7% | 10.7% | 8.2% | 10.1% | 12.1% | 14.6% |
| Vegetation | 16.8% | 8.8% | 19.9% | 24.9% | 26.7% | 28.1% | 20.4% | 20.8% |
| Sky | 15.8% | 25.4% | 10% | 14.5% | 4.1% | 17.7% | 20.4% | 15.4% |
| Water | 0.1% | 4.5% | 1.6% | 7.2% | 1.3% | 1.0% | 2.3% | 2.6% |
| Natural terrain | 0.4% | 1.3% | 0.9% | 4.5% | 0.7% | 0.5% | 0.7% | 1.3% |
| Pedestrians and animals | 2.1% | 0.5% | 1.4% | 0.1% | 8.6% | 0.2% | 0.6% | 1.9% |

In terms of the subjective evaluation of the visual environment, the results of the VECD, VEND, and VEDD are shown in Figure 6. Five WSMCs' results for the VECD were higher than 3 (moderate), indicating that the overall visual environment of WSMCs was relatively comfortable. However, the VEND and VEDD were generally lower. Additionally, the results of the VEND were lower than those of VEDD at 2.5–3.2, between "a little artificial" and "a little natural". This indicates that the visual environments of WSMCs were still lacking richness in landscape diversity, especially the natural landscape.

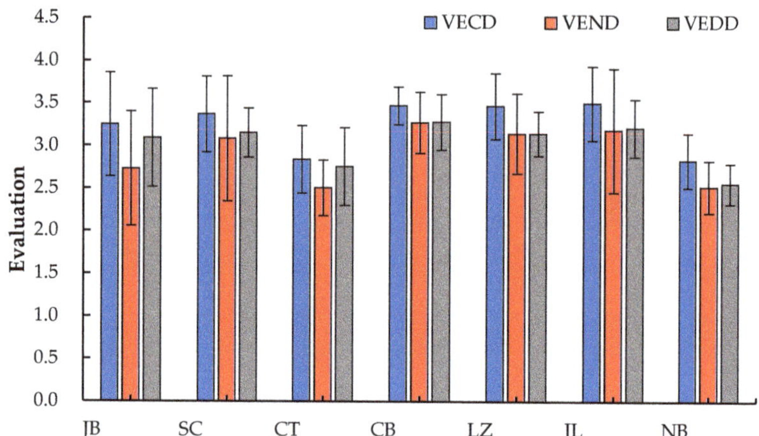

**Figure 6.** Results of VECD, VEND, and VEDD in seven WSMCs. VECD = visual environment comfort degree; VEND = visual environment natural degree; VEDD = visual environment diversity degree.

3.3.2. Smell Environment of WSMCs

According to common odor sources in the city [49], the odors identified by the participants during the sensewalk were classified into five categories: natural odors, emission odors, food odors, building material odors, and human odors (Table 7). It can be seen that, in the WSMCs, the proportion of natural odors was the highest, with an average of 66.2%. The second highest proportion was emission odors, ranging from 7.5% to 32.3%. The proportions of food odors, building material odors, and human odors were lower, no more than 6%. In terms of the subjective evaluation of the smell environment, the results of the SECD in seven WSMCs are shown in Figure 7. It can be seen that the SECDs were higher with a large proportion of natural odors (such as CB and JL) and lower with a large proportion of emission odors (e.g., CT). In general, the scores of the SECD ranged from 2.7 to 3.3, and the five WSMCs' SECD evaluation results were higher than 3 (moderate).

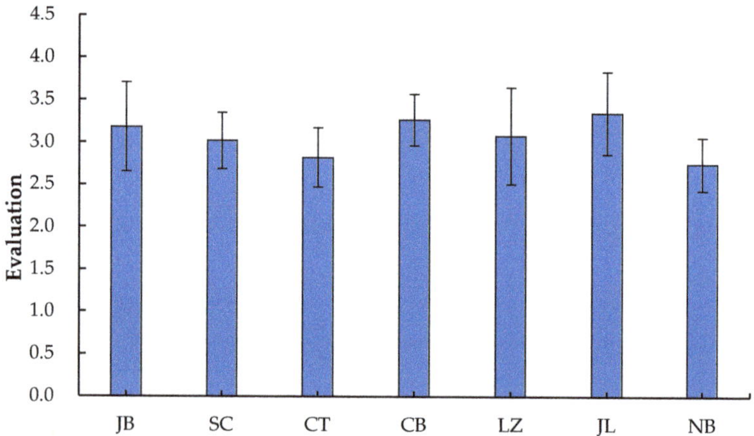

**Figure 7.** Evaluation results of SECD in seven WSMCs. SECD = smell environment comfort degree.

**Table 7.** Average proportions of five categories of odors in seven WSMCs.

|  | JB | SC | CT | CB | LZ | JL | NB | Average |
|---|---|---|---|---|---|---|---|---|
| Natural odors | 63.2% | 56.5% | 43.3% | 82.5% | 70.1% | 81.1% | 66.7% | 66.2% |
| Emission odors | 25.3% | 17.6% | 32.3% | 7.6% | 20.4% | 7.5% | 18.5% | 18.5% |
| Human odors | 3.4% | 5.1% | 7.0% | 9.2% | 4.4% | 0.0% | 11.1% | 5.7% |
| Building material odors | 6.9% | 7.7% | 3.5% | 0.7% | 5.3% | 11.3% | 3.7% | 5.6% |
| Food odors | 1.1% | 11.9% | 13.9% | 0.0% | 0.0% | 0.0% | 0.0% | 3.8% |

### 3.3.3. The Influence of Visual and Smell Environments on Soundscape Evaluation Parameters

The correlation analyses between the proportions of visual elements and soundscape evaluation parameters are shown in Table 8. It was found that the $L_{Aeq\_5min}$ was significantly positively correlated with the proportions of paved ground and pedestrians and animals ($r = 0.276, p < 0.05$; $r = 0.632, p < 0.01$, respectively) and significantly negatively correlated with the proportions of sky and water ($r = -0.460, p < 0.01$; $r = -0.255, p < 0.05$, respectively). In terms of the SCD, it was found that the proportion of buildings was significantly negatively correlated with the SCD ($r = -0.254, p < 0.05$), and the proportions of water and natural terrain were significantly positively correlated with the SCD ($r = 0.262$ and $0.311, p < 0.05$, respectively). This indicates that people prefer the soundscape experience of waterfront space and natural terrain space (such as a natural terrace, slope, or valley), rather than a building-dominated soundscape experience. In general, the results show that the proportions of paved ground, buildings, and pedestrians and animals have negative effects on the soundscape, while the sky, water, and natural terrain have positive effects. In terms of correlation coefficients, pedestrians and animals and natural terrain have the greatest impacts on the $L_{Aeq\_5min}$ ($r = 0.632$) and SCD ($r = 0.311$), respectively.

**Table 8.** Spearman's rho correlation coefficients for the relationship between seven visual elements and soundscape evaluation parameters.

|  | Paved Ground | Buildings | Vegetation | Sky | Water | Natural Terrain | Pedestrians and Animals |
|---|---|---|---|---|---|---|---|
| $L_{Aeq\_5min}$ | 0.276 * | 0.026 | 0.214 | −0.460 ** | −0.255 * | −0.147 | 0.632 ** |
| NDSI | −0.010 | −0.013 | −0.139 | 0.139 | 0.079 | 0.086 | −0.214 |
| SCD | −0.181 | −0.254 * | −0.003 | 0.108 | 0.262 * | 0.311 * | 0.185 |

Notes: * significance at the 0.05 level (2-tailed); ** significance at the 0.01 level (2-tailed).

All subjective evaluations of visual and smell environments were negatively correlated with the $L_{Aeq\_5min}$ and positively correlated with the SCD, and all of the visual environment subjective evaluations were positively correlated with the NDSI ($p < 0.01$), as demonstrated in Table 9. This indicates that visual and smell environments can enhance the soundscape evaluation, which confirms the association between visual and olfactory perceptions in soundscape evaluations [30,50]. Only the SECD was not significantly correlated with the NDSI ($p > 0.05$). This might indicate that the smell environment, as perceived by humans, has little effect on the ecological characteristics of the acoustic environment. In terms of the SCD, it is worth noting that the SCD was more strongly correlated with the SECD ($r = 0.780$) than the VECD ($r = 0.729$). This indicates that the smell environment had a greater impact on the SCD than the visual environment in WSMCs.

From the multiple linear regression analyses (Table 10), it can be seen that subjective evaluations of visual and smell environments accounted for 24.9%, 12.7%, and 69.6% of the variability in the $L_{Aeq\_5min}$, NDSI, and SCD (adjusted $R^2 = 0.249, 0.127,$ and $0.696$, respectively). The adjusted $R^2$ values of subjective evaluations for the $L_{Aeq\_5min}$ and NDSI were lower than 0.3. The impact of the VECD and VEDD on the $L_{Aeq\_5min}$ and NDSI was limited. In a study on the relationship between visual elements and the soundscape, Liu, Kang, Behm and Luo [26] found that the adjusted $R^2$ of the proportions of visual elements was rather low (adjusted $R^2 > 0.1$). The proportions of visual elements account for 21.7% and 4.9% of the variability in the $L_{Aeq\_5min}$ and SCD, respectively (adjusted $R^2 = 0.217$ and 0.049, respectively). The impact of buildings on the SCD was limited. In addition, pedestrians and animals were the most influential variable on the $L_{Aeq\_5min}$ ($\beta = 0.376$), and buildings was the most influential variable on the SCD ($\beta = -0.254$). In the subjective evaluations of visual and smell environments, the SECD was the most influential variable on the SCD ($\beta = 0.553$). The VECD and VEDD were the most influential variables on the $L_{Aeq\_5min}$ and NDSI, respectively ($\beta = -0.511$ and 0.376, respectively).

**Table 9.** Spearman's rho correlation coefficients for the relationships between subjective evaluations of visual and smell environments and soundscape evaluation parameters.

|  | VECD | VEND | VEDD | SECD |
|---|---|---|---|---|
| $L_{Aeq\_5min}$ | −0.525 ** | −0.482 ** | −0.400 ** | −0.506 ** |
| NDSI | 0.328 ** | 0.305 ** | 0.327 ** | 0.195 |
| SCD | 0.729 ** | 0.708 ** | 0.566 ** | 0.780 ** |

Notes: ** significance at the 0.01 level (2-tailed).

**Table 10.** Results of multiple linear regression analyses for the proportions of visual elements and subjective evaluations of visual and smell environments with soundscape evaluation parameters.

| Dimensions | Dependent Variables | Factors | Adjusted $R^2$ | $\beta$ | SE | $t$-Value | $p$ | VIF |
|---|---|---|---|---|---|---|---|---|
| The proportions of visual elements | $L_{Aeq\_5min}$ | Pedestrians and animals | 0.217 | 0.376 | 22.081 | 3.325 | 0.002 | 1.013 |
|  |  | Paved ground |  | 0.278 | 5.272 | 2.453 | 0.017 | 1.013 |
|  | SCD | Buildings | 0.049 | −0.254 | 0.762 | −2.055 | 0.044 | 1.000 |
| Subjective evaluations of visual and smell environments | $L_{Aeq\_5min}$ | VECD | 0.249 | −0.511 | 1.505 | −4.647 | 0.000 | 1.000 |
|  | NDSI | VEDD | 0.127 | 0.376 | 0.080 | 3.170 | 0.002 | 1.000 |
|  | SCD | SECD | 0.696 | 0.553 | 0.122 | 5.920 | 0.000 | 1.778 |
|  |  | VEND |  | 0.365 | 0.093 | 3.906 | 0.000 | 1.778 |

The linear regression curves between the proportions of visual elements and soundscape evaluation parameters are shown in Figure 8a,b. It can be seen that when the proportion of paved ground was lower than 22%, or the proportion of pedestrians and animals was lower than 1%, the $L_{Aeq\_5min}$ could drop below 55 dBA (Figure 8a); when the proportion of buildings was less than 13%, the SCD could reach "moderate" (evaluation score = 3) or above (Figure 8b). In linear regression curves between subjective evaluations of visual and smell environments and soundscape evaluation parameters, when the VECD reached 3.4 or above, the $L_{Aeq\_5min}$ was below 55 dBA (Figure 8c); when the VEDD was above 3.2, the NDSI could reach a positive value (Figure 8d). It should be noted that, in terms of the SCD, only when the VEND and SECD reached 3.1 and 3.2, respectively, could the evaluation of the SCD reach "moderate" (evaluation score = 3) or above (Figure 8e). This indicates that the improvement in soundscape comfort in WSMCs might require a better natural visual environment and a comfortable smell environment.

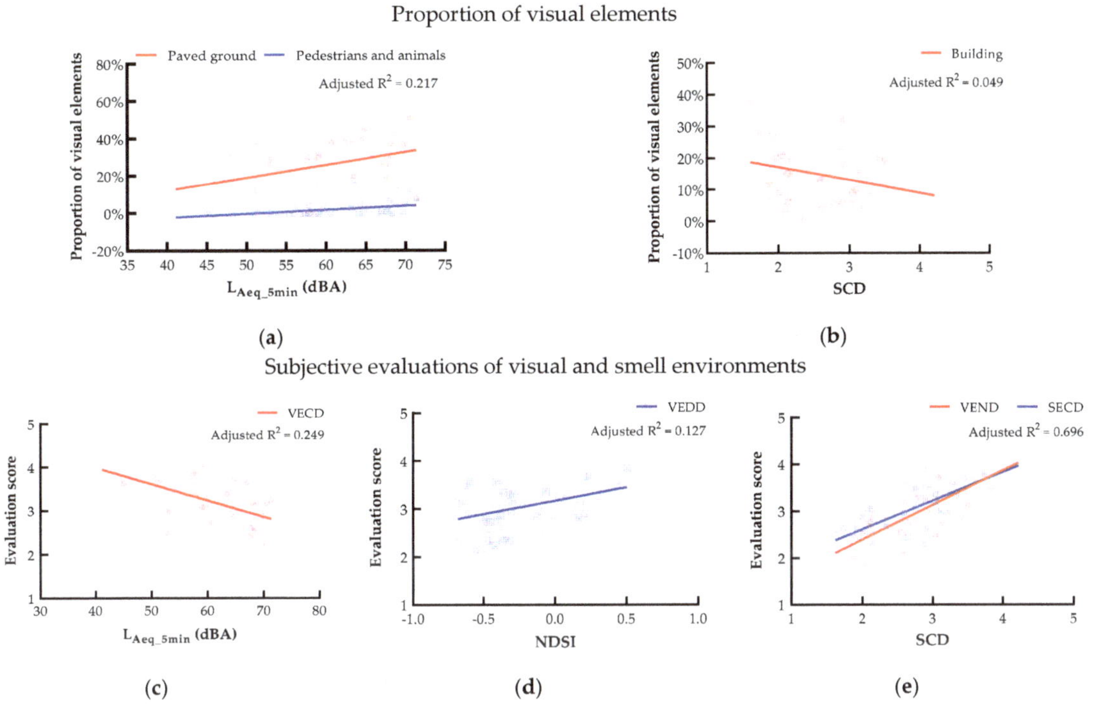

**Figure 8.** Linear regression curves between the proportion of visual elements and soundscape evaluation parameters, and subjective evaluations of visual and smell environments and soundscape evaluation parameters: (**a**) the proportion of visual elements and $L_{Aeq\_5min}$; (**b**) the proportion of visual elements and SCD; (**c**) subjective evaluations of visual and smell environments and $L_{Aeq\_5min}$; (**d**) subjective evaluations of visual and smell environments and NDSI; (**e**) subjective evaluations of visual and smell environments and SCD.

## 4. Discussion

### 4.1. The Influence of Spatial Characteristics on the Soundscape

This study found that the soundscape evaluations of WSMCs could be affected by spatial characteristics, such as elevation, VDS, VDR, and HDR. Due to excessive noise from urban expressways caused by the terrain fluctuation and compact urban structure in mountainous cities [22], regression analyses showed that the position relative to the road (including VDR and HDR) had a greater impact than other spatial indicators on the WSMC soundscape parameters. In a field study of a mountain landscape, Liu, Kang and Meng [42] found that elevation was significantly negatively correlated with sharpness but had no significant effects on the SPL or SCD. In contrast, this study found that elevation was significantly correlated with the $L_{Aeq\_5min}$ and SCD in WSMCs. This may be due to the fact that WSMCs are greatly influenced by urban elements, such as roads and pedestrians. In addition, spatial variations in soundscapes may lead to the distribution of biodiversity along elevation gradients at large spatial scales [51]. This study found that the NDSI was only significantly correlated with the VDR and HDR, but not with elevation, the VDS, or the HDS. This indicates that topography has little impact on biodiversity in small-scale spaces, such as WSMCs, and more attention should be paid to the impact of the spatial distribution of noise sources, such as traffic noise, on the ecological environments of WSMCs. Laboratory studies have shown that a horizontal position near water can significantly improve the evaluation of soundscape comfort by using photographs [52]. However, this study found that, in WSMCs, only the vertical position near water was related to the SCD. Additionally, the correlation coefficient between the VDR and SCD was the highest (r = −0.572). This might suggest that vertical spatial indicators have a greater impact on soundscape comfort in WSMCs.

### 4.2. The Influence of Visual and Smell Environments on Soundscape

In terms of visual elements, this study found that paved ground, buildings, sky, water, natural terrain, and pedestrians and animals were effective landscape elements influencing the soundscape evaluation parameters. Unlike the sound of fountains, this study found that water sounds from rivers have little effect on the SPL in WSMCs [53]. The presence of water can still cause a sense of visual tranquility and improve the comfort of the soundscape [14], but the visual element of the water accounted for only 0.1%–7.2% in the studied WSMCs. This indicates that it is necessary to increase the number of walking points where water can be seen, or build other waterscapes to improve the visibility of water and the proportion of water sounds in WSMCs. In terms of subjective evaluation, the regression analysis found that the diversity of the visual environment is an important factor affecting the proportion of biological sounds. However, this study also found that the visual landscape of WSMCs is still not rich in landscape diversity.

In terms of the smell environment, although the overall evaluation of the SECD tended to be positive, it should be noted that emission odors (such as traffic and waste emissions) accounted for large proportions, ranging from 7.5% to 32.3%. A previous study found that there was a strong similarity between the soundscape and the smell environment evaluation [30]. However, this study found that the SECD was not significantly correlated with the NDSI. This may be due to the fact that biologically emitted odors are less detectable than biological visual elements. Notably, in the regression model, the SECD was found to be the most influential variable on the SCD ($\beta$ = 0.553). This may be because the olfactory experience has a greater impact on subjective feelings (including subjective emotions and environmental and spatial memory) than visual perception [31,54].

### 4.3. Suggestions for Soundscape Improvement in WSMCs

Specifically, this study summarizes the recommended values of specific spatial indicators, the proportions of visual elements, and subjective evaluations of visual and smell environments through linear regression curves so as to achieve a positive soundscape evaluation (Table 11). These data can provide suggestions and references for healthy acoustic environment design and the study of WSMCs in the future. It is worth noting that the adjusted $R^2$ values of the regression models of the $L_{Aeq}$ and NDSI using spatial indicators (0.425 and 0.217, respectively) were higher than those using subjective evaluations of visual and smell environments (0.249 and 0.127, respectively). In contrast, the adjusted $R^2$ value of the regression model of the SCD using subjective evaluations of visual and smell environments (0.696) was higher than that of the regression model using spatial indicators (0.249). This indicates that the objective evaluation of the soundscape is more affected by spatial indicators, and soundscape comfort is more affected by visual and smell environments.

In addition, this study has some limitations that need to be addressed in the future. First, although the linear regression model was effective, it was found that the performance of the models of the VECD and $L_{Aeq\_5min}$, the HDR and VEDD and the NDSI, and buildings and the SCD was relatively low (Table 11). The limitation of photography technology may lead to a low adjusted $R^2$ in the visual element model [26]. In addition, the removal of strong difference points may lead to a higher adjusted $R^2$ coefficient (such as the Grubbs test) [55]. Future studies should discuss how to improve the adjusted $R^2$ coefficient and obtain more accurate recommended values. Secondly, the participants in this study were architecture students. However, participants with different social backgrounds may influence the results of environmental perception [56]. In follow-up studies, randomized participants might be employed across multiple areas to verify the generality of the conclusions of this study.

**Table 11.** The recommended values to achieve positive soundscape evaluation in terms of spatial characteristics and visual and smell environments.

| Soundscape Evaluation Parameters | Objectives | Indicator of Spatial Characteristics, Visual and Smell Environments | | Recommended Values |
|---|---|---|---|---|
| $L_{Aeq\_5min}$ | $\leq$55dBA | Spatial indicators | HDR | $\geq$90 m |
| | | | Elevation | $\geq$220 m |
| | | Visual elements | Paved ground | $\leq$22% |
| | | | Pedestrians and animals | $\leq$1% |
| | | Subjective evaluations of visual and smell environments | VECD * | $\geq$3.4 |
| NDSI | $\geq$0 | Spatial indicators | HDR * | $\geq$90 m |
| | | Subjective evaluations of visual and smell environments | VEDD * | $\geq$3.2 |
| SCD | $\geq$3 | Spatial indicators | VDR | $\leq-$10 m |
| | | | HDR | $\geq$70 m |
| | | Visual elements | Buildings ** | $\leq$13% |
| | | Subjective evaluations of visual and smell environments | SECD | $\geq$3.2 |
| | | | VEND | $\geq$3.1 |

Notes: * adjusted $R^2 < 0.3$; ** adjusted $R^2 < 0.1$.

## 5. Conclusions

This study took Chongqing as an example to investigate the current situation of soundscapes in WSMCs and discussed the influence of spatial characteristics, as well as visual and smell environments, on soundscape evaluation parameters ($L_{Aeq\_5min}$, NDSI, and SCD). The results show that the subjective and objective soundscape evaluations of WSMCs are of poor quality. Traffic sounds are dominant (33%), and natural sounds only account for 27%. Spatial indicators (elevation, VDS, VDR, and HDR) were significantly correlated with soundscape evaluation parameters. Among them, the VDR is the most influential variable on the $L_{Aeq\_5min}$ and NDSI, and the HDR is the most influential variable on the SCD. In addition, elevation and the VDS are positively correlated with the $L_{Aeq}$ and negatively correlated with the SCD. In terms of the proportions of visual elements, paved ground, pedestrians, and buildings in photos have negative effects on the soundscape, while the sky, water, and natural terrain have positive effects. Subjective evaluation results showed that high visual and smell environment quality can enhance soundscape evaluation, although the smell environment had a greater impact on SCD than the visual environment. In general, the $L_{Aeq}$ and NDSI are more affected by spatial characteristics, and the SCD is more affected by visual and smell environments in WSMCs. Finally, this study summarizes the recommended values of spatial characteristics and visual and smell environment indicators to achieve a positive soundscape evaluation. Considering the likely accelerated urban construction process in the future, the results of this study can provide effective data support and references for soundscape design and landscape environment construction in WCMCs in order to improve environmental health and people's happiness. More research is needed to further optimize model performance to improve data accuracy and to discuss the impact of different population experiences in the future.

**Author Contributions:** Conceptualization, H.X. and B.Z.; methodology, B.Z., H.X., T.G., L.Q. and H.L.; validation, H.X., B.Z., T.G. and L.Q.; investigation, B.Z., H.X., H.L. and Z.Z.; writing—original draft preparation, B.Z. and H.X.; writing—review and editing, H.X., B.Z., T.G., L.Q., H.L. and Z.Z.; visualization, B.Z.; supervision, H.X.; project administration, H.X.; funding acquisition, H.X. All authors have read and agreed to the published version of the manuscript.

**Funding:** This research was funded by the National Natural Science Foundation of China (52078077) and Northwest A&F University Youth Talent Cultivation Project (Z1010122001).

**Conflicts of Interest:** The authors declare no conflict of interest.

## Appendix A

**Table A1.** The details of the questionnaire.

| Parts | | Question | |
|---|---|---|---|
| Soundscape | Sound sources identification | Please list sound sources you noticed (limited to 8). | Open question |
| | SCD | How would you rate the comfort of the soundscape? | From 1 = "uncomfortable" to 5 = "comfortable" |
| Visual environment | VECD | How would you rate the comfort of the visual environment? | From 1 = "uncomfortable" to 5 = "comfortable" |
| | VEND | How would you rate the natural of the visual environment? | from 1 = "artificial" to 5 = "natural" |
| | VEDD | How would you rate the diversity of the visual environment? | From 1 = "simple" to 5 = "complex" |
| Smell environment | Odor identification | Please list odors you noticed (limited to 3). | Open question |
| | SECD | How would you rate the comfort of the smell environment? | From 1 = "uncomfortable" to 5 = "comfortable" |

## Appendix B

**Table A2.** Detailed specifications of all measurement equipment and associated data processing software in the study.

| Measurement | | Equipment | Equipment Specifications | Data Processing Software | Software Specifications |
|---|---|---|---|---|---|
| Objective soundscape evaluation parameters | $L_{Aeq}$ | AWA 6228$^+$ Sound Level Meter (Aihua Instruments Co., Ltd., China) | IEC 61672 Class 1 Measurement Range: 20 dB–142 dB (145 dB Peak) Ref.: [57] | – | – |
| | NDSI | PCM-M10 Recorder (Sony Corporation, Japan) | Sampling frequencies: 44.1 kHz Bit rate: 32 kbps–192 kbps Recoding: binaural method Ref.: [58] | Rstudio (RStudio, Inc., Boston, USA) | Packages: tuneR, soundecology Ref.: [59] |
| Spatial indicators | Elevation, VDS, HDS, VDR, HDR | YILI X28 altimeter (Hengyi Technology Co., Ltd., China) | Barometric altimetry: ≤1 m Location accuracy: ≤2 m Ref.: [60] | – | – |
| Identification of the proportions of visual elements | | Mobile phone cameras | Camera: ≥12 megapixels Image size: 4750 × 1080 pixels | The FCN model (GUC. HPSCIL, University of Geosciences, China) | Codes: Java, C++ Accuracy: 67% for actual data Ref.: [43] |

## References


1. Yang, C.; Shi, M.; Geng, H. Study on the pedestrian accessibility of waterfront area based on an understanding of urban fabric level: Taking the estuary area of Suzhou River in Shanghai as an example. *City Plan. Rev.* **2018**, *42*, 104–114.
2. Niu, Y.; Mi, X.; Wang, Z. Vitality evaluation of the waterfront space in the ancient city of Suzhou. *Front. Archit. Res.* **2021**, *10*, 729–740. [CrossRef]
3. Feng, L.; Lei, L. Strategy Studying on Urban Design of Riverfront in Mountainous City: Taking Chongqing for Example. *Inter. Des.* **2009**, 41–45.
4. Leske, S.; Kõlves, K.; Crompton, D.; Arensman, E.; De Leo, D. Real-time suicide mortality data from police reports in Queensland, Australia, during the COVID-19 pandemic: An interrupted time-series analysis. *Lancet Psychiatry* **2021**, *8*, 58–63. [CrossRef]
5. World Health Organization. *Burden of Disease from Environmental Noise: Quantification of Healthy Life Years Lost in Europe*; World Health Organization, Regional Office for Europe: Copenhagen, Denmark, 2011.
6. Aletta, F.; Oberman, T.; Kang, J. Associations between positive health-related effects and soundscapes perceptual constructs: A systematic review. *Int. J. Environ. Res. Public Health* **2018**, *15*, 2392. [CrossRef]
7. Aletta, F.; Oberman, T.; Kang, J. Positive health-related effects of perceiving urban soundscapes: A systematic review. *Lancet* **2018**, *392*, S3. [CrossRef]
8. Escobedo, D.N. Foreigners as gentrifiers and tourists in a Mexican historic district. *Urban Stud.* **2020**, *57*, 3151–3168. [CrossRef]
9. Joo, W.; Gage, S.H.; Kasten, E.P. Analysis and interpretation of variability in soundscapes along an urban–rural gradient. *Landsc. Urban Plan.* **2011**, *103*, 259–276. [CrossRef]

10. Kasten, E.P.; Gage, S.H.; Fox, J.; Joo, W. The remote environmental assessment laboratory's acoustic library: An archive for studying soundscape ecology. *Ecol. Inform.* **2012**, *12*, 50–67. [CrossRef]
11. Bertucci, F.; Parmentier, E.; Lecellier, G.; Hawkins, A.D.; Lecchini, D. Acoustic indices provide information on the status of coral reefs: An example from Moorea Island in the South Pacific. *Sci. Rep.* **2016**, *6*, 33326. [CrossRef]
12. Fuller, S.; Axel, A.C.; Tucker, D.; Gage, S.H. Connecting soundscape to landscape: Which acoustic index best describes landscape configuration? *Ecol. Indic.* **2015**, *58*, 207–215. [CrossRef]
13. Brown, A.; Muhar, A. An approach to the acoustic design of outdoor space. *J. Environ. Plan. Manag.* **2004**, *47*, 827–842. [CrossRef]
14. Watts, G.; Pheasant, R.; Horoshenkov, K. Tranquil spaces in a metropolitan area. In Proceedings of the 20th International Congress on Acoustics, Sydney, Australia, 23–27 August 2010; pp. 23–27.
15. Jeon, J.Y.; Hong, J.Y. Classification of urban park soundscapes through perceptions of the acoustical environments. *Landsc. Urban Plan.* **2015**, *141*, 100–111. [CrossRef]
16. Puyana Romero, V.; Maffei, L.; Brambilla, G.; Ciaburro, G. Acoustic, visual and spatial indicators for the description of the soundscape of waterfront areas with and without road traffic flow. *Int. J. Environ. Res. Public Health* **2016**, *13*, 934. [CrossRef] [PubMed]
17. Watts, G.R.; Pheasant, R.J.; Horoshenkov, K.V.; Ragonesi, L. Measurement and subjective assessment of water generated sounds. *Acta Acust. United Acust.* **2009**, *95*, 1032–1039. [CrossRef]
18. Kang, J. Sound propagation in street canyons: Comparison between diffusely and geometrically reflecting boundaries. *J. Acoust. Soc. Am.* **2000**, *107*, 1394–1404. [CrossRef]
19. Li, H.; Xie, H.; Woodward, G. Soundscape components, perceptions, and EEG reactions in typical mountainous urban parks. *Urban For. Urban Green.* **2021**, *64*, 127269. [CrossRef]
20. Renterghem, T.V.; Botteldooren, D.; Lercher, P. Comparison of measurements and predictions of sound propagation in a valley-slope configuration in an inhomogeneous atmosphere. *J. Acoust. Soc. Am.* **2007**, *121*, 2522–2533. [CrossRef]
21. Feng, K.; Hubacek, K.; Sun, L.; Liu, Z. Consumption-based $CO_2$ accounting of China's megacities: The case of Beijing, Tianjin, Shanghai and Chongqing. *Ecol. Indic.* **2014**, *47*, 26–31. [CrossRef]
22. Li, H.; Xie, H. Noise exposure of the residential areas close to urban expressways in a high-rise mountainous city. *Environ. Plan. B Urban Anal. City Sci.* **2021**, *48*, 1414–1429. [CrossRef]
23. Calvert, G.; Spence, C.; Stein, B.E. *The Handbook of Multisensory Processes*; MIT Press: Cambridge, MA, USA, 2004.
24. Carles, J.; Bernáldez, F.; Lucio, J.d. Audio-visual interactions and soundscape preferences. *Landsc. Res.* **1992**, *17*, 52–56. [CrossRef]
25. Michon, R.; Chebat, J.-C.; Michon, R. The Interaction Effect of Music and Odour on Shopper Spending. Ph.D. Thesis, School of Retail Management, Ryerson University, Toronto, ON, Canada, 2006, *unpublished*.
26. Liu, J.; Kang, J.; Behm, H.; Luo, T. Effects of landscape on soundscape perception: Soundwalks in city parks. *Landsc. Urban Plan.* **2014**, *123*, 30–40. [CrossRef]
27. Shao, Y.; Hao, Y.; Yin, Y.; Meng, Y.; Xue, Z. Improving Soundscape Comfort in Urban Green Spaces Based on Aural-Visual Interaction Attributes of Landscape Experience. *Forests* **2022**, *13*, 1262. [CrossRef]
28. Maffei, L.; Iachini, T.; Masullo, M.; Aletta, F.; Sorrentino, F.; Senese, V.P.; Ruotolo, F. The effects of vision-related aspects on noise perception of wind turbines in quiet areas. *Int. J. Environ. Res. Public Health* **2013**, *10*, 1681–1697. [CrossRef] [PubMed]
29. Puyana-Romero, V.; Ciaburro, G.; Brambilla, G.; Garzón, C.; Maffei, L. Representation of the soundscape quality in urban areas through colours. *Noise Mapp.* **2019**, *6*, 8–21. [CrossRef]
30. Ba, M.; Kang, J. A laboratory study of the sound-odour interaction in urban environments. *Build. Environ.* **2019**, *147*, 314–326. [CrossRef]
31. Michael, G.A.; Jacquot, L.; Millot, J.-L.; Brand, G. Ambient odors modulate visual attentional capture. *Neurosci. Lett.* **2003**, *352*, 221–225. [CrossRef] [PubMed]
32. Adams, M.; Askins, K. Sensewalking: Sensory walking methods for social scientists. In Proceedings of the RGSIBG Annual Conference 2009, Manchester, UK, 26–28 August 2009.
33. Xie, H.; He, Y.; Wu, X.; Lu, Y. Interplay between auditory and visual environments in historic districts: A big data approach based on social media. *Environ. Plan. B Urban Anal. City Sci.* **2021**, *49*, 1245–1265. [CrossRef]
34. Bruce, N.; Condie, J.; Henshaw, V.; Payne, S.R. *Analysing Olfactory and Auditory Sensescapes in English Cities: Sensory Expectation and Urban Environmental Perception*; International Ambiances Network: Grenoble, France, 2015.
35. *ISO/TS 12913-2: 2018*; Acoustics—Soundscape—Part 2: Data Collection and Reporting Requirements. ISO: Geneva, Switzerland, 2018.
36. Berkouk, D.; Bouzir, T.A.K.; Boucherit, S.; Khelil, S.; Mahaya, C.; Matallah, M.E. Evaluation of the soundscapes through the cafe terraces before and after the COVID-19 lockdown in coastal cities in Algeria. *Noise Vib. Worldw.* **2022**, *53*, 377–389. [CrossRef]
37. Liu, J.; Xiong, Y.; Wang, Y.; Luo, T. Soundscape effects on visiting experience in city park: A case study in Fuzhou, China. *Urban For. Urban Green.* **2018**, *31*, 38–47. [CrossRef]
38. Russell, J.A.; Lanius, U.F. Adaptation level and the affective appraisal of environments. *J. Environ. Psychol.* **1984**, *4*, 119–135. [CrossRef]
39. Xiao, J. Smell, smellscape, and place-making: A review of approaches to study smellscape. In *Handbook of Research on Perception-Driven Approaches to Urban Assessment and Design*; IGI Global: Hershey, PA, USA, 2018; pp. 240–258.
40. Henshaw, V. *Urban Smellscapes: Understanding and Designing City Smell Environments*; Routledge: London, UK, 2013.

41. *ISO 1996-2:2017*; Acoustics—Description, Measurement and Assessment of Environmental Noise—Part 2: Determination of Environmental Noise levels. ISO: Geneva, Switzerland, 2007.
42. Liu, F.; Kang, J.; Meng, Q. On the influence factors of audio-visual comfort of mountain landscape based on field surve. *J. Environ. Eng. Landsc. Manag.* **2020**, *28*, 48–61. [CrossRef]
43. Yao, Y.; Liang, Z.; Yuan, Z.; Liu, P.; Bie, Y.; Zhang, J.; Wang, R.; Wang, J.; Guan, Q. A human-machine adversarial scoring framework for urban perception assessment using street-view images. *Int. J. Geogr. Inf. Sci.* **2019**, *33*, 2363–2384. [CrossRef]
44. Hao, Y.; Kang, J.; Krijnders, J.D. Integrated effects of urban morphology on birdsong loudness and visibility of green areas. *Landsc. Urban Plan.* **2015**, *137*, 149–162. [CrossRef]
45. Zhou, Y.; Dai, P.; Zhao, Z.; Hao, C.; Wen, Y. The Influence of Urban Green Space Soundscape on the Changes of Citizens' Emotion: A Case Study of Beijing Urban Parks. *Forests* **2022**, *13*, 1928. [CrossRef]
46. Jeon, J.Y.; Lee, P.J.; You, J.; Kang, J. Perceptual assessment of quality of urban soundscapes with combined noise sources and water sounds. *J. Acoust. Soc. Am.* **2010**, *127*, 1357–1366. [CrossRef]
47. *GB 3096-2008*; Environmental Quality Standard for Noise. Ministry of Environmental Protection of the People's Republic of China: Beijing, China, 2008.
48. Yang, M. *A Review of Regression Analysis Methods: Establishing the Quantitative Relationships between Subjective Soundscape Assessment and Multiple Factors*; Universitätsbibliothek der RWTH Aachen: Aachen, Germany, 2019.
49. Quercia, D.; Schifanella, R.; Aiello, L.M.; McLean, K. Smelly maps: The digital life of urban smellscapes. In Proceedings of the International AAAI Conference on Web and Social Media, Oxford, UK, 26–29 May 2015; pp. 327–336.
50. Lercher, P.; Schulte-Fortkamp, B. The relevance of soundscape research to the assessment of noise annoyance at the community level. In Proceedings of the Eighth International Congress on Noise as a Public Health Problem, Rotterdam, The Netherlands, 29 June–3 July 2003; pp. 225–231.
51. Pijanowski, B.C.; Farina, A.; Gage, S.H.; Dumyahn, S.L.; Krause, B.L. What is soundscape ecology? An introduction and overview of an emerging new science. *Landsc. Ecol.* **2011**, *26*, 1213–1232. [CrossRef]
52. Ren, X.; Kang, J. Effects of the visual landscape factors of an ecological waterscape on acoustic comfort. *Appl. Acoust.* **2015**, *96*, 171–179. [CrossRef]
53. Nilsson, M.E.; Alvarsson, J.; Rådsten-Ekman, M.; Bolin, K. Auditory masking of wanted and unwanted sounds in a city park. *Noise Control Eng. J.* **2010**, *58*, 524–531. [CrossRef]
54. Gottfried, J.A.; Dolan, R.J. The nose smells what the eye sees: Crossmodal visual facilitation of human olfactory perception. *Neuron* **2003**, *39*, 375–386. [CrossRef]
55. Couderc, N. *GRUBBS: Stata Module to Perform Grubbs' Test for Outliers*; Boston College Department of Economics: Boston, MA, USA, 2007.
56. Fang, X.; Gao, T.; Hedblom, M.; Xu, N.; Xiang, Y.; Hu, M.; Chen, Y.; Qiu, L. Soundscape perceptions and preferences for different groups of users in urban recreational forest parks. *Forests* **2021**, *12*, 468. [CrossRef]
57. AWA 6228+ Sound Level Meter Specifications. Available online: https://www.hzaihua.com/public/uploads/admin/file/7bc2 30b46ec5eedb477c3366f1e135b4.pdf (accessed on 7 December 2022).
58. PCM-M10 Recorder Specifications. Available online: https://www.sony.com/electronics/support/res/manuals/4156/fb0e074c4 53380134302281a7d81bfa6/41565411M.pdf (accessed on 7 December 2022).
59. Rstudio Specifications. Available online: https://posit.co/products/open-source/rstudio/ (accessed on 7 December 2022).
60. YILI X28 Altimeter Specifications. Available online: https://mp.weixin.qq.com/s/KHU7q11lUAkEndxYlrCoOA (accessed on 7 December 2022).

*Article*

# Thermal Comfort Analysis and Optimization Strategies of Green Spaces in Chinese Traditional Settlements

Yanyan Cheng [1], Ying Bao [2,3,4,*], Shengshuai Liu [1], Xiao Liu [2,3,4,5,*], Bin Li [2,3,6], Yuqing Zhang [2,3,6], Yue Pei [4], Zhi Zeng [1] and Zhaoyu Wang [1]

1    Department of Architecture, North China University of Water Resources and Electric Power, Zhengzhou 450046, China
2    School of Architecture, South China University of Technology, Guangzhou 510641, China
3    State Key Laboratory of Subtropical Building and Urban Science, South China University of Technology, Guangzhou 510641, China
4    Architectural Design & Research Institute Co., Ltd., South China University of Technology, Guangzhou 510641, China
5    Energy Saving Technology Research Institute, South China University of Technology, Guangzhou 510641, China
6    Department of Architecture and Design, Politecnico di Torino, 10125 Torino, Italy
*    Correspondence: arbaoy@scut.edu.cn (Y.B.); xiaoliu@scut.edu.cn (X.L.)

**Abstract:** The spatial pattern of Weizi settlements features distinct regional characteristics. Moreover, it contains profound wisdom in terms of traditional construction; therefore, studies on its association with the microclimate have important implications for improving the quality of human settlements. In the present study, Guanweizi Village in the Xinyang City of Henan Province was used as an example to analyze and evaluate the thermal comfort of green spaces. The impact of peripheral water bodies on the thermal comfort of outdoor green spaces in the settlement was studied, and the association between the components of outdoor green spaces and physiological equivalent temperature as an indicator of thermal comfort was explored. Further, factors negatively affecting the thermal comfort of green spaces were analyzed through the grid method. Thermal comfort in the Weizi settlement is somewhat correlated with the coverage of water bodies, roads, soil, greening, and buildings. Increasing the water area and creating multi-level greening spaces are effective measures to improve the thermal comfort of green spaces in the settlement. Our findings provide a theoretical basis and a pioneering example for future practices of environment design for human settlements.

**Keywords:** traditional settlements; ENVI-met; green spaces; outdoor environment; thermal comfort; regional characteristics

**Citation:** Cheng, Y.; Bao, Y.; Liu, S.; Liu, X.; Li, B.; Zhang, Y.; Pei, Y.; Zeng, Z.; Wang, Z. Thermal Comfort Analysis and Optimization Strategies of Green Spaces in Chinese Traditional Settlements. *Forests* **2023**, *14*, 1501. https://doi.org/10.3390/f14071501

Academic Editor: Helmi Zulhaidi Mohd Shafri

Received: 2 June 2023
Revised: 10 July 2023
Accepted: 14 July 2023
Published: 22 July 2023

## 1. Introduction

During rapid urbanization, the spatial patterns of traditional settlements in many areas, being unable to adapt to the development needs of modern life, have been seriously damaged by rampant and extensive rural reconstruction [1]. Moreover, many environmental problems are arising from urbanization, such as a rise in temperature and heat waves, a drop in temperature and cold waves, air pollution, meteorological disasters, and ecosystem degradation [2]. These increasingly severe problems [3,4] have seriously compromised the quality and regional characteristics of traditional human settlements [5–7].

As green spaces in settlements are intended for activities closely related to daily life, their safety and comfort are gradually attracting attention. As such, the construction of the living environment is no longer merely intended to create comfortable living spaces inside buildings for people, but rather to create comfortable, safe, and healthy spaces outside of buildings for peoples' interactions, from a human-centered and perception-based perspective [8]. As the outdoor climate directly affects human perception, the thermal environment outside building spaces has garnered much research attention.

Previous studies on factors influencing the thermal environment outside building clusters have primarily focused on the spatial form, building layout, street space, underlying surface, green space, water body, and thermal comfort. Consequently, relevant progress has been made. First, the impact of spatial form on the thermal environment has been explored. For instance, Wang et al. studied thermal environment mitigation strategies for urban buildings of different densities in Toronto, and found that sunshine duration and average radiation play important roles in urban thermal comfort [9]. In another study, Zhou et al. explored the impact of urban spatial form on air temperature and proposed that building shading can effectively improve the thermal environment [10]. Further, Huang et al. studied the relationships among urban form, weather factors, and thermal comfort and observed that weather factors affect thermal comfort in urban environments [11]. Subsequently, Sun et al. studied the impacts of building density, floor area ratio, and green space ratio on the urban thermal environment and showed that an urban form significantly affects the thermal environment; in addition, increases in the building density increased the land surface temperature, while increased floor area and green space ratios decreased the land surface temperature [12]. Recently, Wu et al. studied the correlation between village morphological factors and water-cooling values and proposed that a higher water body rate and green space ratio, and a lower surrounding building density, can improve the cooling effect of water bodies [13]. Second, the impacts of building layout on the thermal environment have been studied. For instance, Jung et al. used an orthogonal experimental design to improve the urban microclimate and showed that the building coverage rate, building interval, and azimuth angle affect the outdoor thermal environment [14]. In another study, Liu et al. evaluated the impact of building layout on the thermal environment in residential districts and demonstrated that the temperature significantly differed across various building layouts [15]. Third, the impacts of street space on the thermal environment have been assessed. For instance, Du et al. studied the impact of street direction, street aspect ratio, and pavement material on the thermal environment in the street area [16], and Shao et al. studied the impact of street landscape on the thermal environment [17]. Fourth, some studies explored the impact of the underlying surface on the thermal environment. Specifically, Tsoka et al. studied different intervention schemes for the thermal environment with urban morphological characteristics in Thessaloniki, including microclimate parameters, such as earth surface and air, and average radiant temperature distribution. The authors reported significant reductions in the surface temperature by replacing conventional coatings with cooling materials featuring a higher albedo and emissivity [18]. Further, Kurazumi et al. studied the impact of the thermal environment in rural and suburban outdoor spaces on the human body, and proposed green space and water surface as natural factors that help reduce air temperature [19]. Recently, Xin et al. studied typical rural settlements in the Guanzhong Region and, using field measurements and ENVI-met data simulation, analyzed the dynamic changes in the thermal environment in rural areas. The authors used the root mean square error (RMSE) and mean absolute percentage error (MAPE) to verify the simulation model and physiological equivalent temperature (PET) to evaluate the outdoor thermal environment. They proposed that trees can significantly improve the outdoor thermal environment. In 1978, Hoppe et al. introduced meteorological factors, clothing, human activities, and human body parameters and proposed PET as an indicator of thermal comfort [20,21]. PET measures thermal comfort according to the energy balance of the human body. It is impacted by air temperature, wind speed, humidity, and average radiation temperature and can reflect thermal comfort more directly and comprehensively. PET is a meteorological indicator in a real sense and a comprehensive outdoor thermal comfort indicator, which can be used to evaluate the thermal environment in different seasons and under diverse climatic conditions [22,23]. Many recent studies have shown that thermal comfort could be greatly improved by optimizing the physical urban environment. For instance, Andreou et al. discussed the impacts of street structure, form, and ground reflectance on thermal comfort and proposed that the thermal comfort of the street microclimate is better in traditional urban areas than in modern urban clusters. Kariminia et al.

studied the thermal comfort and outdoor space utilization of urban squares and proposed that tourists are more sensitive to temperature changes than to other microclimatic factors, and that better air circulation and water evaporation can significantly increase thermal comfort [24,25].

In contrast, recent research findings on the thermal environment of outdoor green spaces in rural areas have been relatively limited, particularly those on the impact of water areas on microclimates in rural settlements. In addition, previous studies have mostly focused on the heat island effect of land settlements or the thermal comfort of settlements neighboring water on only one side. Weizi settlements, as a form of rural settlement surrounded by water on all four sides, are different from land settlements and settlements neighboring water on only one side. They are built based on the production and living experience of local residents. Through the literature review, we noted this research gap in previous studies. Rural areas are likewise subject to environmental-change-induced climatic impact. In some areas, due to the extensive reconstruction of traditional settlements, the spatial characteristics, as well as the ecological and environmental balance of the settlements, are damaged, resulting in frequent occurrences of problems, such as dust and droughts. In the face of modernization, these settlements are facing increasing climate-related challenges. In response to these climatic and environmental problems and their impacts in rural areas, the Fifth Plenary Session of the 19th CPC Central Committee held in 2020 placed strong emphasis on the goal of improving the rural living environment. In this context, it is of great importance and practical significance to rationally reconstruct and revitalize traditional settlements, as well as to evaluate and optimize the thermal environment of existing outdoor green spaces in traditional settlements. Previous studies on the microclimate of rural environments have mostly been based on the influence of natural water bodies on the microclimate, while areas of water within Weizi have been artificially constructed around the settlement. In the present work, we aim to determine whether such purposeful construction can regulate a microclimate. The significance of the present study lies in our finding that areas of water in Weizi do indeed exert a regulatory effect on the microclimate, although there is still room for optimization. The novelty of this research lies in the verification and analysis of thermal comfort in a rural settlement surrounded by water on all four sides.

Based on the impact that areas of water have on the microclimate of traditional villages, the present study evaluated the thermal comfort of outdoor green spaces and used local traditional construction experience to improve and create an outdoor environment with good thermal comfort. Using Guanweizi Village—a typical Weizi settlement in South Henan, China—as an example, we simulated the thermal environment of the settlement through field mapping of the village layout and ENVI-met modeling, and calculated PET with BIO-met. By maintaining the traditional residential building style, we analyzed the impact of peripheral water bodies on thermal comfort in the settlement. Finally, we proposed strategies to improve the thermal environment of Weizi settlements, providing a reference for the ecological revitalization and sustainable development of traditional villages [26–29].

## 2. Materials and Methods

### 2.1. Case Study Area

Guanweizi Village is located in the Yanhe Township of Guangshan County in Xinyang City, Henan Province, China. Surrounded by water bodies, the settlement has a square overall layout, demonstrating a typical settlement form with local characteristics (Figures 1 and 2). The village is surrounded by water bodies, which not only serve irrigation and cultivation purposes, as well as laundry cleaning and rainwater drainage functions, but also play roles in terms of defense and fire protection. More importantly, from the ecological perspective, they protect the whole area against floods and purify the settlement to some extent. In addition, together with plants, streets, and courtyard spaces, these water bodies regulate and improve the thermal environment in the settlement, which was precisely the focus of the present study.

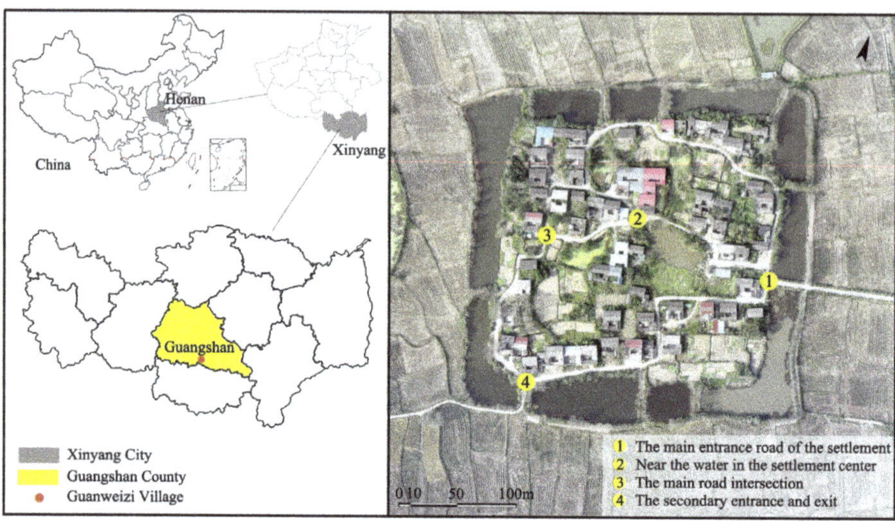



**Figure 1.** Scope of the case study area (map snapshot and photograph obtained by the author).

**Figure 2.** Photographs of the measurement points.

Four measuring points were selected in the present study, which reflected the environmental characteristics of the Guanweizi settlement. Point 1 was located at the main entrance road of the settlement, near peripheral water bodies and surrounded by multi-level greening and many trees. Point 2 was located at the settlement center, near the largest water body inside the settlement. Point 3 was located at the main traffic hub within the settlement, away from the internal water bodies. Point 4 was located at the secondary entrance and exit of the settlement, also near the peripheral water bodies, but with fewer trees and greening clusters than those at point 1. Each point is described in Table 1 and shown in Figures 1 and 2.

**Table 1.** Description of measuring points.

| Measuring Point | Description |
| --- | --- |
| 1 | At the main entrance road of the settlement, near peripheral water bodies and surrounded by multi-level greening and many trees |
| 2 | Near water bodies at the settlement center |
| 3 | At the main road intersection at the settlement center, away from the internal water bodies |
| 4 | At the secondary entrance and exit of the settlement, near peripheral water bodies, but with fewer greening clusters and trees |

### 2.2. Thermal Comfort Research Methodology

### 2.2.1. Low-Altitude Photogrammetry

Six control points for photography were evenly distributed in the survey area of Guanweizi Village, with 3D coordinates measured using RTK and orthophotographs obtained using a DJI UAV (Phantom 4RTK). The aerial photographs were collaged using Pix4Dmapper, and ground elevation was extracted using ArcGIS. The processed orthophotos were then imported into CASS, and ground objects were drawn. Finally, the elevation data were imported into CASS (Figure 3), which showed that the Weizi settlement is on flat land.

**Figure 3.** Elevation map of Guanweizi Village.

### 2.2.2. ENVI-Met Software for Data Simulation

ENVI-met, a piece of software developed by Bruse and Fleer in 1998, is currently updated to version 5 [30]. Based on the principles of thermodynamics and hydromechanics, this software dynamically simulates interactions between the surface and plants in small-scale human settlements with air [26].

In the present study, simulations for Guanweizi Village were conducted using ENVI-met to obtain the distribution of PET, as the thermal comfort indicator, in the green spaces of the settlement [31].

### 2.2.3. PET as the Thermal Comfort Indicator

In the present study, PET was selected as the thermal comfort indicator, which refers to the corresponding temperature when the skin and body temperatures reach the same thermal state as that of the typical indoor environment in a given indoor or outdoor environment [12,15,32]. With reference to the corresponding relationship between PET and human thermal perception evaluation (Table 2) [16], as well as the dynamic ENVI-met-based simulation of the thermal environment in the village, the correlations between various spatial components of the Guanweizi settlement and the thermal comfort indicator PET were comprehensively analyzed [33].

**Table 2.** Physiological equivalent temperature (PET) and the corresponding human thermal sensation.

| PET (°C) | ≤4 | 4–8 | 8–13 | 13–18 | 18–23 | 23–29 | 29–35 | 35–41 | ≥41 |
|---|---|---|---|---|---|---|---|---|---|
| Thermal sensation | Very cold | Cold | Cool | Slightly cool | Neutral | Slightly warm | Warm | Hot | Very hot |
| Physiological stress level | Extreme cold stress | Strong cold stress | Moderate cold stress | Slight cold stress | No | Slight heat stress | Moderate heat stress | Strong heat stress | Extreme heat stress |

### 2.2.4. Grid Analysis

The grid method is primarily applied to urban planning, landscape design, and architectural design. Typically, large-scale grids are used for urban planning, landscape planning, and landscape ecological evaluation, while medium-to-small-scale grids are used for graphic and architectural design [17].

Guanweizi Village is nearly 304 m long from the east to west and 300 m long from the north to south. Therefore, in the present study, the settlement was divided using grids measuring 38 m × 37.5 m into 64 grid units in total (Figure 4), and the coupling relationships between the spatial components of Guanweizi Village (including the water body, road, greening, soil, and building coverage) and PET were discussed in detail on the basis of medium- and small-scale grids.

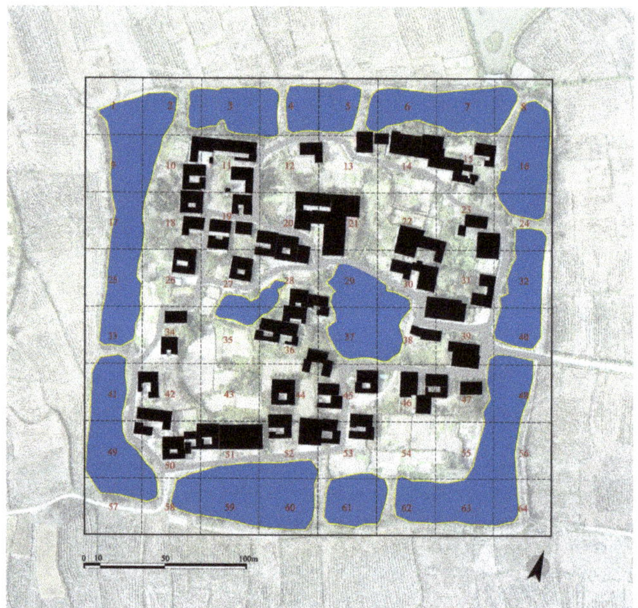

**Figure 4.** Grid of Guanweizi Village.

### 2.3. ENVI-Met-Based Thermal Environment Simulation

### 2.3.1. Measurement of Thermal Environment-Related Indicators in Guanweizi Village

According to the climatic characteristics of Xinyang City, the hottest months are June and July. The data were measured during 08:00 and 17:00 at an interval of 1 h on 21 June 2021, a clear day with moderate wind but no clouds. Air temperature and relative humidity were measured at 1.5 m above the ground level (around the head and neck of an adult) and recorded at an interval of 1 min in accordance with ISO 7726, using the UT332+ temperature humidity meter (air temperature range: 20 °C~70 °C, precision: ±0.2 °C; relative humidity range: 0%~99% RH, precision: ±2% RH).

### 2.3.2. Verification of ENVI-Met Simulation Results

In the present study, RMSE and MAPE were used to verify the simulation results [20,34], as shown in Equations (1) and (2):

$$RMSE = \sqrt{\frac{\Sigma_{i=1}^{n}\left(X_{obs,i}-X_{model,i}\right)^2}{n}} \tag{1}$$

$$MAPE = \frac{1}{n}\Sigma_{i=1}^{n}\frac{\left|X_{obs,i}-X_{model,i}\right|}{X_{obs,i}} \times 100\% \tag{2}$$

where $X_{obs}$ indicates the measured data, $X_{model}$ indicates the simulated data, and $n$ indicates the times of measurement.

Since solar radiation and temperature are closely correlated and air temperature directly affects PET, the correlation between air temperature and PET is the strongest [18]. Therefore, this verification primarily focused on error analysis of the measured and simulation data of air temperature and relative humidity. As is shown in Table 3, the air temperature RMSE ranged between 1.20 °C and 1.51 °C, and the relative humidity RMSE ranged between 2.01% and 2.31% [35]; MAPE ranged between 3.44% and 5.92%, which was less than 10% [36]. This indicates a relatively small error between the measured data and the simulation data, meaning that the model can effectively simulate the thermal environment of the village [37,38].

**Table 3.** Goodness-of-fit analysis of the measured and simulated data.

| Meteorological Parameters | Evaluation Indicators | Point 1 | Point 2 | Point 3 | Point 4 |
|---|---|---|---|---|---|
| Air temperature | RMSE (°C) | 1.51 | 1.34 | 1.40 | 1.20 |
| | MAPE (%) | 5.92 | 4.72 | 5.29 | 4.22 |
| Relative humidity | RMSE (%) | 2.31 | 2.19 | 2.13 | 2.01 |
| | MAPE (%) | 3.70 | 3.60 | 3.47 | 3.44 |

### 2.3.3. ENVI-Met Modeling Parameter

ENVI-met was used to model the Guanweizi Village settlement, with grid resolutions set at dx = 2 m, dy = 2 m, and dz = 2 m (dx and dy are horizontal resolutions, and dz is a vertical resolution). The simulation period was set between 18:00 20 June and 24:00 21 June, with data collected from 13:00 to 15:00, 21 June. According to the actual temperature and seasonal characteristics, the thermal environment in this period varied greatly, with the air temperature reaching the highest at 15:00. Therefore, simulated data during this period were selected for analysis. The simulation parameters are set out in Table 4.

**Table 4.** The basic parameters of the simulation experiment.

| | Parameters | Parameter Values |
|---|---|---|
| Simulation settings | Total simulation time (h) | 30 |
| | Output time interval (min) | 60 |
| | Number of grids (X × Y × Z) | 233 × 231 × 15 |
| Initial parameter setting | Simulation start date | 20 June 2021 |
| | Simulation start time | 18:00 |
| | Initial temperature (°C) | 25.2 |
| | Wind speed at 10 m (m/s) | 2.4 |
| | Wind direction at 10 m (o) | 202.5 S-W |
| | Relative humidity at 2 m (%) | 56 |
| | Specific humidity at 2500 m (g/kg) | 7 |

### 2.3.4. ENVI-Met-Based PET Calculation

The location for the calculation was set at 114.87° E and 31.78° N, and the time was set between 18:00 20 June and 24:00 21 June, 2021. During the calculation, the heat resistance of clothing for a summer outing was set to be 0.5 clo, activity level was 120 W, and human parameters were a standard male body shape (175 cm tall, 75 kg, and aged 35 years). During field measurement, the weather was clear with no clouds; thus, cloud cover was set to 0.

A PET distribution map at 1.5 m above the ground was created using BIO-met. According to the measured data, the temperature peaked at around 15:00; thus, PET distribution at this time was used for analysis. To test the reliability of the simulation data, we compared it with measured data (Figure 5) and used MAPE and RMSE for verification.

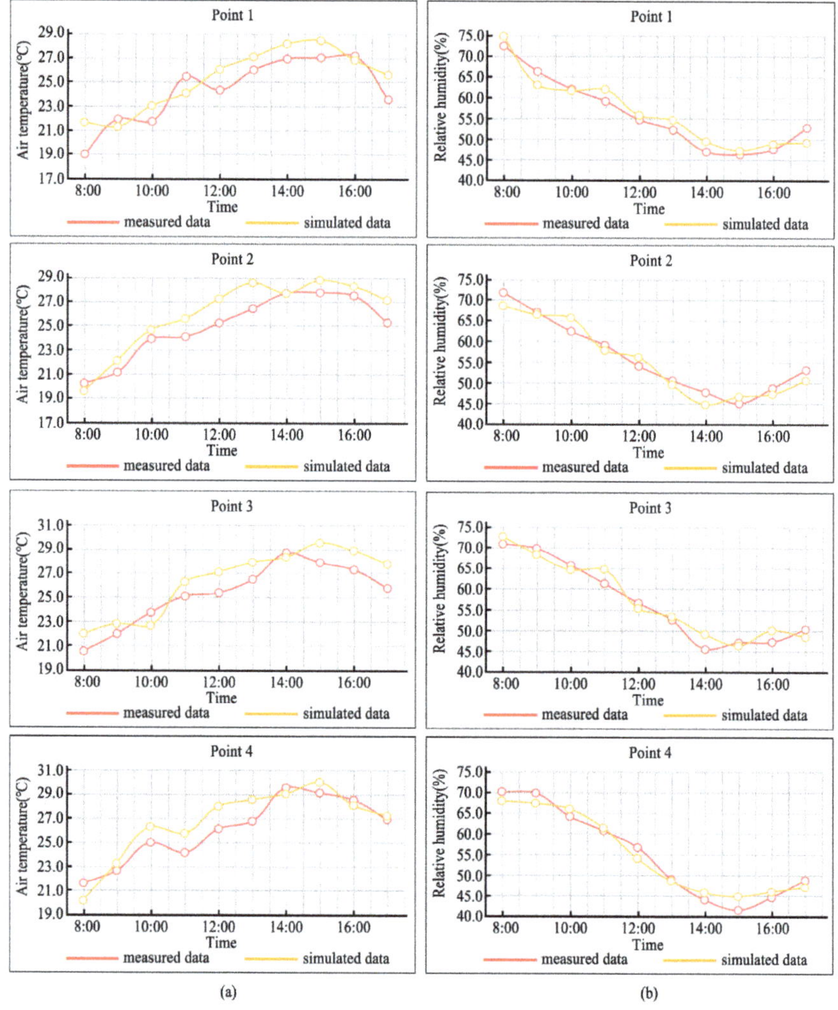

**Figure 5.** Comparison of the simulated and measured data at 1.5 m above the ground at four measuring points used to verify the effectiveness of the ENVI-met software: (**a**) air temperature and (**b**) relative humidity.

## 3. Results

### 3.1. PET Simulation

As is shown in the PET distribution map of Guanweizi Village (Figure 6, left), at 15:00, when air temperature peaked, the PET in green spaces in the village center surrounded by circular water bodies ranged between 34.1 °C and 39.3 °C. This was accompanied by a warm thermal perception, but generally had good thermal comfort. The lowest PET was below 34.1 °C, which appeared in the area surrounded the water bodies and in the shaded region of some buildings, accompanied by a warm thermal perception. The PET values of the hardened road (with a hard concrete pavement) and dense building areas were higher, ranging between 41.9 °C and 47.1 °C, accompanied by a very hot thermal perception; however, the lowest PET ranged between 34.1 °C and 36.7 °C, recorded within a very small area of a shadow of a building.

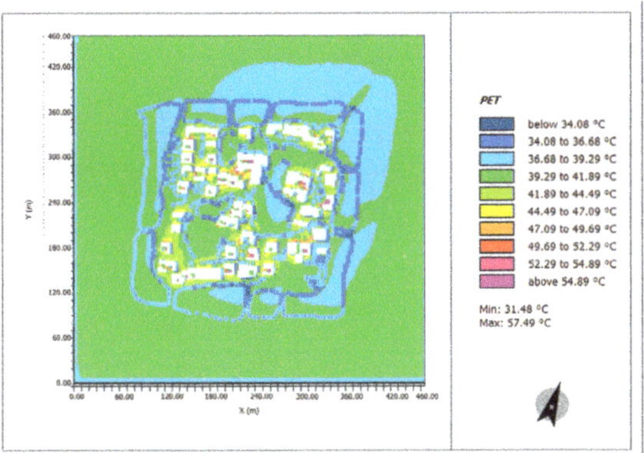
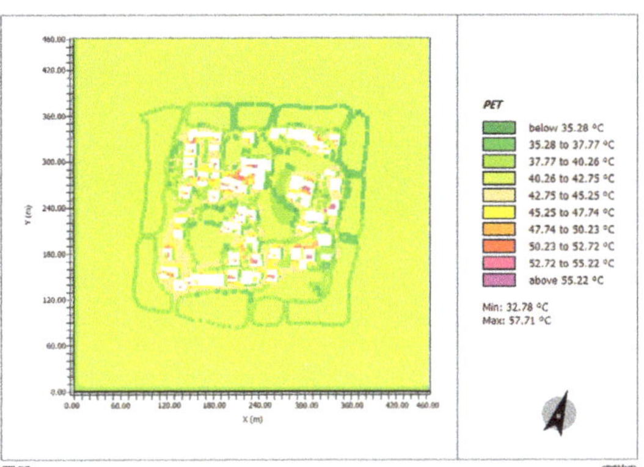

**Figure 6.** Distribution of physiological equivalent temperature (PET) values at 15:00 on 21 June 2021 (with water bodies on the left and without water bodies on the right).

### 3.2. Correlation between Spatial Components and PET

Guanweizi Village was divided into 64 grids, each with a total area of 1425 m². Simultaneously, building, greening, road, water, and soil coverage of each grid were calculated. The summarized data, as well as PET values at 15:00, calculated based on the ENVI-met simulation, were used as the basic database of spatial components and thermal comfort of Guanweizi Village (Table 5) [39].

**Table 5.** Grid database of spatial components and physiological equivalent temperature (15:00) for Guanweizi Settlement.

| Grid No. | Water Body Coverage | Road Coverage | Bare Soil Coverage | Greening Coverage | Building Coverage | PET (°C) Value |
| --- | --- | --- | --- | --- | --- | --- |
| 1 | 35.8% | 0.0% | 64.2% | 0.0% | 0.0% | 39.8 |
| 2 | 49.7% | 0.0% | 29.0% | 21.3% | 0.0% | 36.4 |
| 3 | 70.1% | 0.0% | 22.7% | 7.2% | 0.0% | 37.0 |
| 4 | 64.6% | 0.0% | 35.4% | 0.0% | 0.0% | 37.2 |
| 5 | 60.3% | 0.0% | 39.0% | 0.0% | 0.7% | 37.6 |
| 6 | 56.4% | 0.0% | 35.9% | 6.8% | 0.9% | 36.7 |
| 7 | 69.6% | 0.0% | 29.7% | 0.7% | 0.0% | 36.0 |
| 8 | 46.3% | 0.0% | 47.0% | 6.7% | 0.0% | 37.8 |
| 9 | 53.4% | 0.0% | 46.6% | 0.0% | 0.0% | 36.4 |
| 10 | 22.1% | 0.0% | 23.0% | 34.0% | 20.9% | 36.8 |

**Table 5.** *Cont.*

| Grid No. | Water Body Coverage | Road Coverage | Bare Soil Coverage | Greening Coverage | Building Coverage | PET (°C) Value |
|---|---|---|---|---|---|---|
| 11 | 0.0% | 4.9% | 34.9% | 0.0% | 60.2% | 38.2 |
| 12 | 0.0% | 10.8% | 28.6% | 50.9% | 9.7% | 36.3 |
| 13 | 0.0% | 11.0% | 26.3% | 50.4% | 12.3% | 36.8 |
| 14 | 0.0% | 9.9% | 32.2% | 22.7% | 35.2% | 39.8 |
| 15 | 0.0% | 2.6% | 21.3% | 34.7% | 41.4% | 37.8 |
| 16 | 72.9% | 0.0% | 13.0% | 13.1% | 1.0% | 36.7 |
| 17 | 51.1% | 0.0% | 48.9% | 0.0% | 0.0% | 37.5 |
| 18 | 1.6% | 0.0% | 14.2% | 63.8% | 20.4% | 36.2 |
| 19 | 0.0% | 17.2% | 42.6% | 6.0% | 34.2% | 38.9 |
| 20 | 0.0% | 12.9% | 34.8% | 13.5% | 38.8% | 38.6 |
| 21 | 0.0% | 0.0% | 28.4% | 25.2% | 46.4% | 39.1 |
| 22 | 0.0% | 0.0% | 8.4% | 68.9% | 22.7% | 37.3 |
| 23 | 0.0% | 8.8% | 32.4% | 38.6% | 20.2% | 38.4 |
| 24 | 49.4% | 5.4% | 38.7% | 1.3% | 5.2% | 36.7 |
| 25 | 55.2% | 0.0% | 44.8% | 0.0% | 0.0% | 36.5 |
| 26 | 0.0% | 8.9% | 20.1% | 55.0% | 16.0% | 36.0 |
| 27 | 10.2% | 13.0% | 44.8% | 18.0% | 14.0% | 38.4 |
| 28 | 10.7% | 12.3% | 45.1% | 1.5% | 30.4% | 38.9 |
| 29 | 48.4% | 10.8% | 10.4% | 17.9% | 12.5% | 37.0 |
| 30 | 12.0% | 11.7% | 32.1% | 8.7% | 35.5% | 37.9 |
| 31 | 0.0% | 1.6% | 4.1% | 67.3% | 27.0% | 36.5 |
| 32 | 56.0% | 7.5% | 35.4% | 0.0% | 1.1% | 36.5 |
| 33 | 45.1% | 1.7% | 47.3% | 5.9% | 0.0% | 37.4 |
| 34 | 0.0% | 10.3% | 27.8% | 45.6% | 16.3% | 38.2 |
| 35 | 18.7% | 0.0% | 9.1% | 70.1% | 2.1% | 37.0 |
| 36 | 2.9% | 0.0% | 21.5% | 35.1% | 40.5% | 38.6 |
| 37 | 70.1% | 0.0% | 19.9% | 4.0% | 6.0% | 36.8 |
| 38 | 36.5% | 7.9% | 31.0% | 11.0% | 13.6% | 37.2 |
| 39 | 0.8% | 24.2% | 32.7% | 4.2% | 38.1% | 37.4 |
| 40 | 55.9% | 11.4% | 32.7% | 0.0% | 0.0% | 37.9 |
| 41 | 56.9% | 9.3% | 28.1% | 0.0% | 5.7% | 36.6 |
| 42 | 0.0% | 0.0% | 20.2% | 51.6% | 28.2% | 36.2 |
| 43 | 0.0% | 0.0% | 14.2% | 85.8% | 0.0% | 36.0 |
| 44 | 0.0% | 0.0% | 52.3% | 5.9% | 41.8% | 39.9 |
| 45 | 0.0% | 13.6% | 41.2% | 11.9% | 33.3% | 37.7 |
| 46 | 0.0% | 9.5% | 30.2% | 28.6% | 31.7% | 37.5 |
| 47 | 16.6% | 12.3% | 22.8% | 28.8% | 19.5% | 37.8 |
| 48 | 54.8% | 0.0% | 45.2% | 0.0% | 0.0% | 37.5 |
| 49 | 79.8% | 5.7% | 11.0% | 0.0% | 3.5% | 37.4 |
| 50 | 6.3% | 7.3% | 41.3% | 0.0% | 45.1% | 38.6 |
| 51 | 21.9% | 14.4% | 26.9% | 0.0% | 36.8% | 38.3 |
| 52 | 24.6% | 9.5% | 41.2% | 0.0% | 24.7% | 38.0 |
| 53 | 2.5% | 9.8% | 23.9% | 42.4% | 21.4% | 38.2 |
| 54 | 1.5% | 0.0% | 0.0% | 95.5% | 3.0% | 35.7 |

**Table 5.** *Cont.*

| Grid No. | Water Body Coverage | Road Coverage | Bare Soil Coverage | Greening Coverage | Building Coverage | PET (°C) Value |
|---|---|---|---|---|---|---|
| 55 | 32.4% | 0.0% | 6.8% | 60.8% | 0.0% | 35.9 |
| 56 | 33.2% | 0.0% | 66.8% | 0.0% | 0.0% | 36.6 |
| 57 | 24.2% | 0.0% | 75.8% | 0.0% | 0.0% | 39.2 |
| 58 | 36.6% | 0.0% | 63.4% | 0.0% | 0.0% | 39.1 |
| 59 | 74.1% | 0.0% | 25.9% | 0.0% | 0.0% | 37.4 |
| 60 | 75.7% | 0.0% | 24.3% | 0.0% | 0.0% | 36.5 |
| 61 | 72.9% | 0.0% | 27.1% | 0.0% | 0.0% | 36.6 |
| 62 | 56.1% | 0.0% | 43.9% | 0.0% | 0.0% | 37.2 |
| 63 | 76.6% | 0.0% | 23.4% | 0.0% | 0.0% | 36.8 |
| 64 | 26.8% | 0.0% | 73.2% | 0.0% | 0.0% | 37.2 |

In the present study, the basic grid data (Table 5) were imported into SPSS, and Pearson correlation analysis was conducted between the PET and the basic spatial components of the Guanweizi settlement, including the water, road, bare soil, greening, and building coverage (Table 6). The PET was negatively correlated with water and greening coverage, and positively correlated with road, soil, and building coverage. By comparing the correlation coefficients, the order of correlation strength was as follows: greening coverage > water coverage (slight difference) in the case of negative correlations, and building coverage > bare soil coverage > road coverage in the case of positive correlations.

**Table 6.** Pearson correlation analysis between grid analysis data of settlement space components and the corresponding physiological equivalent temperature (PET).

| | | Water Body Coverage | Road Coverage | Bare Soil Coverage | Greening Coverage | Building Coverage |
|---|---|---|---|---|---|---|
| Pearson correlation between | r | −0.328 ** | 0.281 * | 0.481 ** | −0.331 ** | 0.494 ** |
| settlement space components and PET | Sig. | 0.008 | 0.024 | 0.000 | 0.007 | 0.000 |

** Correlation significant at the 0.01 level (two-tailed); * correlation significant at the 0.05 level (two-tailed).

### 3.3. Impact of Water Bodies on the Thermal Environment

To verify the impact of water bodies around the village on the thermal environment, along with the original model, another copy was created. The copy had only one variable (excluding water bodies around the village), while the other parameters remained consistent with those in the original model. The simulation results were as follows.

According to the data measured at 15:00 on 21 June 2021, during summer, the PET in green spaces at the village center with water bodies ranged between 36.7 °C and 39.3 °C, accompanied by a warm thermal perception. In contrast, that in the green spaces without water bodies ranged between 40.3 °C and 42.8 °C, accompanied by a very hot thermal perception. From the comparison between scenarios with and without water bodies, the PET values around the water bodies and inside the settlement were reduced by approximately 5 °C, on average, and the air temperature was reduced by approximately 3 °C, on average, in the presence of water bodies (Figures 6 and 7). However, with water bodies around the settlement, the PET values outside the water bodies increased (south and west sides in figures on the left). As is shown in Figure 6, the PET in greening areas at the edge of the water bodies ranged between 34.1 °C and 36.7 °C, while those in areas without greening ranged between 36.7 °C and 39.3 °C, reaching the high values shown in blue.

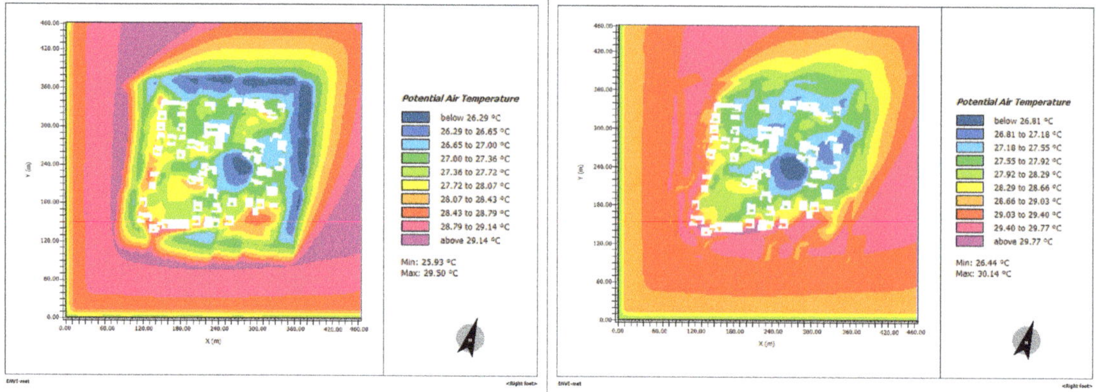

**Figure 7.** Air temperature distribution at 15:00 on 21 June 2021 (with water bodies on the left and without water bodies on the right).

## 4. Discussion

Our literature review revealed that previous studies have mostly focused on the heat island effect of land settlements or the thermal comfort of settlements neighboring water on only one side. For example, Kurazumi et al. studied the relationship of effective temperature (ET), an indicator for the evaluation of outdoor space thermal environments, with human physiological and psychological responses, through subject experiments in 2019. By studying the influence of the thermal environment on the human body in rural and suburban outdoor spaces, natural elements such as green space and water surfaces have been proposed as conducive to reducing air temperature [19]. In 2013, Theeuwes et al. mitigated the heat island effect by introducing open surface water into urban design. By using the WRF mesoscale meteorological model, an idealized circular city was designed. It was concluded that the cooling effect of a water body depends nonlinearly on the partial water coverage, area and distributed wind direction of individual lakes in a city [40]. Sayad et al. proposed two solutions to improve urban thermal comfort through an ENVI-met simulation experiment in 2021: increasing the vegetation ratio relative to air flow and increasing the water surface area within the outdoor space [41]. The combined effect of the increased vegetation ratio and linear water bodies provides the best solution for achieving the optimal thermal comfort level in an outdoor space. In 2022, Wu et al. studied the spatial and temporal distribution characteristics of the water temperature cooling value in Zhoutie Town of Yixing City through field measurement, numerical simulation, etc. They analyzed the correlation between the village morphology and water-cooling value, proposing that the increase in water and greening ratio and the decrease in surrounding building density was conducive to enhancing the water-cooling effect [13]. The above literature only employed a limited number of methods, such as effective temperature evaluation, ideal model construction, field measurement, and digital simulation, to study the influence of an area of water on an outdoor thermal environment. However, this study more systematically adopted the actual measurement, digital simulation, PET outdoor thermal environment evaluation, and grid-based correlation analysis of the experimental data. In addition, the above literature mainly focused on the study of natural water areas, while we focused on artificial areas of water in the particular form of Weizi settlements. This was based on our years of team research on Weizi settlement morphology, since there have been very few studies on Weizi settlements in China.

Based on ENVI-met simulation results, the impact of peripheral water bodies on the thermal comfort of settlements was analyzed. This clearly showed a direct regulatory effect of water bodies on PET and air temperature. Therefore, water bodies can effectively improve the local microclimate and, consequently, PET [42]. Accordingly, an increased water body area can favorably improve the PET and air temperature of Weizi settlements. This is because, although water bodies can effectively reduce the surrounding temperature and increase relative humidity and wind speed, due to heat ventilation, air temperature drops to some extent, thereby increasing PET. With less greening in the south and west and more in the north and east, PET did not increase in the north and east, indicating that waterfront spaces combined with green shade can significantly reduce PET [43].

However, simulation results also indicated that the local temperature around water bodies may increase due to solar refraction. Moreover, Pearson's correlation analysis revealed that, under these circumstances, PET is negatively correlated with greening coverage [44,45]. Therefore, PET may be reduced by combining waterfront spaces and green shade to avoid a local high temperature around the water bodies.

Furthermore, the simulated data showed relatively higher PET values in hard concrete pavement, soil, and dense building areas. This is a result of an increased air temperature due to increased heat radiation from the hard concrete pavement, soil, and building surfaces. Therefore, building density should be adjusted through the reasonable use of hard paving, and the high building density in some areas should be appropriately reduced to improve the local microclimate [46].

Based on grid data and Pearson correlations between PET and the coverage of water bodies, greening, roads, soil, and buildings, a PET ranging between 34 °C and 36 °C is typically accompanied by good thermal comfort due to the relatively higher water body and greening coverage. However, a PET exceeding 36 °C is generally accompanied by poor thermal comfort due to the relatively higher road, soil, and building coverage. Hence, the rational design of water bodies and greening coverage would be an effective means to increase the thermal comfort of Weizi settlements.

The present study has certain limitations as it only focused on a Weizi settlement surrounded by water on all four sides without analyzing partially enclosed settlements, such as those surrounded by water on only two or three sides. According to the current research findings, it is expected that an uneven distribution of the cooling effect will be seen in Weizi settlements surrounded by water on only two or three sides, compared with those surrounded by water on all four sides. This incomplete selection of typical settlement samples is a limitation of this study. More settlements featuring different enclosure types will be selected to conduct further research on the influence of the degree of water enclosure on thermal comfort. In addition, the air speed and average radiation temperature will also be incorporated into field measurements in our subsequent study to obtain more scientific data.

## 5. Conclusions

In the present study, fundamental data of air temperature, relative humidity, and solar radiation were obtained through field measurements in Guanweizi Village. Further, ENVI-met was used for simulation, and the distribution results of PET, a thermal comfort indicator, were obtained for Guanweizi Village on the summer solstice.

Based on the results of fundamental data analysis, grid units with better thermal comfort were concentrated near water bodies or greening clusters. Thus, water and greening coverage are the major factors affecting the thermal comfort of spaces in Guanweizi settlement. Water bodies in Guanweizi Village can regulate air temperature and relative humidity in surrounding areas, thereby playing pivotal roles in improving the thermal comfort of the settlement.

Furthermore, the coupling relationships between the components of settlement spaces and PET values were summarized. On this basis, three suggestions regarding the strategy for improving the thermal comfort of Weizi settlements are summarized below:

1.  Increasing air humidity and reducing air temperature: the area of water bodies and greening should be increased, and the area of bare soil should be reduced within the settlement area, where thermal comfort is poor.
2.  Reducing thermal radiation: green plants should be grown on both sides of hardened roads in the settlement to increase shading and reduce solar thermal radiation; hard concrete pavement should be minimized and planting or permeable bricks should be used [47]. Green plants should be grown on both sides of roads, and surface phytoplankton should be increased to reduce the reflection of solar radiation [48,49].
3.  Reducing high temperatures at the boundary of water bodies: multi-level greenery should be added around water bodies inside and outside the settlement, such as tall trees to shade the water bodies and thus reduce high temperatures at their boundary.

The significance of the present study lies in the site-specific consideration of the relationship between human settlements and water when designing settlements to create spaces with relatively comfortable microclimates. This work can provide a reference and inspiration for future research on the design of water-adaptive spaces in the context of human settlements.

**Author Contributions:** Conceptualization, Y.C. and X.L.; methodology, Y.C., Y.B. and X.L.; software, S.L., Y.B. and Z.Z.; validation, B.L., Y.Z., S.L. and Z.W.; investigation, Y.C., S.L., Z.W. and X.L.; data curation, S.L. and Z.Z.; writing—original draft preparation, Y.C., S.L., Z.Z., Z.W. and X.L.; writing—review and editing, Y.B., Y.P., Y.C., B.L., X.L. and Y.Z.; supervision, X.L. and Y.C.; funding acquisition, X.L. and Y.B. All authors contributed to the development of this work and elaboration of this article. All authors have read and agreed to the published version of the manuscript.

**Funding:** This research was supported by the 2022 Guangdong Philosophy and Social Science Foundation (grant no. GD22XGL02); Guangdong Basic and Applied Basic Research Foundation (grant no. 2023A1515011137); Guangzhou Philosophy and Social Science Planning 2022 Annual Project (grant no. 2022GZQN14); Fundamental Research Funds for the Central Universities (grant no. QNMS202211); Department of Education of Guangdong Province (grant no. 2021KTSCX004); Department of Housing and Urban–Rural Development of Guangdong Province (grant no. 2021-K2-305243); Guangzhou Basic and Applied Basic Research Foundation (grant no. SL2024A04J00890); and State Key Laboratory of Subtropical Building and Urban Science, South China University of Technology (grant no. 2022ZA01).

**Data Availability Statement:** Data are available within the article.

**Conflicts of Interest:** The authors declare no conflict of interest.

## References

1.  Wei, G.; Bi, M.; Liu, X.; Zhang, Z.; He, B.-J. Investigating the impact of multi-dimensional urbanization and FDI on carbon emissions in the belt and road initiative region: Direct and spillover effects. *J. Clean. Prod.* **2023**, *384*, 135608. [CrossRef]
2.  Liu, X.; He, J.; Xiong, K.; Liu, S.; He, B.-J. Identification of Factors Affecting Public Willingness to Pay for Heat Mitigation and Adaptation: Evidence from Guangzhou, China. *Urban Clim.* **2023**, *48*, 101405. [CrossRef]
3.  He, B.-J. Green Building: A Comprehensive Solution to Urban Heat. *Energy Build.* **2022**, *271*, 112306. [CrossRef]
4.  Liu, S.; Wang, Y.; Liu, X.; Yang, L.; Zhang, Y.; He, J. How does future climatic uncertainty affect multi-objective building energy retrofit decisions? Evidence from residential buildings in subtropical Hong Kong. *Sustain. Cities Soc.* **2023**, *92*, 104482. [CrossRef]
5.  Wei, D.; Yang, L.; Bao, Z.; Lu, Y.; Yang, H. Variations in Outdoor Thermal Comfort in an Urban Park in the Hot-Summer and Cold-Winter Region of China. *Sustain. Cities Soc.* **2022**, *77*, 103535. [CrossRef]
6.  He, B.-J.; Zhu, J.; Zhao, D.-X.; Gou, Z.-H.; Qi, J.-D.; Wang, J. Co-Benefits Approach: Opportunities for Implementing Sponge City and Urban Heat Island Mitigation. *Land Use Policy* **2019**, *86*, 147–157. [CrossRef]
7.  Song, C.; Yang, J.; Wu, F.; Xiao, X.; Xia, J.; Li, X. Response characteristics and influencing factors of carbon emissions and land surface temperature in Guangdong Province, China. *Urban Clim.* **2022**, *46*, 101330. [CrossRef]
8.  Li, B.; Guo, W.; Liu, X.; Zhang, Y.; Russell, P.J.; Schnabel, M.A. Sustainable Passive Design for Building Performance of Healthy Built Environment in the Lingnan Area. *Sustainability* **2021**, *13*, 9115. [CrossRef]
9.  Wang, Y.; Berardi, U.; Akbari, H. Comparing the Effects of Urban Heat Island Mitigation Strategies for Toronto, Canada. *Energy Build.* **2016**, *114*, 2–19. [CrossRef]
10. Zhou, X.; Chen, H.; Wu, Y.; Xu, H. Impacts of Mobile Survey-based Urban Spatial Forms on Summer Afternoon Thermal Environments. *Landsc. Archit.* **2018**, *25*, 21–26. (In Chinese) [CrossRef]

11. Huang, C.H.; Tsai, H.H.; Chen, H.C. Influence of Weather Factors on Thermal Comfort in Subtropical Urban Environments. *Sustainability* **2020**, *12*, 2001. [CrossRef]
12. Sun, Z. Impact of urban morphology factors on thermal environment in high density urban areas: A case of Beijing within 5th ring road. *Ecol. Environ. Sci.* **2020**, *29*, 2020–2027. (In Chinese) [CrossRef]
13. Wu, Y.; Peng, C. Influence analysis of rural morphological factors on microclimate cooling effect of water body. *J. Southeast Univ. (Nat. Sci. Ed.)* **2022**, *52*, 179–188. (In Chinese) [CrossRef]
14. Sukjin, J.; Seonghwan, Y. Changes in Sunlight and Outdoor Thermal Environment Conditions Based on the Layout Plan of Flat Type Apartment Houses. *Energies* **2015**, *8*, 9155–9172. [CrossRef]
15. Liu, Z.; Zhao, X.; Jin, H. Thermal environment of riverside residential areas at Harbin in winter. *J. Harbin Inst. Technol.* **2017**, *49*, 164–171. (In Chinese)
16. Du, X.; Shi, Y.; Zhang, Y. Numerical Study and Design of Thermal Environment of Living Street Canyons in Guangzhou. *Build. Sci.* **2015**, *31*, 78–87. (In Chinese) [CrossRef]
17. Shao, Y.; Liu, B. A Study on Microclimate Parameters of Urban Street Space and Its Influential Factors. *Landsc. Archit.* **2016**, *135*, 98–104. (In Chinese) [CrossRef]
18. Tsoka, S. Investigating the Relationship Between Urban Spaces Morphology and Local Microclimate: A Study for Thessaloniki. *Procedia Environ. Sci.* **2017**, *38*, 674–681. [CrossRef]
19. Kurazumi, Y.; Kondo, E.; Fukagawa, K.; Yamato, Y.; Tobita, K.; Tsuchikawa, T. Effects of Outdoor Thermal Environment upon the Human Responses. *Engineering (1947-3931)* **2019**, *11*, 475–503. [CrossRef]
20. Xin, K.; Zhao, J.; Wang, T.; Gao, W.; Zhang, Q. Architectural Simulations on Spatio-Temporal Changes of Settlement Outdoor Thermal Environment in Guanzhong Area, China. *Buildings* **2022**, *12*, 345. [CrossRef]
21. Wang, H.; Hu, S. Analysis on the Applicability of PMV Thermal Comfort Model. *Build. Sci.* **2009**, *25*, 108–114. (In Chinese) [CrossRef]
22. Fiorillo, E.; Brilli, L.; Carotenuto, F.; Cremonini, L.; Gioli, B.; Giordano, T.; Nardino, M. Diurnal Outdoor Thermal Comfort Mapping through Envi-Met Simulations, Remotely Sensed and In Situ Measurements. *Atmosphere* **2023**, *14*, 641. [CrossRef]
23. Yang, F.; Qian, F.; Liu, S. Evaluation of Measurement and Numerical Simulation of Outdoor Thermal Environmental Effect of Planning and Design Strategies of High-rise Residential Quarters. *Build. Sci.* **2013**, *29*, 28–34+92. (In Chinese) [CrossRef]
24. Kariminia, S.; Ahmad, S.S. Dependence of Visitors' Thermal Sensations on Built Environments at an Urban Square. *Procedia Soc. Behav. Sci.* **2013**, *85*, 523–534. [CrossRef]
25. Kariminia, S.; Motamedi, S.; Shamshirband, S.; Petković, D.; Roy, C.; Hashim, R. Adaptation of ANFIS Model to Assess Thermal Comfort of an Urban Square in Moderate and Dry Climate. *Stoch. Environ. Res. Risk Assess.* **2016**, *30*, 1189–1203. [CrossRef]
26. Guo, C.; Fu, J.; Guo, Z.; Zhao, J. Rssearch on the Thermal Comfort Evaluation Framework of Vernacular Architecture under the Subjective and Objective Judgment Mechanism. *Archit. J. (Acad. Issue)* **2020**, *21*, 22–26. (In Chinese)
27. Cheng, Y.; Guo, W. An Analysis of the Principles That Create the Microclimate of Courtyards in Vernacular Houses Surrounded by Embankment. *Archit. J.* **2015**, *557*, 70–73. (In Chinese)
28. Guo, W.; Wang, X.; Zhang, G.; Li, W. Research on the Climate Adaptability of Traditional Villages and Houses: A Case Study of Hengtang Village in Dongyang City. *J. Hum. Settl. West China* **2021**, *36*, 134–140. (In Chinese) [CrossRef]
29. Guo, W.; Ding, Y.; Yang, G.; Liu, X. Research on the Indicators of Sustainable Campus Renewal and Reconstruction in Pursuit of Continuous Historical and Regional Context. *Buildings* **2022**, *12*, 1508. [CrossRef]
30. Bruse, M.; Fleer, H. Simulating Surface–Plant–Air Interactions inside Urban Environments with a Three Dimensional Numerical Model. *Environ. Model. Softw.* **1998**, *13*, 373–384. [CrossRef]
31. Nikolova, I.; Janssen, S.; Vos, P.; Vrancken, K.; Mishra, V.; Berghmans, P. Dispersion Modelling of Traffic Induced Ultrafine Particles in a Street Canyon in Antwerp, Belgium and Comparison with Observations. *Sci. Total Environ.* **2011**, *412–413*, 336–343. [CrossRef]
32. Cheng, Y. Weizi Folk House in Western Anhui—On the Construction Strategy of Traditional Architecture Adapting to the Climate. *Sichuan Build. Sci.* **2009**, *35*, 244–249. (In Chinese)
33. Fang, Z.; Lin, Z.; Mak, C.M.; Niu, J.; Tse, K.-T. Investigation into Sensitivities of Factors in Outdoor Thermal Comfort Indices. *Build. Environ.* **2018**, *128*, 129–142. [CrossRef]
34. Shi, D.; Song, J.; Huang, J.; Zhuang, C.; Guo, R.; Gao, Y. Synergistic Cooling Effects (SCEs) of Urban Green-Blue Spaces on Local Thermal Environment: A Case Study in Chongqing, China. *Sustain. Cities Soc.* **2020**, *55*, 102065. [CrossRef]
35. Ouyang, W.; Sinsel, T.; Simon, H.; Morakinyo, T.E.; Liu, H.; Ng, E. Evaluating the thermal-radiative performance of ENVI-met model for green infrastructure typologies: Experience from a subtropical climate. *Build. Environ.* **2022**, *207*, 108427. [CrossRef]
36. Kumar, S.; Kaur, T. Development of ANN Based Model for Solar Potential Assessment Using Various Meteorological Parameters. *Energy Procedia* **2016**, *90*, 587–592. [CrossRef]
37. Rosso, F.; Golasi, I.; Castaldo, V.L.; Piselli, C.; Pisello, A.L.; Salata, F.; Ferrero, M.; Cotana, F.; de Lieto Vollaro, A. On the Impact of Innovative Materials on Outdoor Thermal Comfort of Pedestrians in Historical Urban Canyons. *Renew. Energy* **2018**, *118*, 825–839. [CrossRef]
38. Salata, F.; Golasi, I.; Petitti, D.; de Lieto Vollaro, E.; Coppi, M.; de Lieto Vollaro, A. Relating Microclimate, Human Thermal Comfort and Health during Heat Waves: An Analysis of Heat Island Mitigation Strategies through a Case Study in an Urban Outdoor Environment. *Sustain. Cities Soc.* **2017**, *30*, 79–96. [CrossRef]

39. Xin, Y.; Shuang, H.; Qi, Z. Thermal Comfort Measurement and Landscape Reconstruction Strategy of Street Space Based on Decomposition Grid Method:A Case Study of Yangmeizhu Street. *J. Landsc. Res.* **2020**, *12*, 33–38+46. [CrossRef]
40. Theeuwes, N.E.; Solcerová, A.; Steeneveld, G.J. Modeling the Influence of Open Water Surfaces on the Summertime Temperature and Thermal Comfort in the City. *J. Geophys. Res. Atmos.* **2013**, *118*, 8881–8896. [CrossRef]
41. Sayad, B.; Alkama, D.; Ahmad, H.; Baili, J.; Aljahdaly, N.H.; Menni, Y. Nature-Based Solutions to Improve the Summer Thermal Comfort Outdoors. *Case Stud. Therm. Eng.* **2021**, *28*, 101399. [CrossRef]
42. Manteghi, G. A Field Investigation on the Impact of the Wider Water Body On-Air, Surface Temperature and Physiological Equivalent Temperature at Malacca Town. *Int. J. Environ. Sci. Dev.* **2020**, *11*, 286–291. [CrossRef]
43. Zheng, S.; Guldmann, J.-M.; Wang, Z.; Qiu, Z.; He, C.; Wang, K. Experimental and Theoretical Study of Urban Tree Instantaneous and Hourly Transpiration Rates and Their Cooling Effect in Hot and Humid Area. *Sustain. Cities Soc.* **2021**, *68*, 102808. [CrossRef]
44. Davtalab, J.; Deyhimi, S.P.; Dessi, V.; Hafezi, M.R.; Adib, M. The Impact of Green Space Structure on Physiological Equivalent Temperature Index in Open Space. *Urban Clim.* **2020**, *31*, 100574. [CrossRef]
45. Morakinyo, T.E.; Lam, Y.F. Simulation Study on the Impact of Tree-Configuration, Planting Pattern and Wind Condition on Street-Canyon's Micro-Climate and Thermal Comfort. *Build. Environ.* **2016**, *103*, 262–275. [CrossRef]
46. Chen, H.; Ooka, R.; Huang, H.; Tsuchiya, T. Study on Mitigation Measures for Outdoor Thermal Environment on Present Urban Blocks in Tokyo Using Coupled Simulation. *Build. Environ.* **2009**, *44*, 2290–2299. [CrossRef]
47. Huang, W.; Liu, X.; Zhang, S.; Zheng, Y.; Ding, Q.; Tong, B. Performance-Guided Design of Permeable Asphalt Concrete with Modified Asphalt Binder Using Crumb Rubber and SBS Modifier for Sponge Cities. *Materials* **2021**, *14*, 1266. [CrossRef] [PubMed]
48. Hong, X.-C.; Cheng, S.; Liu, J.; Dang, E.; Wang, J.-B.; Cheng, Y. The Physiological Restorative Role of Soundscape in Different Forest Structures. *Forests* **2022**, *13*, 1920. [CrossRef]
49. Hong, X.-C.; Liu, J.; Wang, G.-Y. Soundscape in Urban Forests. *Forests* **2022**, *13*, 2056. [CrossRef]

*Article*

# Effect of Landscape Elements on Public Psychology in Urban Park Waterfront Green Space: A Quantitative Study by Semantic Segmentation

Junyi Li [1,2], Ziluo Huang [2], Dulai Zheng [2], Yujie Zhao [2], Peilin Huang [2], Shanjun Huang [2], Wenqiang Fang [2], Weicong Fu [2] and Zhipeng Zhu [1,*]

[1]   College of Architecture and Urban Planning, Fujian University of Technology, Fuzhou 350118, China
[2]   College of Landscape Architecture and Art, Fujian Agriculture and Forestry University, Fuzhou 350000, China
*   Correspondence: 19912151@fjut.edu.cn

**Abstract:** Urban park waterfront green spaces provide positive mental health benefits to the public. In order to further explore the specific influence mechanism between landscape elements and public psychological response, 36 typical waterfront green areas in Xihu Park and Zuohai Park in Gulou District, Fuzhou City, Fujian Province, China, were selected for this study. We used semantic segmentation technology to quantitatively decompose the 36 scenes of landscape elements and obtained a public psychological response evaluation using virtual reality technology combined with questionnaire interviews. The main results showed that: (1) the Pyramid Scene Parsing Network (PSPNet) is a model suitable for quantitative decomposition of landscape elements of urban park waterfront green space; (2) the public's overall evaluation of psychological responses to the 36 scenes was relatively high, with the psychological dimension scoring the highest; (3) different landscape elements showed significant differences in four dimensions. Among the elements, plant layer, pavement proportion, and commercial facilities all have an impact on the four dimensions; and (4) the contribution rate of the four element types to the public's psychological response is shown as spatial element (37.9%) > facility element (35.1%) > natural element (25.0%) > construction element (2.0%). The obtained results reveal the influence of different landscape elements in urban park waterfront green spaces on public psychology and behavior. Meanwhile, it provides links and methods that can be involved in the planning and design of urban park waterfront green space, and also provides emerging technical support and objective data reference for subsequent research.

**Keywords:** urban park waterfront; psychological response; semantic segmentation

Citation: Li, J.; Huang, Z.; Zheng, D.; Zhao, Y.; Huang, P.; Huang, S.; Fang, W.; Fu, W.; Zhu, Z. Effect of Landscape Elements on Public Psychology in Urban Park Waterfront Green Space: A Quantitative Study by Semantic Segmentation. *Forests* **2023**, *14*, 244. https://doi.org/10.3390/f14020244

Academic Editors: Xinhao Wang, Xin-Chen Hong and Jiang Liu

Received: 20 November 2022
Revised: 21 December 2022
Accepted: 26 January 2023
Published: 28 January 2023

## 1. Introduction

According to the "World Urbanization Prospects" prepared by the United Nations, 55% of the population will live in cities by 2018, and this number is expected to rise to nearly 70% by 2050, indicating that human beings are gradually concentrating in cities. Studies from various sources mention that urbanized life actually affects the health of residents on a physical and psychological level. At the physical level, respiratory problems [1], cardiovascular diseases [2,3], immune diseases [4], and kidney diseases [5] brought about by changes in the living environment pose a challenge to human health. Urban life in densely populated areas may also have a number of psychological effects on residents. One study showed that people living in the most densely populated areas had a 68%–77% higher risk of mental illness and a 12%–20% higher risk of depression than those living in areas with low levels of urbanization [6], with urbanized environments placing varying degrees of cognitive and emotional stress on residents [7], along with a reduction in their sense of well-being [8]. In the current context of COVID-19 gripping the world, urban residents are facing dramatic changes in their lives, with disruptions in daily life, the risk of unemployment, and social isolation sounding the alarm for a range of emotional stresses and mental illnesses [9–11]. The emergence of these mental illnesses is a reminder that people are becoming aware of the various stresses associated with urban life and that mental health issues need to be taken seriously.

Wilson, a proponent of the Biophilia Hypothesis, pointed out that human beings come from nature, instinctively such as nature, adore nature, and are attracted to nature [12], and that contact with the natural environment can greatly reduce stress and help with mental recovery [13], as well as enhancing feelings of pleasure [14]. Further research indicates that people show higher preference, restoration, and motivation for landscapes that contain water [15,16]. Compared with other green spaces, spending time in or around waterfront spaces will significantly reduce people's negative emotions and mental stress [17] and effectively increase their sense of well-being [18]. Researchers from environmental psychology and environmental health have pointed to the potential of waterfront spaces to become health-promoting "therapeutic landscapes" [19]. Many studies have confirmed the positive effects of exposure to waterfront environments. For example, epidemiological evidence suggests that urban residents living close to blue spaces report more positive general health conditions compared to residents in other areas [20], and that they are significantly less likely to suffer from mental illness [21]; Gascon M et al. [22,23] based on a study of the association between waterfront landscapes and public psychology indicated that there are mechanisms of influence between the two that reduce stress, downplay anxiety, enhance restorative qualities, and provide evidence of effective positive benefits, as well as positive alleviating effects on specific psychological disorders such as mood disorders and psychological distress; Pasanen et al. noted that waterfront spaces can encourage people to engage in sports and exercise to regulate their lifestyles [24] and promote positive social interactions [25]; Gao et al. [26] argue that forested water spaces, especially dynamic water features, contribute to the public's emotional attitude towards the environment; and Britton et al. [27] further emphasized that an emphasis on the exploration of the mechanisms of influence between waterfront space and human health is key to future research. For this reason, urban park waterfront green space has become an excellent vehicle for urban residents to engage with nature. It provides a variety of positive benefits to residents, and it is also essential to focus on the psychological healing benefits that this environment provides. However, it is worth noting that the presence of an element in the environment and the differences in its visual representation may provide different perceptual effects. For example, studies have mentioned that changes in the sky openness, building height, vegetation area, water proportion, etc. in the scene may also change the public's perception attitude to a certain extent [28–30]. This means that, in addition to the qualitative evaluation of the structure and form of landscape elements, the quantitative visual effects conveyed by the landscape are also an important part of measuring the public's perception attitude. However, there is still a lack of research on urban park waterfront green space, and systematic evaluation indexes and quantitative methods have not been developed yet.

With the continuous development of the computer field, the interdisciplinary application of emerging technologies is gradually gaining academic attention. In recent years, deep learning for semantic segmentation in artificial intelligence has become a trending topic. The principle of semantic segmentation lies in the automatic analysis of a large amount of data and feature learning, which effectively extracts the low, medium, and high level information from the image and achieves pixel-level predictive classification of unknown images depending on the expressed semantics. Its powerful image processing capabilities make automated processing of large volumes of image data a reality [31]. In addition to applications in medical imaging and artificial driving, there is a new trend to use semantic segmentation of images to quantify landscapes and to combine public perception to provide effective guidance for future planning and design. Currently, semantic segmentation is more often applied in studies related to streetscapes in landscape gardening habitat research, such as the measurement of urban street composition [32], pedestrian space [33], quality assessment [34], and aesthetic judgement [35]. We note that there are few examples of this technique being applied to other landscape types, such as the urban park waterfront that is the subject of this study. Urban park waterfront green space has different and more complex components than street landscapes, with common elements such as sky, trees, and buildings, as well as unique landscape elements such as landscape vignettes and landscape structures. However, the application of semantic segmentation to urban park waterfront green space has not yet been discussed or investigated empirically. With the continuous development of deep learning algorithms, more and more deep learning-based segmentation models are gradually developed. The selection of models and datasets applicable to urban park waterfront green space will help expand the scenario of landscape gardening applications of semantic segmentation, and also provide new technical support for the quantitative evaluation of urban park waterfront green space, and there is some room for exploration in this area of research.

Virtual Reality (VR) in the field of computer simulation bridges the gap between traditional image studies. As a method of maximizing the simulation of the perceived spatial environment [36,37], VR provides a convenient and rapid representation of the natural environment for the public without having to leave home. Recent studies have shown that in addition to visual appreciation, the immersive nature scenes offered by VR can have a near-realistic emotional and psychological healing effect on people [38–41]. Subjects experiencing virtual natural scenes within a safe period of time (5–10 min) can significantly improve physiological stress [42,43], become more "fascinating" and "coherent" in psychological recovery [44], and further enhance human emotions and awareness [45]. Based on a large number of existing studies, the application of virtual reality in scene presentation is considered novel, effective, and feasible, and has a wide range of application prospects.

The information conveyed by the objective environment provides multisensory stimulation and subsequently has an impact on human physical and mental health; this interaction indicates a potential relationship between the two [46]. How to change the objective physical environment characteristics to enhance public mental health while incorporating feedback and application of public mental health to landscape design is the key to the study of urban park waterfront green space and public mental health. In addition, the emergence of new technologies has helped us conduct scientific research more efficiently and objectively. Based on this, this paper takes Xihu Park and Zuohai Park in Gulou District, Fuzhou City, Fujian Province, China, as the study subjects and explores the following:

(1)  Analyze the differences between different semantic segmentation models and datasets applied to urban park waterfront green space images, and find out which semantic segmentation model and dataset are capable of efficiently and accurately obtaining quantitative data on urban park waterfront green spaces;

(2)  Using semantic segmentation and virtual reality as technical support, analyze what impact urban park waterfront green spaces have on the public psyche;

(3)  To further explore whether different landscape elements in urban park waterfront green spaces have an impact on public psychology, and what the specific mechanisms of impact are.

This study aims to explore the impact of urban park waterfront green spaces on public psychology from the perspective of public response, to identify landscape elements in urban park waterfront green spaces that may have positive or negative psychological benefits to the public, and to further examine the specific mechanisms and the degree of importance of these landscape elements in influencing public psychology. Through this study, it is possible to provide links and ways to intervene in the landscape construction and optimal design of urban park waterfront green spaces and to provide effective and targeted solutions for the development of urban park waterfront environments with psychological healing benefits.

## 2. Materials and Methods

### 2.1. Study Sites

The study site, Fuzhou, Fujian Province, is located in the southeastern hills, the first of the three major hills in China, and is a typical city in the hilly region. Xihu Park and Zuohai Park are located in the Gulou District of Fuzhou, in the heart of the bustling city, surrounded by famous attractions such as Three Lanes and Seven Alleys, and Fuway. The park is surrounded by many offices, such as the Fujian Provincial Government and the Fujian Provincial Forestry Bureau, and many residential communities, serving a wide range of people and belonging to a typical urban park.

Xihu Park is a classical garden park with a long history and the most complete preservation in Fuzhou and is known as the "Pearl of Fujian Gardens". It was first built in the third year of Jin's Taikang dynasty and was renovated in 1914 to become the present-day Xihu Park. The park is classically beautiful and is a citywide comprehensive park combining history and culture with natural landscape. With a total area of about 0.4251 $km^2$ and a water surface of 0.303 $km^2$, Zuohai Park is one of the first and largest parks in Fuzhou. In 1990, local farmers converted the northern part of the site belonging to the former Xihu to create the present Zuohai Park. Zuohai Park was established with Xihu Park as a reference and incorporates more characteristic parks and modern facilities, covering an area of 0.3547 $km^2$ and a water surface of 0.1814 $km^2$. In 2015, based on the concept of "Great Xihu", Fuzhou City implemented a wooden walkway around the lake and the water system to connect the two parks, making the Xihu and Zuohai Park into one, creating a landscape planning pattern of "one water, two parks, three peaks and four shores", thus forming a representative urban waterfront park in Fuzhou City, which is also an important part of the city's landscape pattern and is regarded as one of the top ten scenic spots in Fuzhou. Both parks are well equipped with functional zones and various facilities that can meet the needs of Fuzhou citizens for various types of leisure and recreational activities. Based on this, the study takes Xihu Park and Zuohai Park as the research objects, which are typical and representative.

The completed landscape nodes of the park usually play the role of guiding visitors to gather. In this study, the distribution map of landscape nodes in the park was used as the basis and further combined with fieldwork to select sample sites; 36 scenes were finally identified as sample sites for this study (Figure 1), including waterfront roads, squares, pavilions, viewing platforms, and other important landscape nodes. The principles of sample site selection are as follows: (1) the waterfront landscape as a linear tour space; the experimental selection of sample scenarios with established landscape nodes that can simulate the walking route of visitors along the waterfront trail. (2) Select sample scenarios with a high frequency of viewing, which refers to a certain amount of human traffic observed in the actual survey in the early stages. (3) The sample scenarios should be able to objectively and comprehensively reflect the characteristics of the park and should have natural (soil, vegetation, water bodies, etc.) and artificial (buildings, squares, paths, facilities, etc.) landscape elements, with the water environment in the field of view.

### 2.2. Virtual Reality Image Acquisition

To restore the objective integrity of the scenes, VR images were used as visual stimuli to initiate public perception assessment. The study used the "insta 360 ONE" panoramic camera for sample scenario scenes collection from 10:00 to 14:30 on 27 December 2020. The camera was fixed at a level of 1.6 m above the ground with a tripod, which ensures that the shooting height is consistent with the level of the human eye and reduces the error caused by manual hand-held operation. A total of 98 VR images were taken, with an output resolution of 6912 × 3456 pixels. After eliminating images with similar angles, large differences in light perception, and more irrelevant influencing factors, 36 VR images that could restore the whole picture of the park were selected as the research samples for this study.

**Figure 1.** Sample scenarios selection.

### 2.3. Participants and Procedure

A total of 40 volunteers, 21 males and 19 females, aged between 19 and 30 years old (M = 23.7, SD = 2.9), were recruited as subjects for this experiment among university teachers and students. Young students and teachers are considered to have diverse professional backgrounds, as well as a certain level of environmental perception and a wide range of preferences [26,47]. The method of using young students and teachers as respondents has been widely used in previous studies [48–51] and is considered to have low cost and high effectiveness, so the selection of this group as subjects is feasible and representative. Considering the purpose of the experiment, all 40 subjects were reported to have good mental health or no mental disorders (as judged by the results of university psychological tests); no history of cardiovascular and neuropsychiatric disease; normal or corrected visual acuity reported in both eyes; and were able to wear VR devices continuously for a short time without strong discomfort. The evaluation process includes the following three steps (Figure 2): (1) stress experiment: before the experiment, participants were invited to complete the short Spielberger State-Trait Anxiety Inventory (STAI-S), which is used to effectively assess participants' current psychological status. The test consists of 20 items, each assessed on a 4-point scale, with scores from 1 to 4 corresponding to a stepwise scale of "not at all" to "fully". Then, the auditory continuous addition test (PASAT) proposed by neuropsychologist Deary was used to stimulate the anxiety and stress levels of the subjects, who were asked to complete random number calculation questions without any computing equipment, and an alarm would be triggered if they answered incorrectly. Lastly, participants were asked to complete the STAI-S test again after the experiment; (2) VR experience: subjects were asked to wear a head-mounted display device and experience 36 VR scenes one by one by turning their heads or bodies. To prevent dizziness, fatigue, and other discomforting reactions caused by prolonged viewing of VR scenes, the length of continuous experience should be 5–10 min [41,52,53]. Given the large number of samples in this experiment, the samples were divided equally into three groups to start the virtual scene experience, with each group experiencing for 8–9 min and being allowed to take a 3-min break at the end of each group experience; (3) questionnaire completion: with the help of the short version of the recovery scale (SRRS) proposed by Han [54], public psychological responses were collected from four dimensions: emotional, physical, cognitive, and behavioral, as shown in Table 1. The questionnaire was scored on a 7-point scale with scores 1 to 7 corresponding to a stepped scale of strongly agree to strongly disagree. All 40 subjects completed the above experiment according to the procedure, which showed that the design and time control of the VR experiment were reasonable.

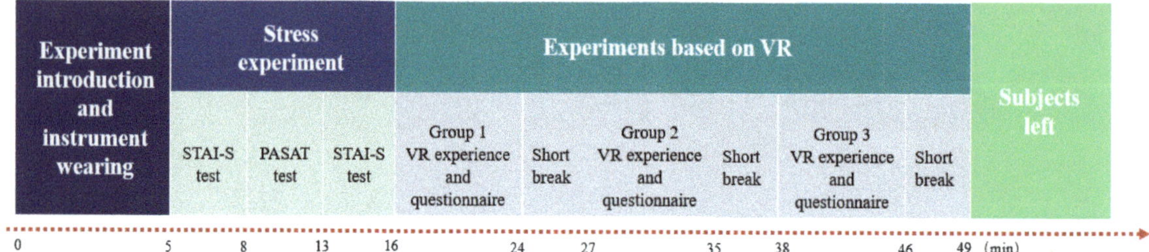

**Figure 2.** Experimental process.

**Table 1.** Short-version revised restoration scale.

| Dimensions | Item | Score |
|---|---|---|
| (F1) Emotional dimensions | V1, Grouchy–Good natured | 1, 2, 3, 4, 5, 6, 7 |
| | V2, Anxious–Relaxed | 1, 2, 3, 4, 5, 6, 7 |
| (F2) Cognitive dimensions | V3, I am interested in the presented scene | 1, 2, 3, 4, 5, 6, 7 |
| | V4, I feel attentive to the presented scene | 1, 2, 3, 4, 5, 6, 7 |
| (F3) Physiological dimensions | V5, My breathing is becoming faster | 1, 2, 3, 4, 5, 6, 7 |
| | V6, My hands are sweating | 1, 2, 3, 4, 5, 6, 7 |
| (F4) Behavioral dimensions | V7, I would like to visit here more often | 1, 2, 3, 4, 5, 6, 7 |
| | V8, I would like to stay here longer | 1, 2, 3, 4, 5, 6, 7 |

### 2.4. GSA-Based Landscape Element Screening

Based on Wang's classification of park landscape elements [55], by consulting experts in landscape architecture and related fields, the environmental characteristics were initially decomposed into 20 pre-selected landscape elements from four aspects: spatial elements, natural elements, construction elements, and facility elements.

Grey Statistic Analysis (GSA) is a fuzzy statistical method based on "little" information and "uncertainty" [56], which can effectively solve the limitations of researchers' knowledge structures and is considered to be effective in improving the scientificity of element selection. For further screening of primary selection elements with GSA, the process is as follows: (1) the questionnaire was sent to 20 landscape architecture experts to solicit their opinions on the importance of the indexes of the elements, and the questionnaire used a 7-level scale, with levels 1 to 7 indicating a stepped scale from "very unimportant" to "very important"; (2) the pre-selected elements are classified into three levels: high, medium, and low. To construct a gray category whitening function, the calculation formula is shown in (1)–(3); (3) to calculate the decision coefficients of gray categories, the formula is shown in (4). Obtain the decision coefficients of high, medium, and low gray categories of each element, and select the elements with "high" importance to start the subsequent study. The final selection of landscape elements is shown in Table 2.

$$f_1(ab) = \begin{cases} 1 & h_{ab} \geq 7 \\ \frac{h_{ab}-4}{7-4} & 4 < h_{ab} < 7 \\ 0 & h_{ab} \leq 4 \end{cases} \tag{1}$$

$$f_2(ab) = \begin{cases} 0 & h_{ab} \geq 7 \\ \frac{7-h_{ab}}{7-4} & 4 < h_{ab} < 7 \\ 1 & h_{ab} = 4 \\ \frac{h_{ab}-1}{4-1} & 1 < h_{ab} < 4 \\ 0 & h_{ab} \leq 1 \end{cases} \tag{2}$$

$$f_3(ab) = \begin{cases} 0 & h_{ab} \geq 4 \\ \frac{4-h_{ab}}{4-1} & 1 < h_{ab} < 4 \\ 1 & h_{ab} \leq 1 \end{cases} \tag{3}$$

$$\eta k(b) = \sum L(ab) \times f_k(ab) \tag{4}$$

In the formula:

$f_k(ab)$—denotes the value of the whitening function, with the importance of the $b$th indicator as $a$, and $k$ denotes the number of gray classes as 1, 2, and 3.

$h_{ab}$—denotes the importance of the $b$th indicator as the value of the assignment of $a$.

$\eta k(b)$—indicates that the $b$th indicator is the decision coefficient of the $k$th gray class.

$L(ab)$—indicates the number of experts whose importance factor of the $b$th indicator is an assignment of $a$.

**Table 2.** Landscape element index selection.

| Type | Landscape Indicators | Calculation Method | Quantification Methods | No. |
|---|---|---|---|---|
| Spatial elements | Sky openness | Proportion of the sky in the view | Semantic segmentation | K1 |
| | Visual complexity | Space composition complexity index | Matlab | K2 |
| | Colorfulness of space | Colorful index of space elements | | K3 |
| Natural elements | Green viewing ratio | Proportion of vegetation in the view | Semantic segmentation | Z1 |
| | Blue viewing ratio | Proportion of water in the view | | Z2 |
| | Plant layers | Number of tree, shrub, and herb strata in the plant landscape | Counting statistics | Z3 |
| | Plant colorfulness | Number of colored foliage and flowering plant species | | Z4 |
| | Soil exposure | Proportion of bare soil in the view | Semantic segmentation | Z5 |
| | Plant growth condition | Condition of decaying and dead plants in the plant landscape, with = 0, without = 1 | Assignment statistics | Z6 |
| Building elements | Building proportion | Proportion of buildings in the view | Semantic segmentation | J1 |
| | Pavement proportion | The proportion of paving in the view | | J2 |
| | Pavement form | Masonry = 0, wood = 1, pebbles = 2 | Assignment statistics | J3 |
| | Vignette proportion | Proportion of vignettes in the view | | J4 |
| | Humanistic atmosphere | Proportion of traditional buildings in the view | Semantic segmentation | J5 |
| Facility elements | Commercial facilities | The proportion of commercial facilities, such as cruise ships and amusement facilities | | S1 |

*2.5. Scene Quantization Decomposition Based on Semantic Segmentation*

We used semantic segmentation as an auxiliary technique to quickly obtain objective and accurate quantitative data from the scenes, which can quickly achieve pixel-level classification based on the semantic information contained in the image. It is usually based on both CNN models and datasets. In order to improve the segmentation accuracy of waterfront green space, we intend to compare the current excellent model algorithms and build a training dataset with urban green space as the main part, so as to select the optimal model for subsequent analysis.

2.5.1. Model Selection and Semantic Segmentation Accuracy Improvement

Using PASCAL VOC, a world-class challenge in computer vision, three models (DeepLabV3+ [57], PSPNet [58], and HRNet [59]) with excellent performance on this dataset were initially selected. The three models are widely used in the current study. The models used scenes that contained streets, natural landscapes, remote sensing images, etc., and showed a high level of accuracy (Table 3).

**Table 3.** Model overview.

| Name of Model | Model Network Architecture |
| --- | --- |
| DeepLabV3+ | |
| PSPNet | |
| HRNet | |

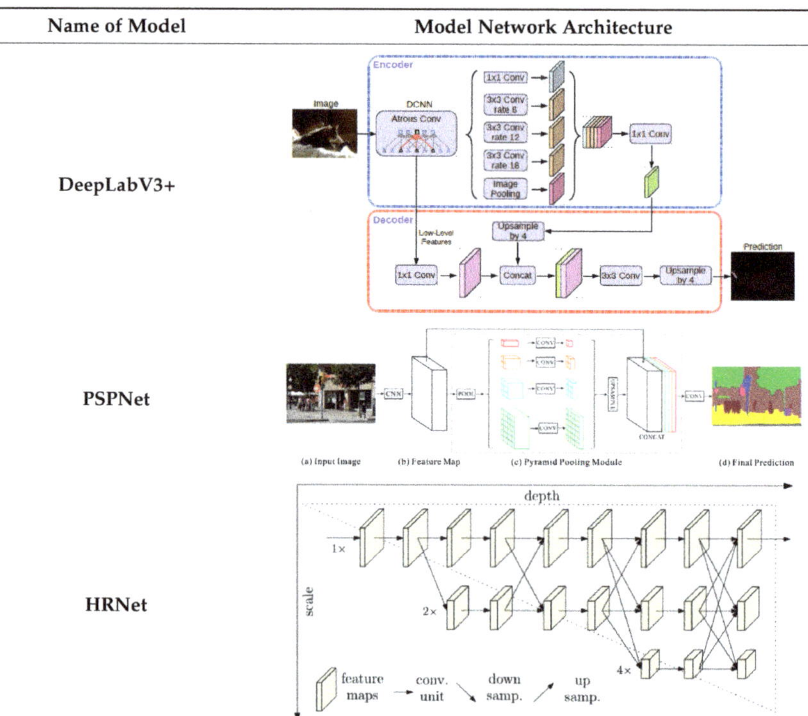

Note: The DeepLabV3+ model refers to the paper: Encoder-Decoder with Atrous Separable Convolution for Semantic Image Segmentation, doi:10.1007/978-3-030-01234-2_49; the PSPNet model refers to the paper: Pyramid Scene Parsing Network, doi:10.1109/CVPR.2017.660; and the HRNet model refers to the paper: Deep High-Resolution Representation Learning for Human Pose Estimation, doi:10.1109/CVPR.2019.00584.

Semantic segmentation requires learning and obtaining laws from scenes with a large number of annotated objects (a training dataset). In addition to the structural conditions of the model itself, the quantity, quality, and diversity of the training dataset will have a significant impact on the segmentation accuracy. To improve the recognition accuracy of the model on urban park waterfront green space images, this study expanded the annotated samples of urban green space images on the basis of ADE20K dataset in a targeted manner to establish the training dataset of this paper, and the specific steps are as follows: (1) image acquisition: a large number of images were acquired using panoramic cameras, DSLR cameras, and other devices, and the acquisition image conditions cover different weather, lighting, etc.; (2) sample labeling: the image labeling software LabelMe, developed by the Computer Science and Artificial Intelligence Laboratory (CSAIL) at the Massachusetts Institute of Technology (MIT), was used to label the acquired images with polylines. The label categories were based on the ADE20K dataset labels; and (3) dataset establishment: sample data were stored in JSON files with annotation information and output.

### 2.5.2. Semantic Segmentation Accuracy Calculation

Semantic Segmentation accuracy is usually measured in terms of Mean Intersection over Union (MIoU) [60], which describes the overlap ratio of predicted and real pixels of an image.

$$MIoU = \frac{1}{k+1} \sum_{i=0}^{k} \frac{p_{ii}}{\sum_{j=0}^{k} p_{ij} + \sum_{j=0}^{k} p_{ji} - p_{ii}} \tag{5}$$

In the formula:

$k$—indicates total number of landscape tag categories.

$p_{ii}$—indicates the total number of pixels of real pixel category $i$ that are predicted to be of category I.

$p_{ij}$—indicates the total number of pixels of real pixel category $i$ that are predicted as category $j$.

$p_{ji}$—indicates the total number of pixels of real pixel category $j$ predicted to be of category I.

### 2.6. Statistical Analysis

Different analysis methods were used for different data: (1) SPSS 20 statistical software was used to conduct multiple reliability and validity tests on the subjective questionnaire to verify its scientific and accuracy; (2) a paired sample t-test was conducted to compare STAI-S data (pre-PASAT vs. post-PASAT) to ensure the PASAT increased the stress level of subjects significantly; (3) partial correlation analysis was used in analysis of subject–object correlations, using partial correlation coefficient as a criterion, the factors with low correlation were gradually removed; (4) conduct multiple linear regression analysis. The adjusted coefficient of determination $R^2$ was used to test the fit of the multiple regression line to the observed values and used analysis of variance (ANOVA) to test whether the multiple regression estimation equation was statistically significant. The significance level ($p < 0.0001$) and the $p$-value indicated the significant level of the single factor ($p < 0.05$). The statistical analysis was completed in SPSS Statistics 22.0.

## 3. Results

### 3.1. Results of Semantic Segmentation

The semantic segmentation results show that the three models trained by the dataset of this paper can correctly identify and classify the elements with a larger pixel share, such as sky, greenery, roads, and pavements, along the contour of the image (Figure 3). PSPNet, compared with DeepLabV3+, is less likely to miss identifying small-scale landscape elements, such as distant buildings, low shrubs, and scenic rocks, and it can also achieve better accurate label classification. HRNet's recognition performance was relatively inaccurate. There were many misclassifications, omissions, and multiple classifications of the same object for both near and distant landscape elements, which were subject to the influence of light and other factors in the segmentation process.

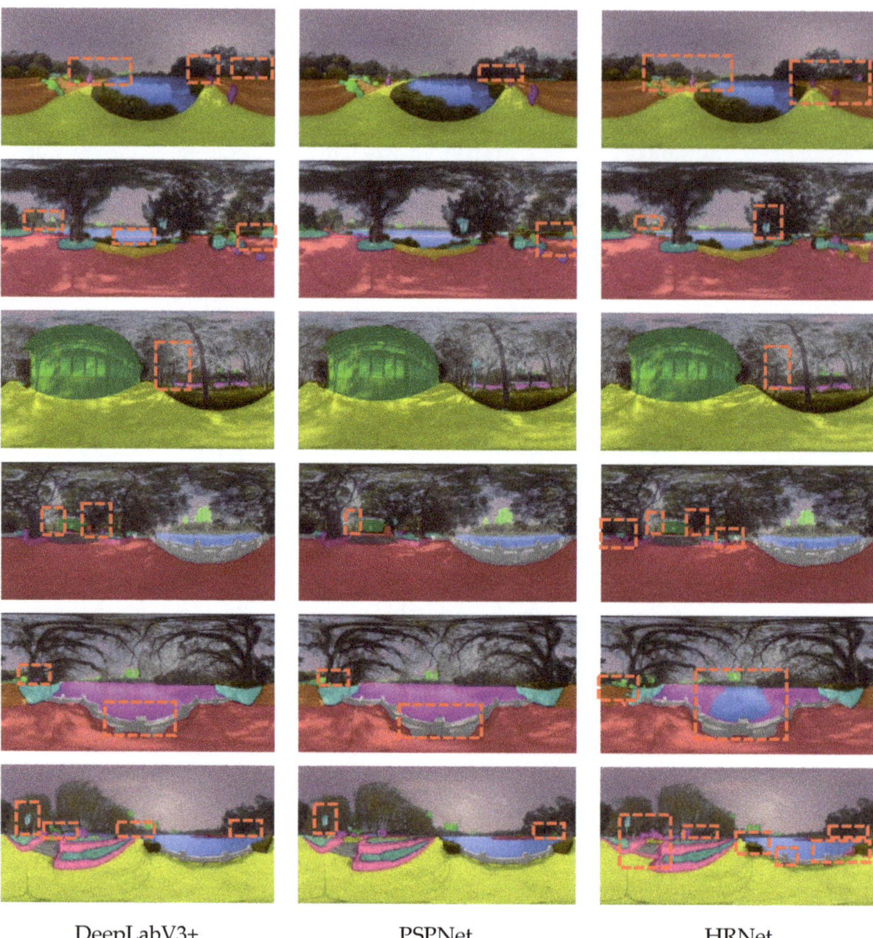

DeepLabV3+        PSPNet        HRNet

**Figure 3.** Segmentation results comparison.

The semantic segmentation accuracy of the three models on the urban park waterfront green space was obtained after excluding the indoor landscape categories such as "Sofa" and "Chair" in the ADE20k dataset (Figure 4). Using MIoU as the criterion, the ranking performance of the three models is PSPNet (0.6865) > DeepLabV3+ (0.6816) > HRNet (0.5179).

To confirm the validity of the dataset constructed in this study, the segmentation results of PSPNet trained by the ADE20k dataset and our dataset were compared. The PSPNet trained by the ADE20k dataset on the scene was unsatisfactory, as shown in Table 4, with fragmented and disordered segmentation, uneven recognition and classification of objects, low edge fit, and almost no recognition of small-scale elements such as distant scenes and facilities. The PSPNet model trained by our dataset greatly improved on the above failures, and most of the elements can be segmented and correctly classified independently. The edge recognition was finer, while the segmentation effect of small-scale elements had been improved, and in general, the segmentation integrity and completeness were significantly better than the previous model. The data show that the PSPNet model trained by the dataset of this paper has significantly improved in both PA and MIoU test criteria, and the MIoU is improved by about 26.8%, which verifies the advantageousness of the dataset of this paper.

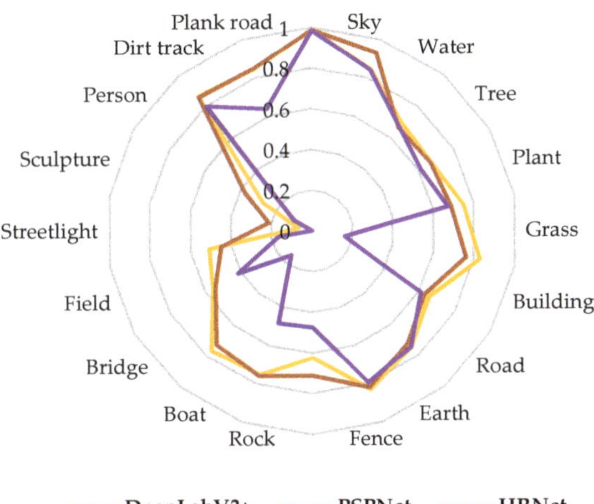


**Figure 4.** Semantic segmentation results.

**Table 4.** PSPNet segmentation results based on different datasets.

| | PA | MIoU | Schematic Diagram 1 | Schematic Diagram 2 |
|---|---|---|---|---|
| PSPNet-trained by ADE20k dataset | 0.8039 | 0.4189 | | |
| PSPNet-trained by the dataset in this paper | 0.8783 | 0.6865 | | |

### 3.2. The Public Psychological Response of Urban Park Waterfront Green Space

#### 3.2.1. Test on Reliability and Validity

The questionnaire reliability test showed that the overall Cronbach's $\alpha$ value of the study scale was 0.741, which was greater than 0.6, indicating that the scale has a relatively good internal consistency and a high degree of reliability; the Kaiser-Meyer-Olkin (KMO) was 0.758, and the significance of the Barlett's spherical test statistic was 0.000, indicating that the questionnaire has a high degree of association and good structural validity.

#### 3.2.2. Public Psychological Response Results

The question items of the psychological dimension (F3) of the SRRS measure psychological arousal [54], so the scores of this dimension were calculated upside down. The results of the questionnaire analysis and cluster analysis showed that the distribution of the public's psychological response levels to the 36 scenarios tended to be moderate to high, with the highest score being sample 4 (5.05), and the lowest being sample 7 (3.89).

The scores for each dimension showed that F3 > F1 > F2 > F4 (Figure 5). The sample scenarios convey a higher restorative benefit to the public in F3, while it still falls short in F4. In terms of standard deviation, the subjects' perceptions were more consistent in F3 (0.17), and the perceived differences were greater in F4 (0.43).

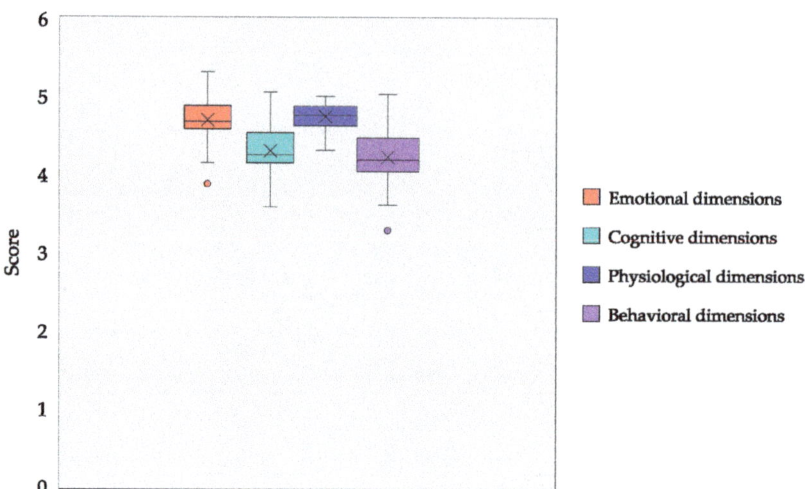


**Figure 5.** Differences in psychological response across four dimensions.

The two scenarios with the highest and lowest psychological response scores were explored further. As shown in Tables 5 and 6, the scores of psychological dimensions (4.89) and cognitive dimensions (4.97) in scenario 4 were lower than those of behavioral dimensions (5.00) and emotional dimensions (5.32). The highest scoring emotional dimension was particularly demonstrated by the switch from anxiety to relaxation (E2 = 5.38), which was the highest mean score for the subfeature factor in this scenario. In scenario 7, the behavioral dimensions (3.30) and the cognitive dimensions (3.59) scored lower than the emotional dimensions (4.19) and the psychological dimensions (4.49). The psychological impact of the scenario on the public is particularly evident in the reduction of people's dwell time (B2 = 3.24) and frequency of visits (B1 = 3.35); in addition, two sub-trait factors of the cognitive dimensions (C1 = 3.57 and C2 = 3.62) contributed to the lower mental response scores in scenario 7.

**Table 5.** Psychological response results of scenario 4.

| | Emotional Dimensions (E) | | Cognitive Dimensions (C) | | Physiological Dimensions (P) | | Behavioral Dimensions (B) | |
|---|---|---|---|---|---|---|---|---|
| | E1 | E2 | C1 | C2 | P1 | P2 | B1 | B2 |
| Mean score of each subconstruct | 5.27 | 5.38 | 5.16 | 4.78 | 4.73 | 5.05 | 4.92 | 5.08 |
| Mean score of each construct | 5.32 | | 4.97 | | 4.89 | | 5 | |

**Table 6.** Psychological response results of scenario 7.

| | Emotional Dimensions (E) | | Cognitive Dimensions (C) | | Physiological Dimensions (P) | | Behavioral Dimensions (B) | |
|---|---|---|---|---|---|---|---|---|
| | E1 | E2 | C1 | C2 | P1 | P2 | B1 | B2 |
| Mean score of each subconstruct | 4.11 | 4.21 | 3.57 | 3.62 | 4.32 | 4.65 | 3.35 | 3.24 |
| Mean score of each construct | 4.16 | | 3.59 | | 4.49 | | 3.30 | |

*3.3. How Do Landscape Elements Affect Public Psychology*

3.3.1. Psychological Response Model Construction

Four dimensions (emotional dimension, cognitive dimension, psychological dimension, and behavioral dimension) were used as dependent variables, and the quantified values of landscape elements were used as independent variables to carry out the partial correlation analysis. The elements with low bias correlation coefficients were gradually eliminated according to the results, and the linear regression model of the four dimensions was established with the final retained elements.

The multiple linear regression model shows that colorfulness of space (K3) plays the greatest positive influence in F1 and F3 ($p < 0.05$), commercial facilities (S1) is the factor with the greatest negative influence in the F1 ($p < 0.05$), and visual complexity (K2) shows the greatest negative influence in F3 ($p < 0.05$); in F2 and F4, those showing the greatest positive as well as negative influence are plant layers (Z3) and commercial facilities (S1) ($p < 0.05$).

The results of the goodness of fit and significant level tests showed that the adjusted $R^2$ for the 4 models were 0.570, 0.452, 0.480, and 0.421, respectively. The established regression equations were considered to have a good degree of explanation, corresponding to a $p \leq 0.001$, indicating a high level of significance and a statistically significant equation (Table 7).

**Table 7.** Multiple linear regression models for the psychological dimensions and the significant influences obtained.

| Model | | B | Std. Error | Standardized Coefficients | t | Sig. |
|---|---|---|---|---|---|---|
| F1<br>(R = 0.795 [a], $R^2$ = 0.632, Adjusted $R^2$ = 0.570; F = 10.296, Sig. = 0.000 [b]) | (constant) | 4.855 | 0.218 | | 22.272 | 0.000 |
| | K3 | 0.420 | 0.183 | 0.302 | 2.300 | 0.029 |
| | Z3 | 0.158 | 0.065 | 0.282 | 2.412 | 0.022 |
| | Z5 | −0.077 | 0.022 | −0.463 | −3.418 | 0.002 |
| | J2 | −0.013 | 0.005 | −0.367 | −2.851 | 0.008 |
| | S1 | −0.653 | 0.111 | −0.690 | −5.901 | 0.000 |
| F2<br>(R = 0.739 [a], $R^2$ = 0.546, Adjusted $R^2$ = 0.452; F = 5.808, Sig. = 0.000 [b]) | (constant) | 5.991 | 0.552 | | 10.852 | 0.000 |
| | K2 | −0.462 | 0.176 | −0.398 | −2.622 | 0.014 |
| | Z1 | −0.013 | 0.004 | −0.614 | −3.163 | 0.004 |
| | Z3 | 0.234 | 0.092 | 0.348 | 2.545 | 0.017 |
| | Z5 | −0.075 | 0.027 | −0.379 | −2.743 | 0.010 |
| | J2 | −0.032 | 0.008 | −0.748 | −4.160 | 0.000 |
| | S1 | −0.558 | 0.148 | −0.490 | −3.768 | 0.001 |
| F3<br>(R = 0.783 [a], $R^2$ = 0.614, Adjusted $R^2$ = 0.480; F = 4.591, Sig. = 0.001 [b]) | (constant) | 5.839 | 0.438 | | 13.321 | 0.000 |
| | K2 | −0.291 | 0.113 | −0.521 | −2.575 | 0.016 |
| | K3 | 0.331 | 0.132 | 0.410 | 2.498 | 0.019 |
| | Z1 | −0.012 | 0.004 | −1.193 | −3.122 | 0.004 |
| | Z2 | −0.033 | 0.010 | −0.773 | −3.305 | 0.003 |
| | Z3 | 0.217 | 0.054 | 0.668 | 4.050 | 0.000 |
| | Z6 | −0.114 | 0.053 | −0.293 | −2.138 | 0.042 |
| | J1 | −0.015 | 0.004 | −0.936 | −3.675 | 0.001 |
| | J2 | −0.015 | 0.006 | −0.706 | −2.597 | 0.015 |
| | S1 | −0.172 | 0.078 | −0.314 | −2.214 | 0.036 |
| F4<br>(R = 0.721 [a], $R^2$ = 0.520, Adjusted $R^2$ = 0.421; F = 5.240, Sig. = 0.001 [b]) | (constant) | 5.776 | 0.684 | | 8.448 | 0.000 |
| | K2 | −0.551 | 0.218 | −0.394 | −2.525 | 0.017 |
| | Z1 | −0.012 | 0.005 | −0.454 | −2.275 | 0.030 |
| | Z3 | 0.320 | 0.114 | 0.393 | 2.802 | 0.009 |
| | Z5 | −0.107 | 0.034 | −0.444 | −3.130 | 0.004 |
| | J2 | −0.032 | 0.010 | −0.611 | −3.307 | 0.003 |
| | S1 | −0.583 | 0.183 | −0.426 | −3.183 | 0.003 |

[a], [b]: Predictors

### 3.3.2. Specific Mechanisms of Influence of Landscape Elements on Public Psychology

From the coefficients of the established regression model, it was clear that there were differences in the contribution of each factor indicator in different dimensions, with the overall performance of spatial elements (average = 37.9%) > facility elements (average = 35.2%) > natural elements (average = 25.0%) > construction elements (average = 2.0%).

Among the types of elements that have an impact on the public's psychological response, the spatial elements include K2 and K3; the facility element is S1; the natural elements include Z1, Z2, Z3, Z5, and Z6; and the construction elements include J1 and J2. Among them, we note that the 3 elements Z3, J2, and S1 provide some contribution to all 4 dimensions of the public's psychological response (Figure 6).

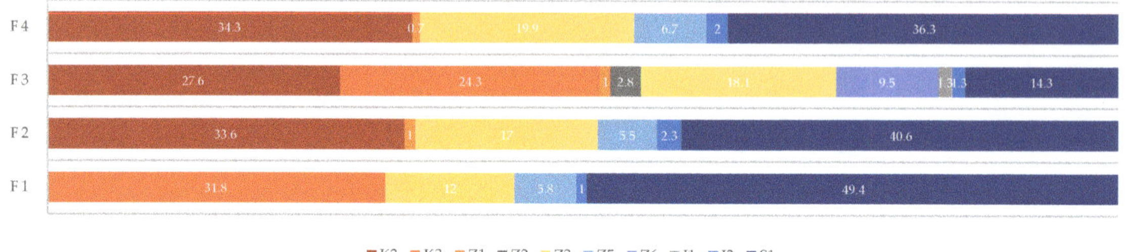


**Figure 6.** Contribution rate of spatial elements.

## 4. Discussion

### 4.1. Semantic Segmentation Model and Dataset for Urban Park Waterfront Green Space

Through the comparison of the three models, the PSPNet trained by the dataset of this paper is considered to be a more suitable semantic segmentation model for urban park waterfront green space, and its accuracy exceeds 0.60, which was considered to reach a good level of semantic segmentation accuracy [61]. The single-element data show that the model performs well (MIoU > 0.60) for 14 of the 18 element categories, including the sky, water, and trees, and achieves an accuracy level of more than 0.90 for 2 elements in particular: the sky and water. The reason was that the morphological and color characteristics of these elements were more fixed in this study [62], which made it easy for the model to determine and recognize them. However, the model is still deficient in identifying small-scale elements, such as sculptures, people, and streetlights. This was due to the fact that small-scale landscape elements contain a relatively low percentage of pixels in the scene representation, and their information, such as color, texture, and semantic features, were weakened in a series of pooling and convolution operations, which increased the difficulty of detection [63]. On the other hand, there were spatial sequences of the foreground and background in the image, and the depth setting of the dataset's annotated objects makes the segmentation process usually recognize the complete outline of the foreground landscape (e.g., trees) as a unit and ignore the pores between trunks, branches, leaves, etc. [64]. The background elements that are heavily obscured by the foreground landscape were difficult to accurately recognize in the complex environment due to the incompleteness of the appearance outline, which leads to a bias in recognition.

Semantic segmentation is based on a combination of two aspects: the model and the dataset. In terms of datasets, the current one that is often used is usually collected from streetscapes and does not include annotated samples of elements specific to park scenes. This makes the model unable to learn and summarize the semantic patterns, which leads to the disadvantage of identifying urban park images. In this paper, we increase the sampling rate and annotation amount of elements specific to park landscapes (e.g., landscape streetlights, scenic stones, pavilions) based on the ADE20k dataset. Comparing the results of the PSPNet trained by the ADE20k dataset and the dataset of this paper, it is noted that the model trained by our dataset shows a significant advantage (a 26.8% improvement in MIoU), confirming that this initiative can effectively improve the recognition accuracy of the model for urban park images, which is consistent with the previously mentioned approach [65].

In addition, the distortion caused by the conversion of VR scenes to flat projection can degrade the segmentation performance of the model [66]. This is due to a certain degree of distortion or stretching at the edges and bottom of the converted image, which leads to a reduction in segmentation accuracy due to non-compliance with the semantic laws of the model. Semantic enhancement inputs can mitigate the distortion of target detection based on this condition [66]. We added a large number of annotated samples taken by panoramic cameras to this study, and a comparative study shows that the initiative effectively helps the model learn the patterns of panoramic images and shows a significant improvement in recognition accuracy.

*4.2. The Impact of Virtual Reality-Based Urban Park Waterfront Green Space on Public Psychological Response*

The public's psychological response ratings for these 36 sample sites tended to be moderate to high, in line with Kaplan's concept of a restorative environment, which refers to the public's ability to recover and relax physically and mentally after spending time in such an environment [67]. The public's psychological response reflects differences in four dimensions: first, the public showed higher restorative benefits and higher consistency on the psychological dimension, which may be due to the fact that green spaces were usually places of relaxation, and subjects were less experience excessive physiological arousal when exposed to natural landscapes [68]; second, the public harvested less recovery in the behavioral dimension and showed more variation. It is possible that this is related to the season, as people are less likely to stay outdoors In the cold winter. In addition, subjects may express different behavioral intentions when facing the same scenario due to their personalities and emotional differences; for example, Scenario 26 was a more remote corner in a park, and some subjects believed that a quiet place to get away from the noise and enjoy solitude could bring better healing effects [69]. Additionally, some subjects would feel insecure due to the remoteness of the scenario, which led to a reluctance to stay for a long time.

Ulrich, a proponent of environmental stress relief theory, argues that the physical environment can be a source of stress or relief and that a healing natural environment should have the following characteristics: an abundance of natural elements, especially greenery and water; a complexity of environmental content with focal points in the spatial structure; a medium to high level of visual depth; winding view corridors; and the absence of things that could pose a danger [70]. Analysis of the scenarios with the highest and lowest public psychological response scores revealed that the scene (scenario 4) with the highest psychological response is a pavilion environment to the left of the entrance to Zuohai Park. The overall style here tends to be classical with a quiet environment, which is thought to be effective in regulating public mood. Through the interviews, we know that the public focuses on resting and other experiences. In addition to visual experiences when visiting, the presence of the pavilion seats in Scenario 4 enhances the public's willingness to act. In the psychological dimension, the public felt slightly faster breathing (P1 = 4.73), which may be related to the scenario being close to water and without a safety guardrail. The results of the scenario analysis found that Scenario 7 has a significantly higher percentage of commercial facilities (0.65%) than other scenarios, with the top 3 percentages. Based on the significant impact factors, we know that commercial facilities have a negative impact on all four dimensions to varying degrees, which is likely the biggest reason for the low score of this scenario. In addition, the scene space is monotonous in color, the plant growth condition is poor, and the overall lack of visual appeal makes it difficult for the public to find attention and interest when viewing. Poor environmental shade may also be responsible for reducing the length of sustained exploration and the frequency of visits by the public.

This study introduced virtual reality technology in the acquisition of public psychological data, and the results of the study showed convergence with the results of the field experiment. That is, near-natural environments in cities, including urban park waterfront green space, provide an effective restorative environment in which experiences can bring positive benefits such as fatigue relief, emotional recovery, and improved cognitive functioning [49,71], while increasing their willingness to act [72]. In previous studies, virtual reality has been used to compare landscape representations with real environments, pictures, videos, and other methods. The results show that virtual reality provides a three-dimensional environment that closely resembles the real environment and gives the viewer a more "immersive" experience than traditional media such as pictures [36,37]. The subjects' view of VR was largely consistent with the live view, and the psychological evaluations yielded feedback that was significantly closer to the live scene [73].

*4.3. Mechanisms of Influence of Urban Park Waterfront Green Space Landscape Elements on Different Psychological Dimensions of the Public*

The four element types have different degrees of influence and mechanisms of action on the four dimensions of public psychological response. Spatial elements show the highest contribution rate (37.9%), while construction elements show the lowest contribution rate (2.0%).

### 4.3.1. Spatial Element

Among the spatial elements, K2 and K3 were two factors that affected the public's psychological response. Scenes with high visual complexity reduced the public's psychological perception of F2, F3, and F4, and showed a greater negative effect in F2 (33.6%) and F4 (34.3%). This is consistent with previous research findings, where dealing with complex scenarios often requires greater cognitive effort on the part of the public [74], which may also further influence the public's willingness to behave. K3 as the element with the greatest positive contribution in both F1 (31.8%) and F3 (24.3%), the color experiment in the last century pointed out that when people are looking at things, color will first attract 80% of attention, and at the same time, color will have a lot of complex effects on people's psychology and physiology through visual contact [75]. This paper further suggests, from a psychological perspective, that rich spatial colors can liberate people from negative emotions and enhance the level of psychological recovery.

### 4.3.2. Facility Element

S1 affects the public's psychological changes to a large extent, which are reflected in F1 (49.4%), F2 (40.6%), and F4 (36.3%), especially in F1, which shows a contribution of nearly 50%, indicating that with the increase in the proportion of commercial facilities, the public is very prone to negative emotions such as anxiety and irritability. In addition, the frequent appearance of S1 greatly weakens the attraction of the scene; the degree of interest and psychological calming effect of the crowd decrease; the frequency of visits and the intention to stay for a long time also decrease; and the overall psychological feeling develops in a negative direction. This is a distinctive feature; the public was repulsed by messy and disorderly commercial facilities [76]. This is reflected in this study by the presence of commercial signs, brightly colored boats, etc. in the sample, which form a clear difference in the park landscape. This suggests that future planning should focus on the location of large commercial facilities and a coordinated and uniform style of commercial elements.

### 4.3.3. Natural Element

Linear regression models indicated that Z3 was the natural element with the largest contribution (average = 16.8%) to public psychology, and showed positive correlation with all 4 dimensions of psychological response. Previous studies have noted that greenery is the element that receives the most attention in visual evaluation and can have positive psychological benefits [77]. We further found in this study that Z3 in particular stimulates and prolongs the frequency and duration of public visits, contributing to the public's willingness to visit the space for extended periods of healthy activity. Z5 is the element with the second largest contribution (4.5%) to public psychology, as reflected in F1, F2, and F4. Previous studies often ignore the impact of surface vegetation on people's visual and psychological response. One study mentioned that when people viewed the natural environment with restoration needs, shrubs are the longest viewed component of the environment [78]; spending time observing trees, bushes, and lower ground vegetation resulted in a higher likelihood of restoration [79]. This suggests that the creation of low-level plant landscapes, such as grass and flowers, also need attention and more soil exposure can have a negative psychological impact. Plant landscapes with conditions such as dead branches and fallen leaves were usually considered to reduce the restorative qualities [51], which in this study were specifically shown to have a psychological arousing effect on the public. There was a negative correlation between Z1 and the public's psychological response, which is contrary to the existence of previous studies. The green-visibility environment in this study mostly consisted of tall trees and less grass and shrubs; this type of planting may create a sense of psychological unease due to the strong sense of enclosure, and may make it difficult to find a visual focus due to the monotony of the landscape, which in turn reduces the desire to stay. Z2 showed a negative effect on F3 but has no obvious effect on other dimensions. This may be related to the fact that the sample sites in this study are all watered environments; however, great blue ratings means that it is closer to the water, while the lack of safety fences also leads to a sense of insecurity.

### 4.3.4. Construction Element

In comparison, the construction element provides a much lower contribution (2.0%) than the other three element types. This may be due to the fact that the sample parks contained relatively few built elements (mean building proportion = 4.46%, mean paving proportion = 29.6%) or were better integrated with the park style and, therefore, less likely to provide a greater impact on public psychology. Both two elements included in construction element show a negative correlation with the public's psychological response; this supports a well-established understanding [77]. The increase in the proportion of the hard landscape leads to the lack of naturalness of the scene, which breaks the continuity and immersion of people enjoying the natural landscape, making it impossible for people to obtain a higher psychological healing effect [80,81].

## 5. Limitations

There were still some limitations in this study: (1) the public's perception of the scene and a series of complex psychological and behavioral activities are related to each other's perception indicators. Focusing on other rich sensory experiences and the public's dynamic perception and recreational behavior in the park will further help us collect more comprehensive feedback data; (2) in terms of research objects, urban waterfronts usually include multiple types, and other different types of waterfront green spaces should be included in the follow-up research to summarize a sound urban waterfront green space characteristic system with a psychological healing effect; (3) there was still room for PSPNet to improve the recognition of small-scale landscape elements. We should target to improve the sampling rate and sample annotation of small-scale elements, and further enhance the detection rate and recognition accuracy of small-scale targets in order to provide more accurate and effective environmental data. In addition, the subjects selected in this study were all university faculty and students, which did not take into account the variability of social groups, and the population sample should be expanded in the future to obtain more comprehensive and in-depth findings.

## 6. Conclusions

As an important place that provides urban residents with access to nature, research on the association between urban park waterfront green space and public psychology is necessary. However, the specific mental health benefits provided by urban park waterfronts and the role of each of the microscopic landscape elements on public psychology remain to be further explored. Based on the above, this study selected 36 sample sites in Xihu Park and Zuohai Park in Gulou District, Fuzhou City, Fujian Province, China, and introduced semantic segmentation and virtual reality as the technical support for obtaining subjective and objective quantitative data to conduct a study on the association between urban park waterfront green space and the public's psychological response. The following conclusions were eventually drawn:

(1) In terms of the application of semantic segmentation, the results indicate that PSPNet is a more suitable semantic segmentation model for urban park waterfronts. In terms of dataset, compared to the model trained by the ADE20k dataset, the PSPNet trained by the dataset in this paper shows a higher level of accuracy, which can be used as a technical support to obtain quantitative environmental data of urban park waterfront green space efficiently and accurately;

(2) In terms of the results of the public's psychological response to urban park waterfront green spaces, urban park waterfront green spaces provide a psychologically healing environment for the public, with the psychological dimension > emotional dimension > cognitive dimension > behavioral dimension in the different psychological response dimensions. The public's psychological relief is better in urban park waterfronts, while the effects of the behavioral dimensions, such as frequency of visits and length of stay, need to be improved;

(3) In terms of the specific role played by landscape elements in urban park waterfront green spaces, among the four types of landscape elements, the spatial element is the element type with the greatest contribution to public psychology, followed by the facility element and the natural element, with the construction element producing the lowest impact. Specifically, rich spatial color, complex plant community forms, and good plant growth can provide positive effects on public psychology, while complex spatial composition, a higher proportion of construction, and facility elements can reduce people's psychological mitigation effects.

These results confirm the effectiveness of the emerging technologies in the application of urban park waterfronts and the specific ways in which they can be applied. In addition, the findings suggest the psychological healing effects of waterfront green spaces in urban parks. In future planning and design, attention to the construction of spatial and facility elements will have a significant effect on the construction of urban park waterfront green spaces with psychological healing as the main theme.

**Author Contributions:** Conceptualization, J.L., Z.Z. and W.F. (Weicong Fu); methodology, J.L., Z.Z. and W.F. (Weicong Fu); software, J.L., D.Z. and P.H.; validation, J.L., Y.Z. and S.H.; formal analysis, J.L. and Z.H.; investigation, J.L., Z.H., D.Z., Y.Z., P.H., S.H. and W.F. (Wenqiang Fang); resources, Z.Z.; data curation, J.L.; writing—original draft preparation, J.L.; writing—review and editing, J.L. and Z.Z. All authors have read and agreed to the published version of the manuscript.

**Funding:** The Natural Science Foundation of Fujian Province of China: Research on the influence mechanism and optimization strategy of urban green space spatial differentiation on air quality. Grant number: 2022J01233348.

**Data Availability Statement:** The data used to support the findings of this study are available from the corresponding author upon request.

**Conflicts of Interest:** The authors declare no conflict of interest.

## References

1. Liu, Q.; Xu, C.; Ji, G.; Liu, H.; Shao, W.; Zhang, C.; Gu, A.; Zhao, P. Effect of exposure to ambient PM2.5 pollution on the risk of respiratory tract diseases: A meta-analysis of cohort studies. *J. Biomed. Res.* **2017**, *31*, 130–142.

2. Mustafić, H.; Jabre, P.; Caussin, C.; Murad, M.H.; Escolano, S.; Tafflet, M.; Périer, M.C.; Marijon, E.; Vernerey, D.; Empana, J.P. Main air pollutants and myocardial infarction: A systematic review and meta-analysis. *Jama J. Am. Med. Assoc.* **2012**, *307*, 713–721. [CrossRef]

3. Kramer, C.K.; Leitão, C.B.; Viana, L.V. The impact of urbanisation on the cardiometabolic health of Indigenous Brazilian peoples: A systematic review and meta-analysis, and data from the Brazilian Health registry. *Lancet* **2022**, *400*, 2074. [CrossRef]

4. Danielsson, M.; Heimerson, I.; Lundberg, U.; Perski, A.; Stefansson, C.G.; Akerstedt, T. Psychosocial stress and health problems: Health in Sweden: The National Public Health Report 2012. Chapter 6. *Scand. J. Public Health* **2012**, *40*, 121–134. [CrossRef] [PubMed]

5. Liang, Z.; Wang, W.; Wang, Y.; Ma, L.; Liang, C.; Li, P.; Yang, C.; Wei, F.; Li, S.; Zhang, L. Urbanization, ambient air pollution, and prevalence of chronic kidney disease: A nationwide cross-sectional study. *Environ. Int.* **2021**, *156*, 106752. [CrossRef] [PubMed]

6. Kristina, S.; Gölin, F.; Jan, S. Urbanisation and incidence of psychosis and depression. *Br. J. Psychiatry* **2004**, *184*, 293–298.

7. Peng, G.; Song, L.; Elizabeth, J.C.; Qingwu, J.; Jianyong, W.; Lei, W.; Justin, V.R. Urbanisation and health in China. *Lancet* **2012**, *379*, 843–852.

8. Okulicz-Kozaryn, A.; Valente, R.R. Urban unhappiness is common. *Cities* **2021**, *118*, 103368. [CrossRef]

9. Tang, F.; Liang, J.; Zhang, H.; Kelifa, M.M.; He, Q.; Wang, P. COVID-19 related depression and anxiety among quarantined respondents. *Psychol. Health* **2020**, *36*, 164–178. [CrossRef]

10. Zabini, F.; Albanese, L.; Becheri, F.R.; Gavazzi, G.; Giganti, F.; Giovanelli, F.; Gronchi, G.; Guazzini, A.; Laurino, M.; Li, Q.; et al. Comparative Study of the Restorative Effects of Forest and Urban Videos during COVID-19 Lockdown: Intrinsic and Benchmark Values. *Int. J. Environ. Res. Public Health* **2020**, *17*, 8011. [CrossRef]

11. Pfefferbaum, B.; North, C.S. Mental Health and the COVID-19 Pandemic. *N. Engl. J. Med.* **2020**, *383*, 510–512. [CrossRef]

12. Wilson, E.; Massachusetts, U. *Biophilia*; Harvard University Press: Cambridge, MA, USA, 1984; pp. 1–157.

13. Guo, X.; Liu, J.; Albert, C.; Hong, X. Audio-visual interaction and visitor characteristics affect perceived soundscape restorativeness: Case study in five parks in China. *Urban For. Urban Green.* **2022**, *77*, 127738. [CrossRef]

14. Hong, X.; Wang, G.; Liu, J.; Song, L.; Wu, E.T.Y. Modeling the impact of soundscape drivers on perceived birdsongs in urban forests. *J. Clean. Prod.* **2021**, *292*, 125315. [CrossRef]

15. Mathew, W.; Amanda, S.; Kelly, H.; Sabine, P.; Deborah, S.; Michael, D. Blue space: The importance of water for preference, affect, and restorativeness ratings of natural and built scenes. *J. Environ. Psychol.* **2010**, *30*, 482–493.

16. Jesper, J.A.; Stefan, W.; Mats, E.N. Stress Recovery during Exposure to Nature Sound and Environmental Noise. *Int. J. Environ. Res. Public Health* **2010**, *7*, 1036–1046.

17. Volker, S.; Kistemann, T. Developing the urban blue: Comparative health responses to blue and green urban open spaces in Germany. *Health Place* **2015**, *35*, 196–205. [CrossRef]

18. Garrett, J.K.; White, M.P.; Huang, J.; Ng, S.; Hui, Z.; Leung, C.; Tse, L.A.; Fung, F.; Elliott, L.R.; Depledge, M.H.; et al. Urban blue space and health and wellbeing in Hong Kong: Results from a survey of older adults. *Health Place* **2019**, *55*, 100–110. [CrossRef]

19. Volker, S.; Kistemann, T. The impact of blue space on human health and well-being—Salutogenetic health effects of inland surface waters: A review. *Int. J. Hyg. Environ. Health* **2011**, *214*, 449–460. [CrossRef]

20. Hooyberg, A.; Roose, H.; Grellier, J.; Elliott, L.R.; Lonneville, B.; White, M.P.; Michels, N.; De Henauw, S.; Vandegehuchte, M.; Everaert, G. General health and residential proximity to the coast in Belgium: Results from a cross-sectional health survey. *Environ. Res.* **2020**, *184*, 109225. [CrossRef]

21. Garrett, J.K.; Clitherow, T.J.; White, M.P.; Wheeler, B.W.; Fleming, L.E. Coastal proximity and mental health among urban adults in England: The moderating effect of household income. *Health Place* **2019**, *59*, 102200. [CrossRef]

22. Nieuwenhuijsen, M.J.; Kruize, H.; Gidlow, C.; Andrusaityte, S.; Antó, J.M.; Basagaña, X.; Cirach, M.; Dadvand, P.; Danileviciute, A.; Donaire-Gonzalez, D.; et al. Positive health effects of the natural outdoor environment in typical populations in different regions in Europe (PHENOTYPE) a study programme protocol. *BMJ Open* **2014**, *4*, e4951. [CrossRef]

23. Mireia, G.; Wilma, Z.; Cristina, V.; Mathew, P.W.; Mark, J.N. Outdoor blue spaces, human health and well-being: A systematic review of quantitative studies. *Int. J. Hyg. Environ. Health* **2017**, *220*, 1207–1221.

24. Tytti, P.P.; Mathew, P.W.; Benedict, W.W.; Joanne, K.G.; Lewis, R.E. Neighbourhood blue space, health and wellbeing: The mediating role of different types of physical activity. *Environ. Int.* **2019**, *131*, 105016.

25. Katherine, J.A.; Sabine, P.; Paul, W.; Mathew, P.W. The beach as a setting for families' health promotion: A qualitative study with parents and children living in coastal regions in Southwest England. *Health Place* **2013**, *23*, 138–147.

26. Gao, Y.; Zhang, T.; Zhang, W.; Meng, H.; Zhang, Z. Research on visual behavior characteristics and cognitive evaluation of different types of forest landscape spaces. *Urban For. Urban Green.* **2020**, *54*, 126788. [CrossRef]

27. Britton, E.; Kindermann, G.; Domegan, C.; Carlin, C. Blue care: A systematic review of blue space interventions for health and wellbeing. *Health Promot. Int.* **2020**, *35*, 50–69. [CrossRef]

28. Yohan, S.; Céline, C.; Jean-Christophe, F. Spatial modelling of landscape aesthetic potential in urban-rural fringes. *J. Environ. Manag.* **2016**, *181*, 623–636.

29. Gungor, S.P.A.T. Relationship between visual quality and landscape characteristics in urban parks. *J. Environ. Prot. Ecol.* **2018**, *19*, 939–948.

30. Li, L.; Riken, H.; Kazuhisa, I. Preferences for a lake landscape: Effects of building height and lake width. *Environ. Impact Asses.* **2018**, *70*, 22–33.

31. LeCun, Y.; Bengio, Y.; Hinton, G. Deep learning. *Nature* **2015**, *521*, 436–444. [CrossRef] [PubMed]

32. Gong, F.Y.; Zeng, Z.C.; Zhang, F.; Li, X.; Ng, E.; Norford, L.K. Mapping sky, tree, and building view factors of street canyons in a high-density urban environment. *Build. Environ.* **2018**, *134*, 155–167. [CrossRef]

33. Nagata, S.; Nakaya, T.; Hanibuchi, T.; Amagasa, S.; Kikuchi, H.; Inoue, S. Objective scoring of streetscape walkability related to leisure walking: Statistical modeling approach with semantic segmentation of Google Street View images. *Health Place* **2020**, *66*, 102428. [CrossRef] [PubMed]

34. Fan, Z.; Bolei, Z.; Liu, L.; Yu, L.; Helene, H.F.; Hui, L.; Carlo, R. Measuring human perceptions of a large-scale urban region using machine learning. *Landsc. Urban Plan.* **2018**, *180*, 148–160.

35. Ye, Y.; Zeng, W.; Shen, Q.; Zhang, X.; Lu, Y.; Yang, P.P.; Yamagata, Y.; Badrinarayanan, V.; Kendall, A.; Cipolla, R.; et al. The visual quality of streets: A human-centred continuous measurement based on machine learning algorithms and street view images. *Environ. Plan. B Urban Anal. City Sci.* **2019**, *46*, 1439–1457. [CrossRef]

36. Amit, B.; Martin, D.; Dick, E.; Joost, D.K.; Geert, D.L.; Nico, D. The utilization of immersive virtual environments for the investigation of environmental preferences. *Landsc. Urban Plan.* **2019**, *189*, 129–138.

37. Chamilothori, K.; Wienold, J.; Andersen, M. Adequacy of Immersive Virtual Reality for the Perception of Daylit Spaces: Comparison of Real and Virtual Environments. *Leukos J. Illum. Eng. Soc. N. Am.* **2019**, *15*, 203–226. [CrossRef]

38. Li, C.; Sun, C.; Sun, M.; Yuan, Y.; Li, P. Effects of brightness levels on stress recovery when viewing a virtual reality forest with simulated natural light. *Urban For. Urban Green.* **2020**, *56*, 126865. [CrossRef]

39. Riva, G.; Bernardelli, L.; Castelnuovo, G.; Di Lernia, D.; Tuena, C.; Clementi, A.; Pedroli, E.; Malighetti, C.; Sforza, F.; Wiederhold, B.K.; et al. A Virtual Reality-Based Self-Help Intervention for Dealing with the Psychological Distress Associated with the COVID-19 Lockdown: An Effectiveness Study with a Two-Week Follow-Up. *Int. J. Environ. Res. Public Health* **2021**, *18*, 8188. [CrossRef] [PubMed]

40. Anna, F.; Oswald, D.K.; Mareike, S.; Anna-Katharina, H.; Leon, B.; Helmut, H.; Ilse, K. Is virtual reality emotionally arousing? Investigating five emotion inducing virtual park scenarios. *Int. J. Hum.-Comput. Stud.* **2015**, *82*, 48–56.

41. Yu, C.; Lee, H.; Luo, X. The effect of virtual reality forest and urban environments on physiological and psychologi-cal responses. *Urban For. Urban Green.* **2018**, *35*, 106–114. [CrossRef]

42. Qiuyun, H.; Minyan, Y.; Hao-ann, J.; Shuhua, L.; Nicole, B. Trees, grass, or concrete? The effects of different types of environments on stress reduction. *Landsc. Urban Plan.* **2020**, *193*, 103654.

43. Jie, Y.; Jing, Y.; Nastaran, A.; Paul, J.C.; Joseph, G.A.; John, D.S. Effects of biophilic indoor environment on stress and anxiety recovery: A between-subjects experiment in virtual reality. *Environ. Int.* **2020**, *136*, 105427.

44. Osmo, M.; Arto, K.; Essi, P.; Kaisa, H.; Jani, H.; Petri, P. Restoration in a virtual reality forest environment. *Comput. Hum. Behav.* **2020**, *107*, 106295.

45. Li, W.; Zhu, J.; Fu, L.; Zhu, Q.; Xie, Y.; Hu, Y. An augmented representation method of debris flow scenes to improve public perception. *Int. J. Geogr. Inf. Sci.* **2021**, *35*, 1521–1544. [CrossRef]

46. Hu, F.; Wang, Z.; Sheng, G.; Lia, X.; Chen, C.; Geng, D.; Hong, X.; Xu, N.; Zhu, Z.; Zhang, Z.; et al. Impacts of national park tourism sites: A perceptual analysis from residents of three spatial levels of local communities in Banff national park. *Environ. Dev. Sustain.* **2022**, *24*, 3126–3145. [CrossRef]

47. Van, D.; Joye, Y.; Koole, S.L. Why viewing nature is more fascinating and restorative than viewing buildings: A closer look at perceived complexity. *Urban For. Urban Green.* **2016**, *20*, 397–401.

48. Jo, H.I.; Jeon, J.Y. The influence of human behavioral characteristics on soundscape perception in urban parks: Subjective and observational approaches. *Landsc. Urban Plan.* **2020**, *203*, 103890. [CrossRef]

49. Deng, L.; Li, X.; Luo, H.; Fu, E.; Ma, J.; Sun, L.; Huang, Z.; Cai, S.; Jia, Y. Empirical study of landscape types, landscape elements and landscape components of the urban park promoting physiological and psychological restoration. *Urban For. Urban Green.* **2020**, *48*, 126488. [CrossRef]

50. Li, D.; Sullivan, W.C. Impact of views to school landscapes on recovery from stress and mental fatigue. *Landsc. Urban Plan.* **2016**, *148*, 149–158. [CrossRef]

51. Zhao, J.; Xu, W.; Ye, L. Effects of auditory-visual combinations on perceived restorative potential of urban green space. *Appl. Acoust.* **2018**, *141*, 169–177. [CrossRef]

52. Calogiuri, G.; Litleskare, S.; Fagerheim, K.A.; Rydgren, T.L.; Brambilla, E.; Thurston, M. Experiencing Nature through Immersive Virtual Environments: Environmental Perceptions, Physical Engagement, and Affective Responses during a Simulated Nature Walk. *Front. Psychol.* **2018**, *8*, 2321–2335. [CrossRef] [PubMed]

53. Zhang, Z.; Xu, L.Q. The Healing Benefits and Application of the Virtual Natural Environment. *South Archit.* **2020**, *04*, 34–40.

54. Han, K. A reliable and valid self-rating measure of the restorative quality of natural environments. *Landsc. Urban Plan.* **2003**, *64*, 209–232. [CrossRef]

55. Ning, X.; Wang, J. Study of Urban Public Space Based on Civic Daily Life: Case Study of Three Kinds of Public Space in Nanjing Old City. *Archit. J.* **2008**, *8*, 45–48.

56. Julon, D. *Gray Theory Foundation*; Huazhong University of Science and Technology Press: Wuhan, China, 2002; pp. 40–42.

57. Chen, L.; Zhu, Y.; Papandreou, G.; Schroff, F.; Adam, H. Encoder-Decoder with Atrous Separable Convolution for Semantic Image Segmentation. In Proceedings of the European Conference on Computer Vision (ECCV), Munich, Germany, 8–14 September 2018; Volume 11211, pp. 833–851.

58. Zhao, H.; Shi, J.; Qi, X.; Wang, X.; Jia, J. Pyramid Scene Parsing Network. In Proceedings of the Conference on Computer Vision and Pattern Recognition, IEEE Computer Society, Honolulu, HI, USA, 21–26 July 2017.

59. Sun, K.; Xiao, B.; Liu, D.; Wang, J. Deep High-Resolution Representation Learning for Human Pose Estimation. In Proceedings of the IEEE/CVF Conference on Computer Vision and Pattern Recognition (CVPR), Long Beach, CA, USA, 16–20 June 2019; pp. 5686–5696.

60. Garcia-Garcia, A.; Orts-Escolano, S.; Oprea, S.; Villena-Martinez, V.; Garcia-Rodriguez, J. A Review on Deep Learning Techniques Applied to Semantic Segmentation. *arXiv* **2017**, arXiv:1704.06857.

61. Dang, H.; Li, J. The integration of urban streetscapes provides the possibility to fully quantify the ecological landscape of urban green spaces: A case study of Xi'an city. *Ecol. Indic.* **2021**, *133*, 108388. [CrossRef]

62. Zhang, W.; Yang, M.; Zhou, Y. Visible Street Space Analysis Based on Geo-tagged Panoramic Images. *J. Wuhan Univ. Technol.* **2020**, *42*, 70–78.

63. Lin, Y. *Research on Small-Scale Pedestrian Real-Time Detection for Complex Road Scenes*; Southwest Jiaotong University: Chengdu, China, 2020.

64. Zhou, B.; Zhao, H.; Puig, X.; Xiao, T.; Fidler, S.; Barriuso, A.; Torralba, A. Semantic Understanding of Scenes Through the ADE20K Dataset. *Int. J. Comput. Vis.* **2019**, *127*, 302–321. [CrossRef]

65. Kisantal, M.; Wojna, Z.; Murawski, J.; Naruniec, J.; Cho, K. Augmentation for small object detection. *arXiv* **2019**, arXiv:1902.07296.

66. Wang, M.; Lyu, X.; Li, Y.; Zhang, F. VR content creation and exploration with deep learning: A survey. *Comput. Vis. Media* **2020**, *6*, 3–28. [CrossRef]

67. Kaplan, S.; Talbot, J.F. *Psychological Benefits of a Wilderness Experience*; Springer: Boston, MA, USA, 1983; Volume 6, pp. 163–203.

68. Magdalena, V.D.B.; Mireille, V.P.; Graham, S.; Margarita, T.; Sandra, A.; Irene, V.K.; Willem, V.M.; Christopher, G.; Regina, G.; Mark, J.N.; et al. Does time spent on visits to green space mediate the associations between the level of residential greenness and mental health? *Urban For. Urban Green.* **2017**, *25*, 94–102.

69. Staats, H.H.T. Alone or with a Friend: A Social Context for Psychological Restoration and Environmental Preferences. *J. Environ. Psychol.* **2004**, *24*, 199–211. [CrossRef]

70. Ulrich, R.S. *Aesthetic and Affective Response to Natural Environment*; Springer: Boston, MA, USA, 1983.

71. Chen, Z.; He, Y.; Yu, Y. Enhanced functional connectivity properties of human brains during in-situ nature experience. *PeerJ* **2016**, *4*, e2210. [CrossRef] [PubMed]

72. Yao, W.; Yun, J.; Zhang, Y.; Meng, T.; Mu, Z. Usage behavior and health benefit perception of youth in urban parks: A case study from Qingdao, China. *Front. Public Health* **2022**, *10*, 923671. [CrossRef]

73. Chen, Z.; Fu, W.; Zhu, Z.; Dong, J. Analysis of advantages and disadvantages of panorama presentation technology in landscape visual evaluation. *J. Xi'an Univ. Archit. Technol. (Nat. Sci. Ed.)* **2021**, *53*, 584–593.

74. Yannick, J.; Agnes, V.D.B. Is love for green in our genes? A critical analysis of evolutionary assumptions in restorative environments research. *Urban For. Urban Green.* **2011**, *10*, 261–268.

75. Fujisawa, H. *Color Psychology*; Scientific and Technical Literature Press: Beijing, China, 1989; pp. 1–103.

76. Lü, X.; Li, H. The Layout of Micro Commercial Facilities in Street Space. *Planners* **2019**, *35*, 57–62.

77. Liu, L.; Qu, H.; Ma, Y.; Wang, K.; Qu, H. Restorative benefits of urban green space: Physiological, psychological restoration and eye movement analysis. *J. Environ. Manag.* **2022**, *301*, 113930. [CrossRef]

78. Nordh, H.; Hagerhall, C.M.; Holmqvist, K. Exploring view pattern and analysing pupil size as a measure of restorative qualities in park photos. *Acta Hortic.* **2010**, *881*, 767–772. [CrossRef]

79. Kaplan, R.; Kaplan, S. *The Experience of Nature: A Psychological Perspective*; Cambridge University Press: Cambridge, UK, 1989.

80. Kjell, N. Forests, trees and human health and wellbeing. *Urban For. Urban Green.* **2006**, *5*, 109.

81. Velarde, M.D.; Fry, G.; Tveit, M. Health effects of viewing landscapes-Landscape types in environmental psychology. *Urban For. Urban Green.* **2007**, *6*, 199–212. [CrossRef]

*Article*

# Effects of Audio–Visual Interaction on Physio-Psychological Recovery of Older Adults in Residential Public Space

**Shan Shu *, Lingkang Meng, Xun Piao, Xuechuan Geng and Jiaxin Tang**

College of Architecture and Urban Planning, Qingdao University of Technology, Qingdao 266033, China; menglk1997@163.com (L.M.); piaoxun@qut.edu.cn (X.P.); gengxuechuan@qut.edu.cn (X.G.); tangjiaaaxin@163.com (J.T.)
* Correspondence: shushan@qut.edu.cn; Tel.: +86-15054156771

**Abstract:** It is now well established that everyday interaction with nature has a restorative potential on the elderly population's health and well-being. However, empirical evidence on the restorative effects of neighborhood greenspace is still lacking, and scant attention has been given to the cross-effect of the visual–audio experience. The present study examined the restorative effects of audio–visual interactions on older adults in typical residential public spaces in Chinese cities. A pretest–post-test design was used to measure changes in participants' physiological responses, mood states, and mental restoration. Participants (mean age = 68.88 years) were asked to experience six simulated audio–visual conditions (3 scenes × 2 sounds) of residential public space. The results showed that: (1) A green scene combined with nature sounds showed the most restorative effect on the elderly participants' psycho-physiological health. (2) Viewing green scenes facilitated the most psycho-physiological recovery for the elderly, followed by viewing the activity scene. (3) Compared to the traffic noise, adding nature sounds could promote many more benefits in HR recovery, positive mood promotion, and perceived restorative effects, and the advantage of nature sounds over traffic noise was mainly demonstrated in the green scene. (4) Visual scenes demonstrated a greater impact on the elderly participants' psycho-physiological recovery than the sounds. Our findings suggested the necessity of providing residential nature and activity spaces, encompassing both sound and vision, to promote healthy aging in Chinese residential contexts.

**Keywords:** older adults; restorative environment; residential public space; physio-psychological recovery

**Citation:** Shu, S.; Meng, L.; Piao, X.; Geng, X.; Tang, J. Effects of Audio–Visual Interaction on Physio-Psychological Recovery of Older Adults in Residential Public Space. *Forests* **2024**, *15*, 266. https://doi.org/10.3390/f15020266

Academic Editors: Jiang Liu, Xinhao Wang and Xin-Chen Hong

Received: 30 December 2023
Revised: 16 January 2024
Accepted: 27 January 2024
Published: 30 January 2024

## 1. Introduction

The world is facing the severe challenge of an aging population. The population of people aged 60 years and older will rise to 2.1 billion by 2050, with 80% of the aging population living in developing countries [1]. In China, the proportion of citizens who are aged 60 or above is expected to increase beyond 30% by 2035. Furthermore, projected fertility rates and expected population age structures show that, in the decades to come, most older people in China will live in urban areas [2]. Meanwhile, urban aging populations have been widely shown to face serious physical and mental health problems, which pose greater challenges for themselves, their family, and the whole society [3]. The question of how to foster healthy aging has, therefore, attracted global attention. It is essential to consider the elderly population's well-being, including physiological and psychological health, in urban environmental design. In view of this, urban greenspace, which provides ideal nature-based open public spaces for recreational activities, is a critical environmental resource that benefits the health and well-being of elderly residents [4]. An increasing body of research has shown that urban residential greenspace can reduce stress-related depression symptoms [5,6], promote recovery of ischemic heart disease [7], reduce dementia risk [8], foster physical activities [9], and increase longevity of urban elderly dwellers [10].

The health benefits of greenspace have been proposed by diverse possible mechanisms, and one of the leading hypotheses is the theory of restoration, indicating the process of renewing, recovering, or re-establishing diminished physical, psychological, and social resources or capabilities in ongoing efforts to meet adaptive demands [11]. Research on the restorative environments has been guided by two main theoretical frameworks: Stress Recovery Theory (SRT) and Attention Restoration Theory (ART). SRT claims that restoration is derived from the release of stress and the reduction of negative moods through being exposed to restorative environments [12]. ART explains how natural environments capture involuntary attention and allow restoration of cognitive capacities [13]. Based on these theories, an increasing amount of research evidence has shown the restorative benefits of greenspace on older adults. One of the pilot studies was conducted by Berto, who recruited 50 older adults and asked them to rate the restorative value of 10 pictures of environments, ranging from natural to built, using the Perceived Restorativeness Scale [14]. Likewise, more research has been performed to examine the restorative benefits of nature on older adults based on their self-reported restorativeness [15–18].

Despite growing studies on the relationship between greenspace exposure and health, empirical evidence for restorative effects remains quite ambiguous. For instance, a field experiment performed in an elderly care institution in China showed that a five-minute VR forest experience can bring immediate psychological improvements, but was unable to significantly decrease blood pressure [19]. Similarly, a study was conducted with 34 middle-aged elderly adults, showing them VR experiences of natural and urban settings. The results showed significant restorative effects in mood levels rather than physiological responses [20]. On the contrary, a lab experiment used bamboo and urban images as stimuli to offer indirect contact with nature and found it enhanced both psychological and physiological conditions in the elderly [21]. Field experiments also showed a significant restorative effect of green walking on the psycho-physiological health of the elderly [22,23]. Taken together, the restorative effects of greenspace on the elderly population's psycho-physiological health have not been conclusively backed up by the existing empirical research. More evidence-based studies are still needed to confirm whether psychological and physiological restoration could be induced by greenspace exposure in older adults.

In addition, most existing studies linking greenspace and restoration focused on the comparison between natural and urban scenarios [20], and suggested the restorative benefits of urban natural environments [16]. However, the restorative designation of small greenspaces in urban residential contexts has yet to be fully examined. With rapid urbanization in China, elderly city dwellers generally live in high-density residential areas and have less opportunity to visit large nature parks due to their limited physical mobility, along with other barriers (e.g., low income). They usually spend a lot of time in peripheral small residential public spaces to rest, socialize, or exercise. An accessible natural environment within walking distance from home is more likely to be used daily. Given this, planning and design considerations for residential public spaces for the elderly require more attention because the location advantage makes it livable and accessible for its high-density aged population. Overall, there is a need to create residential greenspaces that could foster restorative experiences for older adults.

Moreover, extant studies are limited by solely validating and quantifying the restorative benefits of urban greenspace, and the mechanism of possible influential factors on the generation of restorativeness is still unclear. In recent years, much research recognized that restorativeness emerges from a diverse, complex, and integrated process of multi-sensory information about the environment, especially the visual and auditory sources [24]. To date, considerable research has focused on the restorative effects of visual elements, showing that more visible greenery is related to better restorativeness [25]. It is noteworthy that sight is not the only sense that affects the restorative response of humans, which can be partially or even largely attributed to the acoustic environment. The restorative effects of nature soundscapes have been validated from multiple perspectives. Generally, natural sounds (e.g., birdsong and water sounds) were found to be positively associated with psychological and physiological benefits, while urban sounds (e.g., traffic noise) were negatively associated with those benefits [26,27]. It is also important to bear in mind that visual landscapes and sounds never exist and work alone, and the interaction of audio–visual experiences is omnipresent in urban and nature environments [28]. Many studies have shown that sounds influence the evaluation of visual stimuli, and vice versa [29]. Moreover, their coherence has been exemplified to be associated with restorative benefits by some previous studies. For example, a study highlighted that landscapes containing natural water and high plant cover matched with birdsong can produce a higher restorative potential [30]. Another study indicated that naturally related visual and auditory stimuli can promote better restorative benefits than single visual stimuli [31]. Notwithstanding the restorative benefits reported elsewhere, the integration effect of visual and auditory stimuli on the elderly population's restoration has remained largely unexplored. This leads to a lack of clear theoretical guidance on how audio–visual design could be integrated into the restorative environment design practice for the elderly.

This study aimed to find reliable evidence to inform public space design of audio–visual combinations to improve the restorative quality of residential public space for older residents. To this end, a pre–post-test experimental design was employed to measure elderly participants' physio-psychological recovery and mental restoration after the reproduction of common audio–visual scenarios of residential public spaces in a laboratory setting. Specifically, this study aimed to address the following two main questions:

(1) Can residential public space foster psychological and physiological restoration for older adults?
(2) Which visual and acoustic factors could facilitate better restorative effects on older adults in residential public spaces?

## 2. Materials and Methods

### 2.1. Participants

A total of 62 older adults (42 males and 20 females) participated in the experiment. They were recruited via social networking platforms (e.g., WeChat) and snowball sampling in Qingdao, China. People who were at least 60 years old and living in an urban setting were eligible to participate. We opted to select mainly young older adults (aged from 60 to 80, with an average age of 68.88 ± 5.19) in good health, in order to avoid that cardiovascular diseases, cognition disorders, or specific visual and hearing problems might affect their psycho-physiological experience and a reliable evaluation of the environments. All participants were instructed to avoid consumption of alcohol and caffeine for at least 24 h before the experiment.

The study was conducted in accordance with the Declaration of Helsinki. Ethical approval was obtained from the Academic Committee of the College of Architecture and Urban Planning at Qingdao University of Technology. Before the experiment, the participants were informed about the study protocol and gave written informed consent to the researchers. All the elderly participants voluntarily participated in the study, and they were free to quit the experiment if they felt uncomfortable during the process.

### 2.2. Research Design

A pre–post-test experimental design was applied to investigate the restorative effects of audio–visual combinations on the psycho-physiological health of older adults in residential public spaces. Three typical visual scenes of residential public space (square scene, activity scene, and green scene) were separately combined with two sound types (traffic noise and nature sounds) to form six different audio–visual combinations: (1) square scene + traffic noise, (2) square scene + nature sound, (3) activity scene + traffic noise, (4) activity scene + nature sound, (5) green scene + traffic noise, and (6) green scene + nature sound (Figure 1). In addition, in order to control the experiment within an hour to avoid fatigue and impatience effects on health indices, as well as due to the limited available sample size, each participant experienced three or four audio–visual stimuli, which were randomly selected from the total six stimuli and presented in random order.

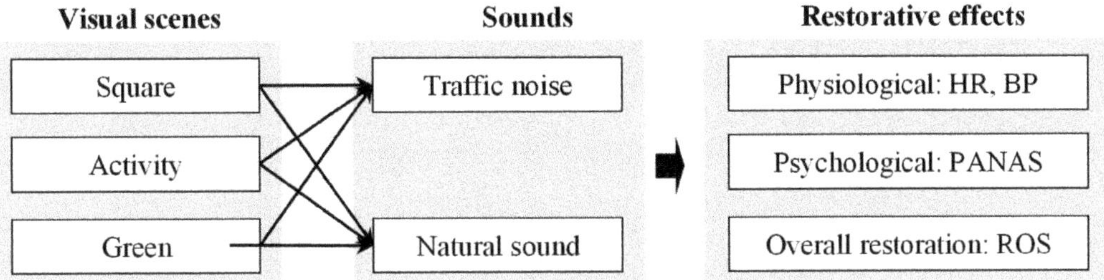


**Figure 1.** Diagram of the study design.

*2.3. Environmental Setting*

2.3.1. Visual Scenes

In this study, a typical high-density residential district in China was modeled in SketchUp 2018 to represent the outdoor contexts in residential areas. The central square, a 5000 m$^2$ public space surrounded by high-rise residential buildings, was selected as the original prototype of the experimental setting. It was used as the "square" without any additional restorative visual landscape.

To test participants' responses in public spaces with different visual characteristics, we categorized visual elements into two types, "activity scene" and "green scene", mainly because we considered two types of visual elements that have shown restorative potential: well-designed man-made features and natural green elements. Specifically, the activity scene incorporated fitness facilities, pavilions, and seats, which are expected to promote the intentions of physical exercise and social interaction and further foster the restorative experience. The green scene incorporated large areas of lawn, various tree species, as well as waterscapes (fountain and ponds). It was designed to foster mental health for older adults. It is important to note that the presence of people was excluded in all environments since it was indicated to have different impacts on the restorative experience of different scenarios [15,32]. We also maintained the same size and a similar layout for the three scenes to maximize the comparability. Except for the restorative design interventions, all three scenes were identical in terms of all other elements.

Regarding the visual scenes, SketchUp software was first used to establish the three-dimensional geometric model, and Lumion 10.0 software was then used to render the videos for screen representation. Video is a valid display medium that older adults are familiar with and it could provide an immersive experience of the scene without dizziness or fear issues that could possibly affect the elderly participants' psycho-physiological fluctuations [33]. Specifically, the video was recorded to simulate walking around the public space, and the walking routes were identical in the three visual scenes for comparability. The visual point of the video was at a height of 1.6 m to simulate the view of participants when they were walking around. The duration of each video was set at 6 min, which was thought to be enough to foster restorative benefits in a laboratory simulation [34]. The videos were rendered at a 4 K resolution at 30 frames per second (fps) and presented on a 27-inch LCD monitor with a 3840 × 2160 pixel resolution and a 60 Hz refresh rate. The aerial views and screenshots of each visual scene are shown in Table 1.

**Table 1.** Aerial views and screenshots of the visual scenes and corresponding detailed descriptions.

| Visual Features | Aerial View | Screenshots of the Video |
| --- | --- | --- |
| **Square scene**<br>Large square surrounded by lawns |  |  |
| **Activity scene**<br>Adding walking trails, fitness equipment, and amenities, seats, and pavilions | |  |
| **Natural scene**<br>Adding more grasslands, colorful flowers, various plant species, and waterscapes |  |  |

### 2.3.2. Sounds

In this study, two types of sounds were considered: traffic noise and natural sounds. Since road traffic is one of the major sources of noise pollution in urban residential areas, it was designated as the non-restorative sound. On the contrary, nature sounds were widely acknowledged to have restorative effects in previous studies; thus, it was designated as the restorative sound. The traffic noise was downloaded from open sources on the Internet, and predominantly contained low-frequency components below 2.0 kHz, as shown in Figure 2a. The nature sounds were recorded in a typical urban park in Qingdao using a Sony PCM-100 digital recorder equipped with two stereo microphones, and featured birdsong and flowing water sounds. Figure 2b shows that the nature sounds were dominated by mid- to high-frequency components from 500 Hz to 8 kHz. In terms of temporal characteristics, the traffic noise was constant with low temporal variability, whereas the nature sounds were intermittent with high temporal variability. All the acoustic stimuli were stored in uncompressed wave digital format and converted into the same sample rate and size of 44.1 kHz, 16 bit. In addition, the sound pressure levels were all normalized to 55 dBA for two reasons: (1) the upper limit of the environmental sound level in residential areas is 55 dBA according to the sound quality standard in China, and (2) people can hear the acoustic stimuli clearly but do not feel annoyed at this sound level, as indicated by previous studies [27]. The calibration was conducted using a AWA6270+ sound level meter coupled to a Beyerdynamic DT 900 Pro headphone, which was driven by a soundcard (YAMAHA Steinberg UR242, Steinberg Media Technologies GmbH, Hamburg, Germany). Besides, the duration of the acoustic stimuli was set at 6 min to match with the visual stimuli. All the editing and calibration of the acoustic stimuli were performed using the Adobe Audition software (version 2022). The acoustic stimuli were played back to the participants via the same Beyerdynamic DT 900 Pro headphones.

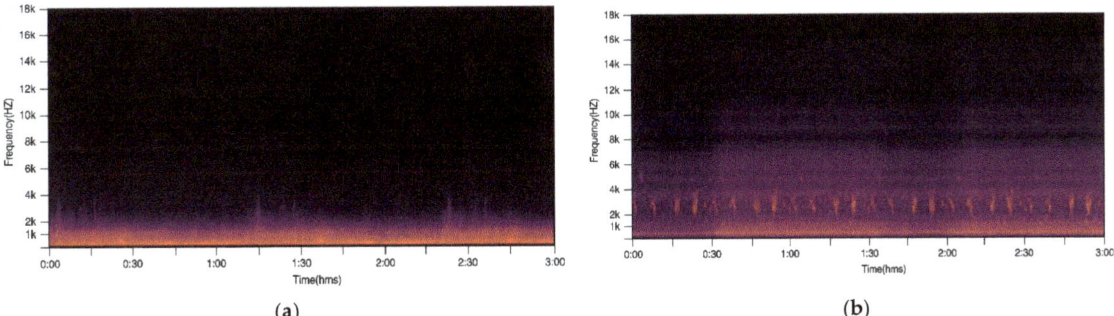

**Figure 2.** Spectrogram of the acoustic stimuli: (**a**) traffic noise and (**b**) nature sounds, including birdsong and water sounds.

*2.4. Measures*

2.4.1. Physiological Measurements

To measure older adults' physiological recovery, two simple measures were applied: mean arterial pressure (MAP) and heart rate (HR). These indices are commonly suggested as sensitive indicators for stress recovery in restorative environment research, and they are widely used to assess cardiovascular effects in older adults [23]. Notably, the MAP was calculated based on systolic blood pressure (SBP) and diastolic blood pressure (DBP), as: $((DBP \times 2) + SBP)/3$. During the experiment, SBP, DBP, and HR were measured using a portable cuff-based electronic sphygmomanometer (HEM-7137, Omron, Tokyo, Japan). Measurements were performed in a relaxed sitting position with the instrument placed on the right arm at the heart level. Any tight-fitting or thick clothing was removed from the arm before measurement. Two measurements were conducted with a 20 s interval. In the analysis, the average of the two measurements was used. Since every individual has a unique MAP and HR baseline, all the measured data after the audio–visual exposure were further calculated as a percentage change (%) from the stress period to accommodate individual differences.

2.4.2. Psychological Measurements

The Positive and Negative Affect Scale (PANAS) was employed to measure the emotional outcomes in this study [35]. The original PANAS consists of 20 items and is widely used to evaluate changes in positive and negative emotions [36,37]. We finally extracted 6 items from the original PANAS since they are easily understood by the older adults and highly matched to the experience in public space, as indicated in a preliminary experiment. Therefore, the shortened version of the scale consisted of six items divided into two sub-scales: the positive mood state was calculated from three items (i.e., interested, excited, or attentive), and the negative mood state was calculated from three items (i.e., nervous, upset, or irritable). The order of these items was randomized to avoid an order effect. Participants were asked to indicate the extent to which they felt a series of different emotions "at the present moment" on a 5-point Likert scale (1 = not at all; 5 = extremely).

2.4.3. Mental Restoration

In this study, participants' overall mental restoration was measured using the Restoration Outcome Scale (ROS) [38]. The ROS scale consisted of six items, three of which reflect relaxation and calmness ("I feel restored and relaxed", "I feel calm", and "I have enthusiasm and energy for my everyday routines"), one reflects attention restoration ("I feel focused and alert"), and two reflect clearing one's thoughts ("I can forget everyday worries" and "My thoughts are clear"). Each item was assessed on a 5-point scale (1 = not at all; 5 = extremely). The scale has been widely used in previous studies and confirmed with high reliability and validity [39].

### 2.5. Procedure

The experiments were conducted in a quiet laboratory from October to December 2022. The background noise in the laboratory was measured as below 35 dB(A) with no distinguished background noise, and the indoor temperature was kept between 18 and 22 °C. Each participant took part in the experiment individually, supervised by a staff member.

After arrival at the laboratory, participants were guided to sit in a comfortable chair and the experimental procedure was explained. The first physiological measure was conducted after sitting quietly for 3 min without any intervention, as the baseline level, which was used to ensure no differences across audio–visual groups. For each audio–visual stimulus, there were two periods: (1) A 3 min stressor period, during which participants were asked to accurately perform continuous oral subtraction from a number with a step of 7 (or 13), as soon as possible. The arithmetic task could effectively induce stress and fatigue, as indicated in previous studies [27,40]. (2) A 6 min recovery period with exposure to one audio–visual condition. Physiological measurements (MAP and HR) and psychological questionnaires (PANAS) were completed immediately at the end of each period. Thereafter, the participants were asked to fill out the restoration scale (ROS) based on their experience of the video. The same process was repeated for the next stimulus, as shown in Figure 3.

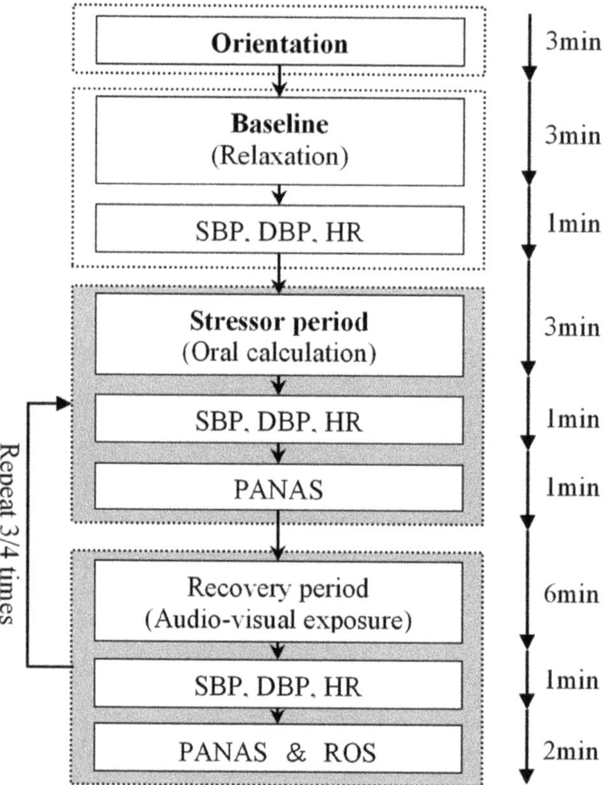


**Figure 3.** Diagram of the experimental procedure.

### 2.6. Statistical Analysis

All the data analyses were performed using SPSS software (version 26). Non-parametric tests were chosen for the raw data analysis due to the non-normality, as examined by a

Kolmogorov–Smirnov test. First, the individual differences across audio–visual conditions were checked using a non-parametric Kruskal–Willis test (scale variables) and chi-square test (nominal variables). Then, the Wilcoxon signed-rank test was conducted to examine the effectiveness of the stress induction. Both of the above steps were the prerequisites to allow testing psycho-physiological recovery differences. Additionally, to verify the restorative effects of the audio–visual conditions on the physio-psychological data, pretest values and post-test values were compared for each audio–visual condition using the Wilcoxon signed-rank test.

To determine whether the audio–visual conditions were different in psycho-physiological recovery, the recovery (i.e., increase or decrease) of the measures was considered, which was calculated as the relative difference: (post-test value – pretest value)/pretest value × 100%. Due to the normality of the recovery values, analysis of variance (ANOVA) tests were performed, with visual stimuli (square, activity, and green) and acoustic stimuli (traffic noise and nature sounds) considered as the between-subjects factors. In all ANOVA tests, the least significant difference (LSD) post-hoc tests were conducted for pairwise comparisons. Additionally, partial eta-squared values ($\eta^2_p$) were determined to measure the effect size. In all analyses, a $p$-value less than 0.05 was used as the criterion to determine significant differences.

## 3. Results

### 3.1. Descriptive Data

In this study, 62 older adults were randomly assigned to 3 (or 4) of the 6 audio–visual conditions. Table 2 shows the demographics and baseline characteristics of the elderly participants in different conditions. No significant difference was found in any of the psycho-physiological measures among the six audio–visual groups.

**Table 2.** Comparison of sample characteristics among the audio–visual conditions (N = 62).

| | Square | | Activity Park | | Green Park | | |
| --- | --- | --- | --- | --- | --- | --- | --- |
| | **Traffic** **(N = 37)** | **Nature** **(N = 33)** | **Traffic** **(N = 33)** | **Nature** **(N = 33)** | **Traffic** **(N = 31)** | **Nature** **(N = 39)** | $p$ |
| Gender (N; % Male) | 22; 59.5% | 26; 78.8% | 25; 75.8% | 21; 63.6% | 24; 77.4% | 24; 61.5% | 0.284 |
| Age (years) | 68.16 ± 4.54 | 68.30 ± 5.15 | 68.79 ± 5.33 | 69.76 ± 6.02 | 68.68 ± 5.24 | 69.56 ± 5.02 | 0.792 |
| Baseline value | | | | | | | |
| MAP (mm Hg) | 92.09 ± 8.07 | 98.16 ± 12.77 | 99.01 ± 12.62 | 95.74 ± 12.85 | 98.08 ± 12.92 | 93.68 ± 8.65 | 0.074 |
| HR (bpm) | 75.30 ± 7.91 | 73.58 ± 10.07 | 74.30 ± 10.81 | 74.03 ± 8.21 | 74.58 ± 10.22 | 75.59 ± 7.71 | 0.879 |
| PA (0–15) | 9.22 ± 1.46 | 10.00 ± 2.10 | 10.12 ± 2.22 | 9.88 ± 1.91 | 10.26 ± 2.14 | 9.54 ± 1.64 | 0.527 |
| NA (0–15) | 3.46 ± 0.77 | 3.36 ± 0.60 | 3.27 ± 0.57 | 3.36 ± 0.74 | 3.29 ± 0.53 | 3.49 ± 0.76 | 0.748 |

MAP = mean arterial pressure, HR = heart rate, PA = positive affect, and NA = negative affect.

Regarding the effect of the stress induction, significant changes from baseline to the pretest were detected for MAP (mean = 3.49%, 95% CI = 2.90–4.08, increase), HR (mean = 1.60%, 95% CI = 1.18–2.03, increase), PA (mean = 25.65%, 95% CI = 19.75–31.55, decrease), and NA (mean = 25.05%, 95% CI = 21.94–28.16, increase). The results proved that the oral calculation could effectively induce elderly participants' stress levels in terms of psychological and physiological measures.

Table 3 shows the descriptive statistics, mean values with standard deviations (SD), of the psycho-physiological measures during the pretest and post-test for each audio–visual condition. The Wilcoxon signed-rank test showed that participants' MAP, HR, and PA strongly recovered after exposure to both the activity scenes and green scenes. Although a tendency of recovery was also indicated in the square scenario, the changes were rarely significant. Notably, older adults' negative mood significantly decreased in the post-test compared to the pretest in all conditions, no matter what audio–visual stimuli the participants were exposed to (Figure 4).

**Table 3.** Descriptive statistics of psycho-physiological measurements in the pretest and post-test for each visual and sound combination.

| Visual | Square | | Activity | | Green | |
|---|---|---|---|---|---|---|
| **Sound** | **Traffic** | **Nature** | **Traffic** | **Nature** | **Traffic** | **Nature** |
| MAP (mm Hg) | | | | | | |
| Pretest | 95.55 ± 8.31 | 101.20 ± 11.16 | 103.71 ± 13.04 | 97.92 ± 12.76 | 100.78 ± 14.94 | 97.62 ± 8.87 |
| Post-test | 95.10 ± 7.04 | 99.78 ± 10.22 | 101.38 ± 10.91 | 95.35 ± 10.21 | 96.81 ± 13.13 | 93.20 ± 8.73 |
| $p$ | 0.081 | 0.017 * | 0.000 ** | 0.000 ** | 0.000 ** | 0.000 ** |
| HR (bpm) | | | | | | |
| Pretest | 77.86 ± 8.35 | 75.79 ± 10.56 | 75.55 ± 11.28 | 75.48 ± 8.83 | 75.97 ± 11.13 | 78.38 ± 8.10 |
| Post-test | 77.46 ± 8.19 | 74.61 ± 10.97 | 74.64 ± 10.66 | 74.33 ± 8.73 | 74.87 ± 10.39 | 75.62 ± 7.81 |
| $p$ | 0.158 | 0.077 | 0.058 | 0.027 * | 0.009 ** | 0.000 ** |
| PA | | | | | | |
| Pretest | 8.00 ± 1.68 | 8.91 ± 2.48 | 8.73 ± 2.45 | 8.70 ± 2.04 | 8.71 ± 2.38 | 8.44 ± 1.87 |
| Post-test | 6.92 ± 2.95 | 9.36 ± 2.69 | 10.39 ± 2.51 | 11.48 ± 2.49 | 11.00 ± 2.88 | 13.23 ± 1.87 |
| $p$ | 0.057 | 0.116 | 0.000 ** | 0.000 ** | 0.000 ** | 0.000 ** |
| NA | | | | | | |
| Pretest | 5.19 ± 1.35 | 4.06 ± 1.17 | 4.36 ± 1.22 | 4.67 ± 1.19 | 4.13 ± 1.18 | 5.33 ± 1.44 |
| Post-test | 3.92 ± 0.76 | 3.15 ± 0.44 | 3.09 ± 0.38 | 3.12 ± 0.70 | 3.03 ± 0.18 | 3.21 ± 0.89 |
| $p$ | 0.000 ** | 0.000 ** | 0.000 ** | 0.000 ** | 0.000 ** | 0.000 ** |
| ROS | 2.48 ± 0.77 | 3.45 ± 0.94 | 3.69 ± 0.79 | 3.62 ± 0.82 | 3.87 ± 0.89 | 4.23 ± 0.71 |

* $p < 0.05$; ** $p < 0.01$ (2-tailed). MAP = mean arterial pressure, HR = heart rate, PA = positive affect, NA = negative affect, and ROS = Restorative Outcome Scale.

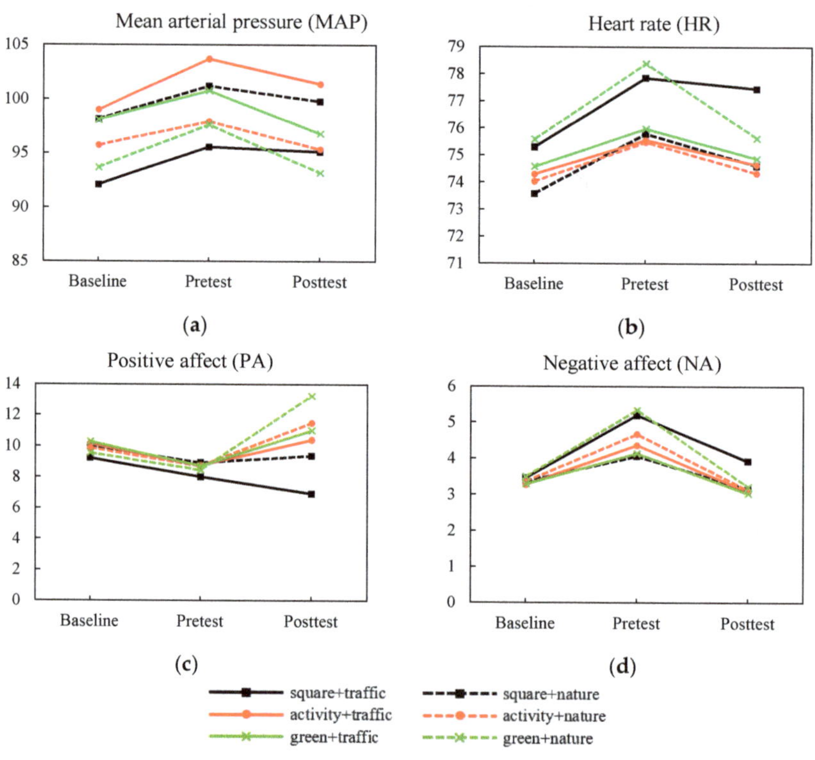


**Figure 4.** Average physiological and psychological outcomes at baseline, pretest (after stressor), and post-test (after recovery) among different audio–visual conditions: (**a**) mean arterial pressure, MAP; (**b**) heart rate, HR; (**c**) positive affect, PA; (**d**) negative affect, NA.

*3.2. Differences between Audio–Visual Conditions in the Recovery*

Table 4 provides an overview of the tested differences in recovery values between visual scenes, sounds, and audio–visual combinations, with the effect sizes indicated by ($\eta^2_p$). Overall, visual scenes showed a significant influence on the recovery of all psycho-physiological measures, while acoustic conditions only showed a significant influence on the recovery of HR, PM, and overall POS. Meanwhile, interactions between audio and visual stimuli were only detected in the POS. Detailed multiple comparison results for the recovery outcomes between different visual and audio conditions are presented in the following.

**Table 4.** Summary of ANOVA results for psycho-physiological recovery.

| | Visual | | | Audio | | | Audio * Visual | | |
|---|---|---|---|---|---|---|---|---|---|
| | F | $p$ | $\eta^2_p$ | F | $p$ | $\eta^2_p$ | F | $p$ | $\eta^2_p$ |
| ΔMAP | 16.39 | 0.000 ** | 0.141 | 2.27 | 0.134 | 0.011 | 0.12 | 0.888 | 0.001 |
| ΔHR | 4.16 | 0.017 * | 0.040 | 8.53 | 0.004 ** | 0.041 | 1.66 | 0.194 | 0.016 |
| ΔPA | 32.81 | 0.000 ** | 0.247 | 17.54 | 0.000 ** | 0.081 | 2.37 | 0.096 | 0.023 |
| ΔNA | 3.55 | 0.031 * | 0.034 | 2.31 | 0.130 | 0.011 | 2.70 | 0.070 | 0.026 |
| ROS | 30.81 | 0.000 ** | 0.236 | 13.63 | 0.000 ** | 0.064 | 6.91 | 0.001 ** | 0.065 |

* $p < 0.05$; ** $p < 0.01$ (2-tailed). Δ = Recovery values; $\eta^2_p$ = partial eta-squared (effect size). MAP = mean arterial pressure, HR = heart rate, PA = positive affect, NA = negative affect, and ROS = Restorative Outcome Scale.

3.2.1. Physiological Recovery

The ANOVA results on MAP change values showed a main effect of visual scenes ($p < 0.001$, $\eta^2_p = 0.141$), but no significant effect of acoustic stimuli ($p > 0.05$, $\eta^2_p = 0.011$). Additionally, no significant interaction between visual and acoustic stimuli was found ($p > 0.05$, $\eta^2_p = 0.001$). As shown in Figure 5a, the green scene facilitated a higher MAP decrease compared to the activity and square scenes despite the acoustic conditions. Nevertheless, slight differences in MAP recovery between sounds were indicated in all the scenes, i.e., the nature sound tended to decrease MAP more than the traffic noise, although not on a statistically significant level ($p > 0.05$).

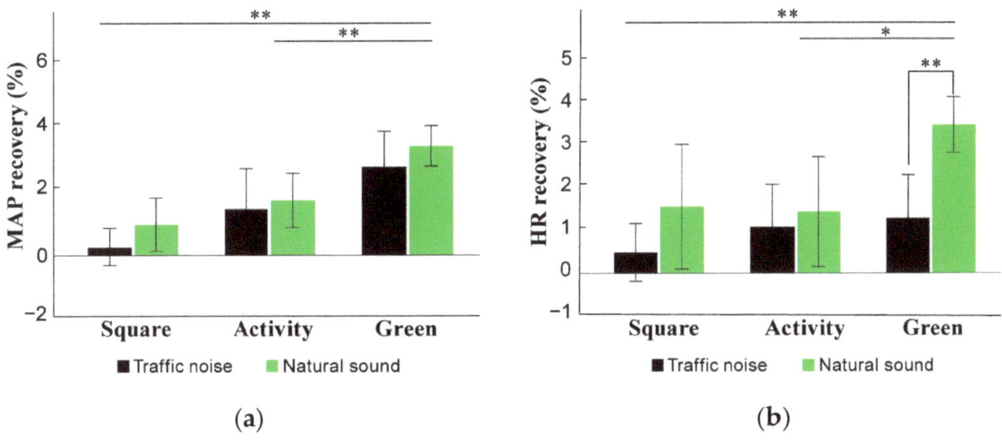


**Figure 5.** Comparison of physiological recovery in different audio–visual conditions: (**a**) MAP recovery and (**b**) HR recovery. * $p < 0.05$; ** $p < 0.01$ (2-tailed).

Regarding HR recovery, ANOVA results showed main effects of both the visual condition ($p < 0.05$, $\eta^2_p = 0.040$) and the acoustic condition ($p < 0.01$, $\eta^2_p = 0.041$). The effect size showed that visual scenes and sounds had a similar influence on the elderly participants' HR recovery. However, no significant acoustic and visual interaction was detected ($p > 0.05$, $\eta^2_p = 0.016$). Further multiple comparisons showed that the green scene facilitated a higher HR reduction than the activity and square scenes. As for the acoustic stimuli, nature sounds facilitated a higher HR reduction than traffic noise, and the advantage of the nature sounds was much more evident in the green scene ($p < 0.001$), as shown in Figure 5b.

### 3.2.2. Psychological Recovery

The ANOVA tests on positive mood promotion showed significant effects of both the visual scene ($p < 0.001$, $\eta^2_p = 0.247$) and sound ($p < 0.05$, $\eta^2_p = 0.034$), but no audio–visual interaction was detected ($p > 0.05$, $\eta^2_p = 0.023$). In particular, the effect size of the visual scene was quite large, while the effect size of the acoustic condition was moderate, indicating a higher impact of visual scenes than sounds on the positive mood promotion. As shown in Figure 6a, the green scene showed the best restorative effect on positive mood, followed by the activity scene, while the square scene exposure showed a detrimental effect. As for the acoustic condition, nature sounds showed more restorative effects compared to the traffic noise in all visual scenes. However, the difference between sounds was not significant in the activity scene.

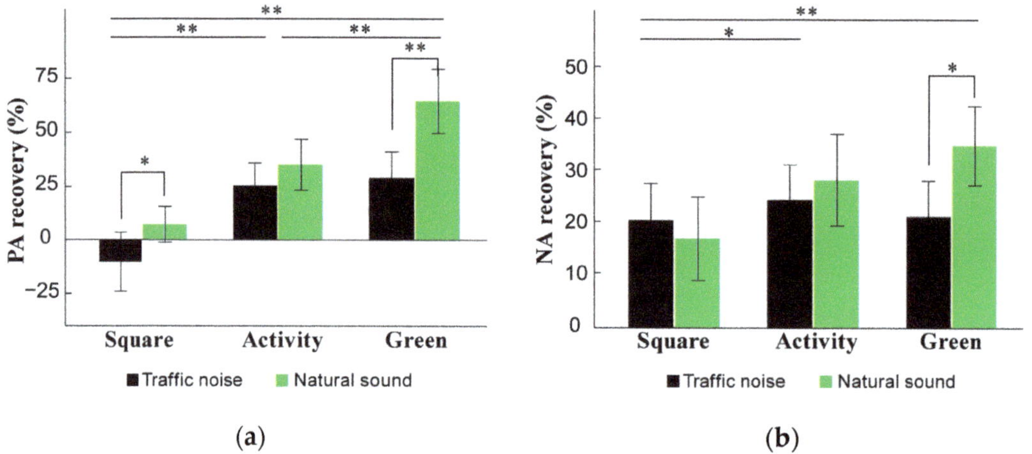


**Figure 6.** Comparison of psychological recovery in different audio–visual conditions: (**a**) positive mood recovery and (**b**) negative mood recovery. * $p < 0.05$; ** $p < 0.01$ (2-tailed).

The ANOVA tests on negative mood recovery showed only a main effect of the visual scene ($p < 0.001$, $\eta^2_p = 0.081$). Compared to the square scene, both the activity and green scenes generated a higher NA decrease. However, no significant effect of the acoustic condition ($p > 0.05$, $\eta^2_p = 0.011$) was identified in terms of NA recovery, nor was an audio–visual interaction revealed ($p > 0.05$, $\eta^2_p = 0.070$). Nevertheless, nature sounds showed a better restorative effect on negative mood than traffic noise in the green scene ($p < 0.01$), although no significant differences between sounds were detected in the square and activity scenes ($p > 0.05$), as shown in Figure 6b.

### 3.3. Overall Perceived Restorativeness

The overall restorativeness was assessed using the Restorative Outcome Scale (ROS), and the results showed main effects of the visual scene ($p < 0.001$, $\eta^2_p = 0.236$), acoustic condition ($p < 0.001$, $\eta^2_p = 0.064$), as well as audio–visual interactions ($p < 0.001$, $\eta^2_p = 0.065$). Notably, the effect size of the visual scene was much larger than that of the sounds. The green scene was perceived as the most restorative scene, followed by the activity scene. As for the audio–visual interaction, Figure 7 shows a comparison of the acoustic performance on the perceived restorativeness under each visual scene. A significant difference between sounds was detected in the square scene, and nature sounds were rated to be more restorative than traffic sounds ($p < 0.001$). However, no significant difference between acoustic stimuli was found in either the activity ($p > 0.05$) or green scenes ($p > 0.05$).

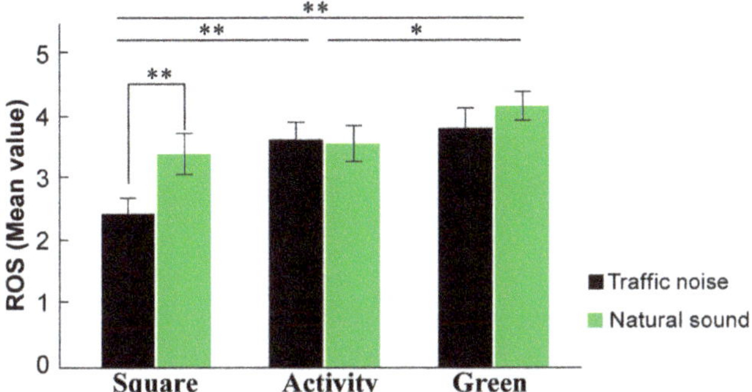


**Figure 7.** Perceived restorativeness of different audio–visual conditions. $*\ p < 0.05$; $**\ p < 0.01$ (2-tailed).

Spearman's correlation analysis showed that older adults' overall perceived restorativeness had a strong positive correlation with MAP ($r = 0.261$, $p < 0.001$) and PA ($r = 0.311$, $p < 0.001$), as well as HR ($r = 0.175$, $p < 0.05$). However, there was no strong relationship between perceived restorativeness and NA ($r = -0.039$, $p > 0.05$). Moreover, the psychological recovery indicators were also significantly correlated with the physiological recovery indicators, except MAP and NA ($r = 0.127$, $p > 0.05$) (Table 5).

**Table 5.** Spearman's relationships between psycho-physiological recovery and perceived restorativeness.

|  | MAP | HR | PA | NA |
|---|---|---|---|---|
| MAP |  |  |  |  |
| HR | 0.210 * |  |  |  |
| PA | 0.284 ** | 0.186 ** |  |  |
| NA | 0.224 * | 0.242 ** | 0.370 ** |  |
| ROS | 0.243 ** | 0.175 * | 0.311 ** | −0.039 |

$*\ p < 0.05$; $**\ p < 0.01$ (2-tailed). MAP = mean arterial pressure, HR = heart rate, PA = positive affect, NA = negative affect, and ROS = Restorative Outcome Scale.

## 4. Discussion

In this study, a laboratory experiment was conducted to examine the effects of audio–visual scenarios on psycho-physiological recovery of older adults in high-density residential public space. Here, we first assess how our findings address our key research questions. We then discuss the implications for planning and design. Finally, we consider the limitations of our study and directions for future research.

### 4.1. Benefits of Residential Public Space on Older Adults' Restoration

The first aim of this study was to confirm whether older adults could recover from a state of stress and fatigue after exposure to a scenario of residential public space. The results of the physiological measures were largely in line with our expectations, indicating that a brief exposure period to the green scenario in residential environments could induce significant positive changes in blood pressure and heart rate in older adults. This finding was also consistent with previous studies suggesting that built-up urban environments with greenery views have positive effects on physiological stress [36]. In addition, we could observe from Figure 4 that the mean values of physiological indicators of post-recovery stress levels went back to, or even lower than, their baseline measures for those elderly participants in the activity and green scenes, indicating complete recovery. Overall, however, the present findings in relation to MAP and HR are consistent with SRT, which suggests that natural landscapes produce immediate affective responses, and that natural views are effective for recovery from stress [41].

Additionally, it is important to note that there was a tendency for significant MAP and HR recovery in the activity scene, indicating that adding activity-based elements could effectively promote the physiological restoration of the residential public space. The restorative effects induced by the activity scene on the blood pressure and heart rate seemed to confirm the outcomes of a previous study, suggesting that well-designed residential environments incorporating fitness facilities and furniture could be perceived by older people as a potential physical activity possibility and, consequently, generate a restorative experience [42].

Results from the positive mood evaluations were similar to the physiological outcomes in that both the green and activity scenes could significantly promote the elderly participants' positive mood, but not the square scene. All the results confirmed that green-based and activity-based public spaces in residential areas could provide restorative effects for older adults. However, it is interesting to note that the negative mood significantly recovered after all experimental conditions, no matter what audio–visual stimuli the participants were exposed to. One possible reason is that the elderly participants in our study generally displayed a low level of negative mood, even after the stress induction, as shown in Table 3. Older people with high levels of negative mood (e.g., anxiety and depression) should be further studied in future research.

*4.2. Different Restorative Effects among Audio–Visual Conditions*

The results from the overall restorative perceptions answered the second research question, indicating that the green scene was perceived as more restorative than other scenes in residential public space. Along the same line, older adults experienced a larger restorative benefit from nature sounds compared to traffic noise. The subjective evaluation outcome was also strongly backed up by physio-psychological evidence, including blood pressure, heart rate, and positive mood.

As for the visual scene, we found that the activity scene was perceived as much more restorative than the square scene by the elderly participants. This result is consistent with a previous study, which suggested that older people preferred colorful parks to squares [43]. However, the actual restorative advantage of the activity scene over the square scene was only identified in terms of the mood recovery, rather than the physiological recovery. It is possible that the activity scene was generally characterized as positive valence but high arousal (e.g., exciting) in the view of older adults. Therefore, although exposure to the activity scene could effectively promote psychological recovery, the physiological measures were still on a rather high level.

Regarding the effects of acoustic stimuli, the restorative advantage of nature sounds over traffic noise was mainly demonstrated in the HR and PA recovery, indicating that natural sounds, including birdsong and water sounds, can effectively reduce autonomic nervous system activation and promote positive mood after acute stress. This conclusion is reinforced by many previous studies, which identified significantly greater HR recovery when participants were exposed to natural sounds rather than urban noise [44,45]. However, the effect sizes of the sounds were generally small ($\eta^2_p < 0.01$), indicating that the effect of sounds on the elderly participants' restorative experience was less strong than the visual scenes. This was different from a prior study, which found that elderly participants assigned more importance to the acoustic characteristics of scenarios than the visual elements [43]. A possible reason is that traffic noise is ubiquitous in high-rise residential areas, and the elderly are quite accustomed to the presence of traffic noise in such residential contexts. Therefore, instead of regarding the traffic sounds as environmental noise, participants rather took them as normal environmental sounds in urban residential contexts, which was quite familiar and relaxing for the elderly, although it was subjectively evaluated to be much less restorative than the natural sounds. Consistent with this interpretation, Alvarsson et al. also noted that the differences between nature sounds (50 dBA) and low traffic noise (50 dBA) did not reach significance in the recovery of SCL [26]. In addition, it is also possible that visual stimuli (composed of trees, plants, water, sky, pavilion, etc.) are more complex than auditory stimuli (only implying a category, i.e., traffic noise, birdsong, and water sounds). Thus, participants may have simply spent more time on visual than acoustic information.

Despite this, it is interesting to note that significantly more recovery was identified in the exposure to nature sounds compared to traffic noise in the green scene. However, the advantage of nature sounds was not pronounced in the activity scene. This might be explained in terms of the congruence between the sound and the visual context [46]. In the activity scene, participants felt confused about the experience when they heard a loud birdsong and water sounds, which indicated trees or a water body were somewhere nearby but they were unable to see them in the video. Alternatively, the nature sounds matched the green scenes with green plants and waterscapes. Human sounds, such as talking, laughing, and children playing, were more common in such an activity context, as reported by some participants after the experiments. This finding supported prior research, which reported that the audio–visual coherence indeed plays an important role in restorative perceptions [47].

### 4.3. Implications for Planning and Design

The findings of this study could have implications for enhancing visual and acoustic design in urban residential areas. First, from the view of psycho-physiological restoration for older adults, green parks should be a priority in the planning of residential public spaces, and where this is not possible, activity parks might represent a viable alternative since they could effectively promote older adults' mood states [43]. However, the existence of a green park or activity park itself may not necessarily be a factor guaranteeing a positive effect on the quality of life. Residents' well-being and soundscape should be taken into account in pre-design as well as re-design of urban parks [48], such as involving the user's experience in co-design processes [49], improving sound environment quality through spatial planning and landscape management [50]. Particularly, congruence between visual scenes and acoustic design should be taken into account. In the case of green parks, nature-based sounds play an important role in enhancing the restorative experience. Thus, residential design should increase and enrich plants to attract birds and insects to live. Meanwhile, moving waterscapes should be considered in the residential landscape design to create water sounds, and they could be artificial (e.g., fountains) or natural (e.g., streams) [51]. Moreover, to improve restorative quality in residential public spaces, acoustic strategies should include not only the addition of nature sounds but also the control of traffic noise. In this study, traffic noise with a sound level of 55 dBA had detrimental effects on older adults' positive mood. Therefore, traffic noise in residential public spaces should be strictly limited by planning methods, such as increasing buffer distances or vegetation barriers [52].

### 4.4. Limitations and Future Research

This study was innovative in examining the restorative effect of residential public spaces on the elderly participants' health. Herein, a differentiation was made in audio versus visual effects, and in physio-psychological and self-reported outcomes. Potential limitations of this study are related to the participants and the experiment setting. As for the participants, only elderly persons in a good state of health and abilities were sampled in the current study. Previous research has shown that healthier older people are more likely to perform physical activities [9]. People with disabilities might be in greater need than their able counterparts for the health benefits from green environments. Besides, their unique embodied biographies (e.g., gender, educational level, pre-retirement occupation, income, and cultural background) were not assessed, which could strongly influence elderly individuals' relationships with potentially restorative environments [5]. It is, therefore, important to consider these factors in future studies to promote the generalization of the results to the entire older population. As for the experiment setting, only a few typical audio–visual stimuli were presented due to the limitation of experiment time. Future studies should consider other environmental factors that might influence human–environment interactions, including common sounds in residential areas (human voices, mechanical noise, etc.) and other visual elements (presence of people, cultural landscape, etc.). Additionally, only limited psycho-physiological measures and short-term effects were explored in this study. To obtain a more complete picture of the restorative effects of residential parks on individuals, more objective measures, such as performing cognitive tasks and long-term effects, should be considered in future studies. Despite these limitations, the results from this study provided more evidence for the actual restorative effects of residential environments as well as extended the understanding of restorative environments to older adults.

## 5. Conclusions

In a simulated situation of residential public space, six (three visual × two audio) scenarios were presented to the elderly participants to examine the restorative effects of audio–visual public space on the elderly population's psycho-physiological health. Based on the elderly participants' MAP and HR responses, as well as emotional ratings and perceived restorativeness, the following conclusions could be drawn:

(1) Viewing green scenes in residential public spaces facilitated the most psycho-physiolo gical recovery for the elderly, followed by viewing the activity scenes, while the square scene only showed restorative benefits on the elderly participants' negative mood.

(2) Compared to the traffic noise, adding nature sounds in residential public spaces could promote many more benefits in HR recovery, positive mood promotion, and perceived restorativeness.

(3) The advantage of nature sounds over traffic noise was mainly demonstrated in the green scene, indicating the importance of the audio–visual congruence in residential design.

(4) Visual elements demonstrated a greater impact on the elderly participants' psycho-physiological recovery than the acoustic elements in residential contexts.

**Author Contributions:** Conceptualization, S.S. and X.P.; methodology, S.S., X.P. and X.G.; software, L.M. and J.T.; validation, S.S. and L.M.; formal analysis, S.S. and J.T.; investigation, L.M. and J.T.; resources, S.S. and L.M.; data curation, L.M. and J.T.; writing—original draft preparation, S.S. and L.M.; writing—review and editing, S.S.; visualization, S.S. and J.T.; supervision, S.S.; project administration, S.S., X.P. and X.G.; funding acquisition, S.S. All authors have read and agreed to the published version of the manuscript.

**Funding:** This research was funded by the National Natural Science Foundation of China (grant number: 52308067), the Ministry of Education Humanities and Social Sciences Research Project (grant number: 23YJCZH185), and the Shandong Natural Science Foundation (grant number: ZR2021QE256).

**Data Availability Statement:** The data in this study are available from the authors upon request.

**Acknowledgments:** The authors would like to express their thanks to all the participants who volunteered to take part in the research.

**Conflicts of Interest:** The authors declare no conflicts of interest.

# References


1. World Health Organization. *Ageing and Health*; WHO: Geneva, Switzerland, 2022.
2. Yang, J.; Siri, J.G.; Remais, J.V.; Cheng, Q.; Zhang, H.; Chan, K.K.Y.; Sun, Z.; Zhao, Y.; Cong, N.; Li, X.; et al. The Tsinghua–Lancet Commission on Healthy Cities in China: Unlocking the Power of Cities for a Healthy China. *Lancet* **2018**, *391*, 2140–2184. [CrossRef]
3. Zhang, T.; Tan, H.; Wu, Y.; Han, B.; Wang, T. Urban Older Adults Becoming Unhealthier in Modern China: A Cross-Temporal Meta-Analysis. *Psychol. Rep.* **2016**, *118*, 737–747. [CrossRef] [PubMed]
4. Wen, C.; Albert, C.; Von Haaren, C. The Elderly in Green Spaces: Exploring Requirements and Preferences Concerning Nature-Based Recreation. *Sustain. Cities Soc.* **2018**, *38*, 582–593. [CrossRef]
5. Lee, H.J.; Lee, D.K. Do Sociodemographic Factors and Urban Green Space Affect Mental Health Outcomes among the Urban Elderly Population? *Int. J. Environ. Res. Public Health* **2019**, *16*, 789. [CrossRef]
6. Zhifeng, W.; Yin, R. The Influence of Greenspace Characteristics and Building Configuration on Depression in the Elderly. *Build. Environ.* **2021**, *188*, 107477. [CrossRef]
7. Zhou, R.; Zheng, Y.J.; Yun, J.Y.; Wang, H.M. The Effects of Urban Green Space on Depressive Symptoms of Mid-Aged and Elderly Urban Residents in China: Evidence from the China Health and Retirement Longitudinal Study. *Int. J. Environ. Res. Public Health* **2022**, *19*, 717. [CrossRef] [PubMed]
8. Slawsky, E.D.; Hajat, A.; Rhew, I.C.; Russette, H.; Semmens, E.O.; Kaufman, J.D.; Leary, C.S.; Fitzpatrick, A.L. Neighborhood Greenspace Exposure as a Protective Factor in Dementia Risk among U.S. Adults 75 Years or Older: A Cohort Study. *Environ. Heal. A Glob. Access Sci. Source* **2022**, *21*, 14. [CrossRef]
9. Xiao, Y.; Miao, S.; Zhang, Y.; Xie, B.; Wu, W. Exploring the Associations between Neighborhood Greenness and Level of Physical Activity of Older Adults in Shanghai. *J. Transp. Health* **2022**, *24*, 101312. [CrossRef]
10. Takano, T.; Nakamura, K.; Watanabe, M. Urban Residential Environments and Senior Citizens' Longevity in Megacity Areas: The Importance of Walkable Green Spaces. *J. Epidemiol. Community Health* **2002**, *56*, 913–918. [CrossRef]
11. Hartig, T. Restorative Environments. *Encycl. Appl. Psychol.* **2004**, *3*, 273–279. [CrossRef]
12. Ulrich, R.S. Effects of Interior Design on Wellness: Theory and Recent Scientific Research. *J. Health Care Inter. Des.* **1991**, *3*, 97–109. [PubMed]
13. Kaplan, R.; Kaplan, S. *The Experience of Nature: A Psychological Perspective*; Cambridge University Press: Cambridge, UK, 1989; ISBN 9780521341394.
14. Berto, R. Assessing the Restorative Value of the Environment: A Study on the Elderly in Comparison with Young Adults and Adolescents. *Int. J. Psychol.* **2007**, *42*, 331–341. [CrossRef]
15. Scopelliti, M.; Giuliani, M.V. Restorative Environments in Later Life: An Approach to Well-Being from the Perspective of Environmental Psychology. *J. Hous. Elder.* **2005**, *19*, 203–226. [CrossRef]
16. Qiu, L.; Chen, Q.; Gao, T. The Effects of Urban Natural Environments on Preference and Self-Reported Psychological Restoration of the Elderly. *Int. J. Environ. Res. Public Health* **2021**, *18*, 509. [CrossRef] [PubMed]
17. Boffi, M.; Pola, L.G.; Fermani, E.; Senes, G.; Inghilleri, P.; Piga, B.E.A.; Stancato, G.; Fumagalli, N. Visual Post-Occupancy Evaluation of a Restorative Garden Using Virtual Reality Photography: Restoration, Emotions, and Behavior in Older and Younger People. *Front. Psychol.* **2022**, *13*, 927688. [CrossRef] [PubMed]
18. Lu, S.; Oh, W.; Ooka, R.; Wang, L. Effects of Environmental Features in Small Public Urban Green Spaces on Older Adults' Mental Restoration: Evidence from Tokyo. *Int. J. Environ. Res. Public Health* **2022**, *19*, 5477. [CrossRef] [PubMed]
19. Yuan, S.; Tao, F.; Li, Y. The Restorative Effects of Virtual Reality Forests on Elderly Individuals during the COVID-19 Lockdown. *J. Organ. End User Comput.* **2022**, *34*, 1–22. [CrossRef]
20. Yu, C.-P.; Lee, H.-Y.; Lu, W.-H.; Huang, Y.-C.; Browning, M.H.E.M. Restorative Effects of Virtual Natural Settings on Middle-Aged and Elderly Adults. *Urban For. Urban Green.* **2020**, *56*, 126863. [CrossRef]
21. Elsadek, M.; Shao, Y.; Liu, B. Benefits of Indirect Contact With Nature on the Physiopsychological Well-Being of Elderly People. *Health Environ. Res. Des. J.* **2021**, *14*, 227–241. [CrossRef]
22. Roe, J.; Mondschein, A.; Neale, C.; Barnes, L.; Boukhechba, M.; Lopez, S. The Urban Built Environment, Walking and Mental Health Outcomes Among Older Adults: A Pilot Study. *Front. Public Health* **2020**, *8*, 528. [CrossRef]
23. Li, H.; Liu, H.; Yang, Z.; Bi, S.; Cao, Y.; Zhang, G. The Effects of Green and Urban Walking in Different Time Frames on Physio-Psychological Responses of Middle-Aged and Older People in Chengdu, China. *Int. J. Environ. Res. Public Health* **2021**, *18*, 90. [CrossRef]
24. Hedblom, M.; Gunnarsson, B.; Iravani, B.; Knez, I.; Schaefer, M.; Thorsson, P.; Lundström, J.N. Reduction of Physiological Stress by Urban Green Space in a Multisensory Virtual Experiment. *Sci. Rep.* **2019**, *9*, 10113. [CrossRef]
25. Yao, Y.; Xu, C.; Yin, H.; Shao, L.; Wang, R. More Visible Greenspace, Stronger Heart? Evidence from Ischaemic Heart Disease Emergency Department Visits by Middle-Aged and Older Adults in Hubei, China. *Landsc. Urban Plan.* **2022**, *224*, 104444. [CrossRef]
26. Alvarsson, J.J.; Wiens, S.; Nilsson, M.E. Stress Recovery during Exposure to Nature Sound and Environmental Noise. *Int. J. Environ. Res. Public Health* **2010**, *7*, 1036–1046. [CrossRef] [PubMed]
27. Ma, H.; Shu, S. An Experimental Study: The Restorative Effect of Soundscape Elements in a Simulated Open-Plan Office. *Acta Acust. United Acust.* **2018**, *104*, 106–115. [CrossRef]

28. Preis, A.; Kociński, J.; Hafke-Dys, H.; Wrzosek, M. Audio-Visual Interactions in Environment Assessment. *Sci. Total Environ.* **2015**, *523*, 191–200. [CrossRef] [PubMed]
29. Carles, J.L.; Barrio, I.L.; de Lucio, J.V. Sound Influence on Landscape Values. *Landsc. Urban Plan.* **1999**, *43*, 191–200. [CrossRef]
30. Zhao, J.; Xu, W.; Ye, L. Effects of Auditory-Visual Combinations on Perceived Restorative Potential of Urban Green Space. *Appl. Acoust.* **2018**, *141*, 169–177. [CrossRef]
31. Deng, L.; Luo, H.; Ma, J.; Huang, Z.; Sun, L.-X.; Jiang, M.-Y.; Zhu, C.-Y.; Li, X. Effects of Integration between Visual Stimuli and Auditory Stimuli on Restorative Potential and Aesthetic Preference in Urban Green Spaces. *Urban For. Urban Green.* **2020**, *53*, 126702. [CrossRef]
32. Herzog, T.R.; Rector, A.E. Perceived Danger and Judged Likelihood of Restoration. *Environ. Behav.* **2009**, *41*, 387–401. [CrossRef]
33. Li, H.; Dong, W.; Wang, Z.; Chen, N.; Wu, J.; Wang, G.; Jiang, T. Effect of a Virtual Reality-Based Restorative Environment on the Emotional and Cognitive Recovery of Individuals with Mild-to-Moderate Anxiety and Depression. *Int. J. Environ. Res. Public Health* **2021**, *18*, 53. [CrossRef]
34. Browning, M.H.E.M.; Saeidi-Rizi, F.; McAnirlin, O.; Yoon, H.; Pei, Y. The Role of Methodological Choices in the Effects of Experimental Exposure to Simulated Natural Landscapes on Human Health and Cognitive Performance: A Systematic Review. *Environ. Behav.* **2020**, *53*, 687–731, ISBN 0013916520906. [CrossRef]
35. Watson, D.; Clark, L.A. Development and Validation of Brief Measures of Positive and Negative Affect: The PANAS Scales. *J. Pers. Soc. Psychol.* **1988**, *54*, 1063–1070. [CrossRef] [PubMed]
36. Tyrväinen, L.; Ojala, A.; Korpela, K.; Lanki, T.; Tsunetsugu, Y.; Kagawa, T. The Influence of Urban Green Environments on Stress Relief Measures: A Field Experiment. *J. Environ. Psychol.* **2014**, *38*, 1–9. [CrossRef]
37. Gaekwad, J.S.; Sal Moslehian, A.; Roös, P.B.; Walker, A. A Meta-Analysis of Emotional Evidence for the Biophilia Hypothesis and Implications for Biophilic Design. *Front. Psychol.* **2022**, *13*, 2476. [CrossRef] [PubMed]
38. Korpela, K.M.; Ylén, M.; Tyrväinen, L.; Silvennoinen, H. Determinants of Restorative Experiences in Everyday Favorite Places. *Heal. Place* **2008**, *14*, 636–652. [CrossRef] [PubMed]
39. Subiza-Pérez, M.; Pasanen, T.; Ratcliffe, E.; Lee, K.; Bornioli, A.; de Bloom, J.; Korpela, K. Exploring Psychological Restoration in Favorite Indoor and Outdoor Urban Places Using a Top-down Perspective. *J. Environ. Psychol.* **2021**, *78*, 101706. [CrossRef]
40. Han, K.T. Psychophysiological Effects of Different Methods of Inducing Restoration Needs. *Psychol. Rep.* **2021**, *124*, 131–162, ISBN 0033294119. [CrossRef]
41. Ulrich, R.S. View through a Window May Influence Recovery from Surgery. *Science* **1984**, *224*, 420–421. [CrossRef]
42. Fu, E.; Zhou, J.; Ren, Y.; Deng, X.; Li, L.; Li, X.; Li, X. Exploring the Influence of Residential Courtyard Space Landscape Elements on People's Emotional Health in an Immersive Virtual Environment. *Front. Public Health* **2022**, *10*, 1017993. [CrossRef]
43. Ruotolo, F.; Rapuano, M.; Masullo, M.; Maffei, L.; Ruggiero, G.; Iachini, T. Well-Being and Multisensory Urban Parks at Different Ages: The Role of Interoception and Audiovisual Perception. *J. Environ. Psychol.* **2023**, *16*, 102219. [CrossRef]
44. Medvedev, O.; Shepherd, D.; Hautus, M.J. The Restorative Potential of Soundscapes: A Physiological Investigation. *Appl. Acoust.* **2015**, *96*, 20–26. [CrossRef]
45. Wang, P.; He, Y.; Yang, W.; Li, N.; Chen, J. Effects of Soundscapes on Human Physiology and Psychology in Qianjiangyuan National Park System Pilot Area in China. *Forests* **2022**, *13*, 1461. [CrossRef]
46. Li, H.; Lau, S.K. A Review of Audio-Visual Interaction on Soundscape Assessment in Urban Built Environments. *Appl. Acoust.* **2020**, *166*, 107372. [CrossRef]
47. Shu, S.; Ma, H. Restorative Effects of Urban Park Soundscapes on Children's Psychophysiological Stress. *Appl. Acoust.* **2020**, *164*, 107293. [CrossRef]
48. Jaszczak, A.; Pochodyła, E.; Kristianova, K.; Małkowska, N.; Kazak, J.K. Redefinition of Park Design Criteria as a Result of Analysis of Well-Being and Soundscape: The Case Study of the Kortowo Park (Poland). *Int. J. Environ. Res. Public Health* **2021**, *18*, 2972. [CrossRef] [PubMed]
49. Fumagalli, N.; Fermani, E.; Senes, G.; Boffi, M.; Pola, L.; Inghilleri, P. Sustainable Co-Design with Older People: The Case of a Public Restorative Garden in Milan (Italy). *Sustainability* **2020**, *12*, 3166. [CrossRef]
50. Jaszczak, A.; Małkowska, N.; Kristianova, K.; Bernat, S.; Pochodyła, E. Evaluation of Soundscapes in Urban Parks in Olsztyn (Poland) for Improvement of Landscape Design and Management. *Land* **2021**, *10*, 66. [CrossRef]
51. Lugten, M.; Karacaoglu, M.; White, K.; Kang, J.; Steemers, K. Improving the Soundscape Quality of Urban Areas Exposed to Aircraft Noise by Adding Moving Water and Vegetation. *J. Acoust. Soc. Am.* **2018**, *144*, 2906–2917. [CrossRef]
52. Joynt, J.L.R.; Kang, J. The Influence of Preconceptions on Perceived Sound Reduction by Environmental Noise Barriers. *Sci. Total Environ.* **2010**, *408*, 4368–4375. [CrossRef]

*Article*

# Expanding the Associations between Landscape Characteristics and Aesthetic Sensory Perception for Traditional Village Public Space

Guodong Chen, Jiayu Yan, Chongxiao Wang and Shuolei Chen *

College of Landscape Architecture, Nanjing Forestry University, 159 Longpan Rd., Nanjing 210037, China; chen35@njfu.edu.cn (G.C.); yussie@njfu.edu.cn (J.Y.); chongxiao@njfu.edu.cn (C.W.)
* Correspondence: shuoleichen@njfu.edu.cn; Tel.: +86-131-8295-0689

**Abstract:** Traditional village landscapes have a cultural and regional significance, and the visual aesthetic quality of the landscape is widely regarded as a valuable resource to benefit the health and well-being of urban residents. Although the literature has analyzed the influential mechanism of landscape features on aesthetic senses, most were from a single dimension. To improve the precision of the landscape aesthetic evaluation method, this study expanded the indicators for landscape characteristics of public spaces in traditional villages by incorporating multiple dimensions, such as landscape visual attraction elements and landscape color. It explored their associations with sensory preferences in a case study in Dongshan (a peninsula) and Xishan (an island) of Taihu Lake. We used multi-source data, a semantic segmentation model, and R language to identify landscape characteristic indicators quantitatively. The research results indicated that the accuracy of the aesthetic sensory assessment model integrating multi-dimensional landscape characteristic indicators was significantly improved; in the open space of traditional villages, the public preferred a scenario with a high proportion of trees, relatively open space, mild and uniform color tones, suitability for movement, and the ability to produce a restorative and peaceful atmosphere. This study can provide a guarantee for the efficient use of village landscape resources, the optimization of rural landscapes, and the precise enhancement of traditional village habitat.

**Keywords:** aesthetic sensory; landscape characteristics; traditional villages; public space

**Citation:** Chen, G.; Yan, J.; Wang, C.; Chen, S. Expanding the Associations between Landscape Characteristics and Aesthetic Sensory Perception for Traditional Village Public Space. *Forests* **2024**, *15*, 97. https://doi.org/10.3390/f15010097

Academic Editors: Jiang Liu, Xinhao Wang and Xin-Chen Hong

Received: 1 December 2023
Revised: 31 December 2023
Accepted: 2 January 2024
Published: 4 January 2024

## 1. Introduction

In comparison to modern cities, there are rich natural and civilized resources in traditional villages, with important cultural and regional significance [1]. As an important part of urban and rural ecosystems, rural landscapes can provide a variety of ecosystem services, such as improving human mental, physical status, and well-being [2,3]. The aesthetic experience of rural landscapes can effectively relieve the psychological pressure of urban residents and provide them with the opportunity to escape the hustle and bustle of the city and enjoy the natural environment in the context of rapid urbanization and expansion [4,5]. Rural landscapes and urban landscapes complement each other. The interaction between the two helps to meet the needs of the growing population, protect the culture and natural resources of rural areas, and realize the sustainable development of both areas [6]. On the other hand, commercial development and industrialization have had fierce impacts on the traditional village landscape in the market economic environment [7]. As a result, the landscapes of traditional villages face a series of problems, such as regional recession, increased homogeneity, and separation of tradition and modernity [8]. The government and all walks of life are deeply concerned about improving the quality of the traditional village landscape environment. The public space of traditional villages, as an essential part of the village landscape, is a key part of the quality improvement of the traditional village landscape [9]. Public space can be interpreted as a gathering place that promotes and facilitates social interaction [9]. The traditional village public space refers mainly to the space where all villagers and tourists can freely enter for daily activities, communication, and relaxation, such as squares, ancient wells, piers, and other spaces [9,10]. The European Landscape Convention defined landscape as an area that is perceived by people, characterized by the interaction of natural and human factors or by the impact of one of them [11,12]. According to the definition of landscape, the core of it is people [13]. Therefore, human sensory perception is an important part of the landscape. Landscape aesthetic sensory perception means observers' good and bad feelings towards the landscape after a series of perceptions, cognition, and some other psychological assessments [14]. The landscape aesthetic sensory judgment is an important tool to assist decision making in landscape planning [15]. Understanding people's aesthetic sensory perception can help planners and designers create a more attractive, sustainable, and culturally appropriate rural landscape [16]. Meanwhile, it can help meet the expectations of residents and visitors and realize the multiple goals of environmental protection, sustainable development, community interaction, and so on [17–19]. Based on the study of the relationship between landscape aesthetic preferences, Arriaza et al., Hernández et al., and Yao et al. have effectively improved urban and rural landscapes and environments [19–21]. Therefore, understanding the spatial landscape elements in the public space of traditional villages that positively influence public aesthetic sensory perception, quantitatively identifying them, and understanding the influence mechanism can provide a basis for the protection and enhancement of the traditional village landscape and construction management.

There are four recognized academic schools for the judgement of landscape aesthetic preference: the expert school, the psychophysical school, the cognitive school, and the empirical school, of which the psychophysical school is the most widely used [9]. This study is based on the theory of the psychophysical school. They believe that landscape aesthetic activity is a visually oriented perception process in which the aesthetic object interacts with the aesthetic subject under the influence of the aesthetic psychological structure. Aesthetic values projected by a subject on an object are culturally conditioned and are subject to intergenerational change [22]. Therefore, aesthetic sensory perception is jointly influenced by the characteristics of the aesthetic subject, such as cultural background, psychological needs, mental state, emotional experience, and other factors, as well as the characteristics of the aesthetic object, such as form, architectural style, color, texture, and other factors [23–25]. Based on the psychophysical paradigm, the aesthetic preference judgment model is mainly divided into three steps: the first step is to construct the landscape feature characterization system; the second step is to evaluate and rate the landscape environment by the aesthetic subject; and the last step is to construct a functional relationship model between the landscape features and the aesthetic preference [26]. Therefore, the construction of a landscape feature characterization system is the premise and foundation of aesthetic preference judgment. The system contains two levels of content. On one hand, it is the identification of the types of elements in the environment. On the other hand, it is the description of the characteristics presented by the landscape elements. Studies by Qin et al. showed that the elements of mountains, trees, and water bodies had a positive impact on the aesthetic preference of road landscape, in which the green view rate (GVR) was significantly related to landscape preference [27]. Li et al. used eye-tracking technology to identify the significant influence of trees, water bodies, and hard paving on the public's aesthetic preferences [28]. López-Martínez explored the public's visual perceptual preferences for Mediterranean landscapes based on landscape photographs. The final results showed that water and vegetation fundamentally contributed to positive evaluation of the overall landscape scene. In summary, the category and type of landscape elements can influence public landscape preferences, with plant elements being a significant factor [29]. Actually, in the study of traditional villages, the plant landscape was not as typical in terms of its material appearance as historical buildings and buildings under the protection of cultural relics. So, the plant landscape was often neglected. At present, meso- and macro-scale landscape element recognition are mainly based on the use of high-resolution satellite images and remote sensing data [30,31]. Surface elements can be automatically selected based on color, texture, shape, edge, spectral reflectance, and other features combined with different classification algorithms [9,32,33]. The recognition accuracy of this method is affected by the resolution of remote sensing data, the selection of classification algorithms, and the accuracy of feature extraction methods [34,35]. The identification of microscale landscape elements is mainly performed by using machine learning and model training based on magnanimous photographs, such as micro-scale landscape element recognition which mainly uses machine learning and model training, such as supporting vector machine (SVM), decision tree, convolutional neural network (CNN), and other models to realize automatic extraction of element categories [36,37]. The recognition accuracy of this class of methods depends on the quality of the features, models, and applied training data.

The material elements of the landscape are the basis of the landscape environment. The attribute characteristics shown by different elements have different effects on landscape aesthetic sensory perception. According to environmental psychology, aesthetic sensory perception is a process in which human vision, hearing, touch, smell, and taste work together, among which the information obtained through vision reaches 87% [38]. Vision is the most direct and effective way of perceiving the landscape. Zhang et al. explored the relationship between four spatial visual attributes, including the openness of visual scale, the richness of composing elements, the orderliness of organization, the depth of view, and landscape preference through realistic photographs. The results showed that the public preferred landscape scenes with high openness and orderliness. Moreover, the high richness of composing elements could positively affect the preference when the landscape is in good order [39]. Rechtman explored the relationship between field size, lot shape, land texture, crop texture built elements, and visual sensory preference of agricultural farming landscape based on photographic works. The results showed that land textures, crop textures, and lot shapes could help explain the visual preference of agricultural farming landscapes [38]. Chen et al. investigated the relationship between public space patterns in traditional villages and landscape aesthetic preference based on radar point cloud data. The research showed that average contour upper height, solid-space ratio, vegetation cover, and comprehensive closure are four indicator factors that significantly correlated with aesthetic preference [9]. Huang et al. used eye-tracking technology to explore the relationship between landscape features, preference, and viewing behavior. Their results showed that more drastic hue variation and chromaticity were conducive to visual fixation. There was a close relationship between landscape preference and the number of gazes in mountainous, aquatic, and forest landscapes [40]. Cao et al. investigated the effect of color block patterns on landscape preference in suburban forests. The research showed that the average area of the color blocks was positively related to landscape preference, and the number of color blocks, maximum patch index, and standard deviation of patch size were negatively related to landscape preference [41]. These above studies showed that landscape color features and spatial form features had a significant effect on landscape aesthetic preference. In these cases, color and spatial form were mostly discussed separately, while the effect of object features on landscape aesthetic preference was explored from a single dimension. Currently, there is no unified standard for identifying plant colors. The previous research mainly focused on qualitative description. Color extraction technology is a method of processing images through computers to identify and extract specific color information. The basic principle is to map the pixels in the image to the color space and extract the colors of interest from it by setting thresholds, clustering algorithms, or color histograms [42]. With the development of color theory and color extraction technology, the colorimetric method, instrumental measurement method, and software extraction method have been widely used. The first colorimetric method is mainly based on the Royal Horticultural Society (RHS) color card, natural color system (NCS) color card, Munsell color card, etc., for color extraction, which is suitable for collecting a large amount of color data; the second instrumental extraction method is mainly performed by using a color measuring instrument such as colorimeter, chromameter, spectroradiometer, etc., which is suitable for color extraction of plant organs nearby; the third software extraction method is mainly based on photo images, using image processing software such as Colorimpact 4, Photoshop CC 2018, and other software for color identification, which are simple and easy to use [42]. Most of the existing studies were aimed at the quantitative identification of a single plant, and few of them involved the color extraction of plant landscape communities. The landscape of plant communities presented irregular three-dimensional spatial patterns. Existing studies generally used tools such as a tape measure, measuring tape, infrared rangefinder, and camera to map and record the sample plots and to describe the flat surface and elevation patterns of the sample plots with the help of AutoCAD 2016, Photoshop CC 2018, SketchUp 2018, and other drawing software [9]. It is difficult to quantitatively describe the three-dimensional spatial pattern indexes. It requires a lot of time and labor costs, and lacks timeliness as well. The Scenic Beauty Evaluation (SBE) proposed based on the psychophysical paradigm proposed by Daniel and Boster is currently the most common method for judging aesthetic preferences [43]. It is widely applied in various types of landscapes, such as rural settlements, roads, waterfronts, settlements, national parks, and so on [19,44,45]. Taking into account time and economic costs, existing studies have mainly used landscape photographs as the evaluation medium, but the content of traditional photos was limited by the angle of view, making it difficult to show the panoramic view. Currently, judging the visual quality of landscapes based on the scenic beauty evaluation method mainly involves calculating the evaluator's composite score for the scene's environment, which is invariably limited by the sample data volume.

Based on the above analyses, landscape features in the three dimensions of landscape components, colors, and spatial forms of plant landscape all have an impact on public landscape aesthetic sensory perception. Existing studies mainly remain on the topic of qualitative description of landscape characteristics and explore the relationship between landscape characteristics and aesthetic preference from a single dimension, which often leads to a low prediction model of landscape aesthetic preference, and makes it difficult to effectively explain the main landscape feature indicators affecting aesthetic sensory perception.

To address these research gaps, this study aims to improve the accuracy of the landscape aesthetic sensory assessment methods from both the construction of the landscape characteristic index system and landscape preference judgment. First, based on the previous single dimension of spatial form features, landscape components and plant color features are added to expand the landscape special index system. A quantitative description of indicators is achieved with the help of digital technology. At the same time, the traditional beauty degree evaluation is improved, and the score of each evaluation subject for the scene environment is calculated, expanding the sample data volume. Finally, the relationship between multi-dimensional landscape characteristics and landscape aesthetic preference is constructed. It will provide a theoretical basis and references for the refined conservation and regeneration of the landscape of traditional villages.

## 2. Materials and Methods

### 2.1. Research Area

Dongshan and Xishan are located in the southwest of Suzhou, on the east bank of Taihu Basin, which is close to Shanghai. Suzhou is located in the southeast of Jiangsu Province, which belongs to the eastern coastal area of China. The region has a long history of congregation which can be traced back to the earliest Spring and Autumn Period in the Wuyue Kingdom. A dozen ancient villages are distributed in Dongshan and Xishan, and most of them lie in front of mountains and boast rivers around, showing distinguishing regional characteristics. Dongshan is a peninsula that extends into Taihu Lake, surrounded by water on three sides, with a total area of 96.6 square kilometers. The existing resident population of it is more than 5300. Xishan Island belongs to the islands in the lake, with an area of about 79.8 square kilometers, and the current population of it is about 45,000 people. Four representative traditional villages at national level with relatively well-preserved historical features, namely Yangwan Village, Wengxiang Village, Dongcun Village, and Zhili, were selected for this study. Yangwan has an area of about 11.86 square kilometers and a resident population of more than 3600; Wengxiang has an area of 3.77 square kilometers and a resident population of approximately 1000; Dongcun has an area of 0.07 square kilometers and a population of about 700; and Zhili covers an area of about 2.1 square kilometers and has a population of approximately 2000. The study selected 31 typical open outdoor spaces in villages where residents and visitors carry out daily communication, activities, and recreation based on field visits (Figure 1) [9], and the boundaries of the sample plots were limited to forest edges, road edges, or corner lines of building side walls.

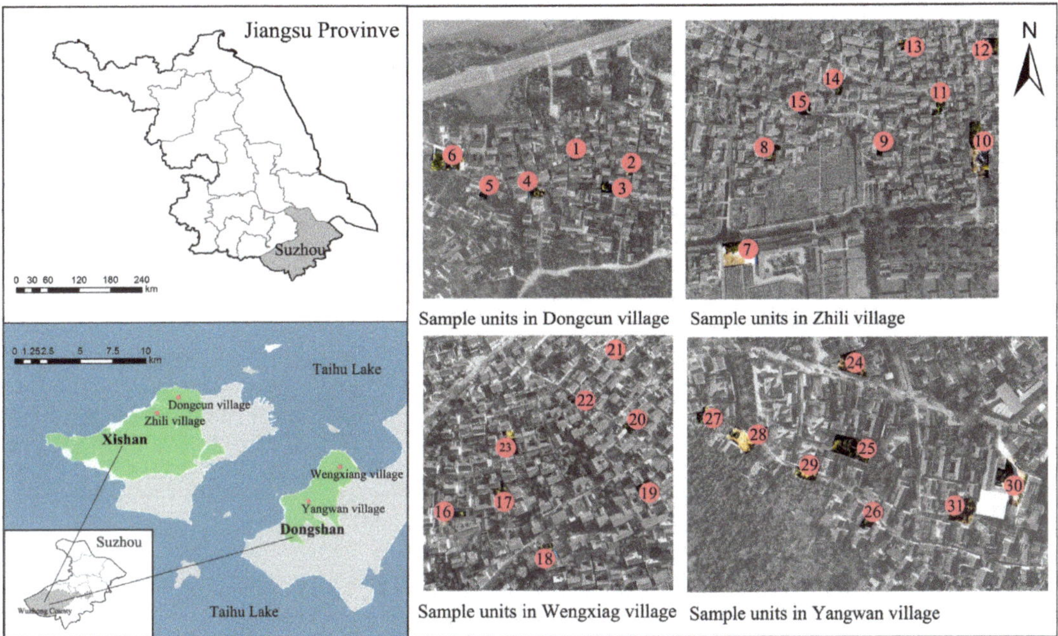


**Figure 1.** Thirty-one sample sites in four traditional villages.

## 2.2. Data Collection

To restore the true feelings of people in open space, this study was based on panoramic pictures to judge aesthetic preferences and to identify landscape elements and color features. Panoramic photographs were taken using a fixed standard of shooting to simulate the human point of view and comprehensively display the landscape features; the cameras were placed in the center of the scene at the height of 1.6 m; and the photographs were taken with the same Insta360 ONE X2, which was connected to the cell phone with the positioning enabled by Bluetooth. It helped to locate the geographic coordinates and the photographs were taken over 3 days from 23–25 November 2021, 9:30–11:30 a.m. and 2:00–4:00 p.m. During this period, the weather conditions were favorable and climatic conditions were similar.

A handheld 3D laser scanner, model GEOSLAM ZEB-HORIZON, collected spatial morphology data. The experimental staff member held the instrument in front of their chest, then started walking from the starting point around the field. Then, he returned to the origin to form a closed loop. Data collection was completed in this way. The spatial collection of morphology data was completed at the same time.

*2.3. Research Methods*

The research was structured into five steps. In the initial stage, a semantic segmentation model was employed to quantitatively identify visual attraction elements within traditional village public space landscapes, utilizing panoramic photos as the primary data source. Subsequently, in the second step, a combination of colorimetric and software methods was applied to extract the color information of plant communities through Colorimpact 4 software (Tiger Color, Akershus, Norway). The color characteristics were then quantitatively described based on the Munsell color card theory. Moving on to the third step, three-dimensional laser scanning technology was introduced. The irregular three-dimensional space of plant landscapes was characterized using the R language 4.1.0 (R Core Team, Vienna, Austria). The fourth step involved an enhancement of traditional beauty evaluation methods, integrating virtual reality technology to assess the aesthetic preferences of the landscape scenes. Lastly, predictive models for aesthetic sensory perception, tailored to different scenarios, were developed.

2.3.1. Identification of Visually Attractive Elements of the Landscape

1.    Image Pre-processing

With color panoramic photographs as a medium, we used the image analysis method during the research to evaluate the quality of rural plant landscapes. As a proxy for real-life scenes, pictures could effectively measure the psychological and aesthetic feedback of the visitors. In addition, with a wide field of view, panoramic images could comprehensively record the study site's visual information and facilitate their quantitative analysis by computer vision techniques. About 28.8% of the periphery of the panoramic images had severe distortions. In contrast, the central portion of the camera lens with a vertical field of view spacing of $\pm 30°$ had less distortion and better matched the visual range of human eyes [46,47]. We referred to the method of Li Yin and Zhenxin Wang to exclude the most distorted part of the image caused by the camera lens by cropping out part of the image. This method could retain the observation content closer to the human perspective [48]. The vision frame showed the view areas reflecting eyelevel equivalent pedestrian experience for three directions: front (A), left (B), and right (C) (e.g., Figure 2). Additionally, the content with the vision frame was low in degree of distortion.

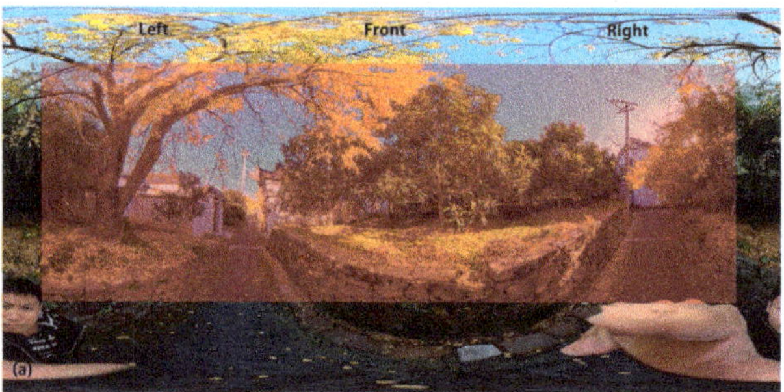

**Figure 2.** Example of Google Street View image and image preprocessing; (**a**) original image (panorama), (**b**) cropped image.

2.  Scene elements identification and extraction

We used the deeplabv3 model trained on the ade20k dataset to extract the scene elements. Ade20k has strong generalization properties, and its extensive use in rural studies verifies its reliability for extracting elements of rural scenes [49]. In addition, the ade20k dataset can identify tree, grass, plant, and sky elements (Table 1), which meets the needs of this study. From the pre-experimental results, image segmentation trained by this dataset was more fine-grained and could accurately outline the countryside plants as well as other elements. As a result, we applied this method to extract each kind of element from all images (e.g., Figure 3).

**Table 1.** Landscape element identification.

| Index | Explanation |
| --- | --- |
| Percentage of structure (Structure) | Structure = $(S_{wall} + S_{building})/S$. In the formula, $S_{structure}$ represents the pixel area of the structure, including walls and buildings; in the scene, S represents the total pixel area of the panoramic image. |
| Percentage of sky (Sky) | Sky = $S_{sky}/S$. In the formula, $S_{sky}$ represents the pixel area of the sky; in the scene, S represents the total pixel area of the panoramic image. |
| Percentage of earth (Earth) | Earth = $S_{earth}/S$. In the formula, $S_{earth}$ represents the pixel area of the earth; in the scene, S represents the total pixel area of the panoramic image. |
| Percentage of grass (Grass) | Grass = $S_{grass}/S$. In the formula, $S_{grass}$ represents the pixel area of the grass; in the scene, S represents the total pixel area of the panoramic image. |

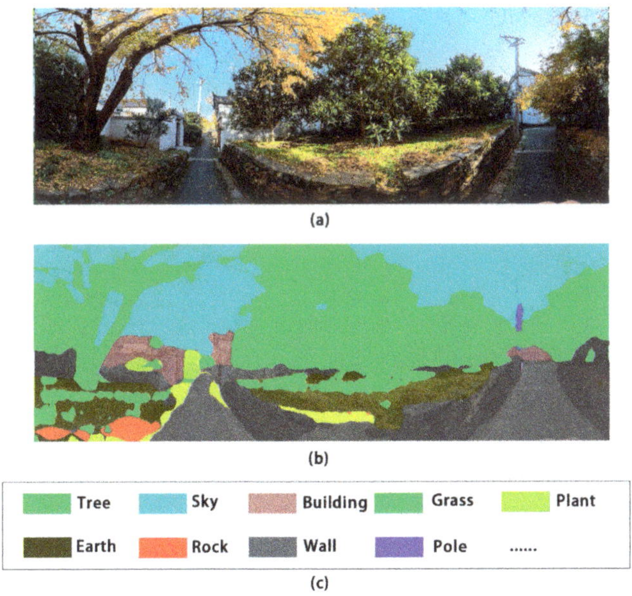


**Figure 3.** Semantic segmentation based on PSPNet. (**a**) Original image before semantic segmentation; (**b**,**c**) the results of semantic segmentation.

2.3.2. Feature Identification of Landscape Colors

Color elements are quantitative indicators that condense most information in the color composition and the color space pattern. The usual means of recording color data in previous studies is to record the RGB values of colors [42]. But, the RGB triple values are in fact not regular. It is challenging to quantify colors from the perspective of the visual sensory characteristics of the human eye, making it difficult to apply the study's results in practice. To solve the above problems, the HSV color model (Hue, Hue, H; Saturation, Saturation, S; Brightness, Value, V) is suitable for the visual characteristics of the human eye. This model was chosen to divide the color threshold interval, which was the one that most closely matched the human eye's ability to perceive colors and can quantify colors non-isotropically and uniformly from the perspective of the human eye's sensory perception of the color characteristics. In conducting this study, the researcher used the ColorImpact 4 software (Tiger Color, Akershus, Norway) to extract the colors present in plant elements. Following this, researchers utilized the quantification method to categorize the non-uniform colors in the HSB color model into intervals of H (Hue): S (Saturation): V (Brightness) = 8:4:4. This process successfully yielded a total of 128 distinct colors (e.g., Figure 4). Representative colors in the same interval range were used to divide the color intervals, which was convenient for quantifying and analyzing the color data in the later stage. The HSV color model was used to describe the detailed color characteristics of plants. According to the non-uniformly quantified color intervals, the colors were divided into different interval ranges, and the three color components were evaluated in a one-dimensional feature vector, that is, L = H × G s × G v + S × G v +V. In the formula, H, S, and V denote the hue, saturation, and luminance, respectively; Gs and Gv denote the number of quantization levels for S and V, which both have 4 levels. Therefore, the final expression was L = 16 H +4 S + V. It could be seen that the weight distribution of the hue is the largest. So, hue was the main factor to distinguish the color characteristics. The related derived indicators were calculated based on the three HSV indicators (e.g., Table 2) [50].

**Figure 4.** Example of color extraction.

**Table 2.** Landscape color recognition.

| Index | Explanation |
|---|---|
| Number of Colors (NC) | NC = SUM(HaSbVc); HaSbVc $\geq$ 1%. NC represents the total number of extracted colors, excluding black, white, and gray, with a pixel percentage of more than 1%. |
| Main Hue Comparison (MHi) | MHi = NMHi/N × 100. NMHiis the pixel of primary color which occupies the largest pixel area in the scene; N is the total number of pixels of the image. |
| Adjacent Hue Comparison (NHi) | NHi = NHi/N × 100. NHi is the pixel of adjacent colors which are within 60 degrees of each other on the left and right of the primary color in the hue circle; N is the total number of pixels of the image. |
| Complementary Hue Comparison (CHi) | CHi = NCH/N × 100. NCH is the pixel of complementary colors which are within 180 degrees of the primary color; N is the total number of pixels in the image N is the total number of pixels of the image. |
| Warm and cool color tone contrast (THi) | THi = NHf/NHw × 100. NHf is the pixel of cool colors; NHw is the pixel of warm colors. |
| Color Diversity Index (H') | $H' = \sum_{i=1}^{S} Pi \ln Pi$; Pi represents the percentage of color i; S represents the total number of extracted colors, S = NC. |
| Color Evenness Index (E') | E' = H'/lnS. H represents the color diversity index; S represents the total number of extracted colors, S = NC. |

### 2.3.3. Identification of the Spatial Morphology Characterization of the Landscape

1.  Data pre-processing

Firstly, the point cloud data were cropped, denoised, and ground points were extracted and normalized to retain the relevant point clouds in the study area while reducing the amount of data. Then, cropped and retained the necessary study objects and separated the hard surfacing point cloud and the tree canopy point cloud for subsequent analyses. Due to the influence of external factors such as human beings or animals, many outliers, namely noisy points, inevitably existed in the point cloud data. To improve the accuracy of subsequent data analysis, noise reduction processing was required. In this study, the distance between points was applied as the primary measure. On the basis of the experience of previous experiments, it was set to be 10 points in surroundings. Then, the median and standard deviation of the average distance of the points in the domain were calculated. Meanwhile, to improve computational efficiency, the box grid filer was set to 0.1 m for filtering based on satisfying the computational accuracy, and the sample point cloud was reduced from ten million orders of magnitude to less than 50 w.

2.  Construction of spatial morphology characterization indexes

This study utilized a traditional index system to characterize the spatial morphology of public areas in the village that have been previously constructed. The system is based on three-dimensions horizontal interface, vertical interface, and three-dimensional spatial level. It included various parameters such as accessible area ratio (AAR), eccentricity rate (E), Spatial Shape Index (SSI), Average Height of Upper Contour (hu), average height of lower contour (hl), solid-space ratio (SVR), contour fluctuation range (FR), fluctuation variance of upper contour (FVU), fluctuation variance of lower contour (FVL), vegetation coverage (VC), plant diversity index (PDI), three-dimensional green visibility (3D-GVI), enclosure degree (ED), and composite closure (CC) [9]. These 14 spatial morphology indicators were used.

3.  Quantitative identification of spatial morphological indicators

Indicators were mainly calculated using Lidar360 v3.2 software combined with R language. The area and length class indicators were calculated by projecting the point cloud of the study area to the XOY plane, and then carrying out edge extraction to identify the edge contour, thus calculating the area within the contour. The height metrics were calculated by outputting the point cloud data as raster data, thus calculating the edge height within the raster. The long and short axes in the site were calculated using the traversal method, which calculated the Euclidean distance from each point to each of the other points, with the maximum value being the long axis and the minimum value being the short axis. The 3D canopy volume was calculated using the $\alpha$-shape method to construct convex packets and accumulate the volume of each convex packet to derive the 3D green volume.

### 2.3.4. Evaluation of Landscape Aesthetic Preference

This research has been approved by the Ethics Review Board of Nanjing Forestry University and the participants have given their informed consent. Scenic beauty estimation (SBE) is widely used to evaluate landscape quality, focusing on visitors' aesthetic feelings for landscape scenes. Considering the evaluator's ability to operate technological products and excluding the influence of utilitarian aesthetics, this study selected a total of 64 students and experts with landscape professional backgrounds, of which 40 were students, and 24 were experts, for scoring. The panorama photos were imported into the Baidu VR platform, and the images were converted to human perspective 360° autonomous rotating VR images, and the evaluators wore VR glasses with the VIVE-VR model for evaluation and scoring. The evaluation rating was divided into 5 levels, with corresponding scores from 1 to 5, indicating dislike very much, dislike, neutral, like, and like very much, respectively. The value of SBE was calculated based on scoring from multi-population for the scenes. In this study, the quality of the 31 rural landscape scenes was audited directly by the results of visitor scoring. As each visitor could be taken into account, compared to calculating the SBE, the method in this study effectively expands the sample size (64 times the SBE).

The validity of visitor scoring has been proved in previous studies related to SBE. This study used one-way ANOVA to test whether there is a difference in the scoring of different scenes. On this basis, this study used the ICC (intraclass correlation coefficient) to evaluate the reliability scale between expert and student scoring on the same scene.

### 2.3.5. Statistics Analysis

1. Data pre-processing

The three types of indicators of the rural plant scene differ greatly in their scale due to their different sources as well as units of measurement. Therefore, they were subjected to maximum–minimum normalization to map the data features into the interval [0, 1] and remove the influence of the scale on the assessment results. The formula is as follows:

$$X_{mmx} = \frac{X - X_{min}}{X_{max} - X_{min}}$$

where $X$ is the original data, $X_{mmx}$ is the normalized data, and $X_{min}$ and $X_{max}$ are the minimum and maximum values of the original data, respectively.

2. Assessment Model Establishment

In previous studies, the excellent assessment capability of a linear model for SBE grade evaluation was validated. This study utilized a multiple linear regression model to predict visitor scoring by selecting indicators such as spatial morphological characteristics, feature composition, and vegetation color characteristics of rural plant scenes as predictors. The parameters of the linear model were solved using the least squares method. The model formula is as follows:

$$\hat{Y} = \beta_1 X_1 + \beta_2 X_2 + \ldots + \beta_n X_n + \beta_0$$

where $\hat{Y}$ is the assessment of visitor scoring, $n$ is the number of predictors included in the model, $X_n$ is the $n$th predictor, and $\beta_n$ is the standardized regression coefficient of the $n$th predictor.

In addition to the full model with all indicators as predictors, the optimized model with streamlined indicators as predictors was established. Based on principal component analysis, indicators with a higher contribution rate were selected for all-subsets regression analysis. Adjusted $R^2$, Mallows' Cp (Cp), and the Bayesian Information Criterion (BIC) were used to determine the optimized model.

### 3. Results

#### 3.1. Result of Public Landscape Preference

Figure 5 illustrates the results of the scores from 40 students and 24 experts. The one-way ANOVA result (F = 147.005, Sig < 0.001) indicated that the difference in scoring across 31 scenes was statistically significant, while the ICC result (ICC = 0.969, $p = 0.00107$) indicated the agreement between experts and students in scoring. In other words, the difference in visitors' aesthetic feelings for 31 scenes, and these differences were not affected by visitors' professional background. Accordingly, this study treated experts' and students' scoring of scenarios consistently.

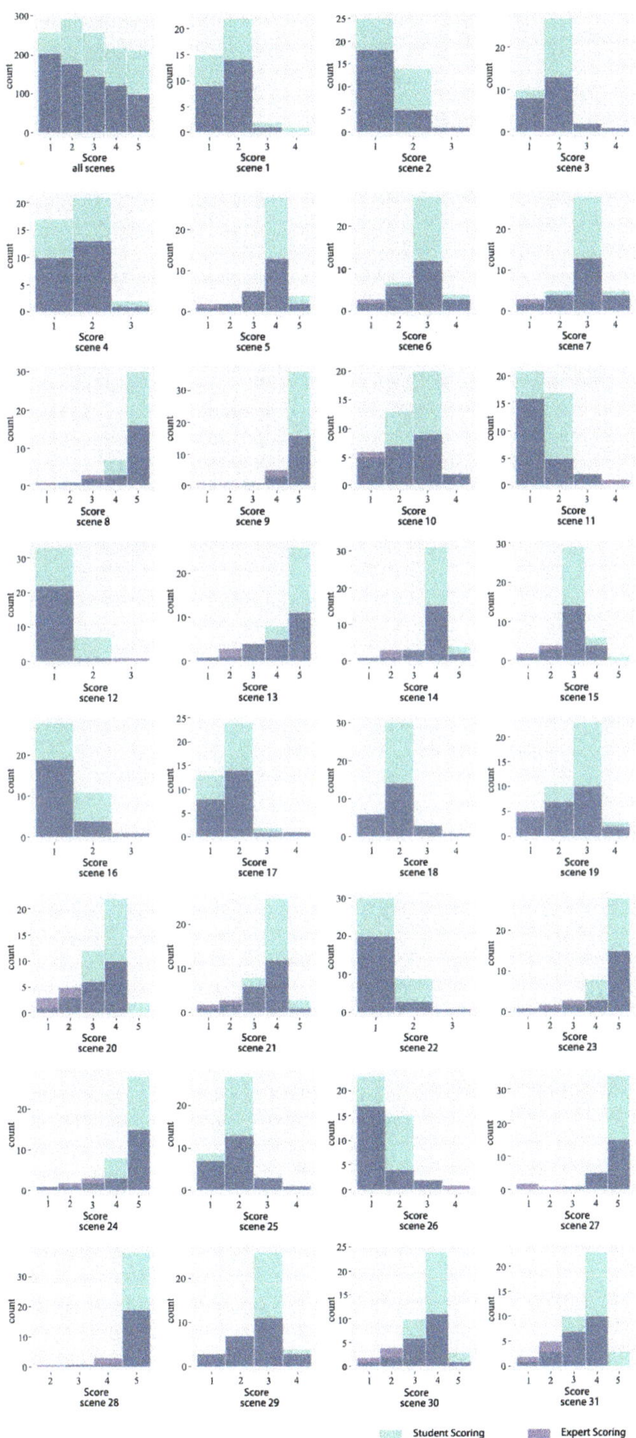

**Figure 5.** Result of visitor scoring.

*3.2. Quantitative Recognition Results of Landscape Features*

3.2.1. Quantitative Results of Spatial Morphology

The quantitative result of spatial morphology is shown in Table 3. The value of the variance of the upper and lower contour fluctuations was in the range of 10–16 levels, indicating that the drastic degree of upper and lower contour fluctuations of traditional villages in the Dongshan and Xishan regions was extremely small. So, the data in the table were not shown. The average of the accessible area ratio (AAR) value of the sample plots in the horizontal interface indicator (HII) reached 31.45%, and the average value of the eccentricity of bottom surface morphology (E) of the sample plots was 1.4489. Most of the sample plots showed a narrow and elongated morphology. The average spatial shape index (SSI) value was 257,269.50, and the standard deviation reached 302,740.87, indicating that the bottom surface morphology of the sample plots was more complex. The vegetation coverage (VC) of the sample plots was generally high, with a mean value of 60%. The average height of the upper layer of vegetation (hu) ranged from 4.23 m to 27.54 m. The average height of the lower layer of vegetation (hl) ranged from 0.50 m to 5.36 m, which indicated that the vegetation in the villages was generally higher, and the lower layer of shrubs was lower. The solid-to-void ratio (SVR) of vertical interface ranges from 0.0015 to 0.0174, with a mean value of 0.0056, indicating that the vertical interface was more open. The three-dimensional morphology index enclosure (ED) indicator ranged from 0.0037–0.2161 with a mean value of 0.0731, and the composite closure (CC) ranged from 0.0753–0.5337 with a mean value of 0.2365, with both indicators indicating high spatial openness. The three-dimensional spatial green visibility (3D-GVA) ranged from 0.0397 to 0.2608, with a mean value of 0.1094. Meanwhile, the plant diversity index revealed that the selection of tree species in villages was relatively unified.

**Table 3.** Results of morphological quantification.

| Spatial Composition | Morphological Characteristics Index | Minimum | Maximum | Mean Value | Standard Deviation |
|---|---|---|---|---|---|
| Horizontal interface | AAR | 0.076 | 0.8563 | 0.3145 | 0.1819 |
| | E | 0.1455 | 2.7862 | 1.4489 | 0.5351 |
| | SSI | 29,816.11 | 1,381,168.18 | 257,269.50 | 302,740.87 |
| | VC | 0.2014 | 0.9202 | 0.6102 | 0.3519 |
| Vertical interface | SVR | 0.0015 | 0.0174, | 0.0056 | 0.0033 |
| | FR | 0.4146 | 1.9128 | 0.9397 | 0.3339 |
| | hu | 4.2324 | 27.5353 | 11.9914 | 5.3382 |
| | hl | 0.4996 | 5.3626 | 2.2667 | 1.1749 |
| Three-dimensional space | 3D-GVA | 0.0397 | 0.2608 | 0.1094 | 0.0548 |
| | ED | 0.0037 | 0.2161 | 0.0731 | 0.0484 |
| | CC | 0.0753 | 0.5337 | 0.2365 | 0.1301 |
| | PDI | 0.1732 | 1.8919 | 1.105 | 0.4049 |

3.2.2. Results of Landscape Element Identification

The results of the identification of landscape elements are shown in Table 4 below. Vegetation and structure were the leading landscape elements that constituted the open space of traditional villages. The structure proportion ranged from 0.84% to 56.93%, with a mean value of 31.14% and a standard deviation of 0.1147. The proportion of bare land in the sample space was low, and the mean value was 5.39%, which indicates that the open space of traditional villages in the region had a high green coverage rate, except for hard paving. The mean value of the proportion of trees in the scene environment reached 14.48%, and the interval range was 1.55%–43.82%, while the proportion of the lower ground cover was lower, with a mean value of 1.94%. The data indicated that the vegetation level in the sample space is more homogeneous, with fewer shrubs in the middle layer.

**Table 4.** Results of landscape element identification.

| Landscape Elements | Minimum | Maximum | Mean Value | Standard Deviation |
|---|---|---|---|---|
| structure | 0.0084 | 0.5693 | 0.3114 | 0.1987 |
| sky | 0.0236 | 0.4923 | 0.2892 | 0.1205 |
| earth | 0 | 0.1881 | 0.0539 | 0.0556 |
| tree | 0.0155 | 0.4382 | 0.1448 | 0.1035 |
| grass | 0 | 0.1488 | 0.0194 | 0.0398 |

### 3.2.3. Results of Landscape Color Recognition

The results of the identification of scene landscape color features are shown in Table 5 below. The number of colors ranged between 3–8, with an average of 5 colors per scene. The main hue of the scene environment was reddish, with low saturation and low value. The distribution of hues in the scene was more dispersed, with fewer neighboring colors, the range of the main hue was 0.57%–22.52%, and the proportion of neighboring colors was 0.22%–17.12%. The scene had almost no complementary colors, and the proportion of complementary colors tended to be 0. The scene environment mainly showed warm tones, and the proportion of warm and cold colors ranged from 0 to 62.23%. The mean value of the color diversity index reached 0.42, with a wide range of colors. However, the color index was not high, with a mean value of 0.25 and a range of 0.04–0.55.

**Table 5.** Results of landscape color recognition.

| Landscape Color Characteristic Index | Minimum | Maximum | Mean Value | Standard Deviation |
|---|---|---|---|---|
| Number of Colors (NC) | 3 | 8 | 5 | 1 |
| Main Hue Comparison (MHi) | 0.0057 | 0.2252 | 0.0853 | 5.6911 |
| Adjacent Hue Comparison (NHi) | 0.0022 | 0.1712 | 0.0496 | 4.3807 |
| Complementary Hue Comparison (CHi) | 0 | 0.0024 | 0.0041 | 0.582 |
| Warm and cool color tone contrast (THi) | 0 | 0.6223 | 0.0757 | 0.1256 |
| Color Diversity Index (H') | 0.0444 | 0.884 | 0.4244 | 0.2119 |
| Color Evenness Index (E') | 0.0404 | 0.5493 | 0.2456 | 0.121 |

### 3.3. Landscape Preference Assessment

### 3.3.1. Indicator Screening

To reduce the parameters and dimensions of the calculation, this study conducted a principal component analysis of the indicators. It calculated the contribution of each indicator in different principal components. Significantly, the first seven principal components and the second principal component presented more than 75% of the information (Table 6) [9]. Therefore, we only took the index contribution rates in the first seven principal components into consideration. Descriptions of the contribution rate of each indicator are in the figure below (Figure 6). Indicators that contributed more than 50% were screened: AAR, VC, FR, hu, hl, FVu, 3D-GVA, CC, structure, tree, MHi, Nhi, CHi, THi, H', and C'.

**Table 6.** Total variance explained.

| Component | Initial Eigenvalues | | | Extraction Sums of Squared Loadings | | |
|---|---|---|---|---|---|---|
| | Total | % of Variance | Cumulative % | Total | % of Variance | Cumulative % |
| 1 | 5.843 | 22.475 | 22.475 | 5.843 | 22.475 | 22.475 |
| 2 | 3.774 | 14.515 | 36.99 | 3.774 | 14.515 | 36.99 |
| 3 | 3.335 | 12.826 | 49.815 | 3.335 | 12.826 | 49.815 |
| 4 | 2.452 | 9.429 | 59.244 | 2.452 | 9.429 | 59.244 |
| 5 | 1.96 | 7.539 | 66.783 | 1.96 | 7.539 | 66.783 |
| 6 | 1.755 | 6.749 | 73.532 | 1.755 | 6.749 | 73.532 |
| 7 | 1.373 | 5.281 | 78.813 | 1.373 | 5.281 | 78.813 |
| 8 | 1.22 | 4.692 | 83.505 | 1.22 | 4.692 | 83.505 |

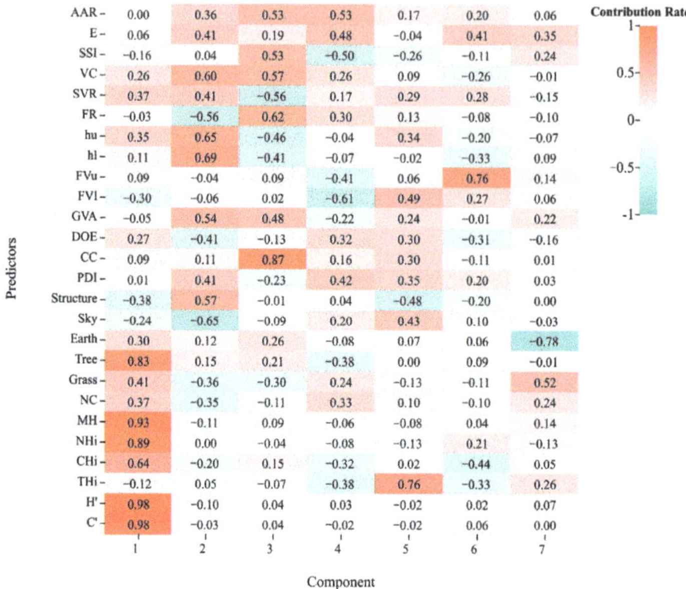

**Figure 6.** PCA component matrix.

The 16 screened indicators were adopted in the all-subsets regression (Figure 7). When the predictors were seven, adjusted $R^2$ peaks, and Cp and BIC were minimized, indicating that the model with streamlined indicators was optimal, and the corresponding predictors were AAR, FR, CC, tree, NHi, CHi, and THi.

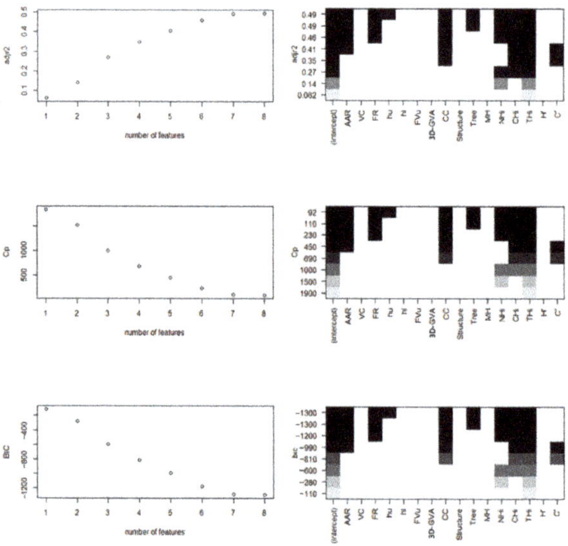

**Figure 7.** All-subsets regression analysis.

### 3.3.2. Result of Public Landscape Preference Assessment

All indicators as predictors were included in the multiple linear regression to build Model 1; the streamlined 7 predictors were included in the multiple linear regression model to build Model 2. As shown in Table 7, the adjusted $R^2 = 0.656$ for Model 1 indicated that all indicators explain 65.6% of the variation in visitor scoring. The adjusted $R^2 = 0.491$ for Model 2 indicated that the 8 streamlined indicators explain 49.1% of the variation in the visitor scoring. The Durbin–Watson values for Model 1 and Model 2 were 0.586 and 0.389, respectively, and were both consistent with independence. The two models' residual histograms and P-P plots were as follows (Figures 8 and 9). The residual histograms obeyed the normal distribution, the mean was close to 0, and the standard deviation was close to 1 (standard normal distribution), which meant that the linear regression was attained at the condition of normality. At the same time, the P–P plots also indicated that the model matches the condition of normality, which thoroughly explained the validity of the models.

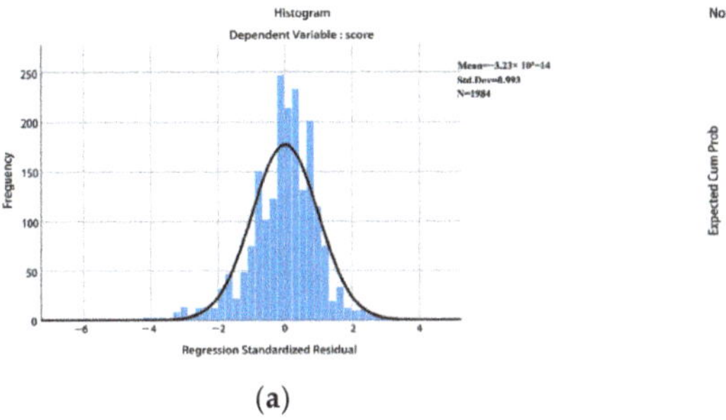

(a)                                    (b)

**Figure 8.** Model 1 (**a**) Regression standardized residual. (**b**) Observed cum prob.

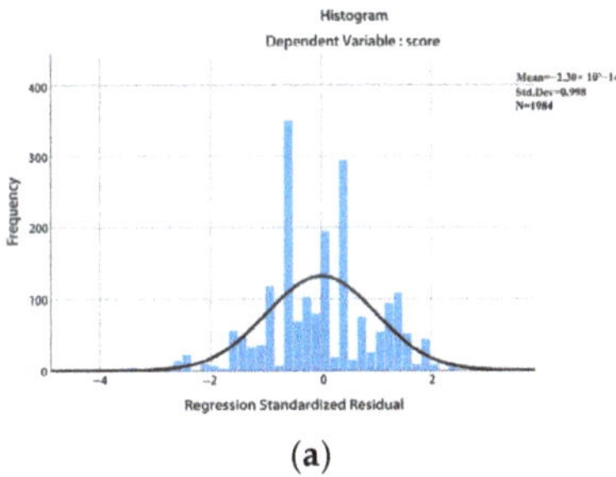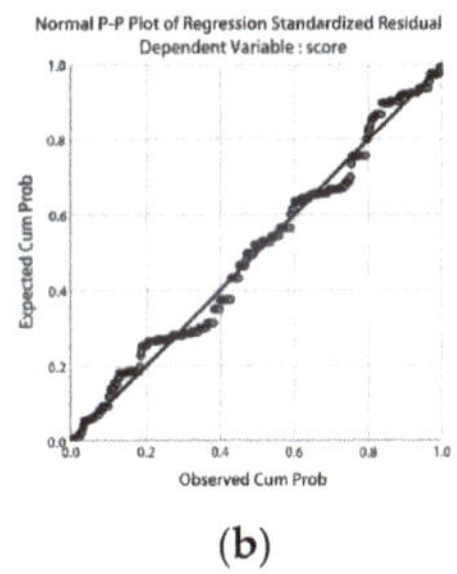

(a)                                    (b)

**Figure 9. Model 2** (**a**) Regression standardized residual. (**b**) Observed cum prob.

**Table 7.** Model summary.

| Model | R | $R^2$ | Adjusted $R^2$ | Std. Error of the Estimate | Durbin-Watson |
|---|---|---|---|---|---|
| 1 | 0.813 | 0.661 | 0.656 | 0.812 | 0.586 |
| 2 | 0.702 | 0.493 | 0.491 | 0.988 | 0.389 |

Table 8 shows the coefficients for Model 1 and Model 2. VC, FVu, structure, grass, NC, Nhi, and H′ in Model 1 were not significant at 95% confidence intervals. This indicated that VC, FVu, structure, grass, NC, Nhi, and H′ hardly predict visitor scoring. Based on the beta coefficients, the accessible area ratio (AAR), spatial shape index (SSi), solid vacancy ratio (SVR), contour fluctuation variance (FVl), sky, tree, main hue percentage (MHi), warm and cool color tone contrast (THi), and color index (C′) can be entered into the Model 1 had a positive effect on landscape aesthetic preferences. Eccentricity (E), upper contour mean height (hu), three-dimensional green volume (3D-GVA), degree of enclosure (DOE), composite closure (CC), plant diversity index (PDI), earth, and complementary color ratio (CHi) had adverse effects on landscape aesthetic preferences. The final model expression for Model 2 was:

$$\hat{Y} = 0.464\text{AAR} + 0.385\text{FR} - 0.819\text{CC} + 0.317\text{TREE} + 0.4\text{NHi} - 0.454\text{CHi} - 0.598\text{THi} + 1.545$$

**Table 8.** Model coefficients.

| Model | Predictors | B | β | t | p |
|---|---|---|---|---|---|
| 1 | (Constant) | 0.362 | | 0.416 | 0.678 |
| | AAR | 2.657 | 0.44 | 11.531 | 0 |
| | E | −1.93 | −0.278 | −10.088 | 0 |
| | SSI | 0.738 | 0.117 | 2.475 | 0.013 |
| | VC | 0.766 | 0.142 | 1.846 | 0.065 |
| | SVR | 5.005 | 0.735 | 15.308 | 0 |
| | FR | 1.774 | 0.281 | 2.428 | 0.015 |
| | hu | −3.222 | −0.524 | −4.124 | 0 |
| | hl | 2.827 | 0.485 | 10.427 | 0 |
| | FVu | −0.531 | −0.072 | −1.837 | 0.066 |
| | FVl | 2.245 | 0.298 | 7.731 | 0 |
| | 3D-GVA | −2.106 | −0.371 | −9.438 | 0 |
| | DOE | −0.942 | −0.152 | −2.417 | 0.016 |
| | CC | −1.312 | −0.265 | −4.416 | 0 |
| | PDI | −1.859 | −0.283 | −9.696 | 0 |
| | Structure | 0.181 | 0.046 | 0.918 | 0.359 |
| | Sky | 1.41 | 0.247 | 4.347 | 0 |
| | Earth | −1.108 | −0.233 | −7.882 | 0 |
| | Tree | 2.081 | 0.364 | 4.209 | 0 |
| | Grass | 0.341 | 0.065 | 0.696 | 0.486 |
| | NC | 0.662 | 0.063 | 0.676 | 0.499 |
| | MHi | 1.258 | 0.232 | 3.531 | 0 |
| | NHi | −0.735 | −0.135 | −1.559 | 0.119 |
| | CHi | −3.13 | −0.551 | −15.23 | 0 |
| | THi | 1.978 | 0.343 | 5.849 | 0 |
| | H′ | −6.68 | −1.198 | −1.863 | 0.063 |
| | C′ | 8.66 | 1.464 | 2.811 | 0.005 |
| 2 | (Constant) | 1.545 | | 24.825 | 0 |
| | AAR | 2.801 | 0.464 | 21.084 | 0 |
| | FR | 2.433 | 0.385 | 17.638 | 0 |
| | CC | −4.062 | −0.819 | −29.062 | 0 |
| | Tree | 1.811 | 0.317 | 11.063 | 0 |
| | NHi | 2.173 | 0.4 | 15.429 | 0 |
| | CHi | −2.578 | −0.454 | −21.334 | 0 |
| | THi | 3.451 | 0.598 | 31.891 | 0 |

## 4. Discussion

To advance the regional and cultural significance of landscape resources in urban-rural areas and promote the sustainable development of villages, tourism, and cultural exchanges, we needed to understand the connection between aesthetic sensory perception and the landscape environment and identify which landscape characteristics could contribute to human mental and physical status and well-being. Many studies have explored landscape characteristics and public aesthetic preferences from a single dimension. However, studies integrating multidimensional features to explore the factors that influence landscape preferences were still scarce. There were many limitations, such as insufficient quantification of landscape feature indicators and low accuracy of landscape preference assessment models. This study aimed to compensate for these deficiencies in two aspects. First, based on the existing landscape spatial morphology dimensions, this study expanded the landscape feature characterization system by introducing two dimensions, namely, landscape visual attraction elements and plant landscape color. At the same time, the traditional beauty degree evaluation was improved and the score of each evaluation subject for the scene environment was calculated, expanding the sample data volume. We focus on two main research questions: (1) identifying the main features that influence landscape aesthetic preference and analyzing the influencing mechanisms; (2) how to improve the progress of the landscape aesthetic preference assessment model. The following discussion will focus on these two aspects.

*4.1. Expanding the Dimensions of Landscape Characteristics Influencing Aesthetic Sensory Perception*

The results of the study showed that seven landscape feature indicators, namely, accessible area ratio (AAR), fluctuating range of contour (FR), comprehensive closure (CC), tree, neighboring color ratio (NHi), complementary color ratio (CHi) and warm and cool color tone contrast (THi), had a significant effect on aesthetic preference. The influence of trees on landscape preference at the level of landscape elements was significantly higher than that of other elements, and it was a positive factor that affected aesthetic sensory perception. Trees were the primary type of vegetation in the public space of traditional villages, and in line with previous studies, an aesthetically pleasing environment was associated with plants [51–53]. The positive influence of plants, as the leading natural element of landscape composition, could be explained by ecological and evolutionary theories. They could create an environment where the provision of nature allowed for humans to retain an affinity for the original natural ecological dynamics and humans could enjoy the natural environment and the relaxing atmosphere [54]. The previous study's explanation for the Chinese people's fondness for village plants was that urbanization in China over the past few decades had caused severe ecological damage and that economic development had intensified the demand for green space, thus making lush vegetation the ideal image of the village landscape for tourists [54]. The type of village vegetation (herbaceous vs. tree) in the study by Arriaza et al. did not produce a statistically significant effect in the regression analysis of the variables, while the percentage of vegetation in the picture was considered an important attribute of the village landscape [20]. Whereas the vegetation cover type trees had a statistically significant effect in this study, herbaceous trees did not provide an adequate explanation for the public aesthetic preference, which may be related to the monoculture vegetation composition of traditional villages because the percentage of herbaceous was relatively low there and was dominated by tall trees and crops, excluding the bare land.

Accessible area ratio (AAR) and contour fluctuation range (FR) in the spatial form dimension were positive factors that significantly affect landscape aesthetic preference. Comprehensive closure (CC) negatively affected landscape aesthetic sensory perception. Accessible area ratio (AAR) somehow indicates the space available to the public per unit of area. The description of popular landscape scene environments in previous studies was large landscape spaces that were naturally open and suitable for transportation and activities, giving a sense of nature [55,56]. It suggested that the accessibility of landscape spaces and the social and recreational activities that they accommodated had a positive effect on landscape preferences. This was consistent with landscape sensory restoration theory, where one of the four characteristics of restorative environments was compatibility. It suggested that good landscape spaces had sufficient content and structure to provide activities that were relatively consistent with individual purposes and preferences, thus satisfying the activity needs of different users [57]. Previous studies have shown that people preferred scenic environments that were open in scale and organized, which was different from the results of this study [58,59]. The contour fluctuation range (FR) indicated the size of the fluctuation range of the highest point of the plant canopy line at the vertical interface, which to some extent destroyed the orderliness of the vertical interface. The study by Zhang et al. showed that the interactions between visual attributes could lead to inconsistencies in the interpretation of the main influences on landscape preference [39]. After careful analysis, it was be found that the range of contour fluctuation (FR) and the variance of upper contour fluctuation (FVu) together reflected the orderliness of the contour on the vertical interface. In this study, the upper contour fluctuation variance (FVu) was infinitely close to 0, which indicated that the intensity of contour fluctuation on the vertical interface in the 31 scene environments was very small. It had already possessed good order in itself, while the increase in the fluctuation range was to strengthen the richness of the whole map surface. Thus, the landscape preference of the scene was enhanced, which validated previous research that richness showed an unstable effect on people's preference, and that richness had a positive effect only if the sense of order was high [60]. Berlyne's study also explained this phenomenon by suggesting that there was an inverted U-shaped functional relationship between scene richness and landscape preference, where landscape preference was initially positively related to scene richness, and then negatively related when a certain threshold was reached [61]. The composite closeness index (CC) measured the degree of closeness of the whole scene environment. It was negatively correlated with landscape aesthetic preference but in line with the explanation of people's preference for scenes with open visual scales in previous studies, while it seemed to contradict the explanation of people's preference for scenes with a high proportion of tree elements. Prospect-refuge theory can provide a reasonable explanation for this: people want to be able to observe others but not be seen by others [62]. The openness of the line of sight provides a good viewing experience, and the increase in the proportion of trees enriches the depth of the field of view, while providing a certain degree of shading.

The proportion of neighboring hues (NHi) and the proportion of warm and cool hues (THi) at the dimension of landscape color characteristics were positively correlated with landscape perceptual preference, and the proportion of complementary hues (CHi) was negatively correlated with it. Previous related studies have shown that large areas of warm-colored plants in autumn can create a warm and peaceful feeling, and the combination of adjacent red and yellow colors makes it more likely to form a harmonious color combination [63]. With a softer and more fluid excess between colors, it helps the public to process visual information more easily, and at the same time creates an emotional experience of calmness, gentleness, and serenity, which is thus preferred by the public [64]. These results explained the positive effect of the proportion of neighboring hues (NHi) on landscape preference in this study, where panoramas were also collected in autumn, and the color of the foliage trees were dominated by reddish-yellow tones. Zhuang et al. showed that increasing the proportion of cool colors and green vegetation in the urban floristic mirror can enhance aesthetic preference [40], which was in line with the results of the present study but contrary to the findings of Luo et al., which were consistent with the results of this study but contrary to the results of Luo et al. A possible explanation for this was that based on the theory of color psychology, warm colors are more likely to produce physiological stimulation than cool colors, and rural landscapes are considered to be rejuvenating or tranquil environments [51,65,66]. The proportion of cool-colored vegetation in the 31 scenes was generally small. Scenes in which the proportion of cool-colored evergreens was slightly higher were more likely to create an atmosphere of calmness, gentleness, and serenity, thus enhancing people's aesthetic preference. Luo et al. showed that the number of experimental colors is between 5 and 7. Among them the color leaf index was high, and the large area of uniform fall warm colors were easily preferred by the public [64]. This can be used to explain why the complementary hue ratio (CHi) hurt landscape aesthetic sensory perception in this study. The sample sites selected in this study had relatively few plant species, of which were colorful foliage plants of similar hues. And the whole picture was relatively uniform, so the increase in complementary colors would destroy the sense of order of the whole picture and make it difficult to obtain uniformity. The study by Kuper showed that attentional restoration was positively correlated with landscape sensory perception, and according to the theory of attentional restoration, exposure to warmer colors was more likely to be preferred by the public [67]. According to the theory of attention restoration, exposure to restorative environments will cause fascination or effortless attention, which can also explain people's preference for scene environments with low complementary hue ratios and relatively uniform hues.

Based on the above analysis, we can describe the ideal public space scene environment of traditional villages, where people can enjoy a wide view, a high proportion of trees, and mild and uniform color tones. The ideal public space is suitable for passages and activities and can produce a restorative and peaceful atmosphere. The way people pay attention to the landscape and the attractiveness of landscape features differs in different landscape sensory perceptions, and landscape features are directly affected by the sensory process, so these sensory perceptions can be strengthened by artificial landscape feature design. Although landscape preference is influenced by many aspects, the most important factor affecting the landscape preference of traditional villages is related to the ideal image of the countryside in people's minds, which is based on the symbolic experience of the perceiving subject for the space. In the subsequent renewal and protection of traditional villages, government authorities and planning and design departments should encourage public participation, considering the physical structure of the landscape in conjunction with its value and meaning, to create a more inclusive environment, thus promoting the sustainable development of rural habitat.

### 4.2. Improvement of the Effectiveness of the Aesthetic Sensory Assessment

Compared with previous studies, this study mainly improved the accuracy of model assessment accuracy from in various aspects. On the one hand, we analyzed the previous landscape scores and the scores of each evaluation subject separately. The sample number was expanded to 64 times of that of the original, which helped to obtain more stable assessment results. On the other hand, based on the previous spatial morphology single-dimension indicators, two-dimensional feature indicators of landscape visual elements and landscape color were introduced, which helped refine the influence dimensions of aesthetic sensory perception. The conditions were conducive to a more comprehensive understanding of the landscape characteristics of the public space of traditional villages. Previous studies have shown that the multiple linear regression model was more advantageous in the assessment of landscape rating, with the coefficient of determination $R^2 = 33.2\%$ and root mean square error RMS = 64.774, but only four significantly related morphology index factors were selected to participate in the construction of the model, making it difficult to understand the influence of other index factors on the assessment of landscape preference. In this study, the multiple linear regression model was also chosen to predict landscape preference, and all indicators of the three dimensions were included in the construction of the model, with the adjusted $R^2$ reaching 65.6%. The study streamlined the indicators based on the comprehensive indicator multiple linear regression model, screening seven significantly related indicators through principal component analysis and full subset regression analysis to participate in the model construction, and the $R^2$ of the streamlined model reached 49.1%, which was also higher than that of 33.2% in the previous study. Both models demonstrated the scientific rationality of the increase in indicator dimensions, and the assessment accuracy of landscape preference was effectively improved based on the previous single dimension. At the same time, the two models could be adapted to different scenarios. The full-indicator assessment model contains all the available indicators, comprehensively considers the landscape characteristics, and better captures the diversity and comprehensiveness. They paid attention to the interrelationships between the indicators of different dimensions, which contributed to the fine management and enhancement of the landscape of the public space of traditional villages. The streamlined predictive model was suitable for rapid judgment of environmental aesthetic sensory perception and could quickly improve the quality of spatial landscape environment in a short period.

### 4.3. Limitations and Future Work

Aesthetic sensory perception is a multidimensional perception process, which is jointly influenced by the aesthetic object and the perception subject. Although this study added two dimensions of landscape visual attraction elements and landscape color characteristics based on previous studies, the influence of other dimensions of landscape characteristics cannot be excluded. Subsequent studies should further improve the indicator system of landscape features and explore how the mutual combinations of visual attributes affect aesthetic sensory perception, as well as how to better control confounding factors to facilitate inter-comparisons between studies. Moreover, the morphological and color characteristics of plant landscapes are time-sequential. The landscape characteristics show great differences in different seasons or different periods of the same season. Therefore, subsequent research should further strengthen the identification of landscape characteristics at different times. At the same time, in the study, buildings, walls, and some other artificial facilities, including poles, were uniformly classified into structures without making further distinctions. Follow-up research will subdivide the composition of the structure and explore the impact of various visual attraction elements on aesthetic sensory perception. Although visual sensory perception is the main way of landscape aesthetics, the role of auditory, tactile, olfactory, and gustatory senses in the process of aesthetic sensory perception should not be ignored [68,69]. As a result, subsequent research should comprehensively consider the influence of human perceptual organs on aesthetic sensory perception. In addition, this study was based on the construction of aesthetic sensory perception assessment model for 31 scenarios of traditional villages in Dongshan and Xishan of Taihu Lake, Suzhou, and the values of each index have a certain range. Therefore, it is difficult to explore the relationship between landscape features and landscape preference beyond the range of the values, and the subsequent study needs to further increase the sample capacity to reveal the general pattern of the public's aesthetic preference for the landscapes of the traditional villages. The following research needs to further increase the sample volume to reveal the general public aesthetic sensory perception of traditional village landscape.

### 5. Conclusions

This study expanded the landscape characterization system for the public space of the traditional village by integrating multiple dimensions: landscape spatial form, visually attractive elements of the landscape, and their colors. It quantitatively identified each index feature based on machine learning and LiDAR scanning technology. The traditional scenic beauty evaluation (SBE) method was improved to construct the aesthetic sensory perception assessment model with all indicators and indicators of significant influence. The accuracy of the full-indicator aesthetic sensory assessment model ($R^2$ = 65.6%) is higher than that of the significant influence indicator aesthetic sensory assessment model (49.1%). The assessment accuracy of both models is greatly improved compared with that of the assessment model of the previous study ($R^2$ = 33.2%). The results showed that the accessibility area ratio (AAR), spatial shape index (SSI), solid vacancy ratio (SVR), contour fluctuation range (FR), the average height of lower contour (hl), variance of lower contour fluctuation (FVl), sky, tree, main color hue (MHI), warm/cold hue (THi), and color index (C') were able to enhance the public's preference for public space in traditional villages. Eccentricity (E), average height of upper contour (hu), three-dimensional green volume (3D-GVA), degree of enclosure (DOE), comprehensive closure (CC), plant diversity index (PDI), earth, and complementary colors (CHi) reduced the public's aesthetic preferences for the public space in traditional villages. Among them, the significant impact factors were AAR, FR, CC, tree, NHi, Chi, and THi. The study revealed the public aesthetic sensory perception of the public space of traditional villages, providing scientific and theoretical guidance and a basis for relevant decision-making departments and planning and design companies. Thus, it promoted the sustainable development of the rural living environment and provided a good relaxation environment for the physical and mental health of urban and rural residents.

**Author Contributions:** Conceptualization, G.C. and S.C.; methodology, G.C., C.W. and S.C.; software, J.Y. and C.W.; validation, G.C., J.Y., C.W. and S.C.; formal analysis, J.Y.; investigation, G.C.; resources, G.C.; data curation, J.Y. and S.C.; writing—original draft preparation, G.C.; writing—review and editing, G.C., J.Y. and S.C.; visualization, G.C.; supervision, J.Y. and S.C.; project administration, S.C.; funding acquisition, S.C. All authors have read and agreed to the published version of the manuscript.

**Funding:** This research was funded by the "National Key Research and Development Program of China" (grant number 2019YFD1100404), "National College Students Innovation and Enterpreneurship Training Program" (grant number 202210298045Z), "Anhui Province Key Research Project for Universities" (grant number 2023AH050968), "A Project Funded by the Priority Academic Program Development of Jiangsu Higher Education Institutions" (PAPD).

**Informed Consent Statement:** Informed consent was obtained from all subjects involved in the study.

**Data Availability Statement:** The data presented in this study are available on request from the author. The data are not publicly available due to privacy. Images employed for the study will be available online for readers.

**Conflicts of Interest:** The authors declare no conflicts of interest.

# References


1. Gao, J.; Wu, B. Revitalizing Traditional Villages through Rural Tourism: A Case Study of Yuanjia Village, Shaanxi Province, China. *Tour. Manag.* **2017**, *63*, 223–233. [CrossRef]
2. Grêt-Regamey, A.; Altwegg, J.; Sirén, E.A.; Van Strien, M.J.; Weibel, B. Integrating Ecosystem Services into Spatial Planning—A Spatial Decision Support Tool. *Landsc. Urban Plan.* **2017**, *165*, 206–219. [CrossRef]
3. Gebre, T.; Gebremedhin, B. The Mutual Benefits of Promoting Rural-Urban Interdependence through Linked Ecosystem Services. *Glob. Ecol. Conserv.* **2019**, *20*, e00707. [CrossRef]
4. Van Zanten, B.T.; Verburg, P.H.; Scholte, S.S.K.; Tieskens, K.F. Using Choice Modeling to Map Aesthetic Values at a Landscape Scale: Lessons from a Dutch Case Study. *Ecol. Econ.* **2016**, *130*, 221–231. [CrossRef]
5. Tveit, M.S. Indicators of Visual Scale as Predictors of Landscape Preference; a Comparison between Groups. *J. Environ. Manag.* **2009**, *90*, 2882–2888. [CrossRef] [PubMed]
6. Almeida, M.; Loupa-Ramos, I.; Menezes, H.; Carvalho-Ribeiro, S.; Guiomar, N.; Pinto-Correia, T. Urban Population Looking for Rural Landscapes: Different Appreciation Patterns Identified in Southern Europe. *Land Use Policy* **2016**, *53*, 44–55. [CrossRef]
7. Ecological Landscape Resource Management and Sustainable Development of Traditional Villages-All Databases. Available online: https://webofscience.clarivate.cn/wos/alldb/full-record/WOS:000588763500039 (accessed on 31 December 2023).
8. Park, J.J.; Selman, P. Attitudes Toward Rural Landscape Change in England. *Environ. Behav.* **2011**, *43*, 182–206. [CrossRef]
9. Chen, G.; Sun, X.; Yu, W.; Wang, H. Analysis Model of the Relationship between Public Spatial Forms in Traditional Villages and Scenic Beauty Preference Based on LiDAR Point Cloud Data. *Land* **2022**, *11*, 1133. [CrossRef]
10. Qi, L.; Liang, W.; Xi, M. The Public Space Pattern Research of Guangfu Traditional Villages Based on Spatial Syntax: A Case Study of Huangpu Village in Guangzhou City, China. *IOP Conf. Ser. Earth Environ. Sci.* **2019**, *267*, 062033. [CrossRef]
11. Déjeant-Pons, M. The European Landscape Convention. *Landsc. Res.* **2006**, *31*, 363–384. [CrossRef]
12. de la O Cabrera, M.R.; Marine, N.; Escudero, D. Spatialities of Cultural Landscapes: Towards a Unified Vision of Spanish Practices within the European Landscape Convention. *Eur. Plan. Stud.* **2020**, *28*, 1877–1898. [CrossRef]
13. Zeller, K.A.; McGarigal, K.; Cushman, S.A.; Beier, P.; Vickers, T.W.; Boyce, W.M. Sensitivity of Resource Selection and Connectivity Models to Landscape Definition. *Landsc. Ecol.* **2017**, *32*, 835–855. [CrossRef]
14. Nohl, W. Sustainable Landscape Use and Aesthetic Perception–Preliminary Reflections on Future Landscape Aesthetics. *Landsc. Urban Plan.* **2001**, *54*, 223–237. [CrossRef]
15. Kalivoda, O.; Vojar, J.; Skřivanová, Z.; Zahradník, D. Consensus in Landscape Preference Judgments: The Effects of Landscape Visual Aesthetic Quality and Respondents' Characteristics. *J. Environ. Manag.* **2014**, *137*, 36–44. [CrossRef] [PubMed]
16. Lee, C.-H. Understanding Rural Landscape for Better Resident-Led Management: Residents' Perceptions on Rural Landscape as Everyday Landscapes. *Land Use Policy* **2020**, *94*, 104565. [CrossRef]
17. Wang, R.-H.; Zhao, J.-W. Evaluation on Aesthetic Preference of Rural Landscapes and Crucial factors. *J. Shandong Agric. Univ. (Nat. Sci. Ed.)* **2016**, *47*, 231–235.
18. Wang, R.; Jiang, W.; Lu, T. Landscape Characteristics of University Campus in Relation to Aesthetic Quality and Recreational Preference. *Urban For. Urban Green.* **2021**, *66*, 127389. [CrossRef]
19. Yao, Y.; Zhu, X.; Xu, Y.; Yang, H.; Wu, X.; Li, Y.; Zhang, Y. Assessing the Visual Quality of Green Landscaping in Rural Residential Areas: The Case of Changzhou, China. *Environ. Monit. Assess.* **2012**, *184*, 951–967. [CrossRef]
20. Arriaza, M.; Cañas-Ortega, J.F.; Cañas-Madueño, J.A.; Ruiz-Aviles, P. Assessing the Visual Quality of Rural Landscapes. *Landsc. Urban Plan.* **2004**, *69*, 115–125. [CrossRef]
21. Hernández, J.; García, L.; Ayuga, F. Integration Methodologies for Visual Impact Assessment of Rural Buildings by Geographic Information Systems. *Biosyst. Eng.* **2004**, *88*, 255–263. [CrossRef]
22. Spennemann, D.H.R. The Shifting Baseline Syndrome and Generational Amnesia in Heritage Studies. *Heritage* **2022**, *5*, 2007–2027. [CrossRef]
23. Li, X.; Zhang, X.; Jia, T. Humanization of Nature: Testing the Influences of Urban Park Characteristics and Psychological Factors on Collegers' Perceived Restoration. *Urban For. Urban Green.* **2023**, *79*, 127806. [CrossRef]

24. Van Den Berg, A.E.; Jorgensen, A.; Wilson, E.R. Evaluating Restoration in Urban Green Spaces: Does Setting Type Make a Difference? *Landsc. Urban Plan.* **2014**, *127*, 173–181. [CrossRef]
25. Li, S.-Y.; Chen, Z.; Guo, L.-H.; Hu, F.; Huang, Y.-J.; Wu, D.-C.; Wu, Z.; Hong, X.-C. How Do Spatial Forms Influence Psychophysical Drivers in a Campus City Community Life Circle? *Sustainability* **2023**, *15*, 10014. [CrossRef]
26. Misthos, L.-M.; Krassanakis, V.; Merlemis, N.; Kesidis, A.L. Modeling the Visual Landscape: A Review on Approaches, Methods and Techniques. *Sensors* **2023**, *23*, 8135. [CrossRef] [PubMed]
27. Qin, X.; Fang, M.; Yang, D.; Wangari, V.W. Quantitative Evaluation of Attraction Intensity of Highway Landscape Visual Elements Based on Dynamic Perception. *Environ. Impact Assess. Rev.* **2023**, *100*, 107081. [CrossRef]
28. Li, J.; Zhang, Z.; Jing, F.; Gao, J.; Ma, J.; Shao, G.; Noel, S. An Evaluation of Urban Green Space in Shanghai, China, Using Eye Tracking. *Urban For. Urban Green.* **2020**, *56*, 126903. [CrossRef]
29. López-Martínez, F. Visual Landscape Preferences in Mediterranean Areas and Their Socio-Demographic Influences. *Ecol. Eng.* **2017**, *104*, 205–215. [CrossRef]
30. García-Ayllón, S.; Martínez, G. Analysis of Correlation between Anthropization Phenomena and Landscape Values of the Territory: A GIS Framework Based on Spatial Statistics. *ISPRS Int. J. Geo-Inf.* **2023**, *12*, 323. [CrossRef]
31. Schüpbach, B.; Kay, S. Validation of a Visual Landscape Quality Indicator for Agrarian Landscapes Using Public Participatory GIS Data. *Landsc. Urban Plan.* **2024**, *241*, 104906. [CrossRef]
32. Loro, M.; Arce, R.M.; Ortega, E. Identification of Optimal Landforms to Reduce Impacts on the Landscape Using LiDAR for Hosting a New Highway. *Environ. Impact Assess. Rev.* **2017**, *66*, 99–114. [CrossRef]
33. Zhu, B. Garden Landscape Planning Based on Digital Feature Recognition. *Phys. Chem. Earth Parts A/B/C* **2023**, *130*, 103372. [CrossRef]
34. Li, Z.; Cheng, Y. Spatial Structure Optimization Model of Island Port Landscape Pattern. *J. Coast. Res.* **2019**, *93*, 404. [CrossRef]
35. Peng, J.; Wang, Y.; Zhang, Y.; Wu, J.; Li, W.; Li, Y. Evaluating the Effectiveness of Landscape Metrics in Quantifying Spatial Patterns. *Ecol. Indic.* **2010**, *10*, 217–223. [CrossRef]
36. Hu, B. Deep Learning Image Feature Recognition Algorithm for Judgment on the Rationality of Landscape Planning and Design. *Complexity* **2021**, *2021*, 9921095. [CrossRef]
37. Chen, Y.; Chen, L.; Li, Y. Recognition Algorithm of Street Landscape in Cold Cities with High Difference Features Based on Improved Neural Network. *Ecol. Inform.* **2021**, *66*, 101395. [CrossRef]
38. Rechtman, O. Visual Perception of Agricultural Cultivated Landscapes: Key Components as Predictors for Landscape Preferences. *Landsc. Res.* **2013**, *38*, 273–294. [CrossRef]
39. Zhang, G.; Yang, J.; Wu, G.; Hu, X. Exploring the Interactive Influence on Landscape Preference from Multiple Visual Attributes: Openness, Richness, Order, and Depth. *Urban For. Urban Green.* **2021**, *65*, 127363. [CrossRef]
40. Zhuang, J.; Qiao, L.; Zhang, X.; Su, Y.; Xia, Y. Effects of Visual Attributes of Flower Borders in Urban Vegetation Landscapes on Aesthetic Preference and Emotional Perception. *Int. J. Environ. Res. Public Health* **2021**, *18*, 9318. [CrossRef]
41. Cao, Y.; Li, Y.; Li, X.; Wang, X.; Dai, Z.; Duan, M.; Xu, R.; Zhao, S.; Liu, X.; Li, J.; et al. Relationships between the Visual Quality and Color Patterns: Study in Peri-Urban Forests Dominated by *Cotinus coggygria* var. *cinerea* Engl. in Autumn in Beijing, China. *Forests* **2022**, *13*, 1996. [CrossRef]
42. Shen, S.; Yao, Y.; Li, C. Quantitative Study on Landscape Colors of Plant Communities in Urban Parks Based on Natural Color System and M-S Theory in Nanjing, China. *Color Res. Appl.* **2022**, *47*, 152–163. [CrossRef]
43. Daniel, T.C. *Measuring Landscape Esthetics: The Scenic Beauty Estimation Method*; Department of Agriculture, Forest Service, Rocky Mountain Forest and Range Experiment Station: Fort Collins, CO, USA, 1976.
44. Clay, G.R.; Daniel, T.C. Scenic Landscape Assessment: The Effects of Land Management Jurisdiction on Public Perception of Scenic Beauty. *Landsc. Urban Plan.* **2000**, *49*, 1–13. [CrossRef]
45. Clay, G.R.; Smidt, R.K. Assessing the Validity and Reliability of Descriptor Variables Used in Scenic Highway Analysis. *Landsc. Urban Plan.* **2004**, *66*, 239–255. [CrossRef]
46. Tsai, V.J.D.; Chang, C. Three-dimensional Positioning from Google Street View Panoramas. *IET Image Process.* **2013**, *7*, 229–239. [CrossRef]
47. Ki, D.; Lee, S. Analyzing the Effects of Green View Index of Neighborhood Streets on Walking Time Using Google Street View and Deep Learning. *Landsc. Urban Plan.* **2021**, *205*, 103920. [CrossRef]
48. Yin, L.; Wang, Z. Measuring Visual Enclosure for Street Walkability: Using Machine Learning Algorithms and Google Street View Imagery. *Appl. Geogr.* **2016**, *76*, 147–153. [CrossRef]
49. Wang, C.; Zou, J.; Fang, X.; Chen, S.; Wang, H. Using Social Media and Multi-Source Geospatial Data for Quantifying and Understanding Visitor's Preferences in Rural Forest Scenes: A Case Study from Nanjing. *Forests* **2023**, *14*, 1932. [CrossRef]
50. Kaufman, A.J.; Lohr, V.I. Does plant color affect emotional and physiological responses to landscapes? *Acta Hortic.* **2004**, 229–233. [CrossRef]
51. Rogge, E.; Nevens, F.; Gulinck, H. Perception of Rural Landscapes in Flanders: Looking beyond Aesthetics. *Landsc. Urban Plan.* **2007**, *82*, 159–174. [CrossRef]
52. Wilson, J.L. The Influence of Individualist-Collectivist Values, Attitudes Toward Women, and Proenvironmental Orientation on Landscape Preference. Master's Thesis, University of North Florida, Jacksonville, FL, USA, 2009.

53. Kaplan, A.; Taşkın, T.; Önenç, A. Assessing the Visual Quality of Rural and Urban-Fringed Landscapes Surrounding Livestock Farms. *Biosyst. Eng.* **2006**, *95*, 437–448. [CrossRef]
54. Ren, X. Consensus in Factors Affecting Landscape Preference: A Case Study Based on a Cross-Cultural Comparison. *J. Environ. Manag.* **2019**, *252*, 109622. [CrossRef] [PubMed]
55. Cai, K.; Huang, W.; Lin, G. Bridging Landscape Preference and Landscape Design: A Study on the Preference and Optimal Combination of Landscape Elements Based on Conjoint Analysis. *Urban For. Urban Green.* **2022**, *73*, 127615. [CrossRef]
56. Grahn, P.; Stigsdotter, U.K. The Relation between Perceived Sensory Dimensions of Urban Green Space and Stress Restoration. *Landsc. Urban Plan.* **2010**, *94*, 264–275. [CrossRef]
57. Berto, R.; Massaccesi, S.; Pasini, M. Do Eye Movements Measured across High and Low Fascination Photographs Differ? Addressing Kaplan's Fascination Hypothesis. *J. Environ. Psychol.* **2008**, *28*, 185–191. [CrossRef]
58. Palmer, J.F. Using Spatial Metrics to Predict Scenic Perception in a Changing Landscape: Dennis, Massachusetts. *Landsc. Urban Plan.* **2004**, *69*, 201–218. [CrossRef]
59. Kotabe, H.P.; Kardan, O.; Berman, M.G. The Order of Disorder: Deconstructing Visual Disorder and Its Effect on Rule-Breaking. *J. Exp. Psychol. Gen.* **2016**, *145*, 1713–1727. [CrossRef]
60. Stamps, A.E. Mystery, Complexity, Legibility and Coherence: A Meta-Analysis. *J. Environ. Psychol.* **2004**, *24*, 1–16. [CrossRef]
61. Swartz, P. Berlyne on Art: A Review of D. E. Berlyne's Aesthetics and Psychobiology. *Can. Psychol./Psychol. Can.* **1973**, *14*, 297–303. [CrossRef]
62. Mumcu, S.; Düzenli, T.; Özbilen, A. Prospect and Refuge as the Predictors of Preferences for Seating Areas. *Sci. Res. Essays* **2010**, *5*, 1223–1233.
63. Jang, H.S.; Kim, J.; Kim, K.S.; Pak, C.H. Human Brain Activity and Emotional Responses to Plant Color Stimuli. *Color Res. Appl.* **2014**, *39*, 307–316. [CrossRef]
64. Luo, Y.; He, J.; Long, Y.; Xu, L.; Zhang, L.; Tang, Z.; Li, C.; Xiong, X. The Relationship between the Color Landscape Characteristics of Autumn Plant Communities and Public Aesthetics in Urban Parks in Changsha, China. *Sustainability* **2023**, *15*, 3119. [CrossRef]
65. Lei, Z.; Fuzong, L.; Bo, Z. A CBIR Method Based on Color-Spatial Feature. In Proceedings of the IEEE Region 10 Conference. TENCON 99. "Multimedia Technology for Asia-Pacific Information Infrastructure" (Cat. No.99CH37030), Cheju Island, Republic of Korea, 15–17 September 1999; Volume 1, pp. 166–169.
66. Natori, Y.; Chenoweth, R. Differences in Rural Landscape Perceptions and Preferences between Farmers and Naturalists. *J. Environ. Psychol.* **2008**, *28*, 250–267. [CrossRef]
67. Kuper, R. Preference and Restorative Potential for Landscape Models That Depict Diverse Arrangements of Defoliated, Foliated, and Evergreen Plants. *Urban For. Urban Green.* **2020**, *48*, 126570. [CrossRef]
68. Hong, X.-C.; Cheng, S.; Liu, J.; Guo, L.-H.; Dang, E.; Wang, J.-B.; Cheng, Y. How Should Soundscape Optimization from Perceived Soundscape Elements in Urban Forests by the Riverside Be Performed? *Land* **2023**, *12*, 1929. [CrossRef]
69. Guo, L.-H.; Cheng, S.; Liu, J.; Wang, Y.; Cai, Y.; Hong, X.-C. Does Social Perception Data Express the Spatio-Temporal Pattern of Perceived Urban Noise? A Case Study Based on 3,137 Noise Complaints in Fuzhou, China. *Appl. Acoust.* **2022**, *201*, 109129. [CrossRef]

*Article*

# Landscape Design Intensity and Its Associated Complexity of Forest Landscapes in Relation to Preference and Eye Movements

Yuanping Shen, Qin Wang, Hongli Liu, Jianye Luo, Qunyue Liu and Yuxiang Lan *

College of Architecture and Planning, Fujian University of Technology, Fuzhou 350108, China
* Correspondence: 19912325@fjut.edu.cn

**Abstract:** Understanding how people perceive landscapes is essential for the design of forest landscapes. The study investigates how design intensity affects landscape complexity, preference, and eye movements for urban forest settings. Eight groups of twenty-four pictures, representing lawn, path, and waterscape settings in urban forests, with each type of setting having two groups of pictures and one group having four pictures, were selected. The four pictures in each group were classified into slight, low, medium, and high design intensities. A total of 76 students were randomly assigned to observe one group of pictures within each type of landscape with an eye-tracking apparatus and give ratings of complexity and preference. The results indicate that design intensity was positively associated with subjective landscape complexity but was positively or negatively related to objective landscape complexity in three types of settings. Subjective landscape complexity was found to significantly contribute to visual preference across landscape types, while objective landscape complexity did not contribute to preference. In addition, the marginal effect of medium design intensity on preference was greater than that of low and high design intensity in most cases. Moreover, although some eye movement metrics were significantly related to preference in lawn settings, none were found to be indicative predictors for preference. The findings enrich research in visual preference and assist landscape designers during the design process to effectively arrange landscape design intensity in urban forests.

**Keywords:** landscape complexity; fractal dimension; eye tracking; preference; urban forest; forest landscape

**Citation:** Shen, Y.; Wang, Q.; Liu, H.; Luo, J.; Liu, Q.; Lan, Y. Landscape Design Intensity and Its Associated Complexity of Forest Landscapes in Relation to Preference and Eye Movements. *Forests* **2023**, *14*, 761. https://doi.org/10.3390/f14040761

Academic Editor: Chi Yung Jim

Received: 19 January 2023
Revised: 24 March 2023
Accepted: 2 April 2023
Published: 7 April 2023



## 1. Introduction

Chinese urbanization has opened up numerous opportunities to expand urban forests. However, a major challenge to landscape designers in China is designing forest landscapes that meet the demands and preferences of the general public. Understanding public perceptions and preferences is critical for effective landscape planning and design. It is the citizens who will ultimately utilize these landscapes [1], and without their support, proposals for urban forest renewal projects may fail [2]. Therefore, it is necessary to gather information about the general public's landscape perceptions and preferences, so that practitioners may incorporate this vital information in the design process to achieve better outcomes.

Previous research has found that environments preferred by the public increase well-being [3] and mental restoration [4–6], as well as generate further positive health outcomes [6,7]. Research has found that landscape preference predicts how well people will function in a given environment [6], and it should be taken into account as an essential factor in the design process [8]. Furthermore, numerous studies have been conducted on the preference for different types of landscapes [9–11], different cultural backgrounds and landscape preferences [12,13], and preferences for potential physical features [14–18].

Few studies have investigated whether varying landscape design interventions in urban forest settings affect respondents' perceptions of landscape complexity and their ratings on preference, particularly regarding Chinese urban forests. Scarce research is available on how these interventions affect landscape complexity and whether the increased complexity influences preferences. Consequently, research on the public's preferences for different levels of landscape design interventions in various forest settings will help designers examine the design outcomes and subsequently develop effective design guidelines. The extent of design intervention refers to design intensity, which was initially put forward by Xu et al. [19] and described as "the amount of the original landscape changed and the degree of artificiality of added elements to the landscape by design." The current study modified this concept to cover a number of different landscape elements used, the complexity of different landscape elements in a configuration, and the extent of landscape maintenance requirements in the present landscape, as well as take it a step further by investigating the influence of design intensity on landscape complexity (including both objective and subjective measures), and thereby on people's preferences.

Research has found that eye movements tend to indicate individuals' preferences for certain landscapes [20]. However, the question of how eye movements can predict these preferences is relatively unexplored, especially for Chinese urban forest settings. In addition, with few studies conducted on the associations between design intensity and eye movement, how landscape design intensities affect eye movements remains unclear. The present study utilizes eye-tracking technology, which offers a great opportunity to explore these questions and provide valuable insights.

### 1.1. The Association between Landscape Design, Complexity, and Preference

Landscape preference, which refers to "liking" specific scenes or places [21], or finding any place aesthetically pleasing [22], has generated an extensive body of published literature [23]. Several studies suggest that landscape design significantly affects public preference. For example, people prefer built environments with natural landscape elements in comparison with those without natural elements [24,25]. Furthermore, the presence of anthropogenic elements in the landscape, such as wind turbines, could have a negative impact on individuals' perceptions [25]. Previous studies also indicate a significant relationship between physical landscape features and people's preferences [26–28]. Within the discussion on physical features, landscape complexity is an important term used for describing visual character [29], and it has been used to explain landscape preference [30]. It refers to the richness and diversity of the visual formation people receive, as well as a measure index for how much is "going on" and how much to look at in a particular scenario [31]. Although there are many visual indicators related to landscape preference, such as historicity, visual scale, and naturalness [32,33], landscape complexity may serve as a bridge between visual aesthetics and landscape ecology [34,35]. Studies suggest that complexity and coherence, which have been examined as potential predictors of landscape preference [36], are closely associated concepts. Landscape complexity, defined based on the distribution, spatial organization, and variation and shape of landscape elements and patterns, can relate to coherence [29,30]. Thus, the present study considers landscape complexity as a good indicator for describing landscape characteristics in urban forest settings, as well as for linking landscape design intensity to preference. However, previous studies that examined landscape and preference were generally only based on subjective ratings of landscape complexity. Additionally, researchers have criticized the existing studies on environmental perception for often lacking quantitative evidence linking the given landscape metrics to human response [30]. Thus, this study uses both objective and subjective measurements of landscape complexity to offer better insights into studies regarding landscape preference and complexity.

Furthermore, previous studies on landscape perception were generally based on extreme examples (e.g., urban vs. rural) [37], and studies on landscape preference were primarily based on photographs without control over image content. These photographs may cause some bias of individuals' perceptions [29], especially when research seeks to test and validate the indicators systematically. As a result, previous studies offer limited guidance in designing landscapes within an urban context, wherein knowledge is required regarding design interventions and preferred level of complexity. One of the aims of this study is thus to examine the association between landscape complexity and preference by using both objective and subjective measures of complexity and ensuring control over the properties of the landscape images.

*1.2. The Objective and Subjective Measurement of Landscape Complexity*

The recently emerged concept of fractal dimension, a parameter used to describe fractured shapes [38], has been extensively used in urban landscape structure studies. Fractal dimension has good ability to consistently describe objective landscape complexity [39]. For instance, Stamps [40] used fractal dimension to describe the physical features of a skyline and attempted to link fractal dimension to preference. In another instance, Cooper [41] employed fractal dimension to characterize the complexity of street edges. In the area of urban design, Robertson [42] investigated fractal dimension in relation to urban character and urban design qualities. Concerning landscape evaluation, fractal dimension has been used to assess fractal characteristics and has been used as a predictor of landscape preference [38]. These studies generally demonstrated a significant relationship between fractal dimension and landscape preference [43,44]. Concerning fractal dimension and landscape complexity, fractal dimension has been described as a statistical quantification of complexity [45]. In addition, it has a good ability to describe complexity of line and composition in the landscape [46]. Sandau and Kurz [47] showed that fractal dimension could be used as a parameter for complexity, as it is related to surface roughness, classification of textures, and line patterns. Hsieh and Lin [39] investigated landscape complexity in relation to preference by using fractal dimension to measure complexity. They pointed out that fractal dimension not only measures the complexity of line and composition but also characterizes the textural details and structural richness. Furthermore, fractal dimension could also be applied to measure the complexity of built environments [43,44]. Moreover, from the perspective of landscape perception and landscape design, the fractal dimension is particularly interesting, since it can be used directly in design work. Therefore, fractal dimension can be regarded as a comprehensive and objective measure of landscape complexity. This study also measures landscape complexity with fractal dimension, which can be easily calculated using software packages such as Fractalyse 2.4 and Benoit 1.3.

Regarding the subjective approach to measuring landscape complexity, previous studies [4,48,49] have commonly asked participants to rate the level of complexity on a Likert scale. For example, to verify the restorative benefits of green landscapes in their study, Kang and Kim [48] asked respondents to rate their perceived complexity using a seven-point scale. In another study, Han asked 274 undergraduate students to report their evaluations of landscape complexity on a five-point Likert-type scale [4]. The evaluations served as controlling and descriptor variables in exploring the relationships between landscape scenic beauty, preference, and restoration. The current study also uses respondent ratings on a seven-point Likert-type scale to subjectively measure landscape complexity.

*1.3. Using Eye-Tracking Technology in Landscape Perception Research*

Questionnaire surveys and in-depth interviews have long been used as subjective approaches to study landscape perception [50]. However, the objective approach, such as eye tracking, has only recently been utilized in landscape studies. The eye-tracking measure can record observers' eye movements when they look at photographs, making it possible to observe the respondents' visual exploration patterns [51]. Dupont et al. [52] have used the eye-tracking measure to explore differences between expert and laypeople's perceptions of landscape photographs. Their study found that these groups may not observe landscapes in the same way and may not even perceive the same landscape features. By analyzing the metrics of eye movement, Valtchanov and Ellard found a longer fixation time for nature scenes than for urban landscapes [53]. However, Berto et al.'s use of eye-tracking technology indicated a greater eye travel distance and number of fixations for urban landscapes than for nature scenes [54]. Both studies formulated and analyzed landscape perception paradigms with the use of eye-tracking technology. However, following the general trend, these studies focused only on landscape properties. Few studies have investigated the association between landscape design intensity, complexity, eye movements, and landscape preferences. There is a need for research focused on landscape design intensity, landscape complexity, and visual exploration.

### 1.4. The Study Objective

The lack of studies concerning objective measures of landscape complexity, and the availability of the relatively new and promising eye-tracking technology, are the key factors for undertaking this study. We use objective and subjective approaches to investigate the landscape design intensity in relation to landscape complexity, visual preference, and eye movements by using urban forest setting photographs. The study seeks to answer the following research questions:

(1) How do landscape design intensities affect objective and subjective landscape complexity and eye movements?
(2) How dose objective and subjective landscape complexity affect visual preference and eye movements?
(3) What are the relationships between eye movement metrics, landscape complexity, and landscape preference?

## 2. Methods

### 2.1. Study Area

The stimuli for the experiment conducted in this study were photographs representing three types of urban forest settings, namely the settings of lawns, paths, and waterscapes, which can be widely seen in the urban forest of Fuzhou, China. These photos were selected from a photo bank of over 678 photos taken by several of the authors in different parks in Fuzhou in similar weather and seasonal conditions. Based on the following criteria, we initially selected 60 images (20 for 1 type of setting): (a) each image should be commonly seen in urban parks of China; (b) each type of landscape image should have similar landscape structures; (c) each type of landscape should include images with varying design intensities; and (d) in the images, it should be feasible to add or remove certain landscape components to create landscapes with certain design intensities (optional). Each type of setting image was then classified into four categories (slight, low, medium, or high design intensity) according to their artificial landscape components, with each category including five images by ten landscape architects. Following this procedure, another ten landscape architects were asked to select two pictures from the five images for each category according to their representativeness and suitability. As there was only minimal difference in design intensities between some images and their counterparts, these images were further modified following the photomontage method to control the landscape design intensity. The photomontage method allows researchers to integrate landscape components to create new landscape settings [55]. Following Xu et al. [19], we added a few sketches or manmade facilities to strengthen the design intensity, and we removed some of these to weaken the design intensity. In addition, the buildings in the background of the images were also removed. To improve the realistic look of the modified images, the addition or removal of certain landscape elements (such as bench, stone, lamppost, etc.) was based on real urban forest settings widely seen in China, and all added components were derived from actual forest setting photos that had similar landscape structures. To ensure the rationality and suitability of the landscape contexts in the manipulated images and the landscape design intensity they represent, these images were subsequently reviewed by senior researchers with research experience on similar topics, as well as by ten landscape architects with extensive experience in landscape design.

Finally, a total of 24 photos were selected for use in the eye-tracking experiment and preference rating survey (Figure 1). The photos were divided into six groups (each type of setting having two groups, A and B), with each group comprising four photos, which indicated slight, low, medium, and high design intensity. The six groups for the three types of settings are lawn settings A and B; path settings A and B; and waterscape settings A and B.

**Figure 1.** Presentation of the stimuli.

### 2.2. Eye-Tracking Apparatus and Measurements

Eye Link 1000 Plus, which has been shown to have high accuracy and precision in eye-tracking measurement, was used to record eye movements. The eye-tracking technology can record observers' point-of-regard every millisecond, as well as continuously record the observers' eye movements while viewing photographs. The records are displayed on a 19 in color monitor with a resolution of 1280 × 1024 pixels. The technology uses infrared light to reflect off the cornea and retina of the eye using low-power infrared light [56,57]. The reflected signal reveals the precise location of the point-of-regard, which is expressed by horizontal and vertical coordinates [56]. Subsequently, the entire gaze pattern is recorded, including fixations and saccades [57]. Although both eyes are engaged in observing the photo, the movements of only one eye are recorded at a measurement rate of 1000 Hz.

Participants' fixations and saccades were continuously recorded by the system throughout the experiment. According to Poole and Ball [57], a fixation is defined as the period of time when the eyes are relatively stationary, which allows for visual perception. It is recommended to set the lower threshold for determining a fixation t at 100 milliseconds. Consequently, in this study, a fixation was depicted as a stationary eye position lasting a minimum of 100 milliseconds. The fixation count and duration were recorded and used to analyze gaze pattern characteristics. Fixation count is the number of fixations in the photo, and fixation duration is the duration time of a fixation. With respect to the metric of saccade, saccade count and saccade amplitude (degree) were used to observe the main viewing pattern. Saccades are the eye movements that move the fovea rapidly from one point of interest to another. Saccade count is the number of saccades in the photo, and saccade amplitude is the angular distance the eye travels during the movement.

### 2.3. Study Participants

The study recruited 76 people (37 males and 39 females) between 20 and 58 years of age to voluntarily participate in the landscape preference rating and eye-tracking experiment. The sample was divided into two groups, each with 38 participants, which was deemed large enough for eye-tracking research and sufficient to detect major effects [58,59]. All participants had normal or corrected-to-normal vision. The participants were given brief practical information about the eye-tracking experiment. The participants were randomly assigned to view one group of photos from each type of setting, and each participant viewed 4 groups of 12 photos in total in random order.

### 2.4. Research Procedure

The experiment was conducted one subject at a time over a period of fifteen days in September 2020 at a lab of a university. After arriving at the lab, the subjects first read and signed the informed consent letter and then were randomly assigned to view one group of images from each setting, for a total of three groups of photographs. The photographs were displayed on the screen for 10 s each in random order to avoid the effects of a fixed order. This specific timespan for observation was derived from prior similar studies [52,60]. Participants were asked to observe the images freely, without any task in mind, because this is how people usually observe landscapes in real life [50]. During the experiment, the subjects were seated at a distance of approximately 50 cm from the color monitor. Before each test, brief instructions were given to the participant, and a 9-dot calibration procedure was executed to match the pupil-center/corneal reflection relationship with the specific x-, y-coordinates of the fixed dots [51]. As a result, an accurate calibration over the entire screen was achieved. To eliminate the effect of unintentional movements or eye breaks, the same calibration procedure was repeated for every image [50]. Before each trial, the participants were informed to look at a dot shown on the center of a blank screen. This was carried out to reduce measurement errors and ensure consistency in the initial conditions of the observation path of each photo [50].

During the experiment, the participants were prohibited from speaking to ensure their full concentration and were asked to use a chin rest to restrict head movements, thus eliminating any deviations. However, the participants were allowed to take a break at any time to avoid the effects of eye fatigue caused by repeatedly looking at a monitor [61]. Fatigue can cause a decrease in fixation count [62] and in the accuracy of observation [63].

After the eye-tracking experiment, the participants were asked to assign ratings of preference and perceived complexity for the photos on a seven-point scale. Participants were asked to rate the photos after the experiment to avoid influence on their viewing patterns in advance and to ensure a free viewing pattern to increase the accuracy of the eye-tracking measurements. In addition, the participants also provided basic demographic information, such as gender and age.

*2.5. The Measurement of Fractal Dimension*

Objective landscape complexity was measured in terms of fractal dimension. Fractal dimension can be calculated using different methods, including the box counting method, divider method, and area–perimeter method [39]. Of those listed, the box counting method is the most used approach to measure fractal dimension among the studies concerning fractal dimension and landscape preference [39]. The fractal dimension using box counting can be used as a comparative measure of visual complexity to quantify the change in landscape [64]. The box counting method involved dividing the graph to be measured into small grids of equal side length δ and calculating how many small grids the pattern occupied. The current study also employs the box counting method to estimate the fractal dimension of landscapes using Fractalyse 2.4. Developed by Gilles Vuidel in 2006, this software was initially developed to measure fractal dimension of built-up areas of cities [65]. Thus, it is particularly suitable to study the fractal dimension of built urban forest landscapes.

*2.6. Statistical Analysis*

Statistical analysis was conducted with SPSS 23.0 software. Firstly, objective landscape complexity and the mean of subjective landscape complexity for every photograph were computed. Spearman's correlation analysis was conducted to explore the relationship between landscape design intensity and objective and subjective landscape complexity. A one-way analysis of variance (ANOVA) was conducted within each group of images and each type of setting to determine whether participants' preference ratings were significantly different among the four design intensities, followed by a post hoc pairwise comparison using Tukey's Honestly Significant Difference (HSD) test to determine where the difference occurred. As the eye-tracking data were not parametric, a nonparametric Kruskal–Wallis test was conducted to examine whether participants' eye movement metrics were significantly different among different design intensities within each type of setting. Finally, Spearman's correlation analysis and linear regression analysis were employed to explore the associations between objective and subjective landscape complexity, preference, and eye movement metrics. The ratings for each image were calculated by the mean value of all participants, while the ratings for each type of setting were determined by the mean score of two related pictures.

## 3. Results

*3.1. Objective and Subjective Landscape Complexity*

The results of the objective and subjective landscape complexity measurements of each image are presented in Figure 2. All objective landscape complexity values ranged between 1.814 and 1.911, which corresponds with the standard range of fractal dimensions (fractal dimensions normally range between 1 and 2). With respect to subjective landscape complexity, the images that displayed slight design intensity were regarded as having the lowest level of complexity in lawn settings A, path settings A and B, and waterscape settings B. However, the images that displayed high design intensity were regarded as having the highest level of complexity across the six groups of settings.

Additional Spearman's correlation analysis results (Table 1) revealed that except in lawn settings B, landscape design intensity has a significant positive or negative relationship with objective complexity in different groups of settings. The relationships between design intensity and subjective complexity in the six groups of settings were all positive. Additionally, significant positive relationships between design intensity and subjective complexity were found in each type of setting, while design intensity positively associated with objective landscape complexity only in lawn and path settings.

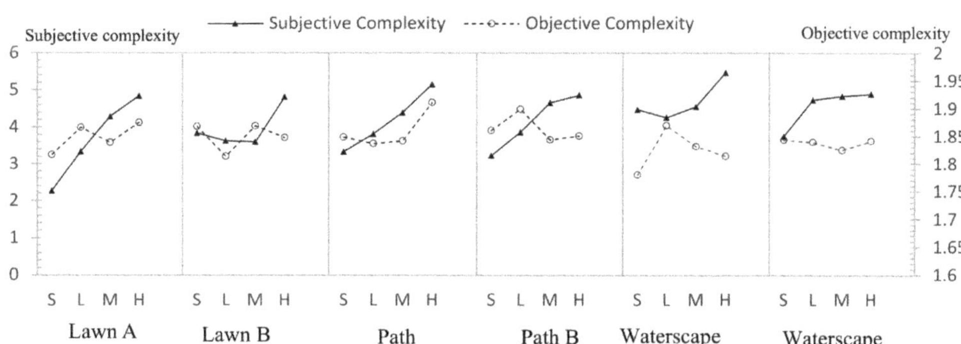



**Figure 2.** Objective and subjective landscape complexity for each image.

**Table 1.** Correlations between landscape design intensity and complexity in each group and type of settings.

| Complexity | Landscape Design Intensity | | | | | | | | |
|---|---|---|---|---|---|---|---|---|---|
| | **Lawn A** | **Lawn B** | **Lawn** | **Path A** | **Path B** | **Path** | **Waterscape A** | **Waterscape B** | **Waterscape** |
| OLC | 0.80 ** | 0.00 | 0.39 *** | 0.40 *** | −0.60 *** | 0.15 * | 0.20 * | −0.40 *** | −0.15 * |
| SLC | 0.71 *** | 0.33 *** | 0.53 *** | 0.63 *** | 0.60 *** | 0.61 *** | 0.34 *** | 0.41 *** | 0.37 *** |

Note: OLC: objective landscape complexity; SLC: subjective landscape complexity; *: $p \leq 0.05$, **: $p \leq 0.01$, and ***: $p \leq 0.001$.

### 3.2. Comparison of Preference for Each Setting

The mean preference rating for each image and each type of setting is shown in Figure 3. Excepting path settings A and the type of path settings, the photos with high design intensity received higher preference scores than their three counterparts. Additional Spearman's correlation analysis of lawn, path, and waterscape settings revealed that the correlation coefficients between landscape design intensity and preference were 0.326, 0.332, and 0.352 respectively. The ANOVA results indicate that all preference ratings were significantly different in each group (excepting path settings A) and in each type of setting (F values ranged between 3.464 and 16.508). Furthermore, the post hoc pairwise comparison results using Tukey's HSD test are presented in Table 2. These results indicate that in terms of all types of settings, the pairwise comparisons of preference scores between slight and high design intensity were significantly different. However, in terms of each group of setting, the pairwise comparisons of preference between slight and high design intensity in lawn settings B was not significant. In all types of settings and in each group of settings, medium vs. high design intensity in terms of preference were nonsignificant. The results seem to indicate that the effect of landscape design intensity on the increase in preference is linked to landscape types.

### 3.3. Marginal Effect of Landscape Design Intensity on Preference Ratings

To further explore the effect of landscape design intensity on increased preference, we calculated the marginal effects of design intensity on landscape preference scores (Figure 4). In terms of the six groups of settings, the marginal effects of low design intensity were only higher than those of medium and high design intensity in lawn settings A. However, the marginal effects of high design intensity were only higher than those of low and medium design intensity in waterscape settings B. In addition, the marginal effects of medium design intensity were higher than those of low and high design intensity in other groups of settings. For the types of lawn and path settings, the marginal effects of medium design intensity were greater than those of low and high design intensity. However, for the type of waterscape settings, the marginal effects of high design intensity were greater than those of low and medium design intensity. These results further indicate that the marginal effect of medium design intensity on preference was greater than that of low and high design intensity in most cases, and the influence of design intensity on improved preference was dependent on the landscape type.

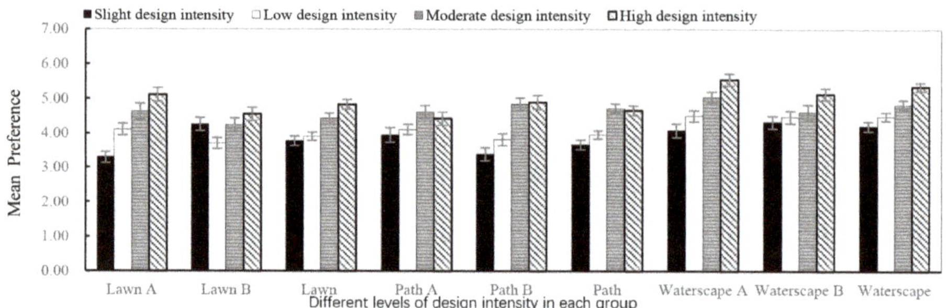


**Figure 3.** The mean of participants' preference ratings for the four images in each group. Error bar: standard error.

**Table 2.** Tukey's HSD test results for individuals' preference in groups.

| I | J | Mean Difference (I–J) | | | | | | | |
|---|---|---|---|---|---|---|---|---|---|
| | | **Lawn A** | **Lawn B** | **Lawn** | **Path B** | **Path** | **Waterscape A** | **Waterscape B** | **Waterscape** |
| S | L | −0.82 ** | 0.55 | −0.13 | −0.42 | −0.29 | −0.42 | −0.16 | −0.29 |
| | M | −1.34 *** | 0.03 | −0.66 ** | −1.45 *** | −1.05 *** | −0.97 *** | −0.29 | −0.63 ** |
| | H | −1.82 *** | −0.29 | −1.05 *** | −1.50 *** | −0.99 *** | −1.47 *** | −0.82 * | −1.14 *** |
| L | M | −0.52 | −0.53 | −0.53* | −1.03 ** | −0.76 *** | −0.55 | −0.13 | −0.34 |
| | H | −1.00 ** | −0.84 ** | −0.92 *** | −1.08 *** | −0.70 ** | −1.05 *** | −0.66 | −0.86 *** |
| M | H | −0.47 | −0.32 | −0.39 | −0.05 | .07 | −0.50 | −0.53 | −0.51 |

Note: S: slight design intensity; L: low design intensity; M: medium design intensity; H: high design intensity; *: $p \leq 0.05$, **: $p \leq 0.01$, and ***: $p \leq 0.001$.

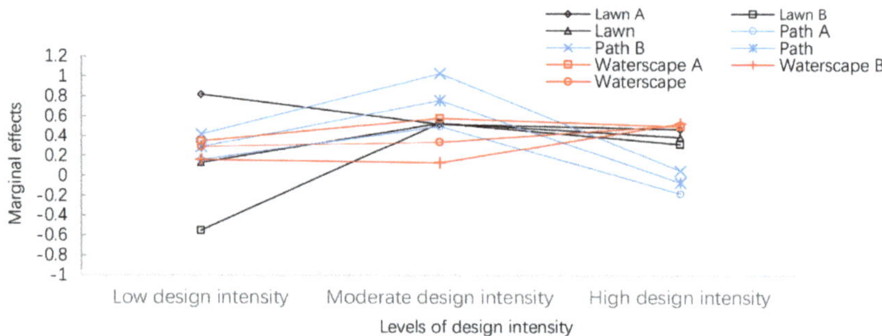


**Figure 4.** Marginal effects of three design intensities on landscape preference in each group and type of landscape.

*3.4. Comparisons of Eye Movement Metrics among Different Levels of Design Intensities within Each Type of Setting*

The Kruskal–Wallis test results are presented in Table 3. They reveal that there were no significant differences in any eye movement metrics between images with different design intensities in the type of path settings. There was also no significant difference in average saccade amplitude between images with different design intensities in the types of lawn ($\chi^2 = 1.165$, $p = 0.761$) and waterscape settings ($\chi^2 = 6.121$, $p = 0.106$). However, significant differences were detected in fixation count, average fixation duration, and saccade count between imagines with various design intensities in the path and waterscape landscapes. Additional pairwise comparison suggests that the fixation count and saccade count for images with high design intensity were significantly more than those with slight design intensity ($p < 0.01$). In addition, the average fixation duration for photographs with high design intensity was significantly shorter than that of images with slight design intensity ($p < 0.01$).

**Table 3.** The Kruskal–Wallis test results of eye movement metrics within each type of setting.

| Settings | Eye Movements | N | $\chi^2$ | df | Mean Rank | | | | $p$ | Real Mean Values | | | |
|---|---|---|---|---|---|---|---|---|---|---|---|---|---|
| | | | | | Sligh | Low | Medium | High | | Sligh | Low | Medium | High |
| Lawn | FC | 76 | 20.079 | 3 | 126.36 | 138.81 | 158.82 | 186.02 | 0.000 | 27.18 | 27.96 | 29.70 | 30.84 |
| | AFD | 76 | 12.310 | 3 | 173.01 | 161.29 | 150.59 | 125.11 | 0.006 | 332.17 | 331.88 | 292.84 | 285.66 |
| | SC | 76 | 19.013 | 3 | 127.59 | 139.33 | 157.28 | 185.81 | 0.000 | 26.50 | 27.29 | 28.92 | 30.14 |
| | ASA | 76 | 1.165 | 3 | 152.07 | 150.30 | 146.40 | 161.23 | 0.761 | 5.83 | 5.70 | 5.66 | 5.94 |
| Path | FC | 76 | 4.322 | 3 | 142.34 | 142.13 | 159.91 | 165.62 | 0.229 | 28.70 | 28.58 | 30.09 | 30.51 |
| | AFD | 76 | 3.202 | 3 | 159.43 | 163.30 | 144.95 | 142.31 | 0.361 | 358.25 | 324.29 | 288.40 | 283.41 |
| | SC | 76 | 4.604 | 3 | 142.16 | 141.95 | 159.07 | 166.82 | 0.203 | 27.92 | 27.83 | 29.33 | 29.86 |
| | ASA | 76 | 5.366 | 3 | 137.80 | 158.80 | 167.98 | 145.41 | 0.147 | 5.65 | 5.98 | 6.18 | 5.80 |
| Waterscape | FC | 76 | 14.179 | 3 | 124.84 | 148.18 | 160.12 | 176.86 | 0.003 | 27.30 | 28.84 | 30.17 | 31.07 |
| | AFD | 76 | 12.494 | 3 | 177.97 | 155.22 | 148.80 | 128.00 | 0.006 | 359.08 | 320.53 | 294.29 | 276.44 |
| | SC | 76 | 13.876 | 3 | 124.83 | 148.74 | 160.09 | 176.34 | 0.003 | 26.54 | 28.11 | 29.41 | 30.30 |
| | ASA | 76 | 6.121 | 3 | 160.94 | 167.16 | 135.06 | 146.84 | 0.106 | 5.71 | 5.90 | 5.45 | 5.51 |

Note: N: number of participants; $\chi^2$: chi-square; df: degrees of freedom; FC: fixation count; AFD: average fixation duration; SC: saccade count; ASA: average saccade amplitude.

### 3.5. Identifying the Correlations between Landscape Eye Movements, Complexity, and Preference

The Spearman correlation analysis was conducted to determine whether preference levels were consistent with landscape complexity levels and whether the eye movement metrics were related to landscape complexity and preference (Table 4). Subjective landscape complexity was positively and significantly related to preference across three types of settings, while objective landscape complexity was not correlated with preference. In addition, both subjective and objective landscape complexity was found to be significantly and positively related to fixation count and saccade count and negatively related to average fixation duration in lawn settings. Moreover, subjective landscape complexity had a positive relationship with average saccade amplitude in path settings, while objective landscape complexity had a positive relationship with average saccade amplitude in waterscape settings.

**Table 4.** Correlations among eye movement parameters, preference andcomplexity.

| | Lawn | | | Path | | | Waterscape | | |
|---|---|---|---|---|---|---|---|---|---|
| | P | OLC | SLC | P | OLC | SLC | P | OLC | SLC |
| Fixation Count | 0.19 ** | 0.14 * | 0.23 *** | 0.09 | 0.04 | 0.08 | 0.09 | −0.11 | 0.07 |
| Average Fixation Duration | −0.13 * | −0.12 * | −0.16 ** | −0.08 | −0.03 | −0.08 | −0.06 | 0.08 | −0.07 |
| Saccade Count | 0.18 ** | 0.14 * | 0.22 *** | 0.09 | 0.04 | 0.08 | 0.09 | −0.10 | 0.07 |
| Average Saccade Amplitude | 0.01 | 0.04 | −0.05 | 0.02 | −0.03 | 0.11 * | −0.09 | 0.11 * | 0.00 |
| OLC | 0.02 | - | 0.19 ** | −0.11 | - | 0.10 | −0.05 | - | −0.28 *** |
| SLC | 0.53 *** | 0.19*** | - | 0.50 *** | 0.10 | - | 0.52 *** | −0.28 *** | - |

Note: P: preference; OLC: objective landscape complexity; SLC: subjective landscape complexity; FC: fixation count; AFD: average fixation duration; SC: saccade count; ASA: average saccade amplitude; *: $p \leq 0.05$, **: $p \leq 0.01$, ***: $p \leq 0.001$.

Significant positive relationships between landscape preference and fixation count and saccade count, and negative association between preference and average fixation duration, were found only in lawn settings. However, further linear regression analysis revealed that none of the eye movement metrics were significant predictors for objective and subjective landscape complexity or for landscape preference. Additionally, linear regression analysis (Table 5) revealed that subjective landscape complexity was a positive predictor for landscape preference in the three types of landscapes. These results indicate that the significant relationship between eye movement parameters, landscape complexity, and preference is largely determined by landscape type.

**Table 5.** Regression results for the effect of landscape complexity on preference.

| | Lawn | | | | Path | | | | Waterscape | | | |
|---|---|---|---|---|---|---|---|---|---|---|---|---|
| | B | SE B | β | *p* | B | SE B | β | *p* | B | SE B | β | *p* |
| Constant | 2.17 | 0.20 | | <0.001 | 1.83 | 0.23 | | <0.001 | 1.86 | 0.28 | | <0.001 |
| SLC | 0.54 | 0.05 | 0.53 | <0.001 | 0.58 | 0.05 | 0.53 | <0.001 | 0.62 | 0.06 | 0.52 | <0.001 |
| Adjusted R2 | | 0.28 | | <0.001 | | 0.27 | | <0.001 | | 0.27 | | <0.001 |

Note: OLC: objective landscape complexity; SLC: subjective landscape complexity.

## 4. Discussion

### 4.1. Landscape Design Intensity, Complexity, and Preference Ratings

This study revealed that variations in landscape design intensity make different contributions to objective and subjective landscape complexity and to landscape preference. As design intensity increased, participants' responses toward subjective complexity also consistently increased in lawn settings A, path settings A and B, and waterscape settings B (see Figure 2). However, this trend was not found in any groups of settings in terms of objective complexity. Previous studies have described landscape complexity arising from the diversity and richness of landscape elements and features [66,67]. In the present study, landscape design intensity was judged based largely on the amount or richness of artificial landscape elements. Thus, with higher design intensity, it is reasonable that the presence of increasing artificial landscape elements enhance participants' feeling of landscape complexity. With regards to the inconsistent trend between design intensity and objective landscape complexity, the reasons may be attributed to the greater amount of artificial landscape elements not being necessarily related to a higher number of occupied small grids. In addition, subjective landscape complexity was positively related to objective landscape complexity in lawn and path settings, while negatively related to objective landscape complexity in waterscape settings. This may further indicate that there are differences between subjective landscape complexity and objective landscape complexity of various green space settings, and participants perception of landscape complexity does not corresponded with fractal dimension. Additional studies using both subjective and objective approaches to measure landscape complexity would provide a more comprehensive insight into this and verify the inference of the present study.

Complexity has been identified as an important characteristic [29] and has been proven to be a good predictor of preference [30]. The current study also found that subjective landscape complexity had a significant positive relationship with landscape preference, regardless of landscape type. This finding supports Kuper's finding of a positive relationship between participants' ratings of complexity and preference [36], as well as the findings of some other research [29,38,68]. Moreover, this finding also echoes [69]'s landscape preference model, which defined complexity as an important construct of preference. However, it is worth noting that subjective landscape complexity positively contributed to preference in all three types of settings, while objective landscape complexity only significantly contributed to preference in lawn settings. This may be attributed to the different effects of design intensity on subjective landscape complexity and objective landscape complexity.

Overall, the study findings suggest that design intensity positively contributes to subjective landscape complexity across landscape types and positively influences respondents' preferences. When landscape architects design forest landscapes that suit public preference, they should prioritize and consider increasing design intensity by arranging artificial landscape elements or organize the landscape contents with a certain level of design intensity to increase individuals' perceived complexity. However, previous studies also indicated an inverted-U-shaped relationship between landscape preference and complexity [70,71], which means a moderate level of landscape complexity may obtain higher preference, and a high level of landscape complexity may lead to a decrease in individuals' preference. In addition, the marginal effect of medium design intensity on preference was greater than that of low and high design intensity in most groups of settings. Thus, creating landscapes with a medium level of complexity may be more suitable and reliable to increase individuals' preference in forest landscape design, which may also help to avoid overdesigning. Although determining a moderate level of complexity or design intensity may be difficult, comparisons among different landscape proposals may help determine which proposal may reflect a medium level of design intensity or complexity. In addition, the present settings with moderate design intensity in the present study would also serve as references.

### 4.2. Design Intensities, Complexity, and Eye Movement

There were significant differences in fixation count, average fixation duration, and saccade count among different design intensities in lawn and waterscape settings. This may largely account for the variations in landscape complexity, as design intensity positively contributed to subjective landscape complexity. Egaña et al. found that landscape complexity is strongly related with eye movement behaviors [72]. Wohlwill argued that as landscape complexity increases, the amount of exploratory activity in terms of eye movement seems to increase linearly [71]. Likewise, Dupont et al. found that the complexity of a given landscape has a great influence on visual exploratory activity, and that the increased complexity of the images can result in a greater fixation and saccade count [51]. In other words, there is a positive relationship between landscape complexity and the metrics of fixation and saccade counts. This is understandable, as studies have claimed that a more complex landscape provides a greater volume of information to process [71] and a greater interest value for the stimulus [70]. Part of our findings indicated there are more fixation and saccade counts for high design intensity settings than slight design intensity in lawn and waterfront settings, which also supports this notion.

However, we also detected nonsignificant differences in any eye movements between different design intensities in path settings. In addition, there was only significant positive relationship between average saccade amplitude and subjective landscape complexity in path settings. We speculated that the inconsistent results in the different settings are because, in the present study, participants' visual exploration was affected not only by landscape design intensity but also by the configuration of landscape components and landscape types. This speculation is in line with the notion that visual exploration is linked with landscape structure [69]. Another possible reason for the inconsistent results is that there may be a threshold for complexity. Perhaps complexity has a significant impact on eye movement metrics only when it reaches a certain threshold. This inference is supported by Dupont et al.'s findings that in heavily urbanized landscapes, if the number of built areas represented in an image reaches a threshold, the built areas will no longer catch the viewer's eyes [51]. In this study, the pictures within each group have a similar landscape structure and may not differ much from their counterparts, which means the complexity between different images within most of the groups may not reach the necessary threshold.

### 4.3. Preference and Eye Movements

It was also found that preference and eye movement patterns were inconsistent across the three settings. Significant relationships between fixation count, average fixation duration, saccade count, and preference were found only in the lawn landscapes. This indicated that the associations between eye movements and preference were determined by landscape types. Supporting our results, a recent study by Huang and Lin [73] revealed strong relationships between landscape preference and fixation count in mountain, aquatic, and forest landscapes, and that people have different preferences and viewing behaviors when observing different landscape types. The present results do not show any significant role of eye movement metrics in predicting preference. Similarly, an earlier study focusing on nightscape preference demonstrated that total fixation duration, total scan path length, and total time spent on images were not good predictors of preference [74]. Perhaps this could also be attributed to the influence of different landscape configuration patterns or image contents. For example, people usually prefer a natural landscape to an urban landscape. However, higher fixations and saccades for urban landscapes were observed despite people's lesser preference for urban landscapes [51].

*4.4. Limitations and Further Research*

While we selected three different but commonly seen urban forest landscape settings and classified four design intensities for each type of setting, the study only used a small sample of landscape scenes. A larger number of settings with different landscape components and a larger sample of respondents should be considered in future research to enhance the validity of the findings. Another limitation is that we only measured the complexity of the landscape. Although landscape complexity has been regarded as a good indicator for describing landscape characteristics, further studies measuring more variables, such as biodiversity, coherence, and naturalness, could be helpful to strengthen understanding of the link between landscape complexity and preference. Following existing studies [52,60], participants' viewing time for one image was limited to 10 s. However, this may skew the eye-tracking results. Additional studies should allow participants to observe the images freely without a time limit. Additionally, although the same group of landscapes were kept the similar colors of water and sky, the brownish color of the water and the white sky may have affected the participants ratings. Finally, we measured objective landscape complexity using fractal dimension with the box counting method, which is recommended by many architects. Additional studies may conduct similar research by using spectral entropy or other approaches to measure objective landscape complexity.

## 5. Conclusions

This study combined objective and subjective landscape complexity to investigate the effects of landscape design intensity on preference and eye movement. Our study suggested that design intensity can positively affect individuals' subjective landscape complexity and promote preferences across various kinds of landscapes, but either significantly or nonsignificantly contribute to objective landscape complexity in different types of landscapes. The significant relationship between objective or subjective landscape complexity, or preference and eye movement metrics, was also dependent on landscape types. However, none of the eye movements were significant predictors for preference in any landscape. These results can enhance our understanding of landscape design intensity in relation to preference and eye movements. They also provided valuable information for urban forestry design to improve public preference. For practitioners, incorporating these findings into the design process to create forest landscapes with medium design intensity could contribute to improved designs for future urban forests in China. For researchers, further research regarding landscape complexity, preference, and eye movements should include more landscape types, as well as both objective and subjective measures of complexity. Thus, the present study not only enriches current research into landscape complexity but also provides a reference for those seeking to promote the outcome of urban forest landscape design. Additionally, it also demonstrates the potential contributions of eye-tracking technology in the visual landscape preference study field.

**Author Contributions:** Conceptualization, Y.S. and Y.L.; methodology, Q.W.; software, Q.W.; validation, Q.L., Y.S. and Y.L.; formal analysis, H.L.; investigation, J.L.; resources, Y.S.; data curation, Y.S.; writing—original draft preparation, Y.S.; writing—review and editing, Y.L.; visualization, Q.W.; supervision, Y.S.; project administration, Q.L.; funding acquisition, Q.L. All authors have read and agreed to the published version of the manuscript.

**Funding:** The authors gratefully acknowledge the financial support from Key Laboratory of New Technology for Construction of Cities in Mountain Area, Ministry of Education, Chongqing University under No. LNTCCMA-20220103, the financial support from The Second Batch of 2021 MOE of PRC Industry-University Collaborative Education Program under No. 202102055016 (Kingfar-CES "Human Factors and Ergonomics"), the financial support from Fujian Provincial Department of Education under No. JAT220220, and from the China Scholarship Council under 202208350037.

**Conflicts of Interest:** The authors declare no conflict of interest.

## References


1. De Groot, R. Function-analysis and valuation as a tool to assess land use conflicts in planning for sustainable, multi-functional landscapes. *Landsc. Urban Plan.* **2006**, *75*, 175–186. [CrossRef]
2. Junker, B.; Buchecker, M. Aesthetic preferences versus ecological objectives in river restorations. *Landsc. Urban Plan.* **2008**, *85*, 141–154. [CrossRef]
3. Suppakittpaisarn, P.; Jiang, B.; Slavenas, M.; Sullivan, W.C. Does density of green infrastructure predict preference? *Urban For. Urban Green.* **2019**, *40*, 236–244. [CrossRef]
4. Han, K.T. An exploration of relationships among the responses to natural scenes: Scenic beauty, preference, and restoration. *Environ. Behav.* **2010**, *42*, 243–270. [CrossRef]
5. Korpela, K.M.; Hartig, T.; Kaiser, F.G.; Fuhrer, U. Restorative experience and self-regulation in favorite places. *Environ. Behav.* **2001**, *33*, 572–589. [CrossRef]
6. Van den Berg, A.E.; Koole, S.L.; van der Wulp, N.Y. Environment preference and restoration: (How) are they related? *J. Environ. Psychol.* **2003**, *23*, 135–146. [CrossRef]
7. Korpela, K.; Hartig, T. Restorative qualities of favorite places. *J. Environ. Psychol.* **1996**, *16*, 221–233. [CrossRef]
8. Lee, S.; Lei, T.C.; Wu, S.C.; Hou, J.S.; Tzeng, G.H. Applying eye tracking technology to examine landscape preference using prospect refuge theory. *Outdoor Recreat. Res.* **2015**, *28*, 1–18.
9. De Groot, W.T.; van den Born, R.J. Visions of nature and landscape type preferences: An exploration in The Netherlands. *Landsc. Urban Plan.* **2003**, *63*, 127–138. [CrossRef]
10. Wang, R.; Zhao, J.; Liu, Z. Consensus in visual preferences: The effects of aesthetic quality and landscape types. *Urban For. Urban Green.* **2016**, *20*, 210–217. [CrossRef]
11. White, E.V.; Gatersleben, B. Greenery on residential buildings: Does it affect preferences and perceptions of beauty? *J. Environ. Psychol.* **2011**, *31*, 89–98. [CrossRef]
12. Buijs, A.E.; Elands, B.H.; Langers, F. No wilderness for immigrants: Cultural differences in images of nature and landscape preferences. *Landsc. Urban Plan.* **2009**, *91*, 113–123. [CrossRef]
13. Yang, B.E.; Brown, T.J. A cross-cultural comparison of preferences for landscape styles and landscape elements. *Environ. Behav.* **1992**, *24*, 471–507. [CrossRef]
14. Gao, T.; Zhu, L.; Zhang, T.; Song, R.; Zhang, Y.; Qiu, L. Is an environment with high biodiversity the most attractive for human recreation? A case study in Baoji, China. *Sustainability* **2019**, *11*, 4086. [CrossRef]
15. Kothencz, G.; Kolcsár, R.; Cabrera-Barona, P.; Szilassi, P. Urban green space perception and its contribution to well-being. *Int. J. Environ. Res. Public Health* **2017**, *14*, 766. [CrossRef]
16. Paul, S.; Nagendra, H. Factors influencing perceptions and use of urban nature: Surveys of park visitors in Delhi. *Land* **2017**, *6*, 27. [CrossRef]
17. Sayadi, S.; González-Roa, M.C.; Calatrava-Requena, J. Public preferences for landscape features: The case of agricultural landscape in mountainous Mediterranean areas. *Land Use Policy* **2009**, *26*, 334–344. [CrossRef]
18. Svobodova, K.; Sklenicka, P.; Vojar, J. How does the representation rate of features in a landscape affect visual preferences? A case study from a post-mining landscape. *Int. J. Min. Reclam. Environ.* **2015**, *29*, 266–276. [CrossRef]
19. Xu, W.; Zhao, J.; Huang, Y.; Hu, B. Design intensities in relation to visual aesthetic preference. *Urban For. Urban Green.* **2018**, *34*, 305–310. [CrossRef]
20. Birmingham, E.; Bischof, W.F.; Kingstone, A. Saliency does not account for fixations to eyes within social scenes. *Vis. Res.* **2009**, *49*, 2992–3000. [CrossRef]
21. Peschardt, K.K.; Stigsdotter, U.K. Associations between park characteristics and perceived restorativeness of small public urban green spaces. *Landsc. Urban Plan.* **2013**, *112*, 26–39. [CrossRef]
22. Hartig, T.; Staats, H. The need for psychological restoration as a determinant of environmental preferences. *J. Environ. Psychol.* **2006**, *26*, 215–226. [CrossRef]
23. Purcell, T.; Peron, E.; Berto, R. Why do preferences differ between scene types? *Environ. Behav.* **2001**, *33*, 93–106. [CrossRef]
24. Sullivan, W.C.; Lovell, S.T. Improving the visual quality of commercial development at the rural–urban fringe. *Landsc. Urban Plan.* **2006**, *77*, 152–166. [CrossRef]
25. Yang, J.; Zhao, L.; Mcbride, J.; Gong, P. Can you see green? Assessing the visibility of urban forests in cities. *Landsc. Urban Plan.* **2009**, *91*, 97–104. [CrossRef]

26. Mundher, R.; Abu Bakar, S.; Maulan, S.; Mohd Yusof, M.J.; Al-Sharaa, A.; Aziz, A.; Gao, H. Aesthetic quality assessment of landscapes as a model for urban forest areas: A systematic literature review. *Forests* **2022**, *13*, 991. [CrossRef]
27. Strumse, E. Environmental attributes and the prediction of visual preferences for agrarian landscapes in Western Norway. *J. Environ. Psychol.* **1994**, *14*, 293–303. [CrossRef]
28. Svobodova, K.; Sklenicka, P.; Molnarova, K.; Salek, M. Visual preferences for physical attributes of mining and post-mining landscapes with respect to the sociodemographic characteristics of respondents. *Ecol. Eng.* **2012**, *43*, 34–44. [CrossRef]
29. Ode, Å.; Miller, D. Analysing the relationship between indicators of landscape complexity and preference. *Environ. Plan. B Plan. Des.* **2011**, *38*, 24–40. [CrossRef]
30. Ode, Å.; Hagerhall, C.M.; Sang, N. Analysing visual landscape complexity: Theory and application. *Landsc. Res.* **2010**, *35*, 111–131. [CrossRef]
31. Kaplan, R.; Kaplan, S. *The Experience of Nature: A Psychological Perspective*; Cambridge University Press: Cambridge, UK, 1989; pp. 50–53.
32. Ode, Å.; Tveit, M.S.; Fry, G. Capturing landscape visual character using indicators: Touching base with landscape aesthetic theory. *Landsc. Res.* **2008**, *33*, 89–117. [CrossRef]
33. Tveit, M.; Ode, Å.; Fry, G. Key concepts in a framework for analysing visual landscape character. *Landsc. Res.* **2006**, *31*, 229–255. [CrossRef]
34. Dramstad, W.E.; Fry, G.; Fjellstad, W.J.; Skar, B.; Helliksen, W.; Sollund, M.L.; Tveit, M.S.; Geelmuyden, A.K.; Framstad, E. Integrating landscape-based values-Norwegian monitoring of agricultural landscapes. *Landsc. Urban Plan.* **2001**, *57*, 257–268. [CrossRef]
35. Fry, G.; Tveit, M.S.; Ode, Å.; Velarde, M.D. The ecology of visual landscapes: Exploring the conceptual common ground of visual and ecological landscape indicators. *Ecol. Indic.* **2009**, *9*, 933–947. [CrossRef]
36. Kuper, R. Evaluations of landscape preference, complexity, and coherence for designed digital landscape models. *Landsc. Urban Plan.* **2017**, *157*, 407–421. [CrossRef]
37. Velarde, M.D.; Fry, G.; Tveit, M. Health effects of viewing landscapes–Landscape types in environmental psychology. *Urban For. Urban Green.* **2007**, *6*, 199–212. [CrossRef]
38. Hagerhall, C.M.; Purcell, T.; Taylor, R. Fractal dimension of landscape silhouette outlines as a predictor of landscape preference. *J. Environ. Psychol.* **2004**, *24*, 247–255. [CrossRef]
39. Hsieh, M.; Lin, Y. The effect of landscape complexity on natural landscape preference. *City Plan.* **2011**, *45*, 427–447. (In Chinese)
40. Stamps, A.E. Fractals, skylines, nature and beauty. *Landsc. Urban Plan.* **2002**, *60*, 163–184. [CrossRef]
41. Cooper, J. Assessing urban character: The use of fractal analysis of street edges. *Urban Morphol.* **2005**, *9*, 96–107. [CrossRef]
42. Robertson, L. A new theory for urban design. *Urban Des. Q.* **1995**, *56*, 11–13.
43. Spehar, B.; Clifford, C.W.G. Chaos and graphics: Universal aesthetic of fractals. *Comput. Graph.* **2003**, *27*, 813–820. [CrossRef]
44. Taylor, R. Architect reaches for the clouds: How fractals may figure in our appreciation of a proposed new building. *Nature* **2001**, *410*, 18. [CrossRef] [PubMed]
45. Ma, L.; He, S.; Lu, M. A measurement of visual complexity for heterogeneity in the built environment based on fractal dimension and its application in two gardens. *Fractal Fract.* **2021**, *5*, 278. [CrossRef]
46. Cooper, J.; Oskrochi, R. Fractal analysis of street vistas: A potential tool for assessing levels of visual variety in everyday street scenes. *Environ. Plan. B: Plan. Des.* **2008**, *35*, 349–363. [CrossRef]
47. Sandau, K.; Kurz, H. Measuring fractal dimension and complexity–an alternative approach with an application. *J. Microsc.* **1997**, *186*, 164–176. [CrossRef]
48. Kang, Y.; Kim, E.J. Difference of restorative effects while viewing urban landscapes and green landscapes. *Sustainability* **2019**, *11*, 2129. [CrossRef]
49. Stamps, A.E. Mystery, complexity, legibility and coherence: A meta-analysis. *J. Environ. Psychol.* **2004**, *24*, 1–16. [CrossRef]
50. Dupont, L.; Antrop, M.; Eetvelde, V.V. Eye-tracking analysis in landscape perception research: Influence of photograph properties and landscape characteristics. *Landsc. Res.* **2014**, *39*, 417–432. [CrossRef]
51. Dupont, L.; Ooms, K.; Duchowski, A.T.; Antrop, M.; Van Eetvelde, V. Investigating the visual exploration of the rural-urban gradient using eye-tracking. *Spat. Cogn. Comput.* **2017**, *17*, 65–88. [CrossRef]
52. Dupont, L.; Antrop, M.; Eetvelde, V.V. Does landscape related expertise influence the visual perception of landscape photographs? Implications for participatory landscape planning and management. *Landsc. Urban Plan.* **2015**, *141*, 68–77. [CrossRef]
53. Valtchanov, D.; Ellard, C.G. Cognitive and affective responses to natural scenes: Effects of low level visual properties on preference, cognitive load and eye-movements. *J. Environ. Psychol.* **2015**, *43*, 184–195. [CrossRef]
54. Berto, R.; Massaccesi, S.; Pasini, M. Do eye movements measured across high and low fascination photographs differ? Addressing Kaplan's fascination hypothesis. *J. Environ. Psychol.* **2008**, *28*, 185–191. [CrossRef]
55. Waldheim, C.; Hansen, A.; Ackerman, J.S.; Corner, J.; Brunier, Y.; Kennard, P. *Composite Landscapes: Photomontage and Landscape Architecture*; Hatje Cantz Verlag: Ostfildern, Germany, 2014.
56. Jacob, R.; Karn, K.S. Eye Tracking in human-computer interaction and usability research: Ready to deliver the promises. *Mind* **2003**, *2*, 573–605.
57. Poole, L.A. Eye tracking in human-computer interaction and usability research: Current status and future. In *Encyclopedia of Human Computer Interaction*; Idea Group Reference: Quebec, QC, Canada, 2005; pp. 211–219.

58. Inhoff, A.W.; Radach, R. Definition and Computation of Oculomotor Measures in the Study of Cognitive Processes. In *Eye Guidance in Reading and Scene Perception*; Elsevier Science Ltd.: Amsterdam, The Netherlands, 1998.

59. Nordh, H.; Hagerhall, C.M.; Holmqvist, K. Tracking restorative components: Patterns in eye movements as a consequence of a restorative rating task. *Landsc. Res.* **2013**, *38*, 101–116. [CrossRef]

60. Dupont, L.; Ooms, K.; Antrop, M.; Eetvelde, V.V. Comparing saliency maps and eye-tracking focus maps: The potential use in visual impact assessment based on landscape photographs. *Landsc. Urban Plan.* **2016**, *148*, 17–26. [CrossRef]

61. Blehm, C.; Vishnu, S.; Khattak, A.; Mitra, S.; Yee, R.W. Computer vision syndrome: A review. *Surv. Ophthalmol.* **2005**, *50*, 253–262. [CrossRef]

62. Van Orden, K.F.; Jung, T.P.; Makeig, S. Combined eye activity measures accurately estimate changes in sustained visual task performance. *Biol. Psychol.* **2000**, *52*, 221–240. [CrossRef]

63. Mcgregor, D.; Stern, J. Time on task and blink effects on saccade duration. *Ergonomics* **1996**, *39*, 649–660. [CrossRef]

64. Tara, A.; Belesky, P.; Ninsalam, Y. Towards managing visual impacts on public spaces: A quantitative approach to studying visual complexity and enclosure using visual bowl and fractal dimension. *J. Digit. Landsc. Archit.* **2019**, *4*, 21–32.

65. Fractalyse-Fractal Analysis Software. 2006. Available online: https://sourcesup.renater.fr/www/fractalyse/ (accessed on 1 April 2022).

66. De Val, G.D.L.F.; Atauri, J.A.; De Lucio, J.V. Relationship between landscape visual attributes and spatial pattern indices: A test study in Mediterranean-climate landscapes. *Landsc. Urban Plan.* **2006**, *77*, 393–407. [CrossRef]

67. Schirpke, U.; Tasser, E.; Lavdas, A.A. Potential of eye-tracking simulation software for analyzing landscape preferences. *PLoS ONE* **2022**, *17*, e0273519. [CrossRef] [PubMed]

68. Van der Jagt, A.P.; Craig, T.; Anable, J.; Brewer, M.J.; Pearson, D.G. Unearthing the picturesque: The validity of the preference matrix as a measure of landscape aesthetics. *Landsc. Urban Plan.* **2014**, *124*, 1–13. [CrossRef]

69. Whitburn, J.; Linklater, W.; Abrahamse, W. Meta-analysis of human connection to nature and pro-environmental behavior. *Conserv. Biol.* **2020**, *34*, 180–193. [CrossRef]

70. Day, H. Evaluations of subjective complexity, pleasingness and interestingness for a series of random polygons varying in complexity. *Percept. Psychophys.* **1967**, *2*, 281–286. [CrossRef]

71. Wohlwill, J.F. Amount of stimulus exploration and preference as differential functions of stimulus complexity. *Percept. Psychophys.* **1968**, *4*, 307–312. [CrossRef]

72. Egaña, J.I.; Devia, C.; Mayol, R.; Parrini, J.; Orellana, G.; Ruiz, A.; Maldonado, P.E. Small saccades and image complexity during free viewing of natural images in schizophrenia. *Front. Psychiatry* **2013**, *4*, 37. [CrossRef]

73. Huang, A.S.; Lin, Y. The effect of landscape color, complexity and preference on viewing behavior. *Landsc. Res.* **2020**, *45*, 214–227. [CrossRef]

74. Kim, M.; Kang, Y.; Abu Bakar, S. A nightscape preference study using eye movement analysis. ALAM CIPTA. *Int. J. Sustain. Trop. Des. Res. Pract.* **2013**, *6*, 85–99.

 *forests*

*Article*

# Spatiotemporal Distribution Analysis of Spatial Vitality of Specialized Garden Plant Landscapes during Spring: A Case Study of Hangzhou Botanical Garden in China

Tian Liu [1], Bingyi Mi [2], Hai Yan [1], Zhiyi Bao [1,*], Renwu Wu [1,*] and Shuhan Wang [1]

[1] College of Landscape Architecture, Zhejiang Agriculture and Forestry University, Hangzhou 311300, China; 2021105042002@stu.zafu.edu.cn (T.L.); yanhai@zafu.edu.cn (H.Y.); 2021105042012@stu.zafu.edu.cn (S.W.)
[2] Huayong Engineering Design Group Co., Ltd., Ningbo 315000, China; nowander777@gmail.com
* Correspondence: 20070007@zafu.edu.cn (Z.B.); wurenwu0034@zafu.edu.cn (R.W.)

**Abstract:** Specialized gardens, as integral components of botanical gardens, bear multiple functions, encompassing plant collection and conservation, scientific research, and public education, as well as serving aesthetic and recreational purposes. Their quality profoundly reflects the landscape artistry of botanical gardens, directly influencing the quality of visitors' enjoyment and the overall experience within the botanical garden. This study aims to investigate the spatial vitality of specialized garden plant landscapes, effectively assessing the usage patterns of plant landscape spaces and promoting the optimal utilization of underutilized spaces. Taking Hangzhou Botanical Garden as a case study, considering the warming climate and suitable temperatures in spring, when most plants enter the flowering period and outdoor visitor frequency increases, the primary observational period focuses on spring to measure the spatial vitality of specialized garden plant landscapes. We obtained data through field measurements and on-site observations. Specifically, We measured and recorded information on plant species, quantity, height, crown width, and growth conditions within the plots. Additionally, we employed ground observations and fixed-point photography to document visitor numbers and activity types. We quantified spatial vitality through four indicators: visitor density, space usage intensity, diversity of age group, and richness of activity type. We explored the spatiotemporal distribution patterns of spatial vitality and investigated the relationship between plant landscape characteristics and spatial vitality using variance analysis and correlation analysis. The results indicate that, in spring, the average spatial vitality index of specialized gardens ranks from highest to lowest as follows: Lingfeng Tanmei (1.403), Rosaceae Garden (1.245), Acer and Rhododendron Garden (0.449), and Osmanthus and Crape Myrtle Garden (0.437). Additionally, the spatial vitality of specialized garden plant landscapes in spring is significantly positively correlated with the ornamental period of specialized plants, characteristics of plant viewing, accessible lawn area, spatial accessibility, and spatial enclosure. Therefore, to create vibrant specialized plant landscapes, managers and planners, when engaging in the planning and design of specialized garden plant landscapes, need to fully consider and respect the visual aesthetics and functional needs of visitors. This study will serve as a theoretical reference for subsequent research on the vitality of plant landscape spaces and other small-scale spaces. It will also provide practical guidance for the construction of plant landscapes in specialized gardens within botanical gardens and other urban green spaces.

**Keywords:** specialized garden; plant landscape space; spatial vitality; visitor behavior; landscape features

**Citation:** Liu, T.; Mi, B.; Yan, H.; Bao, Z.; Wu, R.; Wang, S. Spatiotemporal Distribution Analysis of Spatial Vitality of Specialized Garden Plant Landscapes during Spring: A Case Study of Hangzhou Botanical Garden in China. *Forests* **2024**, *15*, 208. https://doi.org/10.3390/f15010208

Academic Editors: Jiang Liu, Xinhao Wang and Xin-Chen Hong

Received: 28 December 2023
Revised: 14 January 2024
Accepted: 18 January 2024
Published: 20 January 2024

## 1. Introduction

The World Botanical Gardens have long played an active role in conducting scientific research and education, maintaining plant diversity, and monitoring climate change [1–3]. With the establishment of two national botanical gardens in Beijing and South China, along with the emergence of botanical gardens across various regions, the Chinese botanical garden industry is experiencing robust development [4]. Specialized gardens, as the core of botanical gardens, are characterized by specific landscape themes, featuring plants with similar traits as the primary thematic elements [5–7]. These gardens serve as the main venues for plant collection and conservation, scientific research, and education within botanical gardens. Additionally, they encompass functions of leisure and aesthetics, directly influencing the quality of visitors' experiences in botanical gardens. In recent years, some urban parks and scenic areas have increasingly focused on the creation of specialized plant landscapes. Notable examples include the Peony Garden in Heze City in northern China and the Plum Garden in the Taihu Scenic Area in southern China, both renowned tourist destinations, exerting a significant impact on the aesthetic shaping of urban green spaces.

Numerous researchers have conducted research on the plant landscapes of specialized gardens, such as those dedicated to wetland woody plants [8], vine plants [9], and medicinal plants [10]. They have proposed planning and design strategies for specialized gardens, addressing aspects such as plant arrangement, ornamental design, plant culture expression, and regional characteristics prominence. However, existing studies are often confined to individual types of specialized gardens. Moreover, some research has revealed an uneven distribution of visitors in many specialized gardens [11]. This manifests as overcrowding during specific periods and areas, while other periods and areas witness sparse visitor activity, resulting in the underutilization of many spaces. Addressing the issue of improving spatial vitality and creating specialized garden plant landscapes that meet the needs of visitors is a crucial consideration.

Spatial vitality refers to the capacity of urban public spaces to attract human activities [12–14]. Numerous scholars have conducted research on the relationship between environmental factors and spatial vitality in various urban public spaces, such as urban parks [15–17], historic districts [18], streets [19], night markets [20], and urban underground spaces [21]. In the context of this study, spatial vitality refers to the intensity of use in spaces primarily composed of plants. Plant landscape spaces provide the material environment for various outdoor activities, and visitor behavior characteristics reflect the quality and attractiveness of the spaces [22]. Existing vitality studies primarily focus on medium to large-scale urban public spaces, with vitality measurements primarily considering a single indicator, such as population density, lacking the exploration of indicators that can reflect the quality and attractiveness of spaces. Some scholars have studied the vitality distribution of small-scale spaces like plant landscape spaces. However, the majority of these studies have focused on urban park plant landscapes [23–25], with a limited exploration into the plant landscapes of specialized gardens. Moreover, existing research predominantly emphasizes the spatial distribution and variations in vitality, paying relatively less attention to the temporal fluctuations and seasonal changes in vitality.

The research methods for spatial vitality can be broadly categorized into two types: traditional methods and novel methods based on multi-source data analysis [26,27]. Traditional methods include questionnaire surveys, interviews, behavior observations, behavior mapping, etc. [28–30]. With the development and widespread use of Internet big data, new methods based on multi-source data analysis, such as mobile signaling data [31] and social media network data [32], have gradually emerged. These methods enable the collection and visualization of massive sample data in a short period with a low consumption [33,34], providing convenience for vitality research in medium to large-scale spaces. However, current research on the spatiotemporal distribution of vitality in plant landscapes still predominantly relies on traditional methods such as behavior observation and behavior mapping. This is primarily due to the difficulty of meeting the precision requirements for studying vitality in small-scale spaces using current analysis methods based on mobile communication and location navigation data. Additionally, obtaining detailed information about visitor behavior activities, which are integral components of the spatial vitality [35,36], proves challenging with these methods.

In light of the aforementioned, we conducted a study on the spatial vitality of specialized garden plant landscapes in the Hangzhou Botanical Garden, situated in the southeastern coastal region of China. Our objectives were to address the following issues: (1) To clarify the spatiotemporal distribution patterns of spatial vitality of specialized garden plant landscapes in spring based on visitor behavior characteristics. (2) To explore the differences in spatial vitality of specialized garden plant landscapes in spring and the related landscape factors. (3) To provide recommendations on how to stimulate visitor participation, create vibrant specialized garden plant landscapes, and enhance the utilization and participation rates of specialized garden plant landscape spaces.

## 2. Materials and Methods

### 2.1. Study Area

Hangzhou (29°11′–30°34′ N; 118°20′–120°37′ E), situated in the southeastern coastal region of China, serves as the capital of Zhejiang Province and is renowned for its picturesque natural landscapes and rich historical and cultural heritage. Embracing a subtropical monsoon climate, Hangzhou experiences distinct seasons with abundant rainfall and ample sunlight. During spring, the average temperature hovers around 17 °C, accompanied by precipitation ranging from 330 to 450 mm. This pleasant climate marks the prime period for outdoor activities, sightseeing, and flower appreciation as numerous plants enter their blooming phase.

Hangzhou Botanical Garden is situated in the northwest of the West Lake Scenic Area in Hangzhou, covering an area of 284.64 hm$^2$ (Figure 1). The topography within the garden is undulating and adorned with lush vegetation. In the northern region, two primary residential areas coexist, complemented by a surrounding array of dining and office spaces. Established as one of the most influential botanical gardens in China since the founding of the People's Republic, after more than 60 years of transformation and development, it has gradually evolved into a comprehensive botanical garden integrating plant conservation, scientific research, popular science education, tourism, and ecological leisure, holding a significant impact both domestically and internationally.

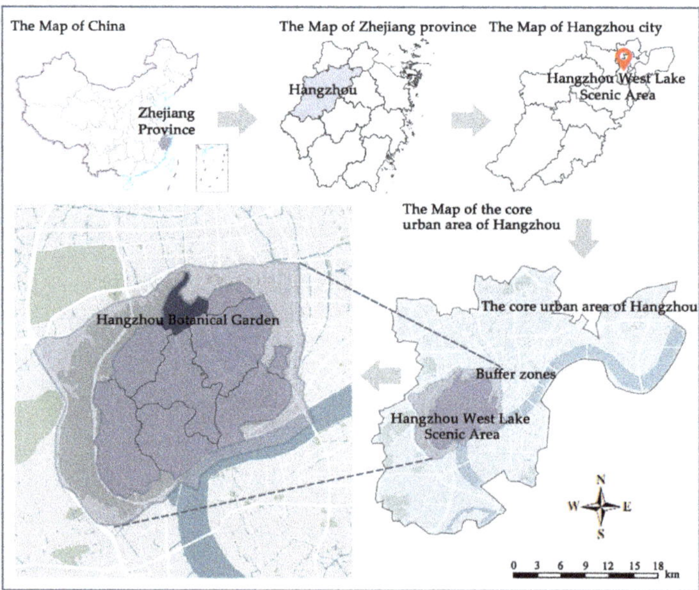


**Figure 1.** Geographical location of Hangzhou Botanical Garden.

### 2.2. Selection of Research Subjects and Plots

Four specialized gardens were selected as research subjects based on well-defined themes, distinctive features, and pronounced seasonal variations. These include Osmanthus and Crape Myrtle Garden, primarily featuring summer and autumn landscapes; Acer and Rhododendron Garden, emphasizing spring and autumn landscapes; Lingfeng Tanmei, showcasing winter and spring landscapes; and Rosaceae Garden (aquatic plants area), predominantly exhibiting spring and summer landscapes.

Osmanthus and Crape Myrtle Garden prominently features various *Osmanthus* varieties, such as *Osmanthus fragrans* var. *thunbergii*, *Osmanthus fragrans* 'Latifolius', *Osmanthus fragrans* var. *aurantiacus* and *Osmanthus fragrans* 'Semperflorens'. Simultaneously, the peripheral areas are landscaped with clusters of *Lagerstroemia indica*, creating a picturesque scene in the summer and autumn seasons. It stands out as an excellent location for enjoying osmanthus blossoms in the autumn. Acer and Rhododendron Garden features primarily plants of the *Rhododendron* and *Acer* in the middle and lower layers. Leveraging the existing upper-layer *Liquidambar taiwaniana* and *Fagaceae* plants, it collaboratively creates a plant landscape for enjoying rhododendron blooms in spring and appreciating red leaves in autumn. Lingfeng Tanmei has a rich history and profound cultural heritage, forming a plant landscape primarily centered around *Prunus mume*. It has earned a reputation as a renowned destination for appreciating plum blossoms. Rosaceae Garden focuses on cultivating plants from the *Rosaceae*, such as *Prunus salicina*, *Prunus serrulata,* and *Malus halliana*, complemented by various aquatic plants, contributing to the landscape during the spring and summer seasons.

To ensure the continuity of plant landscapes and the integrity of their spatial representation, three locations reflecting the themes and features of specialized plant landscapes, possessing typical characteristics of plant landscape spaces and allowing visitor access were chosen within each of the four specialized gardens demarcated by the forest edge and pathway edge [24]. The total area of the plots is 3.42 hm$^2$, with an average size of 2850 m$^2$ (Figure 2). The plots predominantly feature plants and exclude elements such as water bodies, buildings, structures, large paved squares, etc., aiming to minimize interference from non-plant landscape elements on visitor behavior [11].

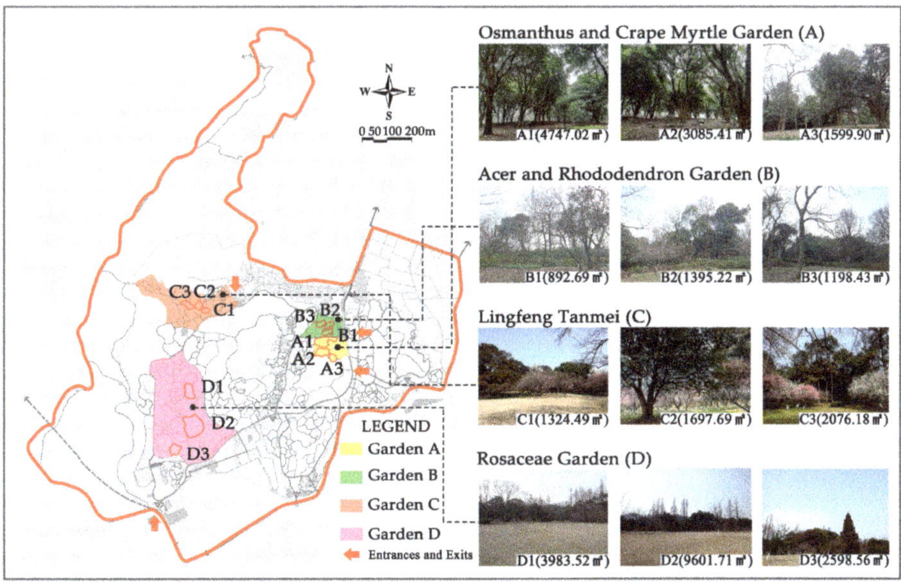


**Figure 2.** Plan and plot distribution of Hangzhou Botanical Garden.

## 2.3. Data Acquisition

### 2.3.1. Plant Landscape Space Characteristics

This study, based on existing research on plant landscape surveys in urban public spaces such as parks [37,38] and water bodies [39], along with studies on individual landscape preference and perception [40,41], identified the specific indicators of specialized garden plant landscape characteristics that need to be collected. A field survey was conducted to investigate the plant landscape space characteristics of each plot. The survey covered a total of 14 landscape variables, comprising both plant and spatial factors. The concepts and quantification methods for each specific indicator are outlined in Table 1 below.

**Table 1.** Spatial characteristics of plant landscape of specialized gardens.

| Types of Indicators | | Number | Indicators | Concepts and Quantification Methods of Indicators |
|---|---|---|---|---|
| Plant Factors | Specialized plant landscape features | X1 | Proportion of specialized plants | The number of specialized plant species as a percentage of total species. To survey the species of specialized plants and other plants and calculate the percentage. |
| | | X2 | Ornamental period of specialized plants | Investigating the main flowering periods of specialized plants on a monthly basis. |
| | | X3 | Color composition of specialized plants | Investigating the richness of color composition of specialized plants and assigning values based on the number of colors. |
| | Plant monomer characteristics | X4 | Plant growth potential | Judging the health status of plant growth and assigning values as 1, 2, 3, and 4 according to the four criteria of poor, medium, good, and excellent. |
| | | X5 | Characteristics of plant viewing | Investigating the ornamental characteristics of plants and assigning values of 1, 2, 3, and 4 based on the number of ornamental parts. |
| | Plant community characteristics | X6 | Types of plant community structure | Investigating the life form composition of plant communities and assigning values of 3, 2, and 1 based on the tree-shrub-herb, tree-herb, and tree-shrub structures, respectively. |
| | | X7 | Canopy closure | The ratio of the total ground projection area of the tree crown in direct sunlight to the total plot area is calculated using AutoCAD. |
| | | X8 | Species richness | The number of species in the plant community. Using the Patrick index to calculate, the formula is $R = S$, where $S$ is the number of species. Due to the difference in plot area, it needs to be converted using the formula $TSR = R/lgA$, where $A$ is the plot area. |
| | | X9 | Species diversity | The richness of species in the plant community. Using the Shannon−Wiener index to calculate, the formula is $H = -\sum_{i=1}^{n} Pilnpi$, where $Pi$ is the ratio of the quantity of each species to the total number of species in the plot. Due to the difference in plot area, it needs to be converted using the formula $TSH = H/lgA$, where $A$ is the plot area. |
| Spatial Factors | Plane surface | X10 | Gross area | Drawing the plot outline by combining the two-step APP and the plan, then using AutoCAD for calculations. |
| | | X11 | accessible lawn area | Drawing the contour of the accessible grassland area within the plot by combining the two-step APP and the plan, then using AutoCAD for calculations. |
| | Accessibility | X12 | Spatial accessibility | Determining the nearest pathway to the entrance/exit by combining the two-step APP and the plan, then using AutoCAD for calculations. |
| | Functional use and psychological feelings | X13 | spatial enclosure | Degree of openness and enclosure in space is calculated using D/H, where D is the distance between people and plants, and H is the plant height. |
| | | X14 | Spatial shape coefficient | Degree of richness in space variation is calculated using the ratio of plot perimeter to the circumference of an equally sized circle, with the formula $S = C/\sqrt[2]{\pi A}$, where $C$ is the plot perimeter and $A$ is the plot area. |

### 2.3.2. Visitor Behavior Characteristics

In this study, visitor behavior primarily refers to recreational activities. As individual behaviors often exhibit variations, the similarities manifested collectively represent visitor behavior characteristics [42]. Behavior observations were conducted to observe and record visitor behavior characteristics within each plot. Visitors' gender, age, and activity type were recorded as a set of data. Based on visitors' behavioral purposes, their activities were categorized into four types: Leisure and Relaxation (chatting, sitting, lying down, picnicking), Sightseeing and Touring (walking, photography, enjoying the scenery), Physical Exercise (Tai Chi, ball sports), and Experience and Entertainment (games, music, chess, picking, kite flying, outdoor education), encompassing a total of 15 visitor behaviors.

Data collection occurred from 4 February 2023 (Start of Spring) to 6 May 2023 (Start of Summer). Observations were conducted on two weekdays and two weekends each month. To ensure sufficient and valid data, observations were carried out on clear weather days conducive to outdoor activities. Observers followed a fixed path and locations, conducting observations from 8:30 a.m. to 16:30 p.m. with a 2-h interval. Each site was observed for 15 min. After 12 rounds of on-site observations, a total of 3367 sets of visitor behavior characteristics data were collected, forming the foundational data for quantifying spatial vitality.

### 2.4. Data Processing and Analysis

#### 2.4.1. Quantification of Spatial Vitality

Through multiple observations of visitor behavior, we acquired a substantial dataset of visitor behavior characteristics. Firstly, we performed preliminary summarization and processing of the data using SPSS 26.0, establishing it as the foundational dataset for quantifying spatial vitality. Subsequently, we employed four indicators, namely visitor density, space usage intensity, diversity of age group, and richness of activity type, as measures of spatial vitality. The first two emphasize the quantity of individuals in space, while the latter two primarily reflect the quality of space. Finally, we utilized formulas to calculate the spatial vitality index for each specialized garden and plot, thereby assessing and evaluating the spatial vitality. The following outlines the concepts and calculation formulas for the four-vitality metrics and the spatial vitality index.

Visitor density refers to the number of visitors per unit area, providing a direct reflection of spatial vitality.

Space usage intensity is calculated using the average area occupied per person during peak hours, expressed as:

$$An = 100/(A/Np) \tag{1}$$

where A is the plot area and Np is the number of visitors during peak hours [43].

Diversity of age group refers to the richness of age group types among visitors, calculated using the Shannon–Wiener diversity index, expressed as:

$$D = -\sum_{i=1}^{n} Pilnpi \tag{2}$$

where $Pi$ is the ratio of visitors in each age group to the total number of visitors [24]. Higher values indicate a more balanced distribution of visitors across different age groups.

Richness of activity type represents the number of simultaneously occurring activity types in space. Higher values indicate that the space can fulfill a more diverse range of functional needs for visitors.

The average values of each indicator on weekdays and weekends for each plot were obtained. These values were standardized, and the entropy method and CRITIC weighting method were employed to determine the weights for each indicator (Table 2). Using the comprehensive weights, the spatial vitality index for each plot was calculated as follows:

$$\text{Spatial Vitality Index} = (\text{Visitor Density} \times 0.271) + (\text{Space Usage Intensity} \times 0.215) + (\text{Diversity of Age Group} \times 0.306) + (\text{Richness of Activity Type} \times 0.208) \tag{3}$$

and then the average spatial vitality index for each specialized garden is presented.

**Table 2.** Weight values of 4 indicators to measure spatial vitality.

| Index | Entropy Method | CRITIC Weighting Method | Comprehensive Weights |
|---|---|---|---|
| Visitor Density | 0.378 | 0.151 | 0.271 |
| Space Usage Intensity | 0.308 | 0.147 | 0.215 |
| Diversity of Age Group | 0.146 | 0.441 | 0.306 |
| Richness of Activity Type | 0.168 | 0.261 | 0.208 |

### 2.4.2. Analysis of Spatial Vitality Differences and Correlations

The spatial vitality index data passed tests for normality and homogeneity of variance. Single-factor analysis of variance (ANOVA) was conducted to compare the differences in spatial vitality among different types of specialized gardens, and Post hoc tests were performed to explore specific differences. A Spearman correlation analysis was conducted to explore the correlation between the 14 plant landscape space characteristics and spatial vitality. All these analyses were carried out using SPSS 26.0.

## 3. Results

### 3.1. Analysis of Plant Landscape Space Characteristics

During the spring, Garden A remains in a non-ornamental phase with minimal changes in plant landscapes. Gardens B, C, and D exhibit significant transformations over time (Figure 3). Field measurements and subsequent calculations were conducted to assess the spatial characteristics of plant landscapes in each specialized garden and plot (Table 3). The specific variations in each indicator are outlined as follows.

**Figure 3.** Photographs of the plots with the highest spatial vitality in each garden.

In Garden A, the average spatial area is 3144.11 m$^2$, with an average accessible grassland area of 2402.14 m$^2$. The average spatial enclosure degree is 5.35, and the average spatial shape coefficient is 1.21. The average length of the nearest pathway to the entrance/exit is 155.32 m. For Plot A1, the community structure is *Magnolia soulangeana—Osmanthus fragrans + Lagerstroemia indica—Ophiopogon bodinieri* (R = 0.816; H = 0.213). For Plot A2, the community structure is *Osmanthus fragrans + Lagerstroemia indica* (R = 0.573; H = 0.066), lacking shrubs and ground cover plants, with the largest grassland area. For Plot A3, the community structure is *Magnolia grandiflora + Ginkgo biloba + Celtis sinensis—Osmanthus fragrans + Podocarpus macrophyllus + Euonymus carnosus—Lycoris radiata* (R = 1.873; H = 0.412), with the smallest grassland area. The D/H values for Plot A1 and A2 are both greater than 5, indicating open spaces, while the D/H value for Plot A3 is much smaller, indicating a stronger sense of spatial enclosure.

**Table 3.** Spatial characteristic indicators for plant landscapes in specialized gardens and plots.

| Garden | Plot | Plant Landscape Space Characteristics | | | | | | | | | | | | | |
|--------|------|------|------|------|------|------|------|------|------|------|------------|------------|-----------|------|------|
| | | X1 | X2 | X3 | X4 | X5 | X6 | X7 | X8 | X9 | X10 (m$^2$) | X11 (m$^2$) | X12 (m) | X13 | X14 |
| A | A1 | 0.67 | 0.00 | 1.00 | 3.00 | 2.00 | 3.00 | 0.07 | 0.82 | 0.21 | 4747.02 | 2896.67 | 99.95 | 6.08 | 1.42 |
| | A2 | 1.00 | 0.00 | 1.00 | 4.00 | 2.00 | 2.00 | 0.09 | 0.57 | 0.07 | 3085.41 | 3085.41 | 225.44 | 6.58 | 1.15 |
| | A3 | 0.17 | 0.00 | 1.00 | 1.00 | 2.00 | 3.00 | 0.07 | 1.87 | 0.41 | 1599.90 | 1224.35 | 140.57 | 3.39 | 1.05 |
| B | B1 | 0.75 | 1.00 | 6.00 | 2.00 | 2.00 | 3.00 | 0.12 | 4.07 | 0.64 | 892.69 | 601.05 | 90.66 | 3.48 | 1.21 |
| | B2 | 0.69 | 1.00 | 8.00 | 3.00 | 2.00 | 3.00 | 0.06 | 4.13 | 0.68 | 1395.22 | 835.47 | 138.49 | 2.94 | 1.15 |
| | B3 | 0.70 | 1.00 | 5.00 | 3.00 | 2.00 | 3.00 | 0.06 | 3.25 | 0.58 | 1198.43 | 932.46 | 149.25 | 3.63 | 1.21 |
| C | C1 | 0.20 | 1.50 | 4.00 | 4.00 | 4.00 | 3.00 | 0.12 | 1.60 | 0.25 | 1324.49 | 1228.94 | 242.41 | 5.75 | 1.07 |
| | C2 | 0.20 | 1.50 | 3.00 | 3.00 | 3.00 | 2.00 | 0.19 | 1.55 | 0.18 | 1697.69 | 1162.44 | 296.79 | 5.31 | 1.11 |
| | C3 | 0.08 | 1.50 | 3.00 | 3.00 | 4.00 | 3.00 | 0.18 | 3.62 | 0.43 | 2076.18 | 1225.43 | 334.59 | 5.46 | 1.29 |
| D | D1 | 0.50 | 2.00 | 6.00 | 2.00 | 3.00 | 3.00 | 0.05 | 2.22 | 0.45 | 3983.52 | 3983.52 | 759.19 | 11.98 | 1.13 |
| | D2 | 0.70 | 1.50 | 8.00 | 2.00 | 3.00 | 3.00 | 0.01 | 2.51 | 0.47 | 9601.71 | 9601.71 | 551.88 | 13.02 | 1.08 |
| | D3 | 0.60 | 1.50 | 7.00 | 3.00 | 2.00 | 3.00 | 0.06 | 2.93 | 0.57 | 2598.56 | 2144.26 | 401.21 | 6.73 | 1.14 |

In Garden B, the average spatial area is 1162.11 m$^2$, with an average accessible grassland area of 789.66 m$^2$. The average spatial enclosure degree is 3.35, and the average spatial shape coefficient is 1.19. The average length of the nearest pathway to the entrance/exit is 126.13 m. The middle and lower layers feature plants from the *Rhododendron* and *Acer*, creating a plant landscape for enjoying rhododendrons in spring and red leaves in autumn. For Plot B1, the upper layer consists of *Liquidambar taiwaniana* (R = 4.067; H = 0.645). For Plot B2, the upper layer includes *Altingia gracilipes, Yulania denudata,* and *Liquidambar taiwaniana* (R = 4.134; H = 0.683). For Plot B3, the upper layer features *Sapindus Saponaria* and *Liquidambar taiwaniana* (R = 3.248; H = 0.583). The canopy closure decreases successively across plots while accessible lawn area increases. D/H values are close, indicating a strong sense of spatial enclosure similar to Garden A.

In Garden C, the average spatial area is 1699.45 m$^2$, with an average accessible grassland area of 1205.60 m$^2$. The average spatial enclosure degree is 5.51, and the average spatial shape coefficient is 1.16. The average length of the nearest pathway to the entrance/exit is 291.26 m. For Plot C1, the community structure is *Elaeocarpus glabripetalus—Prunus mume + Acer palmatum + Illicium henryi—Camellia japonica + Ophiopogon bodinieri* (R = 1.602; H = 0.252). For Plot C2, the community structure is *Elaeocarpus glabripetalus + Magnolia soulangeana—Prunus mume + Michelia figo + Ilex hylonoma* var. *glabra—Ophiopogon bodinieri* (R = 1.548; H = 0.176). For Plot C3, the community structure is *Liquidambar taiwaniana + Pinus elliottii + Magnolia soulangeana—Prunus mume + Michelia figo + Acer palmatum + Podocarpus macrophyllus—Rhododendron* × *pulchrum + Ophiopogon bodinieri + Hedera nepalensis* var. *sinensis + Trachelospermum jasminoides* (R = 3.617; H = 0.426).The D/H values for all plots are close and greater than 5, indicating open spatial configurations.

In Garden D, the average spatial area is 5394.60 m$^2$, with an average accessible grassland area of 5243.16 m$^2$. The average spatial enclosure degree is 10.58, and the average spatial shape coefficient is 1.12. The average length of the nearest pathway to the entrance/exit is 570.76 m. For Plot D1, the plant community structure is *Sapindus saponaria—Prunus mume* 'Danban Xing' *+ Prunus conradinae + Prunus serrulate* var. *lannesiana —Camellia uraku* (R = 2.222; H = 0.449). For Plot D2, the plant community structure is *Ginkgo biloba + Sapindus saponaria—Eriobotrya japonica + Pseudocydonia sinensis + Malus halliana—Rosa chinensis + Chaenomeles cathayensis + Chaenomeles speciosa* (R = 2.511; H = 0.471). For Plot D3, the community structure is *Elaeocarpus glabripetalus + Cryptomeria japonica—Prunus salicina + Eriobotrya japonica + Pyrus pyrifolia + Prunus* 'Yoko'—*Photinia serratifolia + Camellia japonica + Rhaphiolepis umbellate + Liriope spicata* (R = 2.928; H = 0.572). D/H values for Plots D1 and D2 are both greater than 10, suggesting broad visibility and a spacious feel.

### 3.2. Analysis of Visitor Behavior Characteristics

#### 3.2.1. Distribution of Visitor Numbers

Figure 4 illustrates the temporal variations in the number of visitors across various specialized gardens and plots. Each stacked bar in this chart represents data for both weekends and weekdays, progressing from left to right to correspond to various specialized gardens and plots across different months in spring. Generally, Garden A and Garden B show an increasing trend in visitor numbers, while Garden C and Garden D exhibit a decreasing trend. On weekends, the number of visitors is generally higher than on weekdays. From February to early March, Garden C is in its main viewing period, with a significantly higher number of visitors compared to other specialized gardens. Among them, Plot C1 has the highest number of visitors. In mid-March, as the *Prunus mume* completely withers, the number of visitors to Garden C sharply declines. Meanwhile, in Garden D, with *Prunus salicina*, *Pyrus pyrifolia*, *Malus spectabilis*, *Prunus serrulata*, and other *Rosaceae* plants blooming, the number of visitors is significantly higher than in other specialized gardens.

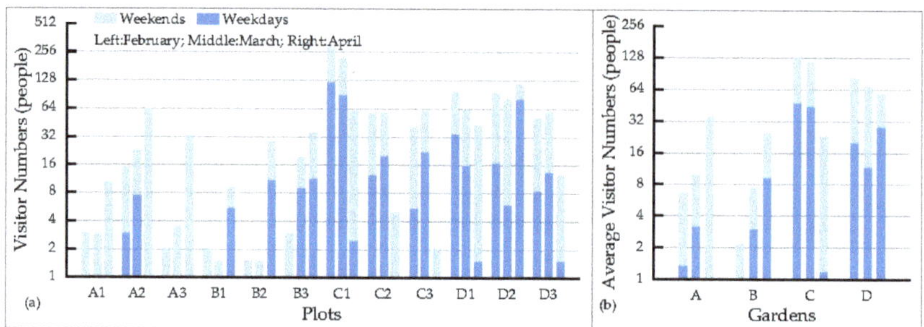

**Figure 4.** Temporal changes in visitor numbers. (**a**) Visitor numbers of plots; (**b**) Visitor numbers of gardens.

Figure 5 presents the temporal variations in visitor density and spatial usage intensity for each specialized garden and plot. Generally, the indicators on weekends are higher than on weekdays and show the same changing trend. Garden A and Garden B show an increasing trend, while Garden C and Garden D exhibit a decreasing trend. In February and March, Garden C has the highest indicators, with Plot C1 having the highest. In April, Garden B has the highest indicators, with Plot B3 having the highest. Although Garden D has the highest number of visitors due to its large area, both visitor density and spatial usage intensity are the lowest.

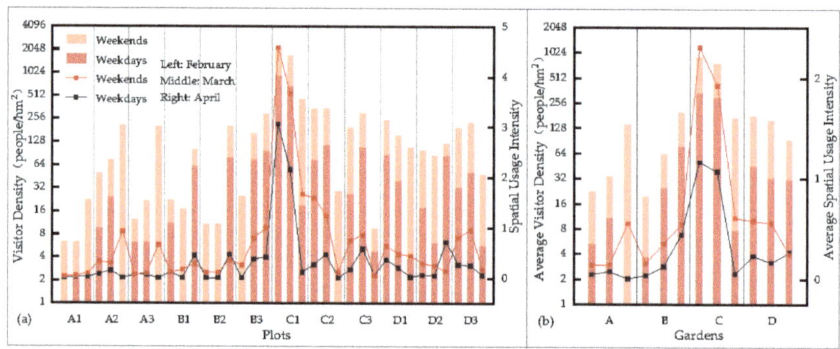

**Figure 5.** Temporal changes of visitor density and space use intensity. (**a**) The indicators of plots; (**b**) The indicators of gardens.

### 3.2.2. Distribution of Visitor Age Composition

Figure 6 illustrates the proportion of visitors in different age groups for each specialized garden and plot. Generally, in spring, the majority of visitors in specialized gardens are young and middle-aged individuals, accounting for over 65% of the total visitors. The next most prominent group is children and teenagers, constituting 5% to 25% of the total visitors, while the elderly make up around 2%. In April, Garden D exhibits a unique distribution in the age composition of visitors. Due to frequent spring activities, children and teenagers constitute over 50% of the total visitors, surpassing the number of young and middle-aged individuals. This is especially concentrated in Plot D2, where over 80% of Garden D's total child visitors gather.

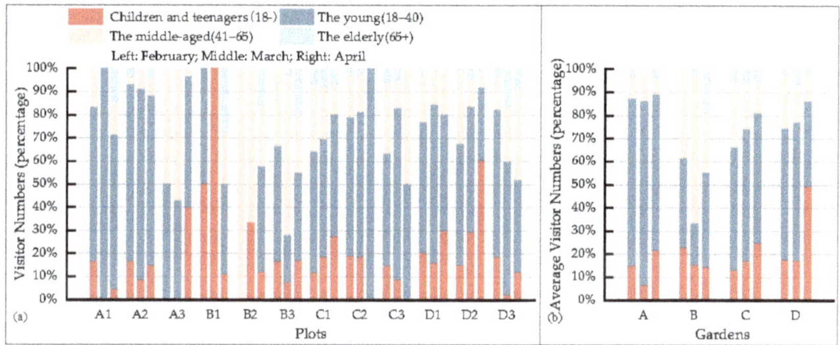


**Figure 6.** Number of visitors of different age groups. (**a**) Number of visitors of different age groups in plots; (**b**) Number of visitors of different age groups in gardens.

Figure 7 illustrates the temporal variations in the diversity of age groups for each specialized garden and plot. In February, Garden C exhibits the highest diversity of age composition. In March, Garden D shows the highest diversity, and in April, Garden B demonstrates the highest diversity. It is evident that the diversity of age groups for each specialized garden peaks during its main viewing period. This suggests that specialized gardens can cater to a broader range of age groups during these peak viewing periods, whereas during non-viewing periods, they are more favored by specific age groups of visitors.

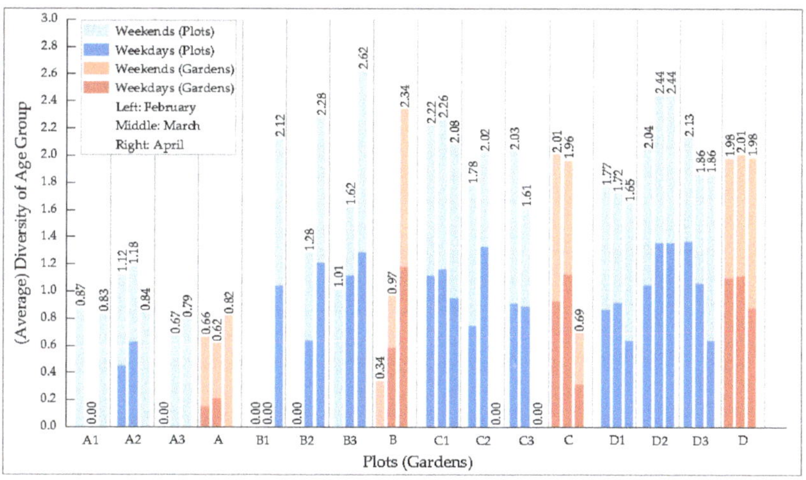


**Figure 7.** Diversity of age groups.

### 3.2.3. Distribution of Visitor Activity Types

Figure 8 presents the common activity types in specialized gardens during spring, along with the participation of visitors of different genders and age groups. Observing the common activity types in each specialized garden, visitors in Garden A predominantly gather around Plot A2, engaging primarily in activities such as sitting and picnicking. In Garden B, visitors are mostly concentrated around Plot B3, participating in activities like photography and scenic appreciation. Garden C attracts visitors mainly to Plot C1, where a diverse range of activities such as sitting, picnicking, chatting, and photography take place. Visitors to Garden D tend to concentrate around Plot D2, engaging in activities such as games, kite flying, and ball sports. The survey indicates that in specialized gardens during spring, female visitors aged 18–40 are the main visitors, engaging primarily in activities such as sitting, chatting, picnicking, and photography, falling under the categories of leisure and relaxation and sightseeing and touring.

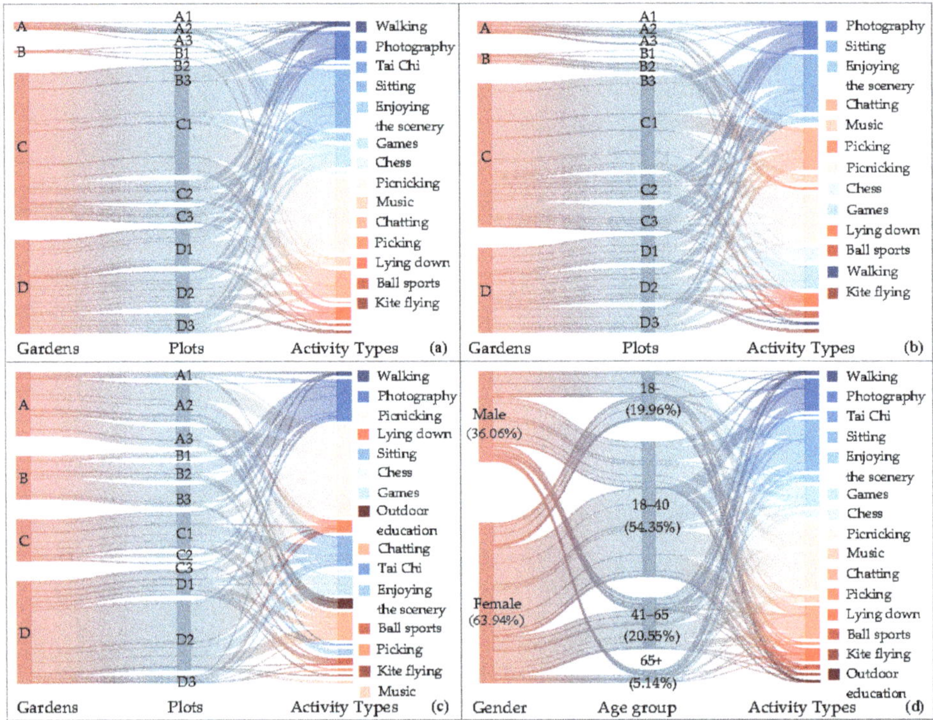

**Figure 8.** Distribution of activity types. (**a**) Common activity types of gardens and plots in February; (**b**) Common activity types of gardens and plots in March; (**c**) Common activity types of gardens and plots in April; (**d**) Activity participation of visitors of different genders and ages in spring.

Figure 9 illustrates the temporal variations in the richness of activity type for each specialized garden and plot. It is evident that the richness of activity type on weekends generally surpasses that on weekdays. In February and March, Garden C exhibits the highest richness of activity type, followed by Garden D, with Plot C1 and Plot D2 having the highest richness, exceeding that of Garden A and Garden B by more than twice. In April, Garden D attains the highest richness of activity type, with Plot D2 exhibiting the highest richness, while the other specialized gardens exhibit similar levels.

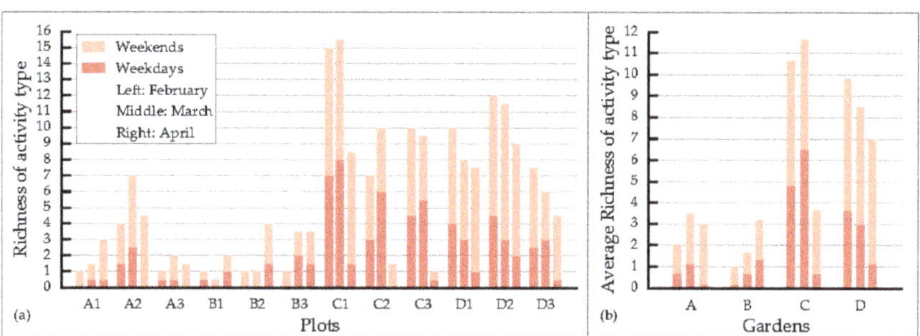

**Figure 9.** Richness of activity type. (**a**) Richness of activity type of plots; (**b**) Richness of activity type of gardens.

*3.3. Spatiotemporal Distribution Characteristics of Spatial Vitality*

3.3.1. Temporal Distribution Characteristics

The temporal distribution characteristics of spatial vitality of specialized garden plant landscapes during spring are primarily reflected in monthly variations. As indicated in Table 4, overall, spatial vitality is higher on weekends than on weekdays. Garden A and Garden B exhibit a monthly increasing trend in spatial vitality. Garden C experienced a slight increase in spatial vitality in March, followed by a sharp decline in April. Meanwhile, Garden D shows a decreasing trend in spatial vitality each month. In February and March, the spatial vitality rankings, from high to low, are Garden C (1.801; 1.845), Garden D (1.271; 1.416), Garden A (0.330; 0.482) and Garden B (0.177; 0.373). The spatial vitality of Garden C and Garden D is more than three times higher than that of Garden A and Garden B. In April, Garden D achieves the highest spatial vitality with an index of 1.048, a slight decrease from February to March. Following is Garden B, with a spatial vitality index of 0.796, more than four times higher than in February and over twice as much as in March. Garden C experiences a sharp decline in spatial vitality with an index of 0.561, slightly higher than Garden A.

**Table 4.** Spatial vitality index of specialized garden and plots in spring.

| Garden | Plot | February | | | March | | | April | | | $\overline{SV}$ |
|---|---|---|---|---|---|---|---|---|---|---|---|
| | | Weekdays | Weekends | Mean | Weekdays | Weekends | Mean | Weekdays | Weekends | Mean | |
| A | A1 | 0.000 | 0.480 | 0.240 | 0.109 | 0.217 | 0.163 | 0.109 | 0.794 | 0.451 | 0.285 |
| | A2 | 0.468 | 0.797 | 0.632 | 0.744 | 1.168 | 0.956 | 0.000 | 1.396 | 0.698 | 0.762 |
| | A3 | 0.118 | 0.118 | 0.118 | 0.118 | 0.539 | 0.328 | 0.000 | 0.702 | 0.351 | 0.266 |
| | A | 0.195 | 0.465 | 0.330 | 0.323 | 0.641 | 0.482 | 0.036 | 0.964 | 0.500 | 0.437 |
| B | B1 | 0.128 | 0.128 | 0.128 | 0.000 | 0.141 | 0.070 | 0.623 | 0.599 | 0.611 | 0.270 |
| | B2 | 0.000 | 0.231 | 0.116 | 0.195 | 0.426 | 0.310 | 0.784 | 0.929 | 0.857 | 0.428 |
| | B3 | 0.000 | 0.572 | 0.286 | 0.841 | 0.638 | 0.739 | 0.800 | 1.044 | 0.922 | 0.649 |
| | B | 0.043 | 0.311 | 0.177 | 0.345 | 0.402 | 0.373 | 0.736 | 0.857 | 0.796 | 0.449 |
| C | C1 | 2.480 | 3.017 | 2.749 | 2.501 | 2.710 | 2.605 | 0.628 | 2.172 | 1.400 | 2.251 |
| | C2 | 0.912 | 1.498 | 1.205 | 1.759 | 1.314 | 1.537 | 0.000 | 0.338 | 0.169 | 0.970 |
| | C3 | 1.251 | 1.651 | 1.451 | 1.545 | 1.243 | 1.394 | 0.114 | 0.114 | 0.114 | 0.986 |
| | C | 1.547 | 2.055 | 1.801 | 1.935 | 1.756 | 1.845 | 0.247 | 0.875 | 0.561 | 1.403 |
| D | D1 | 1.178 | 1.665 | 1.422 | 0.949 | 1.391 | 1.170 | 0.408 | 1.759 | 1.084 | 1.225 |
| | D2 | 1.273 | 1.929 | 1.601 | 1.050 | 2.151 | 1.601 | 0.747 | 1.820 | 1.284 | 1.495 |
| | D3 | 0.997 | 1.452 | 1.225 | 1.003 | 1.081 | 1.042 | 0.311 | 1.245 | 0.778 | 1.015 |
| | D | 1.149 | 1.682 | 1.416 | 1.000 | 1.541 | 1.271 | 0.489 | 1.608 | 1.048 | 1.245 |

3.3.2. Spatial Distribution Characteristics

During spring, there exists variation in spatial vitality across different specialized garden types and within the same garden type's distinct landscape areas. Upon observing the Average Spatial Vitality Index ($\overline{SV}$), it is evident that Garden C has the highest spatial vitality. Specifically, Plot C1 is identified as a high-vitality area, with an average spatial vitality index of 2.251, more than twice that of Plot C2 and Plot C3. Garden D followed closely, with Plot D2 recognized as a high-vitality area, boasting an average spatial vitality index of 1.495, higher than Plot D1 and Plot D3. Garden B demonstrates spatial vitality comparable to Garden A, with average spatial vitality indices for all sites being less than one and decreasing progressively from Plot B3 to Plot B1. Garden A exhibits the lowest spatial vitality, with Plot A2 registering an average spatial vitality index of 0.762, more than 2.5 times that of Plot A1 and Plot A3.

### 3.4. Analysis of Spatial Vitality Differences

In order to explore the difference of spatial vitality of specialized garden plant landscapes in spring, a one-way analysis of variance (ANOVA) was conducted. The results indicate significant differences in spatial vitality of specialized garden plant landscapes (Table 5). Post-hoc tests further revealed pronounced differences, particularly between Garden C and Garden A, as well as Garden B. Additionally, differences were observed between Garden D and Garden A, along with Garden B (Table 6). These findings suggest a potential connection between the distinct plant landscape space characteristics in each specialized garden and the distribution of spatial vitality. Further analysis is required to elucidate this relationship more comprehensively.

**Table 5.** Analysis of variance for spatial vitality.

| Spatial Vitality | Specialized Gardens (Mean ± Standard Deviation) | | | | $F$ | $p$ |
|---|---|---|---|---|---|---|
| | **A** | **B** | **C** | **D** | | |
| Spatial vitality index | 0.44 ± 0.33 | 0.45 ± 0.30 | 1.40 ± 0.70 | 1.24 ± 0.46 | 6.975 | 0.002 ** |

Note: ** indicates a significant difference at the 0.01 level.

**Table 6.** Post hoc tests for spatial vitality.

| Spatial Vitality | (I) Name | (J) Name | (I) Mean | (J) Mean | Difference (I–J) | $p$ |
|---|---|---|---|---|---|---|
| | A | B | 0.437 | 0.449 | −0.012 | 0.967 |
| | A | C | 0.437 | 1.403 | −0.965 | 0.002 ** |
| Spatial Vitality index | A | D | 0.437 | 1.245 | −0.807 | 0.008 ** |
| | B | C | 0.449 | 1.403 | −0.954 | 0.002 ** |
| | B | D | 0.449 | 1.245 | −0.796 | 0.009 ** |
| | C | D | 1.403 | 1.245 | 0.158 | 0.572 |

Note: ** indicates a significant difference at the 0.01 level.

### 3.5. Correlation Analysis between Plant Landscape Characteristics and Spatial Vitality

Table 7 presents the results of the correlation analysis between 14 landscape variables and spatial vitality. The results indicate significant correlations between spring-specialized garden plant landscape spatial vitality and the ornamental period of specialized plants, characteristics of plant viewing, accessible lawn area, spatial accessibility, and spatial enclosure. Except for accessible lawn area, the correlation coefficients for other indicators with spatial vitality index are all higher than 0.7, suggesting a very close positive correlation. In other words, areas with prolonged specialized plant viewing periods, rich plant viewing characteristics, larger accessible lawn areas, greater distance from entrances, and more open and spacious landscape regions are more favored by visitors, resulting in higher spatial vitality. Additionally, the analysis results indicate that there is no significant correlation between indicators such as the proportion of specialized plants, color composition of specialized plants, plant growth potential, types of plant community structure, and canopy closure with spatial vitality. This suggests that the spatial vitality of each specialized garden and plot during the spring season is not influenced by the aforementioned factors.

**Table 7.** Correlation analysis between plant landscape space characteristics and spatial vitality.

| | X1 | X2 | X3 | X4 | X5 | X6 | X7 | X8 | X9 | X10 | X11 | X12 | X13 | X14 | $\overline{SV}$ |
|---|---|---|---|---|---|---|---|---|---|---|---|---|---|---|---|
| X1 | 1.000 | | | | | | | | | | | | | | |
| X2 | −0.377 | 1.000 | | | | | | | | | | | | | |
| X3 | 0.284 | 0.526 | 1.000 | | | | | | | | | | | | |
| X4 | 0.093 | −0.064 | −0.239 | 1.000 | | | | | | | | | | | |
| X5 | −0.564 | 0.747 ** | 0.050 | 0.128 | 1.000 | | | | | | | | | | |
| X6 | −0.163 | 0.136 | 0.426 | −0.419 | 0.000 | 1.000 | | | | | | | | | |
| X7 | −0.372 | −0.074 | −0.563 | 0.430 | 0.296 | −0.453 | 1.000 | | | | | | | | |
| X8 | 0.074 | 0.214 | 0.690 * | −0.320 | −0.079 | 0.583 * | −0.252 | 1.000 | | | | | | | |
| X9 | 0.281 | 0.155 | 0.807 ** | −0.388 | −0.276 | 0.648 * | −0.517 | 0.923 ** | 1.000 | | | | | | |
| X10 | 0.004 | 0.162 | −0.064 | −0.075 | 0.158 | −0.130 | −0.322 | −0.462 | −0.399 | 1.000 | | | | | |
| X11 | 0.021 | 0.265 | −0.057 | 0.041 | 0.296 | −0.065 | −0.350 | −0.545 | −0.441 | 0.888 ** | 1.000 | | | | |
| X12 | −0.312 | 0.817 ** | 0.297 | −0.030 | 0.631 * | −0.065 | −0.245 | −0.098 | −0.126 | 0.531 | 0.643 * | 1.000 | | | |
| X13 | 0.119 | 0.464 | 0.135 | 0.090 | 0.335 | −0.065 | −0.308 | −0.406 | −0.294 | 0.797 ** | 0.923 ** | 0.727 ** | 1.000 | | |
| X14 | 0.326 | −0.302 | −0.152 | 0.264 | −0.296 | 0.065 | 0.126 | 0.224 | 0.140 | 0.000 | −0.182 | −0.371 | −0.084 | 1.000 | |
| $\overline{SV}$ | −0.193 | 0.817 ** | 0.368 | 0.302 | 0.749 ** | 0.000 | −0.105 | −0.119 | −0.126 | 0.357 | 0.587 * | 0.846 ** | 0.713 ** | −0.343 | 1.000 |

Note: * indicates a significant association at the level of 0.05 (two-tailed) and ** indicates a significant association at the level of 0.01 (two-tailed).

## 4. Discussion

### 4.1. Spatiotemporal Distribution Characteristics of Spatial Vitality

During spring, there is a pronounced spatiotemporal fluctuation in the spatial vitality of specialized garden plant landscapes. Temporally, the spatial vitality of Osmanthus and Crape Myrtle Garden (A) and Acer and Rhododendron Garden (B) shows a monthly increasing trend. In contrast, the spatial vitality of Lingfeng Tanmei (C) experiences a slight initial rise followed by a rapid decline to approximately one-third of its initial level. The Rosaceae Garden (D) exhibits a monthly decreasing trend. Spatially, the spatial vitality ranks from high to low as follows: Lingfeng Tanmei, Rosaceae Garden, Acer and Rhododendron Garden, and Osmanthus and Crape Myrtle Garden. The regions with the highest spatial vitality are identified as Plot A2, Plot B3, Plot C1, and Plot D2, respectively.

The renowned tradition of appreciating plum blossoms, both in ancient and modern times, attracts a large number of visitors and promotes various activities. During this period, Lingfeng Tanmei exhibited the highest spatial vitality, with an overall enhancement observed in the spatial vitality of other specialized gardens. It can be inferred that seasonal flower exhibitions significantly captivate visitors, markedly elevating spatial vitality. Research indicates that incorporating a moderate amount of cold-toned flowering plants in urban greenery positively impacts the mental and physical health of the public [44]. Rahnema highlights that flowering plants influence visitors' preferences and emotional perceptions [45]. Ozer also asserts that vibrant flowering plants significantly stimulate activities like photography and appreciation [46]. These findings align with the results of this study.

In comparison to Lingfeng Tanmei, the specialized plants in Rosaceae Garden fail to form a concentrated display during their flowering period. Additionally, some plants exhibit suboptimal maintenance, leading to underutilized spaces. However, the Rosaceae Garden secures the second-highest spatial vitality, likely attributed to its ample grassy areas that cater to recreational and exercise activities such as games, kite-flying, and sports, contributing substantially to spatial vitality enhancement. Previous research has confirmed the attractiveness of grassy spaces to individuals and groups, fulfilling diverse functional needs [47].

The lower level of Acer and Rhododendron Garden predominantly features *Rhododendron* species with rich, vibrant colors during the flowering period, attracting visitors for activities like photography and scenic appreciation. While there is an improvement in spatial vitality, it remains lower than that of Lingfeng Tanmei and Rosaceae Garden, possibly due to the shorter flowering period and comparatively lower promotional efforts. Therefore, a suggestion is to consider conducting extensive and content-rich flower-related promotional activities to prevent visitors from missing the optimal viewing period [11]. Osmanthus and Crape Myrtle Garden undergo minimal changes in its landscape during spring, maintaining the lowest and most stable spatial vitality.

### 4.2. Factors Associated with Spatial Vitality

In spring, there is a significant disparity in spatial vitality among different specialized gardens, indicating an evident uneven distribution of visitor traffic. This phenomenon is influenced by the inherent characteristics of the site [48,49]. The spatial vitality of each specialized garden reaches its peak during its primary viewing period, signifying the ability of these gardens to attract more visitors and stimulate various activities during this time. Mao et al.'s research underscores the importance of the flowering period of specialized plants as a critical determinant of visitor behavior, aligning with the findings of this study [11].

Additionally, visitors tend to engage in various activities in plant landscape spaces that offer rich viewing characteristics, ample access to lawns, greater distance from entrances, and more open and spacious environments, resulting in higher spatial vitality. Interestingly, indicators such as canopy closure, species richness, species diversity, and plant community structure show no correlation with spatial vitality. However, Wang et al.'s investigation into campus landscapes suggests that increasing vegetation coverage promotes recreational activities [50]. Shanahan et al.'s research indicates that species richness and diversity are crucial factors in attracting visitors to parks [51]. People also exhibit preferences for different types of vegetation in green spaces. For instance, compared to tree-shrub-grass composite woodland, individuals prefer single-layer woodland [52].

Furthermore, Ekkel et al.'s study demonstrates that accessibility is a significant factor in increasing the frequency of green space use, with usage decreasing as distance increases [53]. Research by Mao and Liu on the behavior preference of plant landscapes in urban parks suggests that plant landscape spaces with good accessibility are more conducive to visitor participation [25]. However, in contrast to these conclusions, our study indicates that, compared to urban park plant landscapes, visitors place more emphasis on the aesthetic and functional characteristics of specialized plant landscapes rather than accessibility.

### 4.3. Improvement Strategies for Specialized Garden Plant Landscape Construction

Based on the age composition and activity distribution of visitors in specialized gardens and plots during spring, it is advisable to consider the activity needs of different age groups when planning and designing plant landscapes in specialized gardens. For instance, in the spring, the predominant visitors in specialized garden plant landscape spaces are females aged 18–40, engaging primarily in activities such as sitting, picnicking, and photography. Therefore, designers should allocate suitable spaces for lingering, such as shaded areas and lawns, and guide visitors' sightlines through diverse plant arrangements to facilitate better relaxation and scenic enjoyment. Furthermore, considering that children often engage in games, kite-flying, and ball sports, increasing the area of lawns may be beneficial. Additionally, recognizing that the elderly often participate in activities such as Tai Chi and sitting, providing relatively private and quiet spaces for them is recommended.

Significant differences exist in spatial vitality among different specialized gardens and within different landscape areas of the same specialized garden when examining the spatiotemporal distribution characteristics of spatial vitality of specialized garden plant landscapes during spring. It is recommended that the construction of specialized gardens consider factors such as geographical location, historical and cultural features, and visitor interests and needs to maximize the utilization of underutilized spaces and comprehensively enhance spatial vitality. This approach helps avoid low-level redundant construction, which could lead to the wastage of plant and spatial resources. In the case of large-scale botanical gardens with multiple specialized gardens like the Hangzhou Botanical Garden, efforts should be directed not only at enhancing the vitality of individual specialized gardens but also at addressing the connections between these gardens.

Considering the factors related to the spatial vitality of specialized garden plant landscapes during spring, it is crucial for the planning and design of specialized garden plant landscapes to account for seasonal influences. It is essential to allocate and seamlessly connect plants based on their viewing periods, creating specialized garden plant landscapes with distinctive features and personalities while providing year-round attractions. Additionally, the cultivation of plants with rich ornamental characteristics can attract visitor attention and extend the viewing period of the plant landscape. Grassland spaces can fulfill various functional needs of visitors; therefore, expanding accessible lawn areas may be practical. The accessibility and enclosure of plant landscape spaces also influence visitor behavior and distribution. Therefore, careful consideration of these factors is essential in the planning and design of specialized plant landscapes. For specialized gardens with overlapping viewing periods but significant differences in spatial vitality, strategies such as organizing flower exhibitions and promoting floral events can attract visitors, guide the flow of people, and foster a comprehensive and balanced improvement of spatial vitality throughout the entire garden.

### 4.4. Limitations

This study has certain limitations. Firstly, the selected indicators for specialized garden plant landscape spaces are primarily based on existing research on urban park plant landscape spatial vitality, and further adjustments and refinements are needed to explore potential factors specific to specialized garden plant landscape spatial vitality. Secondly, despite the detailed information on individual social attributes, distribution, and behavior obtained through traditional behavior observation methods, they are relatively time and labor-intensive. The investigation into the relationship between plant landscape space characteristics and spatial vitality is limited to correlation without further exploration of other potential relationships between the two. Finally, seasonal variation is a crucial aspect of plant landscape research, and the spatial dynamics formed during this period can undergo significant changes in the short term. The study did not investigate spatial vitality in other seasons, and the limitations of focusing on a single season may result in non-significant outcomes for certain indicators, which nevertheless still hold value for discussion.

### 5. Conclusions

With the vigorous development of specialized garden landscaping forms in urban green spaces, the importance of creating specialized plant landscapes is increasing. Based on the behavior characteristics of visitors, this study investigated the measurement and spatiotemporal distribution patterns of spatial vitality of specialized garden plant landscapes during spring and explored the associated landscape variables. The primary findings are as follows:

1. The findings reveal a hierarchy of spatial vitality, with the order being Lingfeng Tanmei, Rosaceae Garden, Acer and Rhododendron Garden, and Osmanthus and Crape Myrtle Garden. The spatial vitality of Lingfeng Tanmei experiences a slight increase in the mid-term, followed by a sharp decline. Rosaceae Garden exhibits a monthly decreasing trend in spatial vitality, whereas both Acer and Rhododendron Garden and Osmanthus and Crape Myrtle Garden demonstrate a monthly increasing trend.

2. The ornamental period of specialized plants stands out as a pivotal determinant of spatial vitality. Additionally, features such as characteristics of plant viewing, accessible grassland area, spatial accessibility, and enclosure are associated with the spatiotemporal distribution of spatial vitality.

3. The seasonal flower exhibitions and floral event promotions have a significant allure for visitors and concurrently contribute to enhancing the spatial vitality of other specialized gardens.

In future research, exploration and improvement can be undertaken in the following aspects. Regarding indicator selection, consideration could be given to incorporating additional indicators related to plant landscape space characteristics, such as the olfactory features [54], acoustic characteristics [55], and thermal comfort [52]. In terms of method selection, insights from studies on street and urban park spatial vitality can be leveraged to identify other suitable quantifiable vitality indicators. Additionally, exploring the utilization of diverse data sources and investigating more convenient methods for data acquisition and analysis is essential [56,57]. The integration of on-site observations with multiple data sources should be further explored and refined, particularly when applied to the study of small-scale spatial vitality. Finally, given the significant impact of seasonal variations on plant landscape spatial vitality, conducting long-term field observations of visitor behavior in specialized garden plant landscape spaces across different seasons would be beneficial. This approach allows for a comprehensive analysis of the distribution patterns and fluctuation trends of spatial vitality across various time dimensions and spatial scales.

In comparison to existing studies on the vitality of urban public spaces, this research explores the distribution of vitality and related factors in small-scale spaces, specifically focusing on plant landscape spaces. It emphasizes the significance of visitor engagement in enhancing vitality. We believe these findings can assist planners and managers in constructing specialized plant landscapes that align with the preferences and needs of visitors.

**Author Contributions:** Conceptualization, T.L. and R.W.; methodology, T.L.; software, T.L.; validation, B.M., H.Y., and R.W.; formal analysis, T.L.; investigation, T.L. and S.W.; resources, Z.B.; data curation, T.L.; writing—original draft preparation, T.L.; writing—review and editing, B.M., H.Y., and R.W.; visualization, T.L.; supervision, R.W.; project administration, Z.B.; funding acquisition, R.W. All authors have read and agreed to the published version of the manuscript.

**Funding:** This research was funded by the Humanities and Social Science Fund of Ministry of Education of the People's Republic of China (No. 21YJC770026).

**Data Availability Statement:** Data are contained within the article.

**Acknowledgments:** We sincerely thank the editor and anonymous reviewers for their constructive comments and suggestions.

**Conflicts of Interest:** Bingyi Mi was employed by the Huayong Engineering Design Group Co., Ltd. The remaining authors declare that the research was conducted in the absence of any commercial or financial relationships that could be construed as a potential conflict of interest.

## References

1. Chen, G.; Sun, W. The role of botanical gardens in scientific research, conservation, and citizen science. *Plant Divers.* **2018**, *40*, 181–188. [CrossRef]
2. Dunn, C.P. Biological and cultural diversity in the context of botanic garden conservation strategies. *Plant Divers.* **2017**, *39*, 396–401. [CrossRef] [PubMed]
3. Heywood, V.H. The future of plant conservation and the role of botanic gardens. *Plant Divers.* **2017**, *39*, 309–313. [CrossRef] [PubMed]
4. Zhang, Y.; Tang, Y.; Yu, Q.; Bao, Z. Development Prospect of Local Botanical Gardens Under the Background of the Construction of China's National Botanical Garden System: A Case Study of Hangzhou Botanical Garden. *Landsc. Archit.* **2023**, *30*, 52–56.
5. Hu, Y. The Status and Function of Theme Garden in Botanical Garden and its Inspiration for Arrangement of Theme Garden for Chenshan Botanical Garden, Shanghai. *Chin. Landsc. Archit.* **2006**, *7*, 24–31.
6. Zhu, J.; Yuan, L.; Chen, Y.; Liu, H. Discussion on several contents in design of Specified Plant Gardens Design Code. *J. Cent. China Norm. Univ. (Nat. Sci.)* **2022**, *56*, 635–640.
7. Hu, L.; Gu, L.; Zhao, Y.; Chen, T. Research on the Establishment of Open-field Special Gardens in Urban Construction SystemBotanical Gardens—Taking Shaoxing Botanical Garden as an Example. *Chin. Landsc. Archit.* **2022**, *38*, 39–44.
8. Rao, X.; Huang, R.; Shi, Y.; Wu, R.; Bao, Z. Condense Regional Characteristics and Construct the Forest on Water—Analysis of the Planning and Design and Construction of Wet Woody Plants Specialized Garden in Ningbo Botanical Garden. *Chin. Landsc. Archit.* **2020**, *36*, 116–121.
9. Zhang, Y.; Liao, J.; Zhang, R.; Ning, Z. The Planning and Design of the Specialized Garden for Vine Plants in South China Botanical Garden. *Chin. Landsc. Archit.* **2020**, *36*, 92–97.
10. Devkota, H.P.; Watanabe, M. Role of medicinal plant gardens in pharmaceutical science education and research: An overview of medicinal plant garden at Kumamoto University, Japan. *J. Asian Assoc. Sch. Pharm.* **2020**, *9*, 44–52.
11. Mao, Y.; Chen, Y.; Wang, M. Behavioral Preference of Visitors on Special Categorized Plant Landscape. *Chin. Landsc. Archit.* **2021**, *37*, 121–126.
12. Xu, X.; Xu, X.; Guan, P.; Ren, Y.; Wang, W.; Xu, N. The Cause and Evolution of Urban Street Vitality under the Time Dimension: Nine Cases of Streets in Nanjing City, China. *Sustainability* **2018**, *10*, 2797. [CrossRef]
13. Jalaladdini, S.; Oktay, D. Urban Public Spaces and Vitality: A Socio-Spatial Analysis in the Streets of Cypriot Towns. *Procedia—Soc. Behav. Sci.* **2012**, *35*, 664–674. [CrossRef]

14. Liu, S.; Zhang, L.; Long, Y.; Long, Y.; Xu, M. A New Urban Vitality Analysis and Evaluation Framework Based on Human Activity Modeling Using Multi-Source Big Data. *ISPRS Int. J. Geo-Inf.* **2020**, *9*, 617. [CrossRef]

15. Zhai, Y.; Li, D.; Wu, C.; Wu, H. Spatial distribution, activity zone preference, and activity intensity of senior park users in a metropolitan area. *Urban For. Urban Green.* **2023**, *79*, 127761. [CrossRef]

16. Liu, W.; Chen, W.; Dong, C. Spatial decay of recreational services of urban parks: Characteristics and influencing factors. *Urban For. Urban Green.* **2017**, *25*, 130–138. [CrossRef]

17. Zhu, J.; Lu, H.; Zheng, T.; Rong, Y.; Wang, C.; Zhang, W.; Yan, Y.; Tang, L. Vitality of Urban Parks and Its Influencing Factors from the Perspective of Recreational Service Supply, Demand, and Spatial Links. *Int. J. Environ. Res. Public Health* **2020**, *17*, 1615. [CrossRef]

18. Mao, Z.; Chen, X.; Xiang, Z. Research on the Measurement and Influencing Factors of Street Vigour in Historic Districts: A Case Study of Wenming Street Historic District in Kunming. *South Archit.* **2021**, *4*, 54–61.

19. Li, Q.; Cui, C.; Liu, F.; Wu, Q.; Run, Y.; Han, Z. Multidimensional Urban Vitality on Streets: Spatial Patterns and Influence Factor Identification Using Multisource Urban Data. *ISPRS Int. J. Geo-Inf.* **2022**, *11*, 2. [CrossRef]

20. Han, J.; Li, L.; Qiu, B.; Shen, H.; Peng, K. The Vitality of Night Market Space Based on Spatial Data and Influencing Factors: A Case Study Based on the Old City of Nanjing. *South Archit.* **2021**, *6*, 68–75.

21. Xu, Y.; Chen, X. Quantitative analysis of spatial vitality and spatial characteristics of urban underground space (UUS) in metro area. *Tunn. Undergr. Space Technol.* **2021**, *111*, 103875. [CrossRef]

22. Yao, W.; Yun, J.; Zhang, Y.; Meng, T.; Mu, Z. Usage behavior and health benefit perception of youth in urban parks: A case study from Qingdao, China. *Front. Public Health* **2022**, *10*, 923671. [CrossRef] [PubMed]

23. Liu, R.; Xu, X.; Chen, L. Investigation of Plant Landscape Space in Urban Park Based on Users' Behavior—The Case Study of Shenzhen Bay Park. *Chin. Landsc. Archit.* **2019**, *35*, 123–128.

24. Liu, R.; Xu, X. Space Vitality and Environmental Impact of Plant Landscape in Urban Parks. *Chin. Landsc. Archit.* **2018**, *34*, 160–164.

25. Mao, Y.; Zou, Y.; Zheng, Y.; Wang, M. Research on Visitors' Preferences for Plant Landscape Spaces in Ulanqab. *Chin. Landsc. Archit.* **2022**, *38*, 123–128.

26. Riungu, G.K.; Peterson, B.A.; Beeco, J.A.; Brown, G. Understanding visitors' spatial behavior: A review of spatial applications in parks. *Tour. Geogr.* **2018**, *20*, 833–857. [CrossRef]

27. Komossa, F.; Wartmann, F.M.; Kienast, F.; Verburg, P.H. Comparing outdoor recreation preferences in peri-urban landscapes using different data gathering methods. *Landsc. Urban Plan.* **2020**, *199*, 103796. [CrossRef]

28. Zhai, Y.; Li, K.; Liu, J. A Conceptual Guideline to Age-Friendly Outdoor Space Development in China: How Do Chinese Seniors Use the Urban Comprehensive Park? A Focus on Time, Place, and Activities. *Sustainability* **2018**, *10*, 3678. [CrossRef]

29. Lv, G.; Zheng, S.; Hu, W. Exploring the relationship between the built environment and block vitality based on multisource big data: An analysis in Shenzhen, China. *Geomat. Nat. Hazards Risk* **2022**, *13*, 1593–1613. [CrossRef]

30. Do, D.T.; Cheng, Y.; Shojai, A.; Chen, Y. Public park behaviour in Da Nang: An investigation into how open space is used. *Front. Archit. Res.* **2019**, *8*, 454–470. [CrossRef]

31. Korpilo, S.; Virtanen, T.; Saukkonen, T.; Lehvävirta, S. More than A to B: Understanding and managing visitor spatial behaviour in urban forests using public participation GIS. *J. Environ. Manag.* **2018**, *207*, 124–133. [CrossRef] [PubMed]

32. Chen, T.; Hui, E.C.M.; Wu, J.; Lang, W.; Li, X. Identifying urban spatial structure and urban vibrancy in highly dense cities using georeferenced social media data. *Habitat Int.* **2019**, *89*, 102005. [CrossRef]

33. Guan, C.; Song, J.; Keith, M.; Zhang, B.; Akiyama, Y.; Da, L.; Shibasaki, R.; Sato, T. Seasonal variations of park visitor volume and park service area in Tokyo: A mixed-method approach combining big data and field observations. *Urban For. Urban Green.* **2021**, *58*, 126973. [CrossRef]

34. Zhang, S.; Zhou, W. Recreational visits to urban parks and factors affecting park visits: Evidence from geotagged social media data. *Landsc. Urban Plan.* **2018**, *180*, 27–35. [CrossRef]

35. Zhang, A.; Li, W.; Wu, J.; Lin, J.; Chu, J.; Xia, C. How can the urban landscape affect urban vitality at the street block level? A case study of 15 metropolises in China. *Environ. Plan. B Urban Anal. City Sci.* **2021**, *48*, 1245–1262. [CrossRef]

36. Xia, C.; Yeh, A.G.; Zhang, A. Analyzing spatial relationships between urban land use intensity and urban vitality at street block level: A case study of five Chinese megacities. *Landsc. Urban Plan.* **2020**, *193*, 103669. [CrossRef]

37. Wang, R.; Zhao, J. Demographic groups' differences in visual preference for vegetated landscapes in urban green space. *Sustain. Cities Soc.* **2017**, *28*, 350–357. [CrossRef]

38. Wang, X.; Wu, C. An Observational Study of Park Attributes and Physical Activity in Neighborhood Parks of Shanghai, China. *Int. J. Environ. Res. Public Health* **2020**, *17*, 2080. [CrossRef]

39. Nazemi Rafi, Z.; Kazemi, F.; Tehranifar, A. Public preferences toward water-wise landscape design in a summer season. *Urban For. Urban Green.* **2020**, *48*, 126563. [CrossRef]

40. Wang, X.; Rodiek, S. Older Adults' Preference for Landscape Features Along Urban Park Walkways in Nanjing, China. *Int. J. Environ. Res. Public Health* **2019**, *16*, 3808. [CrossRef]

41. Schirpke, U.; Tappeiner, G.; Tasser, E.; Tappeiner, U. Using conjoint analysis to gain deeper insights into aesthetic landscape preferences. *Ecol. Indic.* **2019**, *96*, 202–212. [CrossRef]

42. Zeng, Z.; Wang, Y.; Peng, H. Research on Tourist Recreation Behavior Based on SOPARC and KDE—Taking Wuhan East Lake Greenway as an Example. *Chin. Landsc. Archit.* **2019**, *35*, 58–62.
43. Mu, B.; Liu, C.; Mu, T.; Xu, X.; Tian, G.; Zhang, Y.; Kim, G. Spatiotemporal fluctuations in urban park spatial vitality determined by on-site observation and behavior mapping: A case study of three parks in Zhengzhou City, China. *Urban For. Urban Green.* **2021**, *64*, 127246. [CrossRef]
44. Zhuang, J.; Qiao, L.; Zhang, X.; Su, Y.; Xia, Y. Effects of Visual Attributes of Flower Borders in Urban Vegetation Landscapes on Aesthetic Preference and Emotional Perception. *Int. J. Environ. Res. Public Health* **2021**, *18*, 9318. [CrossRef]
45. Rahnema, S.; Sedaghathoor, S.; Allahyari, M.S.; Damalas, C.A.; Bilali, H.E. Preferences and emotion perceptions of ornamental plant species for green space designing among urban park users in Iran. *Urban For. Urban Green.* **2019**, *39*, 98–108. [CrossRef]
46. Ozer, B.; Baris, M.E. Landscape Design and Park Users' Preferences. *Procedia—Soc. Behav. Sci.* **2013**, *82*, 604–607. [CrossRef]
47. Goličnik, B.; Ward Thompson, C. Emerging relationships between design and use of urban park spaces. *Landsc. Urban Plan.* **2010**, *94*, 38–53. [CrossRef]
48. Zhai, Y.; Baran, P.K. Urban park pathway design characteristics and senior walking behavior. *Urban For. Urban Green.* **2017**, *21*, 60–73. [CrossRef]
49. Chuang, I.; Benita, F.; Tunçer, B. Effects of urban park spatial characteristics on visitor density and diversity: A geolocated social media approach. *Landsc. Urban Plan.* **2022**, *226*, 104514. [CrossRef]
50. Wang, R.; Jiang, W.; Lu, T. Landscape characteristics of university campus in relation to aesthetic quality and recreational preference. *Urban For. Urban Green.* **2021**, *66*, 127389. [CrossRef]
51. Shanahan, D.F.; Lin, B.B.; Gaston, K.J.; Bush, R.; Fuller, R.A. Erratum to: What is the role of trees and remnant vegetation in attracting people to urban parks? *Landsc. Ecol.* **2015**, *30*, 761–762. [CrossRef]
52. Duan, Y.; Li, S. Study of Different Vegetation Types in Green Space Landscape Preference: Comparison of Environmental Perception in Winter and Summer. *Sustainability* **2022**, *14*, 3906. [CrossRef]
53. Ekkel, E.D.; de Vries, S. Nearby green space and human health: Evaluating accessibility metrics. *Landsc. Urban Plan.* **2017**, *157*, 214–220. [CrossRef]
54. Pálsdóttir, A.M.; Spendrup, S.; Mårtensson, L.; Wendin, K.; Food, A.M.I.E.; Department, O.F.A.M.; Högskolan, K.; Kristianstad, U.; Man, A.B.H.M.; Avdelningen, F.M.O.M.; et al. Garden Smellscape–Experiences of Plant Scents in a Nature-Based Intervention. *Front. Psychol.* **2021**, *12*, 667957. [CrossRef]
55. Song, X.; Lv, X.; Yu, D.; Wu, Q. Spatial-temporal change analysis of plant soundscapes and their design methods. *Urban For. Urban Green.* **2018**, *29*, 96–105. [CrossRef]
56. Marquet, O.; Aaron Hipp, J.; Alberico, C.; Huang, J.; Fry, D.; Mazak, E.; Lovasi, G.S.; Floyd, M.F. Park use preferences and physical activity among ethnic minority children in low-income neighborhoods in New York City. *Urban For. Urban Green.* **2019**, *38*, 346–353. [CrossRef]
57. Yue, W.; Chen, Y.; Zhang, Q.; Liu, Y. Spatial Explicit Assessment of Urban Vitality Using Multi-Source Data: A Case of Shanghai, China. *Sustainability* **2019**, *11*, 638. [CrossRef]

MDPI
St. Alban-Anlage 66
4052 Basel
Switzerland
www.mdpi.com

*Forests* Editorial Office
E-mail: forests@mdpi.com
www.mdpi.com/journal/forests